T0254191

LONDON MATHEMATICAL SOCIETY LECTURE NOTE SERIES

Managing Editor: Professor N.J. Hitchin, Mathematical Institute,
University of Oxford, 24–29 St Giles, Oxford OX1 3LB, United Kingdom

The titles below are available from booksellers, or from Cambridge University Press at www.cambridge.org/mathematics.

182 Geometric group theory II, G.A. NIBLO & M.A. ROLLER (eds)
183 Shintani zeta functions, A. YUKIE
184 Arithmetical functions, W. SCHWARZ & J. SPILKER
185 Representations of solvable groups, O. MANZ & T.R. WOLF
186 Complexity: knots, colourings and counting, D.J.A. WELSH
187 Surveys in combinatorics, 1993, K. WALKER (ed)
188 Local analysis for the odd order theorem, H. BENDER & G. GLAUBERMAN
189 Locally presentable and accessible categories, J. ADAMEK & J. ROSICKY
190 Polynomial invariants of finite groups, D.J. BENSON
191 Finite geometry and combinatorics, F. DE CLERCK *et al*
192 Symplectic geometry, D. SALAMON (ed)
194 Independent random variables and rearrangement invariant spaces, M. BRAVERMAN
195 Arithmetic of blowup algebras, W. VASCONCELOS
196 Microlocal analysis for differential operators, A. GRIGIS & J. SJÖSTRAND
197 Two-dimensional homotopy and combinatorial group theory, C. HOG-ANGELONI *et al*
198 The algebraic characterization of geometric 4-manifolds, J.A. HILLMAN
199 Invariant potential theory in the unit ball of C^n, M. STOLL
200 The Grothendieck theory of dessins d'enfant, L. SCHNEPS (ed)
201 Singularities, J.-P. BRASSELET (ed)
202 The technique of pseudodifferential operators, H.O. CORDES
203 Hochschild cohomology of von Neumann algebras, A. SINCLAIR & R. SMITH
204 Combinatorial and geometric group theory, A.J. DUNCAN, N.D. GILBERT & J. HOWIE (eds)
205 Ergodic theory and its connections with harmonic analysis, K. PETERSEN & I. SALAMA (eds)
207 Groups of Lie type and their geometries, W.M. KANTOR & L. DI MARTINO (eds)
208 Vector bundles in algebraic geometry, N.J. HITCHIN, P. NEWSTEAD & W.M. OXBURY (eds)
209 Arithmetic of diagonal hypersurfaces over infite fields, F.Q. GOUVÉA & N. YUI
210 Hilbert C*-modules, E.C. LANCE
211 Groups 93 Galway / St Andrews I, C.M. CAMPBELL *et al* (eds)
212 Groups 93 Galway / St Andrews II, C.M. CAMPBELL *et al* (eds)
214 Generalised Euler-Jacobi inversion formula and asymptotics beyond all orders, V. KOWALENKO *et al*
215 Number theory 1992–93, S. DAVID (ed)
216 Stochastic partial differential equations, A. ETHERIDGE (ed)
217 Quadratic forms with applications to algebraic geometry and topology, A. PFISTER
218 Surveys in combinatorics, 1995, P. ROWLINSON (ed)
220 Algebraic set theory, A. JOYAL & I. MOERDIJK
221 Harmonic approximation, S.J. GARDINER
222 Advances in linear logic, J.-Y. GIRARD, Y. LAFONT & L. REGNIER (eds)
223 Analytic semigroups and semilinear initial boundary value problems, KAZUAKI TAIRA
224 Computability, enumerability, unsolvability, S.B. COOPER, T.A. SLAMAN & S.S. WAINER (eds)
225 A mathematical introduction to string theory, S. ALBEVERIO *et al*
226 Novikov conjectures, index theorems and rigidity I, S. FERRY, A. RANICKI & J. ROSENBERG (eds)
227 Novikov conjectures, index theorems and rigidity II, S. FERRY, A. RANICKI & J. ROSENBERG (eds)
228 Ergodic theory of $\mathbf{Z}^d$ actions, M. POLLICOTT & K. SCHMIDT (eds)
229 Ergodicity for infinite dimensional systems, G. DA PRATO & J. ZABCZYK
230 Prolegomena to a middlebrow arithmetic of curves of genus 2, J.W.S. CASSELS & E.V. FLYNN
231 Semigroup theory and its applications, K.H. HOFMANN & M.W. MISLOVE (eds)
232 The descriptive set theory of Polish group actions, H. BECKER & A.S. KECHRIS
233 Finite fields and applications, S.COHEN & H. NIEDERREITER (eds)
234 Introduction to subfactors, V. JONES & V.S. SUNDER
235 Number theory 1993–94, S. DAVID (ed)
236 The James forest, H. FETTER & B.G. DE BUEN
237 Sieve methods, exponential sums, and their applications in number theory, G.R.H. GREAVES *et al*
238 Representation theory and algebraic geometry, A. MARTSINKOVSKY & G. TODOROV (eds)
240 Stable groups, F.O. WAGNER
241 Surveys in combinatorics, 1997, R.A. BAILEY (ed)
242 Geometric Galois actions I, L. SCHNEPS & P. LOCHAK (eds)
243 Geometric Galois actions II, L. SCHNEPS & P. LOCHAK (eds)
244 Model theory of groups and automorphism groups, D. EVANS (ed)
245 Geometry, combinatorial designs and related structures, J.W.P. HIRSCHFELD *et al*
246 p-Automorphisms of finite p-groups, E.I. KHUKHRO
247 Analytic number theory, Y. MOTOHASHI (ed)
248 Tame topology and o-minimal structures, L. VAN DEN DRIES
249 The atlas of finite groups: ten years on, R. CURTIS & R. WILSON (eds)
250 Characters and blocks of finite groups, G. NAVARRO
251 Gröbner bases and applications, B. BUCHBERGER & F. WINKLER (eds)
252 Geometry and cohomology in group theory, P. KROPHOLLER, G. NIBLO, R. STÖHR (eds)
253 The q-Schur algebra, S. DONKIN
254 Galois representations in arithmetic algebraic geometry, A.J. SCHOLL & R.L. TAYLOR (eds)
255 Symmetries and integrability of difference equations, P.A. CLARKSON & F.W. NIJHOFF (eds)
256 Aspects of Galois theory, H. VÖLKLEIN *et al*
257 An introduction to noncommutative differential geometry and its physical applications 2ed, J. MADORE
258 Sets and proofs, S.B. COOPER & J. TRUSS (eds)
259 Models and computability, S.B. COOPER & J. TRUSS (eds)
260 Groups St Andrews 1997 in Bath, I, C.M. CAMPBELL *et al*
261 Groups St Andrews 1997 in Bath, II, C.M. CAMPBELL *et al*
262 Analysis and logic, C.W. HENSON, J. IOVINO, A.S. KECHRIS & E. ODELL

263 Singularity theory, B. BRUCE & D. MOND (eds)
264 New trends in algebraic geometry, K. HULEK, F. CATANESE, C. PETERS & M. REID (eds)
265 Elliptic curves in cryptography, I. BLAKE, G. SEROUSSI & N. SMART
267 Surveys in combinatorics, 1999, J.D. LAMB & D.A. PREECE (eds)
268 Spectral asymptotics in the semi-classical limit, M. DIMASSI & J. SJÖSTRAND
269 Ergodic theory and topological dynamics, M.B. BEKKA & M. MAYER
270 Analysis on Lie Groups, N.T. VAROPOULOS & S. MUSTAPHA
271 Singular perturbations of differential operators, S. ALBEVERIO & P. KURASOV
272 Character theory for the odd order function, T. PETERFALVI
273 Spectral theory and geometry, E.B. DAVIES & Y. SAFAROV (eds)
274 The Mandelbrot set, theme and variations, TAN LEI (ed)
275 Descriptive set theory and dynamical systems, M. FOREMAN *et al*
276 Singularities of plane curves, E. CASAS-ALVERO
277 Computational and geometric aspects of modern algebra, M.D. ATKINSON *et al*
278 Global attractors in abstract parabolic problems, J.W. CHOLEWA & T. DLOTKO
279 Topics in symbolic dynamics and applications, F. BLANCHARD, A. MAASS & A. NOGUEIRA (eds)
280 Characters and automorphism groups of compact Riemann surfaces, T. BREUER
281 Explicit birational geometry of 3-folds, A. CORTI & M. REID (eds)
282 Auslander-Buchweitz approximations of equivariant modules, M. HASHIMOTO
283 Nonlinear elasticity, Y. FU & R. OGDEN (eds)
284 Foundations of computational mathematics, R. DEVORE, A. ISERLES & E. SÜLI (eds)
285 Rational points on curves over finite fields, H. NIEDERREITER & C. XING
286 Clifford algebras and spinors 2ed, P. LOUNESTO
287 Topics on Riemann surfaces and Fuchsian groups, E. BUJALANCE *et al*
288 Surveys in combinatorics, 2001, J. HIRSCHFELD (ed)
289 Aspects of Sobolev-type inequalities, L. SALOFF-COSTE
290 Quantum groups and Lie theory, A. PRESSLEY (ed)
291 Tits buildings and the model theory of groups, K. TENT (ed)
292 A quantum groups primer, S. MAJID
293 Second order partial differential equations in Hilbert spaces, G. DA PRATO & J. ZABCZYK
294 Introduction to operator space theory, G. PISIER
295 Geometry and Integrability, L. MASON & YAVUZ NUTKU (eds)
296 Lectures on invariant theory, I. DOLGACHEV
297 The homotopy category of simply connected 4-manifolds, H.-J. BAUES
298 Higher operads, higher categories, T. LEINSTER
299 Kleinian Groups and Hyperbolic 3-Manifolds, Y. KOMORI, V. MARKOVIC & C. SERIES (eds)
300 Introduction to Möbius Differential Geometry, U. HERTRICH-JEROMIN
301 Stable Modules and the D(2)-Problem, F.E.A. JOHNSON
302 Discrete and Continuous Nonlinear Schrödinger Systems, M.J. ABLOWITZ, B. PRINARI & A.D. TRUBATCH
303 Number Theory and Algebraic Geometry, M. REID & A. SKOROBOOATOV (eds)
304 Groups St Andrews 2001 in Oxford Vol. 1, C.M. CAMPBELL, E.F. ROBERTSON & G.C. SMITH (eds)
305 Groups St Andrews 2001 in Oxford Vol. 2, C.M. CAMPBELL, E.F. ROBERTSON & G.C. SMITH (eds)
306 Peyresq lectures on geometric mechanics and symmetry, J. MONTALDI & T. RATIU (eds)
307 Surveys in Combinatorics 2003, C.D. WENSLEY (ed)
308 Topology, geometry and quantum field theory, U.L. TILLMANN (ed)
309 Corings and Comdules, T. BRZEZINSKI & R. WISBAUER
310 Topics in Dynamics and Ergodic Theory, S. BEZUGLYI & S. KOLYADA (eds)
311 Groups: topological, combinatorial and arithmetic aspects, T.W. MÜLLER (eds)
312 Foundations of Computational Mathematics, Minneapolis 2002, FELIPE CUCKER *et al* (eds)
313 Transcendantal aspects of algebraic cycles, S. MÜLLER-STACH & C. PETERS (eds)
314 Spectral generalizations of line graphs, D. CVETKOVIC, P. ROWLINSON & S. SIMIC
315 Structured ring spectra, A. BAKER & B. RICHTER (eds)
317 Advances in elliptic curve cryptography, I.F. BLAKE, G. SEROUSSI & N. SMART
318 Perturbation of the boundary in boundary-value problems of Partial Differential Equations, DAN HENRY
319 Double Affine Hecke Algebras, I. CHEREDNIK
320 L-Functions and Galois Representations, D. BURNS, K. BUZZARD & J. NEKOVÁŘ (eds)
321 Surveys in Modern Mathematics, V. PRASOLOV & Y. ILYASHENKO (eds)
322 Recent perspectives in random matrix theory and number theory, F. MEZZADRI, N.C. SNAITH (eds)
323 Poisson geometry, deformation quantisation and group representations, S. GUTT *et al* (eds)
324 Singularities and Computer Algebra, C. LOSSEN & G. PFISTER (eds)
325 Lectures on the Ricci Flow, P. TOPPING
326 Modular Representations of Finite Groups of Lie Type, J.E. HUMPHREYS
328 Fundamentals of Hyperbolic Manifolds, R.D. CANARY, A. MARDEN & D.B.A. EPSTEIN (eds)
329 Spaces of Kleinian Groups, Y. MINSKY, M. SAKUMA & C. SERIES (eds)
330 Noncommutative Localization in Algebra and Topology, A. RANICKI (ed)
331 Foundations of Computational Mathematics, Santander 2005, L. PARDO, A. PINKUS, E. SULI & M. TODD (eds)
332 Handbook of Tilting Theory, L. ANGELERI HÜGEL, D. HAPPEL & H. KRAUSE (eds)
333 Synthetic Differential Geometry 2ed, A. KOCK
334 The Navier-Stokes Equations, P.G. DRAZIN & N. RILEY
335 Lectures on the Combinatorics of Free Probability, A. NICA & R. SPEICHER
336 Integral Closure of Ideals, Rings, and Modules, I. SWANSON & C. HUNEKE
337 Methods in Banach Space Theory, J.M.F. CASTILLO & W.B. JOHNSON (eds)
338 Surveys in Geometry and Number Theory, N. YOUNG (ed)
339 Groups St Andrews 2005 Vol. 1, C.M. CAMPBELL, M.R. QUICK, E.F. ROBERTSON & G.C. SMITH (eds)
340 Groups St Andrews 2005 Vol. 2, C.M. CAMPBELL, M.R. QUICK, E.F. ROBERTSON & G.C. SMITH (eds)
341 Ranks of Elliptic Curves and Random Matrix Theory, J.B. CONREY, D.W. FARMER, F. MEZZADRI & N.C. SNAITH (eds)
342 Elliptic Cohomology, H.R. MILLER & D.C. RAVENEL (eds)
343 Algebraic Cycles and Motives Vol. 1, J. NAGEL & C. PETERS (eds)
344 Algebraic Cycles and Motives Vol. 2, J. NAGEL & C. PETERS (eds)
345 Algebraic and Analytic Geometry, A. NEEMAN
346 Surveys in Combinatorics, 2007, A. HILTON & J. TALBOT (eds)
347 Surveys in Contemporary Mathematics, N. YOUNG & Y. CHOI (eds)

London Mathematical Society Lecture Note Series: 350

Model Theory with Applications to Algebra and Analysis

Volume 2

ZOÉ CHATZIDAKIS
CNRS – Université Paris 7

DUGALD MACPHERSON
University of Leeds

ANAND PILLAY
University of Leeds

ALEX WILKIE
University of Manchester

CAMBRIDGE
UNIVERSITY PRESS

CAMBRIDGE
UNIVERSITY PRESS

University Printing House, Cambridge CB2 8BS, United Kingdom

One Liberty Plaza, 20th Floor, New York, NY 10006, USA

477 Williamstown Road, Port Melbourne, VIC 3207, Australia

314-321, 3rd Floor, Plot 3, Splendor Forum, Jasola District Centre, New Delhi - 110025, India

79 Anson Road, #06-04/06, Singapore 079906

Cambridge University Press is part of the University of Cambridge.

It furthers the University's mission by disseminating knowledge in the pursuit of education, learning and research at the highest international levels of excellence.

www.cambridge.org
Information on this title: www.cambridge.org/9780521709088

First published 2008

A catalogue record for this publication is available from the British Library

ISBN 978-0-521-70908-8 Paperback

Table of contents for Volume 2

Table of contents for Volume 1

Preface

These two volumes contain both expository and research papers in the general area of model theory and its applications to algebra and analysis. The volumes grew out of the semester on "Model Theory and Applications to Algebra and Analysis" which took place at the Isaac Newton Institute (INI), Cambridge, from January to July 2005. We, the editors, were also the organizers of the programme. The contributors have been selected from among the participants and their papers reflect many of the achievements and advances obtained during the programme. Also some of the expository papers are based on tutorials given at the March-April 2005 training workshop. We take this opportunity, both as editors of these volumes and organizers of the MAA programme, to thank the Isaac Newton Institute and its staff for supporting our programme and providing a perfect environment for mathematical research and collaboration.

The INI semester saw activity and progress in essentially all areas on the "applied" side of model theory: o-minimality, motivic integration, groups of finite Morley rank, and connections with number theory and geometry. With the exception of motivic integration and valued fields, these topics are well represented in the two volumes.

The collection of papers is more or less divided into (overlapping) themes, together with a few singularities. Aspects of the interaction between stability theory, differential and difference equations, and number theory, appear in the first six papers of volume I. The first paper, based on Pillay's workshop tutorial, can also serve as a fast introduction to model theory for the general reader, although it quickly moves to an account of Mordell-Lang for function fields in characteristic 0. The "arithmetic of differential equations" figures strongly in Pillay's paper on the Grothendieck-Katz conjecture and its nonlinear generalizations, as well as in Bertrand's paper which initiates the investigation of versions of Ax-Schanuel for nonisoconstant semiabelian varieties over function fields. The Galois theory of difference equations is rather a hot topic and the Chatzidakis-Hardouin-Singer paper compares definitions and concepts that have arisen in algebra, analysis, and model theory.

Interactions of complex analytic geometry with model theory and logic (in the form of stability, o-minimality, as well as decidability issues) appear in papers 7 to 10 of volume 1. The papers by Peterzil-Starchenko and Moosa-Pillay (on nonstandard complex analysis and compact Kähler manifolds respectively) are comprehensive accounts of important projects, which contain new results and set the stage for future research. In the first, o-minimality is the model-theoretic tool. In the second it is stability. Wilkie's paper characterizes the holomorphic functions locally definable from a given family of holomorphic functions, and Macintyre's paper is related to his work on the decidability of Weierstrass functions. They are both set in the o-minimal context.

The o-minimality theme is continued in papers 12 and 13 of volume 1 from a (real) geometric point of view. In particular Rolin's paper is a comprehensive account of the most modern techniques of finding o-minimal expansions of the real field.

In recent years Zilber has been exploring connections between model theory and noncommutative geometry, and in his paper in volume I he succeeds in interpreting certain "quantum algebras" as Zariski structures. Fesenko's short note contains a wealth of speculations and questions, including the use of nonstandard methods in noncommutative geometry.

Definable groups of "finite dimension" in various senses (finite Morley rank, finite SU-rank, o-minimal) figure strongly in papers 1 to 5 of volume II. Papers 1 and 2 contain new and striking general results on groups of finite Morley rank, coming out of techniques and results developed in work on the Cherlin-Zilber conjecture. Paper 3 gives an overview of a model-theoretic approach to asymptotics and measure stimulated by the analogous results and concepts for finite and pseudofinite fields. The article by Hrushovski and Wagner, on the size of the intersection of a finite subgroup of an algebraic group with a subvariety, generalizes a theorem of Pink and Larsen. Otero's paper gives a comprehensive description of work since the 1980's on groups definable in o-minimal structures. This includes an account of the positive solution to "Pillay's conjecture" on definably compact groups which was proved during the Newton semester.

Hilbert's 10th problem and its generalizations, as well as first order properties of function fields, appear in papers 6 to 8 of volume II. The Pheidas-Zahidi and Eistenträger papers are based on tutorials given at INI, and give a comprehensive account of work on Hilbert's 10th problem for the rational field and for various rings and fields of functions. Paper 8 proves among other things definability of the constant field in function fields whose constant field is "large". The three papers together give a good picture of an exciting and very active subject at the intersection of logic and number theory.

The volumes are rounded off by important papers on Hrushovski constructions, ordered abelian groups, and continuous logic. In particular the paper 10 in volume II (based again on a tutorial) is an elementary and self-contained presentation of "continuous logic" or the "model theory of metric structures" which is fast becoming an autonomous area of model theory with links to both stability and functional analysis.

Zoé Chatzidakis
Dugald Macpherson
Anand Pillay
Alex Wilkie

Contributors

Itaï Ben Yaacov
Institut Camille Jordan
Université Claude Bernard (Lyon-1)
43 boulevard du 11 novembre 1918
69622 Villeurbanne cédex
France

Alexander Berenstein
Departamento de Matematicas
Universidad de los Andes
Carrera 1 Nro 18A-10
Bogotá
Colombia

Daniel Bertrand
Institut de Mathématiques de Jussieu
Université Paris 6
Boite 247
4 place Jussieu
75252 Paris cedex 05
France

Alexandre Borovik
School of Mathematics
University of Manchester
Oxford Road
Manchester M13 9PL
UK

Zoé Chatzidakis
UFR de Mathématiques
Université Paris 7 - Case 7012
2 place Jussieu
75251 Paris cedex 05
France.

Gregory Cherlin
Department of Mathematics
Rutgers University
110 Frelinghuysen Rd
Piscataway, NJ 08854
USA

Kirsten Eisenträger
Department of Mathematics
The Pennsylvania State University
109 McAllister Building
University Park, PA 16802
USA

Richard Elwes
School of Mathematics
University of Leeds
Leeds LS2 9JT
UK

Ivan Fesenko
Department of Mathematics
University of Nottingham
Nottingham NG7 2RD
UK

Olivier Frécon
Laboratoire de Mathématiques et Applications
Université de Poitiers
Téléport 2 - BP 30179
Boulevard Marie et Pierre Curie
86962 Futuroscope Chasseneuil cedex
France

Charlotte Hardouin
IWR
Im Neuenheimer Feld 368
69120 Heidelberg
Germany

C. Ward Henson
Department of Mathematics
University of Illinois
1409 W. Green St.
Urbana, IL 61801
USA

Ehud Hrushovski
Einstein Institute of Mathematics
The Hebrew University of Jerusalem
Jerusalem 91904
Israel

Eric Jaligot
Institut Camille Jordan
Université Claude Bernard (Lyon-1)
43 boulevard du 11 novembre 1918
69622 Villeurbanne cédex
France

Angus Macintyre
School of Mathematics
Queen Mary, University of London
Mile End Road
London E1 4NS
UK

Dugald Macpherson
School of Mathematics
University of Leeds
Leeds LS2 9JT
UK

Rahim Moosa
Department of Pure Mathematics
200 University Avenue West
Waterloo, Ontario N2L 3G1
Canada

Margarita Otero
Departamento de Matemáticas
Universidad Autónoma de Madrid
28049 Madrid
Spain

Ya'acov Peterzil
Department of Mathematics
University of Haifa
Haifa
Israel

Thanases Pheidas
Department of Mathematics
University of Crete
Knossos Avenue
71409 Iraklio, Crete
Greece

Anand Pillay
School of Mathematics
University of Leeds
Leeds LS2 9JT
England

Bjorn Poonen
Department of Mathematics
University of California
Berkeley, CA 94720-3840
USA

Florian Pop
Department of Mathematics
University of Pennsylvania, DRL
209 S 33rd Street
Philadelphia, PA 19104
USA

Jean-Philippe Rolin
Université de Bourgogne
I.M.B.
9. Avenue Alain Savary
BP 47870
21078 Dijon Cedex
France

Damian Rössler
Institut de Mathématiques de Jussieu
Université Paris 7 Denis Diderot
Case Postale 7012
2, place Jussieu
F-75251 Paris Cedex 05
France

Thomas Scanlon
University of California, Berkeley
Department of Mathematics
Evans Hall
Berkeley, CA 94720-3840
USA

Philip Scowcroft
Department of Mathematics
and Computer Science
Wesleyan University
Middletown, CT 06459
USA

Michael F. Singer
North Carolina State University
Department of Mathematics
Box 8205
Raleigh, North Carolina 27695-8205
USA

Sergei Starchenko
Department of Mathematics
University of Notre Dame
Notre Dame, IN 46556
USA

Alexander Usvyatsov
UCLA Mathematics Department
Box 951555
Los Angeles, CA 90095-1555
USA

Frank Wagner
Institut Camille Jordan
Université Claude Bernard (Lyon-1)
43 boulevard du 11 novembre 1918
69622 Villeurbanne cédex
France

A.J. Wilkie
School of Mathematics
Alan Turing Building
The University of Manchester
Manchester M13 9PL
UK

Karim Zahidi
Dept of Mathematics, statistics and actuarial science
University of Antwerp
Prinsenstraat 13
B-2000 Antwerpen
Belgium

Martin Ziegler
Mathematisches Institut
Albert-Ludwigs-Universität Freiburg
D79104 Freiburg
Germany

B. Zilber
Mathematical Institute
24 - 29 St. Giles
Oxford OX1 3LB
UK

Conjugacy in groups of finite Morley rank

Olivier Frécon
Université de Poitiers

Eric Jaligot†
Université de Lyon, CNRS and Université Lyon 1

Summary

We survey conjugacy results in groups of finite Morley rank, mixing unipotence, Carter, and Sylow theories in this context.

Introduction

When considering certain classes of groups one might expect conjugacy theorems, and the class of groups of finite Morley rank is not an exception to this. The study of groups of finite Morley rank is mostly motivated by the Algebricity Conjecture, formulated by G. Cherlin and B. Zilber in the late seventies, which postulates that infinite simple groups of this category are isomorphic to algebraic groups over algebraically closed fields. The model-theoretic rank involved appeared in the sixties when M. Morley proved his famous theorem on the categoricity in *any* uncountable cardinal of first order theories categorical in *one* uncountable cardinal [Mor65]. He introduced for that purpose an ordinal valued rank, later shown to be finite by J. Baldwin in the uncountably categorical context [Bal73], and this rank can be seen as an abstract version of the Zariski dimension in algebraic geometry over an algebraically closed field.

In particular, the category of groups of finite Morley rank encapsulates finite groups and algebraic groups over algebraically closed fields. One of the most basic tools for analyzing finite groups is Sylow theory, and in algebraic groups semisimplicity and unipotence theory play a similar role. It is thus not surprising to see these two theories, together with all

† Parts of this work were done while the authors were visiting the Isaac Newton Institute, Cambridge, during the model theory program in the spring 2005.

conjugacy results they suggest, having enormous and close developments in the more abstract category of groups of finite Morley rank. The present paper is intended to give an exhaustive survey on these parallel developments.

In a connected linear algebraic group, the centralizers of maximal tori are conjugate and cover the group generically. In the category of groups of finite Morley rank, these Cartan subgroups are best approximated by *Carter* subgroups, which are defined merely by the outstanding properties of being definable, connected, nilpotent, and of finite index in their normalizers. The main feature of Carter subgroups is their existence in any group of finite Morley rank. They constitute, together with all relevant approximations of semisimplicity and unipotence, the core of our preoccupations in this paper.

Sylow theory, as the study of maximal p-subgroups, is well understood for any p in *solvable* groups of finite Morley rank, and in any group of finite Morley rank for the prime $p = 2$. The second point is the key for a classification program of simple groups of finite Morley rank, suggested by A. Borovik and based on the architecture of the Classification of the Finite Simple Groups. In this process, some specific developments have naturally been needed for groups of finite Morley rank. In this context there is a priori no Jordan decomposition as in the linear algebraic context, and hence no nice distinction between semisimple and unipotent elements. The situation is furthermore enormously complicated by some so-called *bad* fields, as we will see in §1.7. Nevertheless, the finiteness of the Morley rank has allowed J. Burdges to develop a graduated notion of unipotence in this general context. This graduated notion of unipotence leads naturally to a new kind of Sylow theory, not related to torsion elements directly, but rather to the *unipotence degree* of the subgroups involved. In finite groups the study of Carter subgroups mostly boils down to Sylow theory; in groups of finite Morley rank this is replaced by this new kind of Sylow theory.

More precisely, we deal here with $\tilde{p}$-groups, where $\tilde{p} = (p, r)$ and p should be understood as the usual prime, or ∞ when dealing with elements of infinite order or merely divisible groups (which is more or less the same up to saturation). In this theory the unipotence degree r measures simultaneously how much a $\tilde{p}$-group can act on, and be acted upon by, another such group. Our $\tilde{p}$-groups are connected and nilpotent by definition and can really be thought of as the p-groups from finite group theory, incorporating the important unipotence degree parameter in our context. They are of three types depending on the value of $\tilde{p}$, listed below by increasing unipotence degree.

- $(\infty, 0)$-groups, or (abelian) "decent tori",
- (∞, r)-groups, with $0 < r < \infty$, or "nilpotent Burdges' $U_{\tilde{p}}$-groups",
- (p, ∞)-groups, with p prime, or (nilpotent) "p-unipotent groups".

This will be explained in §2. In particular, we will see in §2.4 that these $\tilde{p}$-groups cover in some sense all the "basic" connected groups which can occur in our context.

Imposing maximality on these $\tilde{p}$-groups leads naturally to a notion of Sylow theory, reminiscent of that of finite group theory. These new Sylow $\tilde{p}$-subgroups allow one to show the existence of Carter subgroups in any group of finite Morley rank, and hence to have a good approximation of Cartan subgroups of an algebraic group in any case. The natural question arising then is that of their conjugacy. This remains an open problem in general, but we will see in §3 that conjugacy of Carter subgroups is known in two important cases: under a generosity assumption on the one hand, and in solvable groups on the other. We say that a definable subgroup is *generous* if its conjugates cover the ambient group generically. Generosity appeared over the years to be a weak form of conjugacy, and this is confirmed for Carter subgroups also. More precisely, we will see in §3.3 below that an arbitrary group of finite Morley rank contains at most one conjugacy class of generous Carter subgroups. Using this conjugacy result by generosity, we rework then the theory of Carter subgroups in connected solvable groups of finite Morley rank, which was well developed before the unipotence theory mentioned above came into play.

In the present paper we are mostly concerned with conjugacy of certain *connected* subgroups, except in the parenthetical §6.5 which deals with nonnecessarily connected solvable groups of finite Morley rank. In §3.8 we will also compare the theory of Carter subgroups in groups of finite Morley rank, which relies heavily on connectedness, to its analog in finite group theory, where of course connectedness has no exact analog. Concerning the conjugacy of certain connected subgroups of groups of finite Morley rank, the most challenging conjectures are probably the three following.

Conjugacy Conjectures *In any group of finite Morley rank,*

1.12 *Borel subgroups are conjugate,*

2.6 *Sylow $\tilde{p}$-subgroups are conjugate,*

3.1 *Carter subgroups are conjugate.*

We will never consider Conjecture 1.12 here, but we will see that it is stronger than the two others, as we will see the conjugacy of Carter subgroups and of Sylow $\tilde{p}$-subgroups in connected solvable groups. As visible already, solvable groups satisfy many conjugacy theorems. This is merely because they mesh perfectly well with induction arguments, and for this reason the majority of results surveyed here concern solvable groups of finite Morley rank.

For finite solvable groups, *formation theory* is a general and powerful framework for analyzing the interplay between Sylow subgroups, Carter subgroups, and conjugacy. In §4 below we develop, with new results, a very general subformation theory designed for connected solvable groups of finite Morley rank. This theory encapsulates Carter subgroups and several generalizations of Sylow $\tilde{p}$-subgroups in connected solvable groups of finite Morley rank. All expectable conjugacy theorems are derived in §5 below.

To summarize, the architecture of this paper is as follows. The first section concerns preliminary developments on groups of finite Morley rank, with an emphasis on classical, i.e. involving torsion elements, Sylow theory in solvable groups. Then we develop in §2 the theory of semisimplicity and unipotence. In §3 we consider Carter subgroups, with their existence in general and their conjugacy in two important cases. Then in §4 and §5 we are concerned with subformation theory. In §4 a general and quite formal subformation theory is developed, and in §5 we give applications, with conjugacy and structural theorems in connected solvable groups. Finally, §6 deals with additional structural results in solvable groups of finite Morley rank, which are of a slightly different nature but of a certain interest. To conclude, we give in §7 a few examples of applications of this theory beyond solvable groups.

All developments and related notions presented here are well understood under a *linearity* assumption, thanks to the work of Y. Mustafin [Mus04]. Here we work with no linearity assumption and refer to that paper for linear groups.

1 Preliminary developments

We start with some early developments on groups of finite Morley rank.

1.1 Borovik-Poizat axioms

We consider groups $\langle G, \cdot, ^{-1}, 1, \cdots \rangle$ from a model theoretic point of view and think in G^{eq} throughout. They may carry additional structure, and this is an important issue in this context. For example, a group definable in another may carry extra structure not definable in its own pure group structure. We say that a group is *ranked* if there is a function "rk", assigning to each nonempty definable set an integer (its *rank*, or dimension) and satisfying the following axioms for every definable sets A and B:

Definition: $\text{rk}\,(A) \geq n+1$ if and only if A contains infinitely many pairwise disjoint subsets A_i such that $\text{rk}\,(A_i) \geq n$.

Definability: For every uniformly definable family A_b of subsets of A, with b varying in B, the set of elements $b \in B$ such that A_b is of given rank n is a definable subset of B.

Finite sets: For every uniformly definable family A_b of subsets of A, with b varying in B, there is a uniform bound on the cardinals of the finite sets A_b.

In an arbitrary structure, the existence of such a rank implies superstability [BC02]. In a group theoretic context this is equivalent to the finiteness of Morley rank [Poi87], implying in particular the additivity of the rank. We rather tend to work with these purely combinatorical axioms and the book [BN94a] develops all the theory from them.

1.2 Connectedness

If $X \subseteq Y$ are two definable sets in a group of finite Morley rank, we say that X is *generic* in Y if $\text{rk}\,(X) = \text{rk}\,(Y)$. Each definable set has a finite (Morley) *degree*, the maximal number of disjoint generic subsets. It follows easily that groups of finite Morley rank satisfy the Descending Chain Condition on definable subgroups. In particular, such a group G has a smallest definable subgroup of finite index, the intersection of all of them, its *connected component* denoted by G°. It is of course a definably characteristic subgroup, and G is said to be *connected* if $G = G^\circ$. The main property of connected groups can be stated as follows.

Lemma 1.1 [Che79] *A group of finite Morley rank is connected if and only if it has Morley degree one.*

We are going to deal essentially with connected groups. Most of the time, connectedness allows one to avoid all complications from finite combinatorics. For example, the following corollary of Lemma 1.1 has a striking application in §3.3 below.

Corollary 1.2 *A connected group of finite Morley rank acting definably on a finite set fixes it pointwise.*

By the Descending Chain Condition on definable subgroups again, each (nonnecessarily definable) subset X of a group G of finite Morley rank is contained in a smallest definable subgroup $d(X)$, its *definable closure*, which can be seen as a sharper form of the Zariski closure in the algebraic context. This allows one to define the *generalized* connected component of X as $X^\circ = d(X)^\circ \cap X$. If X is a subgroup of G, one sees easily that X° is still, though not necessarily definable, a normal subgroup of finite index in X, and again one says that X is *connected* if $X = X^\circ$. This generalized connected component is particularly relevant for Sylow theory in the classical sense of the study of torsion subgroups. For example, the torsion subgroup of the multiplicative group $\mathbb{C}^\times$ of the complex numbers is not first-order definable in the pure field structure.

For the sake of future arguments, we include here some corollaries of Zilber's theorem on indecomposable sets and connectedness.

Theorem 1.3 ([Zil77]; [BN94a, Corollary 5.29]) *Let G be a group of finite Morley rank. Then:*

- *a. Any family of definable connected subgroups of G generates a definable connected subgroup of G.*
- *b. If H is a definable connected subgroup of G and X any subset, then the commutator subgroup $[H, X]$ is a definable connected subgroup of G.*

1.3 Classical Sylow theory

If p is a prime, a *p-torus* is a divisible abelian p-subgroup of a group of finite Morley rank. By abelian group theory, such a subgroup is a direct product of copies of the quasicylic Prüfer p-group $\mathbb{Z}_{p^\infty}$. In the finite Morley rank context, the number of copies must be *finite* [BP90], and is called the *Prüfer p-rank.* Typically, if K is an algebraically closed field of characteristic different from p, then the n-dimensional torus $K^\times \times \cdots \times K^\times$ contains a p-torus of Prüfer p-rank n.

At the opposite of p-tori, which are of unbounded exponent and not necessarily definable, *p-unipotent* subgroups are the definable connected nilpotent p-subgroups of bounded exponent of groups of finite Morley rank. Notice nilpotence in our definition. A typical example of a p-unipotent group is the group of strictly upper triangular matrices of the general linear group $\mathrm{GL}_n(K)$, with K an algebraically closed field of characteristic p.

The following result describes locally finite p-subgroups of groups of finite Morley rank, mostly in terms of a p-torus and of a p-unipotent subgroup.

Theorem 1.4 [BP90] *Let P be a locally finite p-subgroup, p prime, of a group of finite Morley rank. Then $P^\circ = T * U$ is a central product, with finite intersection, of a p-torus T and a p-unipotent subgroup U.*

It is well known that torsion subgroups of solvable groups are locally finite, and thus the preceding theorem applies in particular to any p-subgroup of a solvable group of finite Morley rank. As usual, one defines *Sylow* p-subgroups as maximal p-subgroups, or, equivalently in a solvable context, as maximal locally finite p-subgroups. The conjugacy of Sylow p-subgroups is not known in general, except in a solvable context or for the prime $p = 2$. Indeed, the singularity of the prime $p = 2$ yields an absolute control of 2-subgroups.

Theorem 1.5 [BP90] *In any group of finite Morley rank, maximal 2-subgroups are locally finite and conjugate.*

Theorem 1.5 is the origin of many arguments in the presence of nontrivial 2-elements. Here we are going to concentrate on aspects not depending on such a presence, hence on other primes and even (mostly, indeed) elements of infinite order.

As alluded to already, there is a conjugacy theorem for Sylow p-subgroups in a solvable context. This is indeed true for a larger class of torsion subgroups. If π is a set of primes, then a *Hall* π-subgroup of a solvable group G of finite Morley rank is a maximal π-subgroup of G.

Theorem 1.6 [ACCN98] *In any solvable group of finite Morley rank, Hall π-subgroups are conjugate for any set π of primes.*

There is an analog of the structural Theorem 1.4 for Hall π-subgroups.

Theorem 1.7 [Fré00c, Proposition 4.22] *Let G be a solvable group of finite Morley rank and R a Hall π-subgroup of G. Then $R^\circ = TU$ where $U \trianglelefteq G$ is a definable connected subgroup of bounded exponent and T an abelian divisible subgroup.*

Also, as in the algebraic context, Hall π-subgroups of *connected* solvable groups of finite Morley rank are connected.

Theorem 1.8 ([Fré00b, Corollaire 7.15], see also [BN92]) *Let G be a connected solvable group of finite Morley rank. Then Hall π-subgroups of G are connected.*

Finally, there are results of a Schur-Zassenhaus type, due to A. Borovik and A. Nesin.

Theorem 1.9 [BN92, BN94b] *Let G be a solvable group of finite Morley rank and H a normal Hall π-subgroup of G. Then:*

a. H has a complement in G.
b. If H is of bounded exponent, then any subgroup K of G with $K \cap H = 1$ is contained in a complement of H in G, and the complements of H in G are definable and conjugate.

1.4 Generalized Hall π-subgroups

Of course, the preceding theorems depend heavily on the presence of torsion elements. To deal with infinite groups of finite Morley rank, one would also like some kind of similar theory for elements of infinite order, at least in solvable groups again. An attempt in this direction is taken in [Fré00c], leading to the following definition. Denoting by $\mathcal{P}$ the set of all primes together with the ∞ symbol, we consider arbitrary subsets π of $\mathcal{P}$ and $\pi^\perp = \mathcal{P} \setminus \pi$. If G is a solvable group of finite Morley rank and R a subgroup of G, we say that:

- An element $x \in G$ is a *π-element* if, for every $p \in \pi^\perp$, $d(x)$ has no elements of order p.
- R is a *π-subgroup* if each $x \in R$ is a π-element.
- R is a *Hall* π-subgroup of G if R is a maximal π-subgroup of G.

Of course, this definition coincides with the usual one if π consists of finite primes only. The main feature of this definition allowing the infinite prime is that Theorems 1.6 and 1.7 on conjugacy and structure are preserved.

Theorem 1.10 [Fré00c, Théorème 4.18, Proposition 4.22] *Let G be a solvable group of finite Morley rank and π any subset of $\mathcal{P}$. Then:*

a. Hall π-subgroups are conjugate.

b. For any Hall π-subgroup R of G, $R^\circ = UBD$ where $U \trianglelefteq G$ is a definable torsion-free subgroup, $B \trianglelefteq G$ a definable connected subgroup of bounded exponent and D a divisible nilpotent subgroup. Moreover, R is locally closed in the sense of §6.5 below.

If the ambient group G is connected, then it is also shown in [Fré00c] that its Hall π-subgroups in this generalized sense are connected.

1.5 Borel subgroups

As visible already, conjugacy theorems are particularly abundant in solvable groups of finite Morley rank. The following result links a given group of finite Morley rank to its solvable subgroups.

Theorem 1.11 ([Fré00a, Corollaire 3.4.4], see also [ACCN98]) *Every locally solvable subgroup of a group of finite Morley rank is solvable*

In general, we are mostly concerned with definable connected subgroups. A *Borel* subgroup of a group G of finite Morley rank is a maximal definable connected solvable subgroup of G. The following very strong conjecture is widely open.

Conjecture 1.12 *In any group of finite Morley rank, Borel subgroups are conjugate.*

Conjecture 1.12 covers all natural conjugacy conjectures which are formulated here about connected subgroups of groups of finite Morley rank. For example, we will see that it is stronger than both Conjectures 3.1 and 2.6 below, by Theorems 3.11 and 5.10 respectively. In particular, the class of connected solvable groups of finite Morley rank is very well understood, and this is extremely relevant as the analysis of an arbitrary group of finite Morley rank is often done with its Borel subgroups.

1.6 Actions

If X and Y are two definable subgroups of a group of finite Morley rank, then X is *Y-minimal* if it is infinite, normalized by Y, and minimal with respect to these properties. In a solvable context, the study of a serious action of a group of finite Morley rank on another most of the time boils down to the following crucial theorem, which gives in many cases an interpretable field.

Zilber's Field Theorem (cf. [BN94a, Theorem 9.1]) *Let $G = U \rtimes T$ be a group of finite Morley rank, with U and T infinite abelian definable subgroups, $C_T(U) = 1$ and U T-minimal. Then G interprets an algebraically closed field K with U definably isomorphic to K_+, T definably isomorphic to a definable subgroup T_1 of $K^\times$, and*

$$U \rtimes T \simeq K_+ \rtimes T_1 = \left\{ \begin{pmatrix} t & u \\ 0 & 1 \end{pmatrix} : t \in T_1 \ , u \in K_+ \right\}.$$

The *Fitting* subgroup of a group G of finite Morley rank is its maximal normal definable nilpotent subgroup. It is well defined and the unique maximal normal nilpotent subgroup of G [Nes91]. If B is a Borel subgroup of a linear algebraic group over an algebraically closed field, then $B = U \rtimes T$ where U is the maximal unipotent subgroup (strictly upper triangular matrices if B is the standard Borel subgroup) and T is a maximal torus of B (diagonal matrices). If the ambient group is simple, then $U = F(B)$ and, hence, Fitting subgroups are usually a good first approximation of the unipotent radical in the finite Morley rank context. In general, it is not known whether a connected solvable group B of finite Morley rank splits as $F(B) \rtimes T$ for some complement T. But any Carter subgroup Q of B satisfies $B = F(B)Q$ by Corollary 3.13 below, and hence provides a good approximation of maximal tori in this context.

For future references, we record here miscellaneous results around connected solvable groups of finite Morley rank.

Theorem 1.13 ([Nes90]; [BN94a, Ex. 5 p. 98]) *Let G be a connected solvable group of finite Morley rank. Then:*

a. *$G/F^\circ(G)$ is divisible abelian. In particular G' is nilpotent.*
b. *If A is G-minimal in G, then $A \leq Z(F^\circ(G))$*

1.7 Fields

An infinite field of finite Morley rank is always algebraically closed [Mac71]. If it is involved in some action, one might become extremely concerned with the definable subgroups of its multiplicative or additive group. The multiplicative group is particularly sensitive when Zilber's Field Theorem is applied.

A *bad* field is a field $\langle K, +, \cdot, 0, 1, T\rangle$ of finite Morley rank where T is an infinite proper definable subgroup of its multiplicative group $K^\times$. A pure algebraically closed field is strongly minimal, and hence cannot be bad. Also, the multiplicative subgroup T of a bad field cannot contain the multiplicative group of an infinite subfield of K [Wag90, New91, AC99]. It is commonly believed that there are no bad fields of characteristic $p > 0$ [Wag03], but building on [Poi01] the existence of bad fields of characteristic 0 has been announced very recently in [BHMPW06].

In characteristic $p > 0$ there is a deep theorem of F. Wagner which gives precious information on the multiplicative subgroup T.

Theorem 1.14 [Wag01] *Let $\langle K, T\rangle$ be a field of finite Morley rank of characteristic $p > 0$ and where T is a definable subgroup of the multiplicative group $K^\times$. Then*

$$\langle K \cap \tilde{\mathbb{F}}_p, T \cap \tilde{\mathbb{F}}_p\rangle \preceq \langle K, T\rangle.$$

In particular, there is always some torsion in the subgroup T of Theorem 1.14, and this good behaviour will be examined in the more general notion of good torus of §2.1. In characteristic 0 it is rather additive groups which have a good behavior.

Theorem 1.15 [Poi87, Corollaire 3.3] *Let K be a field of finite Morley rank of characteristic 0. Then the additive group K_+ has no nontrivial proper definable subgroups.*

Theorem 1.15 has no analog in characteristic $p > 0$, and indeed a field of characteristic $p > 0$ of Morley rank 2 and with a definable additive subgroup of Morley rank 1 is built in [BMPZ06]. Fortunately, we will see in §2.2 that the behaviour of p-unipotent groups is not really sensitive to such pathologies.

To summarize very informally and loosely, the situation with regards to multiplicative and additive subgroups of fields can be described as follows.

	char $(K) = 0$	char $(K) > 0$
$K^\times$	bad	rather good
K_+	very good	bad

There is unfortunately no analog of Theorem 1.14 in characteristic 0, and in the bad field of characteristic 0 of [BHMPW06] the subgroup T is torsion-free. That's why an abstract unipotence theory for groups of finite Morley rank cannot follow merely from Zilber's Field Theorem; the latter might fail to provide divisible torsion subgroups where one might expect some. However, if $T < K^\times$ in that theorem and if r and r' denote $\text{rk}(T)$ and $\text{rk}(K_+)$ respectively, then

$$r < r'$$

regardless of the presence of torsion in T. This rank asymmetry is the main motivation for the introduction of parameters r in the graduated unipotence theory of §2.3 below for torsion-free groups. In this perspective Theorem 1.15, coupled with Zilber's Field Theorem, is particularly useful in analyzing the action of a torsion-free group on another, as we will see in Proposition 2.10 below.

2 Semisimplicity versus unipotence

In this section we will consider "$\tilde{p}$-groups", with $\tilde{p} = (p, r)$, p prime or ∞, r nonnegative or ∞, and $r = \infty$ if and only if $p < \infty$. These groups are connected and nilpotent by definition, and can be thought of as variations of the usual p-groups from finite group theory. Here $p = \infty$ means that we are dealing with elements of infinite order, or merely divisible groups, which is more or less the same up to saturation. The prime p is not so important for our theory, but it is rather the parameter r which we call the "unipotence degree" and which plays a leading role. This parameter will measure how much a group can act on, or be acted upon by, another. The bigger r will be, the less the group will act on seriously, and the smaller r will be, the less the group will be acted upon seriously.

2.1 Decent tori

We first look for the best approximation to algebraic tori, hence groups which are semisimple and not unipotent at all. If K is a pure algebraically closed field, then an algebraic n-dimensional torus $K^\times \times \cdots \times K^\times$ is divisible abelian and has the property that each Zariski closed subgroup is the Zariski closure of its torsion subgroup. This generalizes naturally to the notion of *good torus*, i.e. of divisible abelian group of finite Morley rank in which each definable subgroup is the definable closure of its torsion subgroup. The following examples are provided as a consequence of Theorem 1.14.

Example 2.1 [AC04, §3] Let K be an infinite field of finite Morley rank of characteristic $p > 0$. Then any finite product $K^\times \times \cdots \times K^\times$ of copies of its multiplicative group $K^\times$ is a good torus.

For our present purpose a weaker property suffices. A *decent torus* is a divisible abelian group of finite Morley rank of the form $d(S)$, where S is its (divisible) torsion subgroup. By divisibility, decent tori are connected. By finiteness of Prüfer p-ranks, for each n there are only finitely many elements of order n in a decent torus. Hence Corollary 1.2 implies that decent tori cannot be acted upon in the following sense.

Proposition 2.2 (Rigidity of decent tori) *Let T be a decent torus definable as a subgroup of a group of finite Morley rank. Then $N^\circ(T) = C^\circ(T)$.*

Soon, we will use the term "$(\infty, 0)$-groups" for decent tori, ∞ designating divisiblity, and 0 designating groups which are not at all unipotent in the sense of Proposition 2.2.

The presence of nontrivial torsion elements with finiteness conditions in decent tori allowed G. Cherlin to prove the following conjugacy theorem suggested by the behaviour of algebraic tori.

Theorem 2.3 [Che05] *Let G be any group of finite Morley rank. Then $C^\circ(T)$ is generous in G for every definable decent subtorus T of G, and maximal definable decent subtori of G are conjugate.*

It is noteworthy in Theorem 2.3 that the conjugacy result is indeed obtained as a consequence of the genericity result. In §3.3 below we will see other conjugacy results obtained via genericity.

2.2 p-unipotent groups

At the opposite of decent tori which are divisible, we now turn our attention to groups of bounded exponent. The "basic" ones are the p-unipotent subgroups. Recall that they are connected and nilpotent in our definition, though there exist potentially infinite simple groups of finite Morley rank of bounded exponent. The following dual of Proposition 2.2 can be deduced from Zilber's Field Theorem [ABC97, Fact 2.36].

Proposition 2.4 (p-unipotent action) *Let HU be a group of finite Morley rank, with H and U definable connected nilpotent subgroups, U p-unipotent for some prime p, and H normal. Then HU is nilpotent.*

With Theorem 1.13 (a), it follows that any connected solvable group G of finite Morley rank has a unique maximal p-unipotent subgroup, its *p-unipotent radical* denoted by $U_p(G)$, which is contained in $F^\circ(G)$ and definably characteristic in G. Such a unipotent radical notion will be defined in a considerable higher level of generality in §2.6 below.

While decent tori cannot be acted upon by Proposition 2.2, p-unipotent groups cannot act on by Proposition 2.4. Hence we have found our most extreme basic subgroups in our unipotence theory, and we will use the term "(p,∞)-groups" for p-unipotent groups, ∞ designating an infinite unipotence degree in the sense of Proposition 2.4.

2.3 Burdges' $U_{\tilde{p}}$-groups

The dual Propositions 2.2 and 2.4 rely heavily on the presence of torsion. But it is not a secret that many groups are torsion-free. It is thus necessary to develop a theory for actions of torsion-free groups, or, equivalently by Zilber's Field Theorem, of unipotence in characteristic 0. All this technology comes from Burdges' Thesis [Bur04a].

A group A of finite Morley rank is *indecomposable* if it is abelian and not the sum of two proper definable subgroups (without any assumption of disjointness on the two subgroups). It is easily seen [Bur06b, Lemma 1.2] that any infinite indecomposable group of finite Morley rank is connected. Given an indecomposable group A, we let $\Phi(A)$ denote the maximal definable connected proper subgroup of A. By indecomposability it is well defined, and $\Phi(A)$ is indeed in this case the Frattini subgroup we will introduce more generally in §2.7.

Given a group G of finite Morley rank and an integer $r \geq 1$, we let $\tilde{p} = (\infty, r)$ and define

$$U_{\tilde{p}}(G) = \left\langle A \leq G \;\middle|\; \begin{array}{c} A \text{ is a definable indecomposable subgroup,} \\ A/\Phi(A) \text{ is torsion-free and of rank } r \end{array} \right\rangle$$

As indecomposable groups are connected, $U_{\tilde{p}}(G)$ is definable and connected by Theorem 1.3, and we say that G is a *Burdges' $U_{\tilde{p}}$-group*, or a *$U_{\tilde{p}}$-group* for short, if $G = U_{\tilde{p}}(G)$. In the literature the notation "$U_{0,r}$" is used instead of "$U_{(\infty,r)}$". We prefer the ∞ symbol here to emphasize that we are dealing with some elements of infinite order and to be coherent with the notation "$U_{\tilde{p}}$" of §2.6 below, incorporating decent tori and p-unipotent groups as well. The parameters r will take on their full significance in Proposition 2.10 below. For the moment we merely mention some stability properties under definable homomorphisms.

Lemma 2.5 [Bur04b, Lemma 2.11] *Let $f : G \longrightarrow H$ be a definable homomorphism between two groups of finite Morley rank. Then*

a. (Push-forward) $f(U_{\tilde{p}}(G)) \leq U_{\tilde{p}}(H)$ is a $U_{\tilde{p}}$-group.
b. (Pull-back) If $U_{\tilde{p}}(H) \leq f(G)$, then $f(U_{\tilde{p}}(G)) = U_{\tilde{p}}(H)$.

In particular, an extension of a $U_{\tilde{p}}$-group by a $U_{\tilde{p}}$-group is a $U_{\tilde{p}}$-group.

2.4 $\tilde{p}$-groups

Now we incorporate decent tori and p-unipotent groups into the classes of *nilpotent* $U_{\tilde{p}}$-groups. We consider couples $\tilde{p} = (p, r)$, with p prime or ∞, r a nonnegative integer or ∞, and with $r = \infty$ if and only if $p < \infty$. We define *$\tilde{p}$-groups* as follows, depending on the value of $\tilde{p}$.

- $(\infty, 0)$-groups: decent tori,
- (∞, r)-groups, with $0 < r < \infty$: nilpotent $U_{\tilde{p}}$-groups,
- (p, ∞)-groups, with p prime: p-unipotent groups.

The parameter p is called the *characteristic*, or *prime exponent*, of the $\tilde{p}$-group considered, and r is called its *unipotence degree*, or *weight*. The couple $\tilde{p} = (p, r)$ is called a *unipotence parameter*. By definition, $\tilde{p}$-groups are connected and nilpotent. Unlike for finite p-groups, being a $\tilde{p}$-group is in general not a notion inherited by passing to definable connected subgroups, and we will consider such homogeneity issues in §2.8.

If G is a group of finite Morley rank and $\tilde{p}$ as above, then by the Ascending Chain Condition on definable connected subgroups the union of a chain of definable $\tilde{p}$-subgroups is also a definable $\tilde{p}$-subgroup, and we naturally call any maximal definable $\tilde{p}$-subgroup of G a *Sylow* $\tilde{p}$-subgroup of G. The conjugacy of such subgroups is a natural conjecture.

Conjecture 2.6 *In any group of finite Morley rank, Sylow $\tilde{p}$-subgroups are conjugate.*

Conjecture 2.6 holds for $\tilde{p} = (\infty, 0)$ by Theorem 2.3. It also holds for any $\tilde{p}$ in connected solvable groups as we will see in Theorem 5.10

below, so that Conjecture 2.6 is weaker than Conjecture 1.12. For the moment, we merely concentrate on properties $\tilde{p}$-groups share with the classical finite p-groups. First, we mention an analog for $\tilde{p}$-groups of the decomposition of finite nilpotent groups as direct product of their Sylow p-subgroups.

Theorem 2.7 [Bur06b, §3] *Any connected nilpotent group of finite Morley rank is the central product of its Sylow $\tilde{p}$-subgroups.*

In other words, a group G as in Theorem 2.7 can be written in our earlier notation as

$$[d(S) * U_{(\infty,1)}(G) * \cdots * U_{(\infty,r_{\max})}(G)] * [U_2(G) \times \cdots \times U_{p_{\max}}(G)]$$

where $d(S)$ denotes the maximal decent torus of G, equivalently S is its maximal divisible abelian torsion subgroup, $r_{\max}$ the maximal integer r such that G contains a nontrivial $U_{(\infty,r)}$-group, and $p_{\max}$ the maximal prime p such that G contains a nontrivial p-unipotent subgroup. Indeed, it is known from a theorem of A. Nesin [BN94a, Theorem 6.8] that G is a central product $G = D * B$ of a definable divisible subgroup D, with central torsion, and of a definable subgroup B of bounded exponent. The decomposition of B as a central product of its Sylow p-subgroups gives the decomposition of B into its maximal p-unipotent subgroups. The decomposition of D as a central product of its maximal $\tilde{p}$-subgroups is shown in [Bur06b, §3].

By Reineke's Theorem [BN94a, Theorem 6.4], minimal infinite groups of finite Morley rank are abelian. Hence, Theorem 2.7 says in particular that $\tilde{p}$-groups cover all the "basic" connected subgroups which can be encountered in a group of finite Morley rank. Furthermore, it follows easily from Theorem 2.7 that a $\tilde{p}$-group cannot be a $\tilde{q}$-group whenever $\tilde{p} \neq \tilde{q}$.

We recall again that $\tilde{p}$-groups are nilpotent by definition. One more piece of strong analogy with finite p-groups is the following version of the normalizer condition.

Proposition 2.8 (Normalizer Condition, [Bur06b, Lemma 2.4]) *Let G be a $\tilde{p}$-group and $H < G$ a proper definable subgroup. If S_1 is the Sylow $\tilde{p}$-subgroup of H and S_2 that of $N_G(H)$, then $S_1 < S_2$.*

Proof. As decent tori are abelian, this is clear when $\tilde{p} = (\infty, 0)$. When $\tilde{p} = (p, \infty)$, this follows from the normalizer condition in connected nilpotent groups [BN94a, Lemma 6.3]. The remaining cases are treated by J. Burdges in [Bur06b, Lemma 2.4]. □

We finish here by mentioning that commutators in $\tilde{p}$-groups tend to provide a good source of $\tilde{p}$-groups.

Lemma 2.9 (compare with [Bur06b, Corollary 3.7]) *Let G be a group of finite Morley rank, S a subset of G, and H a definable $\tilde{p}$-subgroup of G normalized by S. Then $[H, S]$ is a $\tilde{p}$-subgroup of H.*

Proof. This is clear for p-unipotent groups, so we may assume that the unipotence degree r of H is finite. As the central product of two decent tori is a decent torus, we may assume that S is a singleton $\{s\}$. If H is a decent torus, then the map $\gamma(h) = [h, s]$ is a definable homomorphism from H onto $[H, s]$, and its image must be a decent torus. Assume now $0 < r < \infty$. Both H and s normalize $[H, s]$. By Lemma 2.5 $[H, s]$ modulo $U_{\tilde{p}}([H, s])$ has a trivial Sylow $\tilde{p}$-subgroup. Hence, by Theorem 2.7, each of its Sylow $\tilde{q}$-subgroups is central in the $\tilde{p}$-group H modulo $U_{\tilde{p}}([H, s])$, and $[H, s]$ modulo $U_{\tilde{p}}([H, s])$ is indeed in the center of H modulo $U_{\tilde{p}}([H, s])$. Now the map $\gamma(h) = [h, s]$ induces a definable homomorphism from H modulo $U_{\tilde{p}}([H, s])$ onto $[H, s]$ modulo $U_{\tilde{p}}([H, s])$. But its image must be a $\tilde{p}$-group by Lemma 2.5 (push-forward), and $[H, s] = U_{\tilde{p}}([H, s])$. □

2.5 Actions and $\tilde{p}$-groups

In the next proposition we look at actions of $\tilde{p}$-groups on each other, clarifying the interest of the unipotence degree parameter r. We rework here, without invoking the theory of Carter subgroups, proofs which can be found alternatively in [Bur06b, §4] or [FJ05, §3].

Proposition 2.10 *Let U_1U_2 be a group of finite Morley rank, with each U_i a definable nilpotent subgroup and U_1 normal. Assume U_1 is generated by definable $\tilde{p}$-subgroups of unipotence degrees $\leq r$ and U_2 is generated by definable $\tilde{p}$-subgroups of unipotence degrees $\geq r$, where $0 \leq r \leq \infty$. Then U_1U_2 is nilpotent.*

Proof. We assume toward a contradiction that $G = U_1U_2$ is a counterexample of minimal rank. By Propositions 2.2 and 2.4 and Theorem 2.7, we may easily assume that the generating $\tilde{p}$-subgroups are $U_{(\infty,s)}$-groups with $0 < s < \infty$.

Let A be a G-minimal subgroup of U_1. By Lemma 2.5 and the induction hypothesis, G/A is nilpotent, and $A \not\leq Z(G)$. Since U_1 and U_2 are divisible (Theorem 2.7), A is divisible and, since $A \not\leq Z(G)$, A is torsion-free. By Theorem 1.13, $A \leq Z(F^\circ(G))$ and $G/C(A)$ is abelian, so Zilber's Field Theorem applies in $A \rtimes G/C(A)$. It gives an interpretable field of characteristic 0, and

$$(*) \quad \begin{array}{l} \text{either } \mathrm{rk}\,(G/C(A)) = \mathrm{rk}\,(A) \text{ and } G/C(A) \text{ is not torsion-free,} \\ \text{or } \mathrm{rk}\,(G/C(A)) < \mathrm{rk}\,(A), \end{array}$$

depending on whether $G/C(A)$ is, or is not, the full multiplicative group of the field. Besides, the subgroup A, being the additive group of the field of characteristic 0, has no nontrivial proper definable subgroups by Theorem 1.15. In particular A is indecomposable and $\Phi(A) = 1$.

Defining by induction $V_1 = U_1$ and $V_{k+1} = [V_k, G]$, one sees with Theorem 2.7 and Lemma 2.9 that the normal definable connected subgroups V_k are generated, like U_1, by their Sylow $\tilde{p}$-subgroups of unipotence degrees s, with $0 < s \leq r$. By Descending Chain Condition the sequence V_k is stationary at some step ℓ, and $V_\ell \neq 1$ as G is not nilpotent. Since G/A is nilpotent, $V_\ell \leq A$, and $A = V_\ell$ by G-minimality of A. Hence $\mathrm{rk}\,(A) = s$ for some $s \leq r$.

As $U_1 \leq F^\circ(G)$, $U_1 \leq C(A)$. Hence $G/C(A)$ is generated by its Sylow $\tilde{p}$-subgroups of unipotence degrees at least r by Lemma 2.5, and in particular it contains a definable indecomposable subgroup X such that $X/\Phi(X)$ is torsion-free and of rank $\geq r$. In particular, $r \leq \mathrm{rk}\,(G/C(A))$. Now $(*)$ gives $r \leq \mathrm{rk}\,(G/C(A)) \leq \mathrm{rk}\,(A) = s \leq r$, and all these quantities are equal. In particular $G/C(A)$ is not torsion-free by $(*)$. But as $r \leq \mathrm{rk}\,(X/\Phi(X)) \leq \mathrm{rk}\,(G/C(A)) = r$, the only possibility for the definable connected subgroup X of $G/C(A)$ is that $X = G/C(A)$, and the definable connected subgroup $\Phi(X)$ must be trivial. Hence $G/C(A) \simeq X/\Phi(X)$ is torsion-free, a contradiction. □

2.6 The $U_{\tilde{p}}$-radical

Given a group G of finite Morley rank and a unipotence parameter $\tilde{p} = (p, r)$ as above, we define the *$U_{\tilde{p}}$-radical* of G as

$$U_{\tilde{p}}(G) = \langle S \mid S \text{ is a definable } \tilde{p}\text{-subgroup of } G \rangle$$

By Theorem 1.3 (a), $U_{\tilde{p}}(G)$ is definable and connected, and we say that G is a *$U_{\tilde{p}}$-group* if $G = U_{\tilde{p}}(G)$. As it is given by a generation property, this definition is very general and does not imply nilpotence. For example, a connected simple group of finite Morley rank is a $U_{\tilde{p}}$-group for any $\tilde{p}$ for which it contains a nontrivial Sylow $\tilde{p}$-subgroup.

Of course, $\tilde{p}$-groups are $U_{\tilde{p}}$-groups, and nilpotent $U_{\tilde{p}}$-groups are $\tilde{p}$-groups thanks to Theorem 2.7. If $p = \infty$, then $\tilde{p}$-groups are generated by their abelian definable $\tilde{p}$-subgroups by definition, so that the radical $U_{\tilde{p}}(G)$ defined here corresponds to that of §2.3 above if $0 < r < \infty$. For p prime however, one cannot reduce the generation property to abelian subgroups, as p-unipotent groups are not necessarily generated by their abelian p-unipotent subgroups [BN94a, §3.2.3].

In connected solvable groups the $U_{\tilde{p}}$-radical provides for certain unipotence parameters $\tilde{p}$ a good notion of "unipotent radical", which justifies the letter "U" for this radical. Indeed, the "unipotence degree" terminology for the parameters r is entirely clarified by Proposition 2.10 in terms of actions, but it was primarily motivated by the following consequence of that proposition in solvable groups (see [Bur04b, Theorem 2.16] for the original).

Lemma 2.11 *Let G be a solvable group of finite Morley rank, and $\tilde{p}$ a unipotence parameter maximizing the unipotence degree of a nontrivial definable $\tilde{p}$-subgroup of G. Then $U_{\tilde{p}}(G) \leq F^\circ(G)$ is the Sylow $\tilde{p}$-subgroup of $F^\circ(G)$.*

Proof. If U is any definable $\tilde{p}$-subgroup of G, then $F^\circ(G)U$ is nilpotent by Proposition 2.10, and normal in G° by Theorem 1.13. Hence $F^\circ(G)U$ is contained in $F^\circ(G)$, and our claim follows. □

Using Theorem 1.14, J. Burdges also shows in [Bur06b, Lemma 4.3] that if a $\tilde{p}$-group of unipotence degree r, with $0 < r < \infty$, normalizes a p-unipotent group, then the resulting product is nilpotent. It follows then easily that the conclusion of Lemma 2.11 also holds if G contains a nontrivial Sylow $\tilde{p}$-subgroup of unipotence degree r, $0 < r < \infty$, and $\tilde{p}$ maximizes the unipotence degree of these $\tilde{p}$-subgroups.

The following additional result of J. Burdges definitively clarifies links with unipotence in algebraic groups.

Theorem 2.12 [Bur04a] *Let G be a connected solvable group of finite Morley rank whose $U_{(p,r)}$-radicals are trivial when $0 < r \leq \infty$. Then G is a good torus.*

Of course, one can elaborate on the definition of the $U_{\tilde{p}}$-radical as follows. If $\tilde{\pi}$ is any set of unipotence parameters, then one can define the $U_{\tilde{\pi}}$*-radical* of G as

$$U_{\tilde{\pi}}(G) = \langle U_{\tilde{p}}(G) \mid \tilde{p} \in \tilde{\pi} \rangle$$

and all relevant notions have their obvious analog in this slightly more general context. In [Bur04a] J. Burdges refines this definition further by imposing extra limitations on divisible torsion. We won't go into these additional refinements. For $U_{\tilde{\pi}}$-groups, the push-forward and pull-back Lemma 2.5 takes the following general form.

Lemma 2.13 *Let $f : G \longrightarrow H$ be a definable homomorphism between two groups of finite Morley rank. Then*

a. (Push-forward) $f(U_{\tilde{\pi}}(G)) \leq U_{\tilde{\pi}}(H)$ is a $U_{\tilde{\pi}}$-group.
b. (Pull-back) Assume all unipotence degrees involved in $\tilde{\pi}$ are finite, or G is solvable. If $U_{\tilde{\pi}}(H) \leq f(G)$, then $f(U_{\tilde{\pi}}(G)) = U_{\tilde{\pi}}(H)$.

In particular, an extension of a solvable $U_{\tilde{\pi}}$-group by a solvable $U_{\tilde{\pi}}$-group is a $U_{\tilde{\pi}}$-group.

Proof. As images of $\tilde{p}$-groups are $\tilde{p}$-groups, the push-forward part follows readily. The pull back of decent tori is shown in [Fré06b, Lemma 3.1] using Theorem 2.3 and a saturation argument, and that of Burdges' $U_{\tilde{p}}$-groups is Lemma 2.5. Hence each piece can be pulled back component by component, or using Theorem 1.13 in presence of a p-unipotent subgroup in a solvable group. □

Unfortunately, we are not able at the moment to pull back p-unipotent groups in general, except in a solvable context.

2.7 The Frattini subgroup

The Frattini subgroup will eventually become an important actor in the theory we are developing. In general group theory, this subgroup is defined as the intersection of all maximal proper subgroups of the ambient group, and in finite group theory the study of the Frattini subgroup mostly boils down to Sylow theory. We define here a Frattini subgroup more adapted to our connectedness concerns, and where Sylow theory is usually replaced by our theory of Carter subgroups. If G is a group of finite Morley rank, its *Frattini* subgroup, denoted by $\Phi(G)$, is the intersection of all maximal proper *definable connected* subgroups of G. We should maybe call this subgroup the "connected Frattini" subgroup of G, to emphasize that the intersection is taken over definable *connected* subgroups only, but we prefer to keep the terminology lite. As for the Fitting subgroup, the Frattini subgroup is in general not connected with this definition, but many results remain true for its connected component.

Two straightforward lemmas on the Frattini subgroup:

Lemma 2.14 [Fré00b, Lemma 5.7, Lemma 5.8] *Let G be a connected group of finite Morley rank.*

- *a. Let N be a definable normal subgroup of G. If $\Phi(G)$ contains N, then $\Phi(G/N) = \Phi(G)/N$. In particular $\Phi(G/\Phi(G)) = 1$.*
- *b. Let H be a definable subgroup of G. If $G = \Phi(G)H$, then $G = H$.*

Lemma 2.14 (*b*) has an interesting consequence relating in general $U_{\tilde{\pi}}$-groups to their Frattini subgroups.

Corollary 2.15 *Let G be a connected group of finite Morley rank, and $\tilde{\pi}$ a set of unipotence parameters. Assume all unipotence degrees involved in $\tilde{\pi}$ finite, or G solvable. Then G is a $U_{\tilde{\pi}}$ group if and only if $G/\Phi(G)$ is a $U_{\tilde{\pi}}$-group.*

Proof. If G is a $U_{\tilde{\pi}}$-group, then $G/\Phi(G)$ is a $U_{\tilde{\pi}}$-group by Lemma 2.13 (push forward). If $G/\Phi(G)$ is a $U_{\tilde{\pi}}$-group, then $G = \Phi(G)U_{\tilde{\pi}}(G)$ by Lemma 2.13 (pull back) and $G = U_{\tilde{\pi}}(G)$ by Lemma 2.14 (*b*). □

2.8 Homogeneity

Definable connected subgroups of $\tilde{p}$-groups are a priori not necessarily $\tilde{p}$-groups when $p = \infty$. To deal with this aspect, we say that a $U_{\tilde{p}}$-group G of finite Morley rank is

- *ultra $\tilde{p}$-homogeneous* if every definable connected subgroup of G is a $U_{\tilde{p}}$-group.
- *$\tilde{p}$-homogeneous* if every definable connected nilpotent subgroup of G is a $U_{\tilde{p}}$-group, or, equivalently, a $\tilde{p}$-group.

In general, we expect these two notions to coincide.

Conjecture 2.16 *A group is ultra $\tilde{p}$-homogeneous if and only if it is $\tilde{p}$-homogeneous.*

Connected solvable groups of finite Morley rank are generated by their definable connected nilpotent subgroups, by Corollary 3.13 below for example. Hence, in a $\tilde{p}$-homogeneous group any Borel subgroup is a $\tilde{p}$-group by Lemma 2.11, and it follows that Conjecture 2.16 has a positive answer if one considers groups in which each definable connected subgroup is generated by its Borel subgroups. But, unfortunately, the generation of connected groups by their Borel subgroups is an irritating open problem in general, which reduces to the problem of the pull-back of p-unipotent groups out of a solvable context.

Under a nilpotency assumption, where Conjecture 2.16 holds by definition, things are well in hand.

Lemma 2.17 *Depending on the value of $\tilde{p}$, the (ultra) $\tilde{p}$-homogeneity of a $\tilde{p}$-group is equivalent to the following:*

a. if $\tilde{p} = (\infty, 0)$, to be a good torus.
b. if $\tilde{p} = (\infty, r)$, with $0 < r < \infty$, to have all its definable connected abelian subgroups $\tilde{p}$-groups.
c. if $\tilde{p} = (p, \infty)$, with p prime, then a $\tilde{p}$-group is always $\tilde{p}$-homogeneous.

Proof. Cases (a) and (c) follow readily. If G is a $\tilde{p}$-group as in case (b), then G is divisible by Theorem 2.7 and the fact that subgroups of bounded exponent of divisible abelian groups of finite Morley rank are finite. Thus, if H is a definable connected subgroup of G, then H is divisible and, by Theorem 2.7, H is generated by its abelian definable connected subgroups, so H is a $U_{\tilde{p}}$-subgroup. □

After these specific characterisations depending on the value of $\tilde{p}$, we return to considerations independent of $\tilde{p}$. In Corollary 5.3 below we will see higher links between $\tilde{p}$-groups and the $\tilde{p}$-homogeneous ones, namely that being a $\tilde{p}$-group is equivalent to being $\tilde{p}$-homogeneous modulo the Frattini subgroup. For the moment we mention that commutator subgroups tend to provide sources of $\tilde{p}$-homogeneous subgroups.

Theorem 2.18 [Fré06a, Theorem 4.11] *Let G be a connected group of finite Morley rank acting definably by conjugation on a $\tilde{p}$-group H. Then $[G, H]$ is a definable $\tilde{p}$-homogeneous subgroup of H.*

As for the $U_{\tilde{p}}$-radical, one can consider several unipotence parameters. If $\tilde{\pi}$ is any set of such unipotence parameters, one naturally says that a group G of finite Morley rank is *$\tilde{\pi}$-homogeneous* if every definable connected nilpotent subgroup of G is a $U_{\tilde{\pi}}$-group. Indeed, we will see in Corollary 5.3 below that a connected nilpotent group is a $U_{\tilde{\pi}}$-group if and only if it is $\tilde{\pi}$-homogeneous modulo the Frattini subgroup.

3 Carter subgroups

A definable subgroup of a group of finite Morley rank is a *Carter* subgroup if it is connected, nilpotent, and of finite index in its normalizer. By elementary properties of generalized connected components and definable closures, one can check easily that Carter subgroups are exactly the nilpotent subgroups Q satisfying $N^\circ(Q) = Q$. This definition is designed to approximate Cartan subgroups, i.e. centralizers of maximal tori in algebraic groups. In a simple algebraic group a Cartan subgroup is exactly a maximal torus T, hence abelian and almost selfnormalizing, and the finite group $N(T)/T$ is called the *Weyl* group.

We will see shortly that Carter subgroups exist in an arbitrary group of finite Morley rank. Their conjugacy remains a natural conjecture in general.

Conjecture 3.1 *In any group of finite Morley rank, Carter subgroups are conjugate.*

We will see nevertheless that Conjecture 3.1 is verified in two important cases: under a generosity assumption on the one hand, and in connected solvable groups on the other.

3.1 Existence

As promised, the main result concerning Carter subgroups is their existence in any group of finite Morley rank.

Theorem 3.2 [FJ05] *Any group G of finite Morley rank contains a Carter subgroup. Furthermore, any nontrivial Sylow $\tilde{p}$-subgroup of G of minimal unipotence degree is contained in a Carter subgroup of G.*

Carter subgroups are built in [FJ05] from their least unipotent part to their most unipotent part. We first consider a nontrivial Sylow $\tilde{p}$-subgroup U_1 of G of minimal unipotence degree, and we take U_2 a Sylow $\tilde{p}$-subgroup of $N^\circ(U_1)$ of minimal unipotence degree greater than that of U_1. By Proposition 2.10, U_1U_2 is a connected nilpotent group, whose subgroups U_1 and U_2 are Sylow $\tilde{p}$-subgroups by construction and Theorem 2.7. Repeating the same process, we build groups $U_1 \cdots U_i$ which are nilpotent by Proposition 2.10, and whose subgroups $U_1, \ldots, U_i$ are Sylow $\tilde{p}$-subgroups by construction and Theorem 2.7. Also, one sees easily in this process, using Lemma 2.13 (*b*) and Proposition 2.10 again, that the unipotence degree of a nontrivial Sylow $\tilde{p}$-subgroup of $N^\circ(U_1 \cdots U_i)/U_1 \cdots U_i$ can only be greater than that of U_i.

By finiteness of Morley rank, this process is stationary, say at step n. Then one can check that the group $U_1 \cdots U_n$ is almost-selfnormalizing as follows: if $U_1 \cdots U_n < N^\circ(U_1 \cdots U_n)$, then Reineke's Theorem yields an infinite definable abelian subgroup of $N^\circ(U_1 \cdots U_n)/U_1 \cdots U_n$, which might be chosen to be indecomposable, and then by the preceding remark and Lemma 2.13 the above process would not be stationary at step n.

Now $U_1 \cdots U_n$ is nilpotent and of finite index in its normalizer, and hence a Carter subgroup of the ambient group G.

3.2 Carter $\tilde{\pi}$-subgroups

If $\tilde{\pi}$ denotes a nonempty set of unipotence parameters, then there is a natural notion of Carter subgroup relativized to $\tilde{\pi}$. Indeed, one can relax the almost selfnormalization condition to selfnormalization relatively to $\tilde{\pi}$. Technically, a *Carter $\tilde{\pi}$-subgroup* of a group of finite Morley rank is a definable nilpotent subgroup Q such that $Q = U_{\tilde{\pi}}(N(Q))$. This definition implies that Carter $\tilde{\pi}$-subgroups are $U_{\tilde{\pi}}$-groups, and in particular connected, but not necessarily almost selfnormalizing.

Of course, the groups produced by this definition correspond to the usual Carter subgroups when $\tilde{\pi}$ consists of the set of all unipotence parameters, or merely all unipotence parameters involved in the ambient group. By Proposition 2.8, they correspond to Sylow $\tilde{p}$-subgroups when $\tilde{\pi} = \{\tilde{p}\}$. In particular Conjectures 2.6 and 3.1 can be encapsulated in the more general conjecture that Carter $\tilde{\pi}$-subgroups are conjugate, for any $\tilde{\pi}$.

Whereas conjugacy remains an open question, existence is fortunately not a problem. Repeating the same procedure as in §3.1, but restricting the choices of Sylow $\tilde{p}$-subgroups to those $\tilde{p}$ belonging to $\tilde{\pi}$, one can show the existence of Carter $\tilde{\pi}$-subgroups for any $\tilde{\pi}$, which was first seen in Burdges' Thesis [Bur04a, §4.3].

Theorem 3.3 *Let G be a group of finite Morley rank and $\tilde{\pi}$ a set of* *unipotence parameters. Then any nontrivial Sylow $\tilde{p}$-subgroup of G, where $\tilde{p} = (p, r) \in \tilde{\pi}$ and r is minimal with this property, is contained in a Carter $\tilde{\pi}$-subgroup of G.*

In [Bur04a, §4.3] this theory is even developed by imposing additional restrictions to the torsion subgroups of decent tori. As this goes back to torsion elements, we won't do that here, but it exists.

3.3 Generous Carter subgroups

We say that a definable subset X of a group G of finite Morley rank is *generous* in G if the union of its conjugates X^G forms a generic subset of G. Conjecture 3.1 holds for generous Carter subgroups.

Theorem 3.4 [Jal06] *In any group G of finite Morley rank, generous Carter subgroups are generous in G° and conjugate in G°.*

The proof of Theorem 3.4 uses the following preliminary analysis on generous subgroups.

Proposition 3.5 [Jal06, Proposition 2.1] *Let H be a definable connected generous subgroup of a group G of finite Morley rank. Then there is a definable generic subset Y of H such that each element of Y belongs to only finitely many conjugates of H.*

Proposition 3.5 can be proved either by direct rank computations on conjugacy classes of elements, or by looking at the geometry whose points are the elements of H^G, that is elements in H up to conjugacy, and lines are the conjugates of H, with the natural incidence relation. A converse of Proposition 3.5 can also be shown, linking thus strongly generosity to such finiteness conditions. Incidentally, this has the following general corollary of independent interest on generosity.

Lemma 3.6 [Jal06, Lemma 3.9] *Let $L \leq H \leq G$ be groups of finite Morley rank, with L and H definable. Then:*

a. If H is connected, L generous in H, and H generous in G, then L is generous in G.

b. Conversely, if L is generous in G, then L is generous in H and H is generous in G.

In particular, in any of these two cases, L and H are both of finite index in their normalizers in G.

A finiteness condition as in Proposition 3.5 allows one to locate many normalizers with the following important independent lemma, whose proof depends only on Corollary 1.2.

Lemma 3.7 [Jal06, Lemma 3.3] *Let G be a group of finite Morley rank, H a definable subgroup of G, and Y the definable subset of H consisting of those elements of H which are in only finitely many conjugates of H. Then, for any definable subset U of H meeting Y in a nonempty subset, $N^\circ(U) \leq N^\circ(H)$.*

Proposition 3.5 and Lemma 3.7 can be applied to Carter subgroups, using only one time their nilpotence, via the normalizer condition, and the fact that they are almost-selfnormalizing, to get some characterizations of generosity for Carter subgroups.

Theorem 3.8 [Jal06, Corollary 3.8] *Let G be any group of finite Morley rank, and Q a Carter subgroup of G. Then the following are equivalent to the generosity of Q in G:*

a. For each y of a definable generic subset Y of Q, Q is the unique maximal definable connected nilpotent subgroup containing y.
b. Q is generically disjoint from its conjugates, i.e.,

$$Q \setminus \bigcup_{g \in G \setminus N(Q)} Q^g$$

is generic in Q.
c. There exists a definable generic subset Y of Q such that each $y \in Y$ is contained in only finitely many conjugates of Q.

The conjugacy of generous Carter subgroups in Theorem 3.4 follows then easily from item (a) of Theorem 3.8 and Lemma 1.1.

3.4 Genericity Conjectures

Of course, Theorem 3.4 depends on a strong genericity assumption which is inspired by the algebraic case, but not known in general. Nevertheless we believe in one of the following genericity conjectures.

Conjecture 3.9 *One of the following is true in any group of finite Morley rank:*

(1) *Any Borel subgroup is generous.*

(2) *Any Carter subgroup is generous.*
(3) *At least one Carter subgroup is generous.*

We will see shortly, in Theorem 3.11 below, that Conjecture 3.9 (2) holds in connected solvable groups. In particular, by Lemma 3.6, Conjectures 3.9 (1) – (3) are successively weaker, and Conjecture 3.9 (3) is equivalent to the generosity of at least one Borel subgroup.

Beyond the solvable context, Conjecture 3.9 (2) holds in two extreme types of groups of finite Morley rank, in the worst counterexamples to the Algebricity Conjecture on the one hand, and under assumptions pushing heavily toward algebricity on the other.

Consider the worst counterexamples first. A *full Frobenius* group of finite Morley rank is a pair $B < G$, with G a group of finite Morley rank, B a proper subgroup satisfying $B \cap B^g = 1$ for every $g \in G \setminus B$ and $G = B^G$. If G is connected, then B is definable and connected [Jal01a, Proposition 3.3]. If G is connected and nonsolvable and B is nilpotent, then G can be called, in a large sense following the ideology of [Jal01a], a *bad* group (indeed, the ideology of that paper would rather be to call bad group any full Frobenius group with G connected). The existence of none of these groups is known yet, but they constitute a good test for many questions in the subject. It is remarkable that all genericity and conjugacy conjectures formulated here are true in these bad groups, or more generally in connected full Frobenius groups with B solvable.

At the opposite, Conjecture 3.9 (2) holds if one excludes bad fields and extreme configurations such as these bad groups. We say that a group of finite Morley rank is *tame* if it interprets no bad fields naturally, i.e. via Zilber's Field Theorem, and *strongly tame* if in addition it interprets no bad groups naturally, i.e. as a section. With all these pathologies excluded everything works as expected. It follows from Zilber's Field Theorem that a nonnilpotent connected strongly tame group of finite Morley rank contains a nontrivial decent torus T. Then it is shown in [Fré06b] under the same assumption that if T is maximal with this property, then $C^\circ(T)$ is a Carter subgroup of G and Conjecture 3.9 (2) holds.

Theorem 3.10 [Fré06b] *In any strongly tame group of finite Morley rank, Carter subgroups are generous, conjugate, and of the form $C^\circ(T)$ for some maximal definable decent subtorus T.*

A structural question inspired by the algebraic case [Hum75, Ex. 6 p. 142; §22-3] is to know whether Carter subgroups of a *connected* group of finite Morley rank are necessarily maximal nilpotent subgroups. We will see in Corollary 3.12 below that this is true in connected solvable groups. Beyond the solvable case this question might well be doable for generous Carter subgroups.

Finally, it is worth mentioning that in simple algebraic groups Carter subgroups are maximal tori, and thus abelian. Hence one can ask whether Carter subgroups of a simple connected group are abelian, and more generally whether a Carter subgroup Q of a connected group G satisfies $Q' \leq \Phi(G)$. We will show this latter inclusion when G is solvable in Corollary 6.6 below. But in general there might exist bad groups with nonabelian Carter subgroups, and Conjecture 3.9 does not seem to help toward such questions.

3.5 Carter subgroups of solvable groups

After these general considerations in arbitrary groups of finite Morley rank we are going to concentrate heavily, now and for the remainder of this paper, on the case of solvable groups. This will eventually lead to subformation theory for connected solvable groups of finite Morley rank.

The conjugacy of Carter subgroups in connected solvable groups of finite Morley rank was first shown by F. Wagner in [Wag94], and independently by O. Frécon in [Fré00b] using the machinery of §3.6 below.

Theorem 3.11 *In any connected solvable group G of finite Morley rank, Carter subgroups are generous, conjugate, and selfnormalizing.*

We provide here a simplified proof of Theorem 3.11, using the general Theorem 3.4. But before, we mention some corollaries of Theorem 3.11. The selfnormalization in Theorem 3.11, together with properties of definable closures of nilpotent groups, unifies the notion of Carter subgroups in connected solvable groups and in finite solvable groups as follows (see also §3.8 below).

Corollary 3.12 [Fré00b, Théorème 1.1] *In any connected solvable group of finite Morley rank, Carter subgroups are exactly the nilpotent selfnormalizing subgroups.*

As another corollary of Theorem 3.11, a Frattini Argument gives a lifting result.

Corollary 3.13 [Fré00b, Corollaire 5.20] *Let G be a connected solvable* *group of finite Morley rank and N a normal definable subgroup of G. Then Carter subgroups of G/N are exactly of the form QN/N, with Q a Carter subgroup of G. In particular $G = QF^\circ(G)$ by Theorem 1.13.*

Solvability allows arguments by induction, and this is the reason why the theory of Carter subgroups works well in this context. In the induction process, the minimal configuration appearing for Carter subgroups is the following situation reminiscent of Borel subgroups of $SL_2(K)$, first studied by A. Nesin.

Theorem 3.14 [BN94a, Lemma 9.14] *Let G be a connected solvable group of finite Morley rank such that G' is G-minimal and $Z(G)$ is finite. Then Carter subgroups of G are exactly the complements of G' in G. Furthermore, they are selfnormalizing, generous, and conjugate in G.*

Proof. We slightly simplify the original proof by using Theorem 3.4.

Let $x \in G \setminus C(G')$ and $Q = C(x)$. We show that Q is a selfnormalizing generous complement of G' in G containing $Z(G)$. By Zilber's Field Theorem applied to $G' \rtimes G/C(G')$, $C_{G'}(x) = 1$ and $\mathrm{rk}\,(x^G) \geq \mathrm{rk}\,(G')$. On the other hand the commutativity of G/G' gives $x^G \subseteq xG'$, and thus $\mathrm{rk}\,(x^G) = \mathrm{rk}\,(G')$. Hence $\mathrm{rk}\,(Q) = \mathrm{rk}\,(G) - \mathrm{rk}\,(x^G) = \mathrm{rk}\,(G/G')$ and, as $Q \cap G' = 1$, Q is a complement of G' in G and $N(Q) = N_{G'}(Q)Q$. But $[N_{G'}(Q), Q] \leq G' \cap Q = 1$, hence $N_{G'}(Q) = C_{G'}(Q) \leq C_{G'}(x) = 1$, and $N(Q) = Q$. In particular $Z(G) \leq Q$. Moreover, if $y \in Q \cap Q^g$ for some $g \in G \setminus Q$, then, as $Q \simeq G/G'$ is abelian, y centralizes $\langle Q, Q^g \rangle$. But $G = G'Q$ and G' is G-minimal, so $G = \langle Q, Q^g \rangle$ and $y \in Z(G)$. Hence $\mathrm{rk}\,(Q^G) = \mathrm{rk}\,(Q) + \mathrm{rk}\,(G/Q) = \mathrm{rk}\,(G)$ and Q is generous in G.

Let now Q be an arbitrary Carter subgroup of G. We show that $Q = C(x)$ for some $x \in Q \setminus C(G')$. If Q centralizes G', then $G' \leq Q$ and, since G/G' is abelian, $G/G' = N^\circ(Q)G'/G' = Q/G'$, $G = Q$ is nilpotent, $Z(G)$ is infinite by Theorem 1.13 (*b*), a contradiction to our assumption. So there exists $x \in Q \setminus C(G')$ and the preceding paragraph gives $C_{G'}(x) = 1$. In particular $Z(Q) \cap G' = 1$. As Q is nilpotent, $Q \cap G' = 1$ and, as G/G' is abelian, Q is abelian. Thus $Q \leq C(x)$ and as $C(x) \simeq G/G'$ is abelian and connected we obtain $C(x) = Q$.

If Q is a complement of G' in G, then $Q \simeq G/G'$ is abelian and $N_{G'}(Q) \leq Z(G)$. Hence $N^\circ(Q) = N^\circ_{G'}(Q)Q = Q$ and Q is a Carter subgroup of G. Now Theorem 3.4 finishes our proof. □

With the previous preparations, we revisit now the proof of Theorem 3.11 by using the general Theorem 3.4.

Proof of Theorem 3.11. We proceed by induction on $\mathrm{rk}(G)$. Let Q be a Carter subgroup of G and A a G-minimal subgroup of G.

Assume $QA < G$. Then $N(QA) = N(Q)QA = N(Q)A$ by induction hypothesis and a Frattini Argument. In particular $N_{G/A}(QA/A) = N(Q)A/A$ and QA/A is a Carter subgroup of G/A, which gives $N(Q) = Q$ by induction. Also by induction, Q is generous in QA and QA/A is generous in G/A, which implies easily that QA is generous in G. As generosity is transitive by Lemma 3.6 (a), it follows that Q is generous in G.

Assume now $QA = G$. If $A < F^{\circ}(G)$, then A centralizes $F^{\circ}(G)$ by G-minimality and Theorem 1.13, and $Q \cap F^{\circ}(G)$ must contain a G-minimal subgroup B. Hence the preceding paragraph applies as $QB = Q < G$, giving Q selfnormalizing and generous. If $F^{\circ}(G) = A$, then, by G-minimality, $Z(G)$ is finite and $G' = A$ by Theorem 1.13. Hence Theorem 3.14 applies, giving the selfnormalization and generosity of Q.

In any case, Theorem 3.4 yields conjugacy. □

3.6 Approach by abnormal subgroups

We see here that Carter subgroups of connected solvable groups of finite Morley rank satisfy a property much stronger than selfnormalization. We say that a subgroup H of a group G is *abnormal* in G if $g \in \langle H, H^g \rangle$ for every $g \in G$. Of course, abnormal subgroups are selfnormalizing. The approach of Carter subgroups of connected solvable groups in [Fré00b] is essentially based on abnormal subgroups. Their basic properties are the following.

Proposition 3.15 [Fré00b, §3] *Let G be a connected solvable group of finite Morley rank. Then:*

a. *Every abnormal subgroup of G is definable and connected.*
b. *If G is nonnilpotent, then G has a proper abnormal subgroup.*
c. *Abnormality is transitive: if H is abnormal in G and if K is abnormal in H, then K is abnormal in G.*

It follows from Proposition 3.15 that *minimal* abnormal subgroups of a connected solvable group G of finite Morley rank are Carter subgroups of G. In [Fré00b, Proposition 4.9] it is deduced from Proposition 3.15 that minimal abnormal subgroups of connected solvable groups of finite Morley rank are conjugate. This gives an alternative proof of the conjugacy of Carter subgroups in this context with the following characterisation of Carter subgroups.

Theorem 3.16 [Fré00b, Théorème 1.1] *In any connected solvable group of finite Morley rank, Carter subgroups are exactly the minimal abnormal subgroups.*

Without minimality, there might be infinitely many conjugacy classes of abnormal subgroups, as the following example shows.

Example 3.17 Let $G = (\oplus_{i=0}^{n} K_+) \rtimes K^\times$, where $n > 0$, K is an algebraically closed field, and $K^\times$ acts by multiplication on $\oplus_{i=0}^{n} K_+$. Let $\mathcal{H} = (H_i)_{i \in I}$ be an infinite family of vector subspaces of $\oplus_{i=0}^{n} K_+$ and, for each $i \in I$, $U_i = H_i \rtimes K^\times$. Then $(U_i)_{i \in I}$ is an infinite family of nonconjugate abnormal subgroups of G.

3.7 Consequences on the Frattini subgroup

The good understanding of Carter subgroups of a connected solvable group of finite Morley rank G allows one to obtain precious information about $G/\Phi(G)$.

Proposition 3.18 [Fré00b, §5.2] *Let G be a connected solvable group of finite Morley rank. Then $\Phi(G)$ is nilpotent, $F(G)/\Phi(G) = F(G/\Phi(G))$ is abelian, and $G/\Phi(G)$ is 2-solvable. In particular, G is nilpotent if and only if $G/\Phi(G)$ is nilpotent if and only if $G/\Phi(G)$ is abelian.*

Proposition 3.18 becomes particularly interesting with the following precise description of connected 2-solvable groups of finite Morley rank.

Proposition 3.19 [Fré00b, Remarks following Corollaire 7.7] *Let G be a connected 2-solvable group of finite Morley rank. Then G has a definable connected characteristic abelian subgroup A such that $G = A \rtimes Q$ for every Carter subgroup Q of G.*

3.8 Interlude on finite group theory

We conclude this section on Carter subgroups by comparing the situation in finite group theory. Indeed, Carter subgroups are named after the following result of R. Carter concerning finite solvable groups.

Theorem 3.20 [Car61] *Any finite solvable group has a nilpotent selfnormalizing subgroup, and these subgroups are conjugate.*

Of course, Theorem 3.11 and Corollary 3.12 provide an exact analog of Theorem 3.20 for connected solvable groups of finite Morley rank, and hence *nilpotent selfnormalizing* subgroups are the finite cousins of Carter subgroups of groups of finite Morley rank. The main difference is that in the finite Morley rank context Carter subgroups are *connected*, a notion which has no exact analog in the finite case. Indeed, we want a Carter subgroup to behave in any case like a Cartan subalgebra of a finite dimensional Lie algebra. If K is an algebraically closed field of characteristic different from 2, elementary abelian 2-subgroups of order 4 of $PSL_2(K)$ are almost selfnormalizing, but they are certainly not what we expect a Carter subgroup of $PSL_2(K)$ to be.

This major connectedness difference is illustrated by existence. Theorem 3.2 gives the existence of (connected) Carter subgroups in an arbitrary group of finite Morley rank, but at the opposite most finite nonsolvable groups have no nilpotent selfnormalizing subgroups. The alternating group $\mathcal{A}_5$ is such an example. However, Conjecture 3.1 on the conjugacy of Carter subgroups of groups of finite Morley rank has an exact analog in finite group theory.

Conjecture 3.21 *A finite group has at most one conjugacy class of nilpotent selfnormalizing subgroups.*

Work toward Conjecture 3.21 can be found in [DVLT98] and [TV02]. It has been proved in [PTV04], using the Classification of the Finite Simple Groups, that a minimal counterexample to Conjecture 3.21 cannot be simple.

With Theorem 3.20 in hand, W. Gaschütz developed in [Gas63] the *formation theory* of finite solvable groups. In a group, a subgroup H *covers* a section K/L if $K \cap H$ modulo L equals K/L. Formation theory is a global approach to existence and conjugacy results by some covering phenomena, unifying and generalizing considerably Hall and Carter theories in finite solvable groups. This theory has been adapted in [Fré00b] to *connected* solvable groups of finite Morley rank. We will elaborate on formation theory for connected solvable groups of finite Morley rank in §4 below. As for finite groups, the theory unifies several important existence and conjugacy results, and is particularly designed here for *connected* subgroups such as Carter subgroups, Sylow $\tilde{p}$-subgroups, π^*-subgroups, etc. We will see such applications of the theory in §5 below.

We finish the comparison with finite groups by mentioning that there is another way to unify Theorems 3.11 and 3.20 in solvable groups of finite Morley rank which are *not necessarily* connected. We will see this in §6.5 below, more precisely in Proposition 6.13 and Theorem 6.15. This leads eventually in [Fré00a] to another kind of formation theory, rather designed for nonnecessarily connected solvable groups of finite Morley rank, and incorporating to Carter subgroups the theory of Hall π-subgroups instead of the connected groups mentioned above. As this escapes from the world of connected groups, we won't go into these considerations.

4 Subformations: the theory

We elaborate now on a version of formation theory particularly designed for *connected* solvable groups of finite Morley rank. In this section we develop the theory quite formally, and in the next we will apply it. This will incorporate most conjugacy results of connected subgroups of solvable groups of finite Morley rank that interest us, such as the conjugacy of Sylow $\tilde{p}$-subgroups of [Bur06b], but also the conjugacy of π^*-groups of [BN94b]. We recall that [Fré00a] provides a version of formation theory rather designed for nonconnected solvable groups, based on Theorem 6.15 below, but which we are not going to present here

4.1 Definition and examples

A *connected subformation* is a collection $\mathcal{F}$ of connected solvable groups of finite Morley rank, called *$\mathcal{F}$-groups* for short, which is closed under images by definable homomorphisms and satisfies the following two conditions for every connected solvable group G of finite Morley rank:

(S1) If $(N_i)_i$ is a family of definable normal subgroups of G such that $\cap_i N_i$ is finite and G/N_i is an $\mathcal{F}$-group for each i, then G has a definable $\mathcal{F}$-subgroup H such that $G = N_i H$ for each i.

(S2) If $G/\Phi(G)$ is an $\mathcal{F}$-group, then G is an $\mathcal{F}$-group.

Naturally, one would like to impose additionally in clause (S1) that G *is* an $\mathcal{F}$-group. This stronger definition corresponds to the *saturated connected formations* of [Fré00b]. Here we work with a weaker definition, and generalize the results of that paper concerning saturated connected formations.

The following collections of groups of finite Morley rank are connected subformations.

- $\mathcal{N}$: Nilpotent connected groups (Proposition 3.18 for clause (S2)).
- $\mathcal{U}_{\tilde{\pi}}$: Solvable $U_{\tilde{\pi}}$-groups (Lemma 2.13 (*b*) for clause (S1) and Corollary 2.15 for clause (S2)).
- Solvable π^*-groups (cf. §6.1 below).

One can find new connected subformations by taking intersections of these "basic" examples.

Proposition 4.1 *Every intersection of connected subformations is a connected subformation.*

Proof. The intersection $\mathcal{F}$ of connected subformations $(\mathcal{F}_k)_k$ verifies clause (S2). For clause (S1), let $(N_i)_i$ be a family of definable normal subgroups of a connected solvable group G of finite Morley rank such that $\cap_i N_i$ is finite and $G/N_i \in \mathcal{F}$ for each i. Let H be a definable connected subgroup of G which covers all quotients G/N_i, and which is minimal with respect to these properties. For each k, clause (S1) applies with the subformation $\mathcal{F}_k$ in H, with the family of subgroups $(H \cap N_i)_i$, and hence H has a definable $\mathcal{F}_k$-subgroup L such that $H = (H \cap N_i)L$ for each i. This implies $G = N_iH = N_iL$ for each i, and hence $H = L \in \mathcal{F}_k$ by minimality of H. Thus $H \in \mathcal{F}$, proving clause (S1). □

Among the miscellaneous possible combinations of intersections of basic examples, the case which particularly interests us is the following.

Corollary 4.2 *Let $\tilde{\pi}$ be any set of unipotence parameters. Then the collection $\mathcal{N}_{\tilde{\pi}}$ of nilpotent $U_{\tilde{\pi}}$-groups, and in particular the collection $\mathcal{N}_{\tilde{p}}$ of $\tilde{p}$-groups if $\tilde{\pi} = \{\tilde{p}\}$, is a connected subformation.*

4.2 Results

The main theme of connected subformation theory concerns the existence and the conjugacy of certain subgroups, namely $\mathcal{F}$-covering subgroups and $\mathcal{F}$-projectors. These subgroups sharpen up the notion of maximal definable $\mathcal{F}$-subgroups in terms of some covering phenomena. Before the definitions, we wish to summarize our main result about these subgroups as follows.

Theorem 4.3 *Let $\mathcal{F}$ be a connected subformation. Then any connected solvable group of finite Morley rank has $\mathcal{F}$-covering subgroups and $\mathcal{F}$-projectors. Furthermore these subgroups coincide and are all conjugate.*

In this subsection we record the results leading to Theorem 4.3, and in the next we will give proofs not directly adaptable from the case of saturated connected formations of [Fré00b]. In practice we are mostly interested in $\mathcal{F}$-projectors, $\mathcal{F}$-covering subgroups being rather a tool in the development of this theory.

After these general comments, we are now ready to embark on technical definitions and to state the miscellaneous results leading to Theorem 4.3. If G is a connected solvable group of finite Morley rank and $\mathcal{F}$ a collection of connected solvable groups of finite Morley rank, an *$\mathcal{F}$-covering subgroup* of G is a definable $\mathcal{F}$-subgroup H of G satisfying $S = NH$ for every definable connected subgroup S of G containing H and every definable subgroup $N \trianglelefteq S$ such that $S/N \in \mathcal{F}$. In other words, an $\mathcal{F}$-covering subgroup covers all definable $\mathcal{F}$-sections containing it. These subgroups are related to subformations as follows.

Theorem 4.4 (compare with [Fré00b, Théorème 8.5]) *Let $\mathcal{F}$ be a collection of connected solvable groups of finite Morley rank closed under images by definable homomorphisms. Then $\mathcal{F}$ is a connected subformation if and only if every connected solvable group of finite Morley rank has an $\mathcal{F}$-covering subgroup, and in this case $\mathcal{F}$-covering subgroups are conjugate in every connected solvable group of finite Morley rank.*

Given a connected subformation $\mathcal{F}$, Theorem 4.4 yields some properties of $\mathcal{F}$-covering subgroups which are reminiscent of those of Carter subgroups of solvable groups seen in §3.5 and §3.6.

Corollary 4.5 (compare with [Fré00b, 8.6-8.9]) *Let $\mathcal{F}$ be a connected subformation and G a connected solvable group of finite Morley rank.*

a. *If N is a definable normal subgroup of G, then the $\mathcal{F}$-covering subgroups of G/N are exactly the subgroups of the form HN/N, with H an $\mathcal{F}$-covering subgroup of G.*
b. *If H is an $\mathcal{F}$-covering subgroup of G, then $N(H)$ is abnormal in G.*
c. *If G is nilpotent, then G has a unique $\mathcal{F}$-covering subgroup, the unique maximal definable $\mathcal{F}$-subgroup of G.*

d. *If H is a definable $\mathcal{F}$-subgroup of G such that $G = F(G)H$, then H is contained in an $\mathcal{F}$-covering subgroup of G.*

We define now the most interesting subgroups of the theory. If $\mathcal{F}$ is a collection of groups and G a connected solvable group of finite Morley rank, an *$\mathcal{F}$-projector* of G is a definable $\mathcal{F}$-subgroup H of G such that HN/N is a maximal definable $\mathcal{F}$-subgroup of G/N for every normal definable subgroup N of G.

Theorem 4.6 (compare with [Fré00b, Théorème 8.11]) *Let $\mathcal{F}$ be a collection of connected solvable groups of finite Morley rank closed under images by definable homomorphisms. Then $\mathcal{F}$ is a connected subformation if and only if every connected solvable group of finite Morley rank has an $\mathcal{F}$-projector, and in this case the $\mathcal{F}$-covering subgroups are exactly the $\mathcal{F}$-projectors in every connected solvable group of finite Morley rank.*

Obviously, Theorem 4.3 is just a summary of Theorems 4.4 and 4.6 for connected subformations.

When $\mathcal{F}$ is a connected subformation, it is natural to ask when maximal definable $\mathcal{F}$-subgroups of a connected solvable group are conjugate. Since $\mathcal{F}$-projectors are maximal definable $\mathcal{F}$-subgroups, Theorem 4.3 guarantees such a conjugacy when these two classes of subgroups coincide. We provide a new necessary and sufficient condition for this property.

Theorem 4.7 *Let $\mathcal{F}$ be a connected subformation. Then the following are equivalent:*

a. *Maximal definable $\mathcal{F}$-subgroups are conjugate in any connected solvable group of finite Morley rank.*
b. *Maximal definable $\mathcal{F}$-subgroups coincide with $\mathcal{F}$-projectors in any connected solvable group of finite Morley rank.*
c. *The definable product HK is an $\mathcal{F}$-group for any two definable $\mathcal{F}$-groups H and K with H normalizing K.*

When the necessary and sufficient conditions of Theorem 4.7 fail, there exists a connected solvable group of finite Morley rank with nonconjugate maximal definable $\mathcal{F}$-subgroups. Concrete examples are provided in Example 5.6 below. Nevertheless there is still a finiteness result.

Theorem 4.8 *Let $\mathcal{F}$ be a connected subformation such that the product of any two definable normal $\mathcal{F}$-groups is an $\mathcal{F}$-group. Then every connected solvable group of finite Morley rank has only finitely many conjugacy classes of maximal definable $\mathcal{F}$-subgroups.*

Now it remains to give the proofs of the theorems we have just stated.

4.3 Proofs: the minimal configuration

For the remainder of this section on subformation theory, we provide proofs of the results of §4.2 not directly or indirectly available in the literature. We start by elaborating on Theorem 3.14, and consider the minimal configuration which for connected subformations plays exactly the same role as Theorem 3.14 for Carter subgroups.

Theorem 4.9 (compare with [Fré00b, Lemme 8.2]) *Let G be a connected solvable group of finite Morley rank, $\mathcal{F}$ a connected subformation, and N a G-minimal subgroup of G such that $G/N \in \mathcal{F}$ and $G \notin \mathcal{F}$. Then there exists a definable $\mathcal{F}$-subgroup K of G such that $G = NK$ and $N \cap K$ is finite, and such subgroups are conjugate.*

Proof. Since $G/N \in \mathcal{F}$ and $G \notin \mathcal{F}$, $N \not\leq \Phi(G)$ by clause (S2). Therefore there exists a maximal proper definable connected subgroup K of G not containing N. Then $G = NK$ by maximality of K. As N is abelian by G-minimality, $K \cap N$ is normal in G, and finite by G-minimality of N again. Now $K/(K \cap N) \simeq G/N \in \mathcal{F}$, and K is an $\mathcal{F}$-group by clause (S1). Therefore subgroups K satisfying the stated conditions exist, and it remains only to prove their conjugacy.

For that purpose we proceed by induction on $\mathrm{rk}\,(G)$, and consider two definable $\mathcal{F}$-subgroups K_1 and K_2 satisfying the stated conditions, i.e. $G = NK_i$ and $N \cap K_i$ is finite for $i = 1$ and 2. If L is a proper definable connected subgroup of G containing K_i, with $i = 1$ or 2, then the G-minimality of N yields as above the finiteness of $L \cap N$, and $L = K_i(L \cap N) = K_i$ by connectedness. Thus each K_i is a maximal proper definable connected subgroup of G. We are going to distinguish two cases, depending on whether G as several, or just one, G-minimal subgroups.

Suppose first that G has a G-minimal subgroup M distinct from N. Notice that $M \cap N$ is finite. Consider the first subcase $G/M \notin \mathcal{F}$. Then $M \leq K_1 \cap K_2$, as otherwise $M \not\leq K_i$ for some i, and the maximality of K_i

and the connectedness of M would imply that $G = MK_i$, contradicting $G/M \notin \mathcal{F}$. Now the induction hypothesis applied in G/M gives the conjugacy of K_1 and K_2 in this subcase $G/M \notin \mathcal{F}$. Consider now the second subcase $G/M \in \mathcal{F}$. Then by clause (S1) and the finiteness of $M \cap N$, G has a definable $\mathcal{F}$-subgroup H such that $G = LH$ for each normal definable subgroup L of G satisfying $G/L \in \mathcal{F}$. In particular, $G = NH = MH$. Notice also that $H < G$, as $G \notin \mathcal{F}$. Now we let $A = (NM \cap H)^\circ$ and show that A is G-minimal (although A normal and infinite would suffice for our purpose!). Notice first that A is normal in G, as $G = NH = MH$ and both M and N are in $Z(F^\circ(G))$ by G-minimality. Since H covers G/M, $AM/M = (NM)^\circ M/M = NM/M$, and the latter quotient is G-minimal as N is and $N \cap M$ is finite. This proves in particular that A is infinite. Moreover $A \cap M \leq H \cap M$ and the latter subgroup, being normal in $G = MH$, is finite by G-minimality of M and the fact that $H < G$. Now a G-minimal subgroup A_0 of A must cover the G-minimal quotient AM/M, and then $A = A_0$ by connectedness and rank equality. Hence A is G-minimal, as claimed. Notice now that $G/A \notin \mathcal{F}$, as otherwise $G = AH$ by choice of H, contradicting the fact that $AH = H < G$. In particular AK_1 and AK_2 are two proper definable connected subgroups of G. Now by maximality of K_1 and K_2 we obtain $A \leq K_1 \cap K_2$, and the induction hypothesis applied in G/A gives the conjugacy of K_1 and K_2. This completes the proof of the conjugacy of K_1 and K_2 in presence of a G-minimal subgroup distinct from N.

Suppose now that N is the unique G-minimal subgroup of G. If $Z(G)$ is infinite, then $N \leq Z(G)$ by uniqueness of N, and K_1 and K_2 are normal as $G = NK_1 = NK_2$. Hence each K_i is trivial or contains a G-minimal subgroup. By uniqueness of N, and as $N \cap K_i$ is finite for $i = 1$ and 2, we obtain $K_1 = K_2 = 1$, and in particular the desired conjugacy when $Z(G)$ is infinite. So we may assume $Z(G)$ finite, and in particular the connected group G' is nontrivial. By Theorem 1.13, N centralizes G' and as $G = K_1N = K_2N$, both groups K_i' are normal in G. If one of the two connected subgroups K_i' is nontrivial, then it must contain N by uniqueness, a contradiction as then $G = K_iN = K_i$. Hence K_1 and K_2 are abelian. As $G = K_1N$ with K_1 abelian, $G' \leq N$, and it follows that $G' = N$ by connectedness of G' and G-minimality of N. We are now in a position to apply Theorem 3.14, and its conclusions imply that $Z(G) \cap N = 1$ (as the center is in Carter subgroups which are disjoint from G'!). As each K_i is abelian, as well as N, the two finite groups $N \cap K_i$ are central in $K_1N = K_2N = G$, hence contained in

$Z(G) \cap N$, and thus trivial. Now K_1 and K_2 are two complements of $N = G'$ in G, and Theorem 3.14 gives their conjugacy when N is the unique G-minimal subgroup of G. □

4.4 Proofs: $\mathcal{F}$-covering subgroups and $\mathcal{F}$-projectors

We consider now the proofs of Theorem 4.4, Corollary 4.5, and Theorem 4.6 on $\mathcal{F}$-covering subgroups and $\mathcal{F}$-projectors. We first prove some parts of Theorems 4.4 and 4.6 not explicit in [Fré00b]

Lemma 4.10 *Let $\mathcal{F}$ be a collection of connected solvable groups of finite Morley rank closed under images by definable homomorphisms. If every connected solvable group of finite Morley rank has an $\mathcal{F}$-covering subgroup, or if every such group has an $\mathcal{F}$-projector, then $\mathcal{F}$ is a connected subformation.*

Proof. If G is a connected solvable group of finite Morley rank, then G has an $\mathcal{F}$-covering subgroup or an $\mathcal{F}$-projector H by assumption. Now clause (S1) follows easily from the definitions, and if $G/\Phi(G) \in \mathcal{F}$, then $G = \Phi(G)H$ and $G = H \in \mathcal{F}$ by Lemma 2.14 (*b*), proving clause (S2). □

Given a connected subformation $\mathcal{F}$, the proof of the existence and conjugacy of $\mathcal{F}$-covering subgroups in every connected solvable group of finite Morley rank can be adapted from [Fré00b, Théorème 8.5], Theorem 4.9 treating here the minimal configuration for existence and conjugacy exactly as Theorem 3.14 in Theorem 3.11 for Carter subgroups. This, together with Lemma 4.10, proves Theorem 4.4.

Items (*a*) – (*c*) of Corollary 4.5 can be recovered from [Fré00b, Corollaires 8.6-8.8]. For item (*d*) we provide an independent proof, which can be compared to [Fré00b, Proposition 8.9].

Proof of Corollary 4.5 (*d*). By induction on $\text{rk}\,(G)$. We may assume $G \notin \mathcal{F}$, and consider a G-minimal subgroup N of G. Then HN/N is an $\mathcal{F}$-subgroup of $G/N = F(G/N)(HN/N)$. By induction, HN/N is contained in an $\mathcal{F}$-covering subgroup of G/N, which by Corollary 4.5 (*a*) is of the form KN/N for an $\mathcal{F}$-covering subgroup K of G.

Suppose $KN < G$. Since $KN = F(KN)H$, H is contained in an $\mathcal{F}$-covering subgroup M of KN by induction. But M is conjugate to K in KN by Theorem 4.4, so M is an $\mathcal{F}$-covering subgroup of G.

Suppose $KN = G$ and $K \cap F(G)$ infinite. Then, as $N \leq Z(F^\circ(G))$ by Theorem 1.13, $U := (K \cap F(G))^\circ$ is normal in G. By induction applied in G/U, there is a definable connected subgroup L of G containing HU such that L/U is an $\mathcal{F}$-covering subgroup of G/U. By Corollary 4.5 (*a*), L contains an $\mathcal{F}$-covering subgroup of G, and hence Theorem 4.4 applied in L provides the desired conclusion whenever H is contained in an $\mathcal{F}$-covering subgroup of L. As $L = F(L)H$, this is the case by induction if $L < G$, and thus it remains only to consider the case $L = G$. But then $G/U = L/U \in \mathcal{F}$ and, since K is an $\mathcal{F}$-covering subgroup, $G = UK = K \in \mathcal{F}$, contradicting our assumption $G \notin \mathcal{F}$.

Suppose $KN = G$ and $K \cap F(G)$ finite. Then $F^\circ(G) = N$, and $G = HN$. If H_0 is a maximal proper definable connected subgroup of G containing H, then $H_0 \cap N$ is normal in G as N is abelian, hence finite by G-minimality of N, and hence $H_0 = H(H_0 \cap N) = H$ by connectedness. Thus H is a maximal proper definable connected subgroup of G. Besides, every normal infinite definable subgroup of G contains N: it contains a G-minimal subgroup N_0, and as $N_0 \leq F^\circ(G) = N$ it follows that $N_0 = N$. Thus, if a definable connected subgroup S of G contains H and has a normal definable subgroup A such that $S/A \in \mathcal{F}$, then either $S = G$, in which case $N \leq A^\circ$ (the case $A^\circ = 1$ would imply $G \in \mathcal{F}$ by clause (S1)) and $S = F^\circ(G)H = NH = AH$, or $S = H = AH$. Hence H is an $\mathcal{F}$-covering subgroup of G. □

With Lemma 4.10 and the present version of Corollary 4.5 appropriate to connected subformations, especially Corollary 4.5 (*d*) which is particularly crucial for the study of $\mathcal{F}$-projectors, the proof of Theorem 4.6 is similar to [Fré00b, Théorème 8.11].

With Theorems 4.4 and 4.6, the proof of Theorem 4.3 is complete, as noticed in §4.2 already.

4.5 Proofs: maximal definable $\mathcal{F}$-subgroups

To complete the proofs of all results stated in §4.2, it remains now just to prove the conjugacy Theorem 4.7 and the finiteness Theorem 4.8 on maximal definable $\mathcal{F}$-subgroups.

Proof of Theorem 4.7. Since $\mathcal{F}$-projectors are maximal definable $\mathcal{F}$-subgroups, the equivalence between (*a*) and (*b*) follows from Theorem 4.3.

We prove that (*b*) implies (*c*). Let H and K be definable $\mathcal{F}$-subgroups of a group G of finite Morley rank with H normalizing K. Let M be a maximal definable $\mathcal{F}$-subgroup of HK containing K. By hypothesis, M is an $\mathcal{F}$-projector of HK. Hence M/K is a maximal definable $\mathcal{F}$-subgroup of $HK/K \in \mathcal{F}$. Therefore $HK = M \in \mathcal{F}$, as desired.

To prove that (*c*) implies (*b*), we proceed by induction on the Morley rank. It suffices to show that every maximal definable $\mathcal{F}$-subgroup is an $\mathcal{F}$-projector. Let G be a connected solvable group of finite Morley rank, H a maximal definable $\mathcal{F}$-subgroup of G, and N a normal definable subgroup of G. If $G/N \notin \mathcal{F}$, then there is a maximal definable $\mathcal{F}$-subgroup L/N of G/N such that $HN/N \leq L/N < G/N$. The preimage L of L/N satisfies $N^\circ \leq L^\circ$, and $H \leq L^\circ N^\circ = L^\circ < G$. By induction hypothesis, H is an $\mathcal{F}$-projector of L°, hence H covers the $\mathcal{F}$-group $L^\circ/(L^\circ \cap N)$, as well as $L^\circ N/N = L/N$, and $HN/N = L/N$ is a maximal definable $\mathcal{F}$-subgroup of G/N. So we may assume $G/N \in \mathcal{F}$, and $HN < G$. We will get a contradiction in this case.

Suppose first $G'NH < G$. By induction, H is an $\mathcal{F}$-projector of $G'N^\circ H$. By Theorem 4.3 and a Frattini Argument, $G = G'NN(H)$, and by connectedness $N^\circ(H)$ covers $G/G'N$. In particular $N^\circ(H)$ modulo its intersection with $G'N$ is in $\mathcal{F}$. If $N(H) < G$, then H is an $\mathcal{F}$-projector of $N^\circ(H)$ by induction, so it covers $N^\circ(H)$ modulo its intersection with $G'N$, and it follows that $G = G'NN^\circ(H) = G'NH$, a contradiction. Hence H is normal in G. By Theorem 4.4, G has an $\mathcal{F}$-covering subgroup F and $G = NF$. As F normalizes H, $HF \in \mathcal{F}$ by assumption and $F \leq H$ by maximality of H. Then $G = NF = NH$, a contradiction to $G'NH < G$.

Suppose now $G'NH = G$. Under our assumption $G/N \in \mathcal{F}$, this implies that $G'H$ modulo its intersection with N is in $\mathcal{F}$. But $G'H = F(G'H)H$ by Theorem 1.13, and Corollary 4.5 (*d*) and Theorem 4.3 applied in this group imply that H is an $\mathcal{F}$-projector of $G'H$. Thus $G'H = H(G'H \cap N)$ and $G = G'NH = HN$, a contradiction to our current assumptions. □

Proof of Theorem 4.8. We proceed by induction on $\mathrm{rk}(G)$. We may assume $G \notin \mathcal{F}$ and G nonnilpotent by Corollary 4.5 (*c*).

Suppose that G has a G-minimal subgroup A such that $G/A \notin \mathcal{F}$. Then every maximal definable $\mathcal{F}$-subgroup of G is contained in a subgroup H of G containing A and such that H/A is a maximal definable $\mathcal{F}$-subgroup of G/A. Then the induction hypothesis, applied in G/A and H, gives the desired finiteness. Therefore we may assume $G/A \in \mathcal{F}$ for every G-minimal subgroup A of G.

Fix a G-minimal subgroup B of G. Since $G/B \in \mathcal{F}$, $G = BF$ for some $\mathcal{F}$-projector F of G. Suppose $(F \cap F(G))^\circ \neq 1$. Then F has an F-minimal subgroup A in $(F \cap F(G))^\circ$. By Theorem 1.13, B centralizes $F^\circ(G)$, so A is a G-minimal subgroup of G. Then $G/A \in \mathcal{F}$, and $G = AF = F \in \mathcal{F}$, contradicting $G \notin \mathcal{F}$. Hence $(F \cap F(G))^\circ = 1$, and in particular $F' = 1$ and F is abelian. Therefore, as $G = BF$, $G' = B$ and this is the unique G-minimal subgroup of G as the preceding argument holds for any B. Also, $B = F^\circ(G)$. If $Z(G)$ is infinite, then $Z^\circ(G) = B = G'$ and G is nilpotent, a contradiction to our assumptions. So G verifies the assumptions of Theorem 3.14.

By assumption, G has a unique maximal normal definable $\mathcal{F}$-subgroup L. We show that any maximal definable $\mathcal{F}$-subgroup $H \neq L$ must be a Carter subgroup of G. Since $H \neq L$, H is not normal in G, and $G' \not\leq H$. By Corollary 4.5 (d), H is an $\mathcal{F}$-covering subgroup of $G'H$ and, by a Frattini Argument, $G = G'N(H)$. Therefore $N(H) \cap G' \trianglelefteq G$ is finite, as otherwise $G' \leq N(H)$ by G-minimality of G' and $H \trianglelefteq G$. Hence $N(H) \cap G' \leq Z(G) \cap G'$. But by Theorem 3.14 $Z(G)$ is in each Carter subgroup of G, and these are disjoint from G'. Hence $Z(G) \cap G' = 1$ and $N(H) \cap G' = 1$ also. Hence $N(H)$ is a complement of G' in G and by Theorem 3.14 again it must be a Carter subgroup of G. Thus $N(H) \simeq G/G'$ is a definable connected $\mathcal{F}$-subgroup and $N(H) = H$ by maximality of H, and therefore H is a Carter subgroup of G.

Consequently in this case, G has at most two conjugacy classes of maximal definable $\mathcal{F}$-subgroups, its Carter subgroups on the one hand, and its maximal normal definable $\mathcal{F}$-subgroup on the other. □

5 Subformations: applications

Now we merely apply the general machinery of §4 above on connected subformations into specific cases, starting with important remarks on nilpotent groups in connected subformations.

5.1 The subformation of nilpotent groups

The collection $\mathcal{N}$ of connected nilpotent groups of finite Morley rank is a connected subformation by Proposition 3.18. The main link between Carter subgroup theory and subformation theory is then Corollary 3.13, which guarantees that Carter subgroups of a connected solvable group G of finite Morley rank are $\mathcal{N}$-covering subgroups of G. Hence Theorem 4.3 gives the following important result.

Proposition 5.1 [Fré00b, Corollaire 8.14] *In any connected solvable group of finite Morley rank, Carter subgroups are exactly the $\mathcal{N}$-covering subgroups and the $\mathcal{N}$-projectors.*

In the next two subsections we will get more structural information about *nilpotent* $\mathcal{F}$-projectors.

5.2 Nilpotence and homogeneity

If $\mathcal{F}$ is a connected subformation, then a group G of finite Morley rank is *$\mathcal{F}$-homogeneous* if every definable connected subgroup of G is an $\mathcal{F}$-group.

Proposition 5.2 *Let $\mathcal{F}$ be a connected subformation. If G is a nilpotent $\mathcal{F}$-group, then $G/\Phi(G)$ is $\mathcal{F}$-homogeneous.*

Proof. Replacing G by $G/\Phi(G)$, we may assume $\Phi(G) = 1$ by Lemma 2.14 (a) and G abelian by Proposition 3.18, and then we want to show that any definable connected subgroup K of G is an $\mathcal{F}$-group. Assuming K is a minimal counterexemple to this, then $K \neq 1$ and every proper definable connected subgroup of K is an $\mathcal{F}$-group. As $\Phi(G) = 1$, there exists a maximal proper definable connected subgroup M of G such that $K \not\leq M$. For such a subgroup M, $G = MK$ by maximality of M and $K/(M \cap K) \simeq G/M$ is an $\mathcal{F}$-group. Thus the family of definable subgroups $(N_i)_i$ of K, defined by the property that $K/N_i \in \mathcal{F}$, has a trivial intersection. Since $\mathcal{F}$ is a connected subformation, K has a definable $\mathcal{F}$-subgroup H such that $K = N_iH$ for every i.

Since $K \neq 1$, there exists i such that N_i is a proper subgroup of K. In particular N_i° is an $\mathcal{F}$-group. But $K = N_i^\circ H$, so $K/H \simeq N_i^\circ/(N_i^\circ \cap H)$ is an $\mathcal{F}$-group, and by definition of the family $(N_i)_i$ it follows that $H = N_j$ for some j. Hence $K = N_jH = H$ is an $\mathcal{F}$-group. □

Recall from §4.1 that if $\tilde{\pi}$ is a set of unipotence parameters, then the collection $\mathcal{U}_{\tilde{\pi}}$ of solvable $U_{\tilde{\pi}}$-groups is a connected subformation. Corollary 2.15 and Proposition 5.2 in this specific case have the following consequence.

Corollary 5.3 (compare with [Bur06b, Lemma 2.8]) *Let G be a connected nilpotent group of finite Morley rank. Then G is a $U_{\tilde{\pi}}$-group if and only if $G/\Phi(G)$ is $\tilde{\pi}$-homogeneous.*

Applied with $\tilde{\pi} = \{(\infty, 0)\}$, Corollary 5.3 says in particular that G is a decent torus if and only if $G/\Phi(G)$ is a good torus. Remarkably, the proof of Cherlin's Theorem 2.3 was done for good tori first, and then adapted to decent tori after a remark of A. Borovik who noticed this interplay between good tori and decent tori.

5.3 Nilpotence and generation

If G is a group of finite Morley rank and $\mathcal{F}$ a connected subformation, we denote by $\mathcal{F}(G)$ the subgroup of G generated by all definable $\mathcal{F}$-subgroups of G. If G° is nilpotent, then $\mathcal{F}(G)$ is the unique $\mathcal{F}$-projector of G° by Corollary 4.5 (c).

Proposition 5.4 (compare with [Bur06b, Theorem 2.9]) *Let $\mathcal{F}$ be a connected subformation, G a nilpotent $\mathcal{F}$-group, and $(H_i)_i$ a finite family of definable subgroups which generates G. Then G is generated by the subgroups $\mathcal{F}(H_i)$.*

Proof. We use the notation "$\overline{}$" to denote the quotients by $\Phi(G)$ in G. By Proposition 5.2, $\overline{G}$ is $\mathcal{F}$-homogeneous. Hence for each i $\overline{H_i^\circ} = \mathcal{F}(\overline{H_i})$, and it follows that $H_i^\circ/H_i^\circ \cap \Phi(G)$ is covered by $\mathcal{F}(H_i)$ by Corollary 4.5 (c) applied in H_i°, i.e. $\overline{H_i^\circ} = \overline{\mathcal{F}(H_i)}$. But $G = \langle \cup_i H_i \rangle = \langle \cup_i H_i^\circ \rangle$ since the family $(H_i)_i$ is finite and G is connected nilpotent. Thus $\overline{G} = \langle \cup_i \overline{\mathcal{F}(H_i)} \rangle$ and $G = \langle \cup_i \mathcal{F}(H_i) \rangle \Phi(G)$. Now our claim follows from Lemma 2.14 (b). □

Of course, the finiteness of the family of generating subgroups in Proposition 5.4 can always be supposed when the family consists of definable connected subgroups only, by Zilber's theorem on indecomposable sets. Together with the approach of Carter subgroups by abnormal subgroups, Proposition 5.4 has the following structural corollary.

Corollary 5.5 *Let $\mathcal{F}$ be a connected subformation and G a connected solvable group of finite Morley rank. If S is a nilpotent $\mathcal{F}$-projector of G, then there exists a Carter subgroup Q of G such that $\mathcal{F}(Q) \leq S \leq \mathcal{F}(G')\mathcal{F}(Q)$.*

Proof. By Corollary 4.5 (b) $N(S)$ is abnormal, and by Proposition 3.15 and Theorem 3.16 $N(S)$ contains a Carter subgroup. Hence S is normalized by a Carter subgroup of G, say Q. Then S covers $\mathcal{F}(Q)S/S$, and $\mathcal{F}(Q) \leq S$. On the other hand, SG'/G' is the maximal definable $\mathcal{F}$-subgroup of $G/G' = QG'/G' \simeq Q/(Q \cap G')$. But $\mathcal{F}(Q)$ modulo $(Q \cap G')$ is the maximal definable $\mathcal{F}$-subgroup of Q modulo $(Q \cap G')$ by Corollary 4.5 (c), hence $SG' = \mathcal{F}(Q)G'$. As $\mathcal{F}(Q) \leq S$, it follows that $S = \mathcal{F}(Q)(S \cap G')$. Now Proposition 5.4 gives $S = \mathcal{F}(S \cap G')\mathcal{F}(Q) \leq \mathcal{F}(G')\mathcal{F}(Q)$. □

5.4 Maximal nilpotent $U_{\tilde{\pi}}$*-subgroups*

If we fix a set $\tilde{\pi}$ of unipotence parameters as usual, then the collection $\mathcal{U}_{\tilde{\pi}}$ of connected solvable $U_{\tilde{\pi}}$-groups is a connected subformation as seen in §4.1, and clearly any connected solvable group G of finite Morley rank has a unique maximal definable $\mathcal{U}_{\tilde{\pi}}$-subgroup, namely $U_{\tilde{\pi}}(G)$. More interestingly, the collection $\mathcal{N}_{\tilde{\pi}}$ of nilpotent $U_{\tilde{\pi}}$-groups is also a connected subformation by Corollary 4.2. The conditions of Theorem 4.7 fail for the subformation $\mathcal{N}_{\tilde{\pi}}$ if $|\tilde{\pi}| > 1$, and maximal definable nilpotent $U_{\tilde{\pi}}$-subgroups are not exactly the $\mathcal{N}_{\tilde{\pi}}$-projectors of connected solvable groups of finite Morley rank. In particular, there is a connected solvable group of finite Morley rank with several conjugacy classes of maximal definable nilpotent $U_{\tilde{\pi}}$-subgroups, as the following example shows.

Example 5.6 Let K be an infinite field of finite Morley rank and characteristic $p > 0$. In $G = K_{+} \rtimes K^{\times}$, the subgroups K_{+} and $K^{\times}$ are two nonconjugate maximal definable nilpotent $\{(p, \infty), (\infty, 0)\}$-subgroups.

However, the assumption of Theorem 4.8 is clearly verified by the connected subformation $\mathcal{N}_{\tilde{\pi}}$, and hence it gives a finiteness result.

Theorem 5.7 *Let G be a connected solvable group of finite Morley rank. Then, for every set $\tilde{\pi}$ of unipotence parameters, G has only finitely many conjugacy classes of maximal definable nilpotent $U_{\tilde{\pi}}$-subgroups.*

5.5 Carter $\tilde{\pi}$*-subgroups*

Again we fix a nonempty set $\tilde{\pi}$ of unipotence parameters. Recall from §3.2 that a Carter $\tilde{\pi}$-subgroup is a definable connected nilpotent $U_{\tilde{\pi}}$-subgroup Q such that $U_{\tilde{\pi}}(N(Q)) = Q$, instead of $N^{\circ}(Q) = Q$ as for Carter subgroups, and that these subgroups always exist by Theorem 3.3. Of course, a Carter $\tilde{\pi}$-subgroup is a maximal definable $\mathcal{N}_{\tilde{\pi}}$-subgroup by Theorem 2.7 and Proposition 2.8. In Example 5.6, the conjugates of $K^{\times}$ are Carter $\{(p, \infty), (\infty, 0)\}$-subgroups.

The conjugacy part of the following theorem can be found in [Bur04a, Theorem 4.16].

Theorem 5.8 *Let G be a connected solvable group of finite Morley rank and $\tilde{\pi}$ a set of unipotence parameters. Then Carter $\tilde{\pi}$-subgroups are exactly the $\mathcal{N}_{\tilde{\pi}}$-projectors and the $\mathcal{N}_{\tilde{\pi}}$-covering subgroups of G, and are in particular conjugate.*

Proof. If S is an $\mathcal{N}_{\tilde{\pi}}$-projector of G, then $S \leq U \leq N^\circ(S)$, where U denotes $U_{\tilde{\pi}}(N(S))$, and the equivalence between $\mathcal{N}_{\tilde{\pi}}$-projectors and $\mathcal{N}_{\tilde{\pi}}$-covering subgroups of Theorem 4.3 implies that S is also an $\mathcal{N}_{\tilde{\pi}}$-projector of U. In particular S/S is a maximal definable $\mathcal{N}_{\tilde{\pi}}$-subgroup of U/S. Now it follows from the push forward part of Lemma 2.13 that $U = S$, and S is a Carter $\tilde{\pi}$-subgroup of G.

For the reverse inclusion we proceed by induction on $\text{rk}(G)$. Let Q be a Carter $\tilde{\pi}$-subgroup of G. It suffices to show that $H = QN$ for every definable connected subgroup H containing Q and every definable subgroup $N \trianglelefteq H$ such that $H/N \in \mathcal{N}_{\tilde{\pi}}$. We may assume $G = H$ and we let P denote the preimage of $\Phi(G/N)$. Then by Lemma 2.14 (b) it suffices to show that $G = QP$. As G/N is a nilpotent $U_{\tilde{\pi}}$-group, G/P is an abelian $U_{\tilde{\pi}}$-group by Proposition 3.18. In particular $G = N(QP)$. Assume now toward a contradiction $QP < G$. Then the induction hypothesis applied in QP, together with Theorem 4.3 and a Frattini Argument, gives $G = N(QP) = N(Q)P$. Now $N(Q)P/P$ is a $U_{\tilde{\pi}}$-group, as well as $N(Q)/(N(Q) \cap P)$. By Lemma 2.13 (b), the latter quotient is covered by $Q = U_{\tilde{\pi}}(N(Q))$, i.e. $N(Q) = Q(N(Q) \cap P)$. Now $G = N(Q)P = QP$, a contradiction.

Our final claim is just Theorem 4.3. □

Theorem 5.8 says that, in any connected solvable group of finite Morley rank, Carter subgroups relativized to a set of unipotence parameters $\tilde{\pi}$ are $\mathcal{N}_{\tilde{\pi}}$-projectors. Then it gives with Corollary 5.5 the following structural information about these Carter $\tilde{\pi}$-subgroups in terms of (non relativized) Carter subgroups.

Corollary 5.9 *Let G be a connected solvable group of finite Morley rank. Then any Carter $\tilde{\pi}$-subgroup $Q_{\tilde{\pi}}$ of G satisfies $U_{\tilde{\pi}}(Q) \leq Q_{\tilde{\pi}} \leq U_{\tilde{\pi}}(G')U_{\tilde{\pi}}(Q)$ for some Carter subgroup Q of G.*

We will finally see that the upper bound is reached when $\tilde{\pi}$ consists of only one unipotence parameter.

5.6 The case $|\tilde{\pi}| = 1$: Sylow $\tilde{p}$-subgroups

The conjugacy of Sylow $\tilde{p}$-subgroups in connected solvable groups of finite Morley rank was first seen in [Bur04a] and [Bur06b]. When $\tilde{\pi}$ consists of a single unipotence parameter $\tilde{p}$, the subformation $\mathcal{N}_{\tilde{p}}$ satisfies the conditions of Theorem 4.7 by Proposition 2.10, and hence Sylow $\tilde{p}$-subgroups are exactly the $\mathcal{N}_{\tilde{p}}$-projectors of connected solvable groups of finite Morley rank. Hence Theorem 4.3 gives their conjugacy in this context. As we saw in §3.2 that Sylow $\tilde{p}$-subgroups coincide with Carter $\tilde{\pi}$-subgroups when $\tilde{\pi} = \{\tilde{p}\}$, this can also be seen as a special case of Theorem 5.8.

Theorem 5.10 *Let G be a connected solvable group of finite Morley rank and $\mathcal{N}_{\tilde{p}}$ the connected subformation of $\tilde{p}$-groups. Then the Sylow $\tilde{p}$-subgroups of G are exactly the $\mathcal{N}_{\tilde{p}}$-covering subgroups and the $\mathcal{N}_{\tilde{p}}$-projectors of G, and are in particular conjugate.*

Corollary 5.5, or 5.9, together with Proposition 2.10 and conjugacy allow one to describe the structure of Sylow $\tilde{p}$-subgroups of connected solvable groups, as seen originally by J. Burdges.

Corollary 5.11 [Bur06b, Theorem 6.7] *Let G be a connected solvable group of finite Morley rank. Then Sylow $\tilde{p}$-subgroups of G are exactly subgroups of the form $U_{\tilde{p}}(G')U_{\tilde{p}}(Q)$ for some Carter subgroup Q of G.*

6 More on solvable groups

Before concluding, we mention additional structural results of a slightly different nature concerning solvable groups of finite Morley rank. The first one in this motley crew is a direct application to π^*-groups of the general connected subformation theory developed above.

6.1 π^*-groups

Following [BN94b], if π is a set of primes and G a connected solvable group of finite Morley rank, we say that G is a π^**-group* if its definable connected abelian sections are π-divisible. One sees easily that a connected solvable group of finite Morley rank is a π^*-group if and only if its (unique) maximal definable connected π-subgroup of bounded exponent is trivial. These groups have been used in [BN94b] for the study of Hall π-subgroups in solvable groups of finite Morley rank and for the proof of

Theorem 1.9. Even if they are not directly related to our main concern here, we insert them just as a special case of the subformation theory developed above.

Lemma 6.1 [Fré00b, Lemme 8.17] *The collection of π^*-groups is a connected subformation.*

This can be proved exactly as in [Fré00b, Lemme 8.17]. Then the conjugacy of maximal definable π^*-subgroups follows from Theorem 4.7.

Corollary 6.2 ([Fré00b, Théorème 8.18]; see also [BN94b]) *Let G be a connected solvable group of finite Morley rank and $\mathcal{F}$ the subformation of π^*-groups. Then the $\mathcal{F}$-projectors of G are exactly the maximal definable π^*-subgroups of G, which in particular are conjugate.*

From this one can also delineate the structure of maximal definable π^*-subgroups, as seen by A. Borovik and A. Nesin in [BN94b]. The reader can also consult [Fré00b, Corollaire 8.19].

6.2 Uniformly definable families of decent tori

Maximal definable decent tori are conjugate. This is true in a connnected solvable group by the connected subformation theory, more precisely by Theorem 5.10 with the unipotence parameter $\tilde{p} = (\infty, 0)$, or much more generally in an arbitrary group of finite Morley rank by Theorem 2.3. Hence one cannot hope for decent tori for a better finiteness result such as the one obtained in Theorem 5.7 for certain *maximal* subgroups. But one can still look at subgroups, not necessarily maximal, in the class of subgroups considered. In the case of decent tori, there is a finiteness theorem on conjugacy classes of such nonnecessarily maximal subgroups.

More precisely, it was proved in [Che05] that if G is an arbitrary group of finite Morley rank and $\mathcal{F}$ a uniformly definable family of good tori in G, then the tori in $\mathcal{F}$ fall into finitely many conjugacy classes under the action of G. We are going to generalize this to decent tori. Notice first that the uniformity assumption cannot be dropped from such a statement, even in the good tori case. Indeed, if T is a good torus, then subgroups of $T \times T$ of the form $\{(t, t^n) \mid t \in T\}$ are, as n varies, pairwise distinct definable one-dimensional good tori. In particular they are nonconjugate in $T \times T$; the family is not uniformly definable.

We start by looking at uniformly definable families of decent subtori inside a given decent torus.

Lemma 6.3 *Let T be a decent torus, interpretable in an arbitrary group of finite Morley rank. Then any uniformly definable family of decent subtori of T is finite.*

Before the proof, we note that all assumptions are made relatively to the ambient group only, and that the latter is not assumed to satisfy any saturation assumption.

Proof. Assume that a group provides a counterexample T, which may be assumed to be of minimal Morley rank, and denote by $\mathcal{F}$ a uniformly definable family of decent subtori of T with infinitely many pairwise distinct members. Then the family $\mathcal{F}_\Phi$ of images of members of $\mathcal{F}$ in $T/\Phi(T)$ is a uniformly definable family of definable connected subgroups of $T/\Phi(T)$. But the latter is a good torus by Corollary 5.3. Now the proof used in [Che05, Lemma 6] shows that uniformly definable families of subgroups of a good torus interpretable in a group of finite Morley rank have to be finite, and in particular the family $\mathcal{F}_\Phi$ of subgroups of $T/\Phi(T)$ has to be finite.

Hence there is an element $\tilde{T}/\Phi(T)$ of $\mathcal{F}_\Phi$ which is the image of infinitely many elements in $\mathcal{F}$. As $\tilde{T} \trianglelefteq T$ is abelian, it contains a unique maximal definable decent subtorus S by Corollary 4.5 (c). Then S is a decent torus which contains infinitely many elements of $\mathcal{F}$, and $T = S$ by minimality of $\mathrm{rk}\,(T)$. Since $S \leq \tilde{T} \leq T$ we obtain $T = \tilde{T}$, and, shrinking $\mathcal{F}$ if necessary, we may assume that the infinitely many elements of $\mathcal{F}$ all cover $T/\Phi(T)$. But now any such element is equal to T by Corollary 2.14 (b). In particular $\mathcal{F} = \{T\}$ is finite, a contradiction. □

Combined with Theorem 2.3, Lemma 6.3 gives our finiteness theorem for conjugacy classes of uniformly definable families of decent tori.

Theorem 6.4 *Let G be a group of finite Morley rank, and $\mathcal{F}$ a uniformly definable family of decent tori in G. Then the tori in $\mathcal{F}$ fall into finitely many conjugacy classes under the action of G.*

Proof. Each decent torus of the family is contained in a maximal one, and maximal definable decent tori are conjugate in the ambient group by Theorem 2.3. Hence one can apply Lemma 6.3 in such a maximal one. □

6.3 Frattini and Carter subgroups

Two unpublished structural results linking the Frattini subgroup together with Carter subgroups in connected solvable groups:

Proposition 6.5 *Let G be a connected solvable group of finite Morley rank and Q a Carter subgroup of G. Then $F(G) \cap \Phi(Q) \leq \Phi(G)$.*

Proof. Assume toward a contradiction $F(G) \cap \Phi(Q) \not\leq \Phi(G)$. We are first going to show that dividing by $\Phi(G)$ we can assume $\Phi(G) = 1$, without altering our assumption.

For each maximal proper definable connected subgroup $M/\Phi(G)$ of the quotient $Q\Phi(G)/\Phi(G)$, the connected component M° of the preimage M contains $\Phi^\circ(G)$ and is a maximal proper definable connected subgroup of $Q\Phi^\circ(G)$. Now if N denotes the connected component of $Q \cap M^\circ$, then one can see as follows that the proper definable connected subgroup N of Q is maximal for these properties. Indeed, if N_1 is a definable connected subgroup of Q containing N, then either $N_1\Phi(G) < Q\Phi(G)$, in which case the maximality of $M/\Phi(G)$ forces $N_1 = N$, or $N_1\Phi(G) = Q\Phi(G)$, in which case the subgroup N_1 of Q covers $Q/(Q\cap\Phi(G))$, as well as $Q/(Q\cap\Phi^\circ(G))$, and $Q \leq N_1(Q\cap\Phi^\circ(G))^\circ \leq N_1(Q \cap M^\circ)^\circ = N_1N = N_1$. Hence N is a maximal proper definable connected subgroup of Q as claimed. In particular M° contains a maximal proper definable connected subgroup of Q, and M contains $\Phi(Q)$. We have shown in any case that

$$\Phi(Q)\Phi(G)/\Phi(G) \leq \Phi(Q\Phi(G)/\Phi(G)).$$

By Proposition 3.18, $\Phi(G) \leq F(G)$ and $F(G)/\Phi(G) = F(G/\Phi(G))$. By Corollary 3.13, $Q\Phi(G)/\Phi(G)$ is a Carter subgroup of $G/\Phi(G)$. Now our contradictory assumption that $F(G) \cap \Phi(Q) \not\leq \Phi(G)$ implies that in the quotient $G/\Phi(G)$ the intersection $F(G/\Phi(G)) \cap \Phi(Q\Phi(G)/\Phi(G))$ is nontrivial. As $\Phi(G/\Phi(G)) = 1$ by Lemma 2.14 (a), our contradictory assumption is also met in the quotient $G/\Phi(G)$. The moral of this is that, dividing G by $\Phi(G)$, we may assume $\Phi(G) = 1$, and our assumption becomes $F(G) \cap \Phi(Q) \neq 1$.

As G is connected, its hypercenter $Z_\infty(G)$, defined for example as in [Wag97, §0.1], is equal to $Z_n(G)$ for some finite n [Poi87, Corollaire 3.15]. It is thus definable and nilpotent, and hence contained in $F(G)$. An easy induction shows that the hypercenter is also contained in any selfnormalizing subgroup, and this applies to Carter subgroups of G by Theorem 3.11. Thus $Z_\infty(G) \leq F(G) \cap Q$. Since $\Phi(G) = 1$, $F(G)$ is abelian by Proposition 3.18. It follows then easily, as $G = QF(G)$ by Corollary 3.13, that $Z_\infty(G) = F(G) \cap Q$. In particular the conjugacy of Carter subgroups gives $F(G) \cap Q = F(G) \cap R$ for every Carter subgroup R of G.

Since $F(G) \cap \Phi(Q) \neq 1$, there is a nontrivial element x in $F(G) \cap \Phi(Q)$, and x must belong to every Carter subgroup of G. As $\Phi(G) = 1$, x does not belong to a maximal proper definable connected subgroup H of G. As $\Phi(G) = 1$ again, G has by Propositions 3.18 and 3.19 a definable connected abelian characteristic subgroup A such that $G = A \rtimes Q$. We claim that $A \leq H$. Otherwise, $G = AH$ by maximality of H. If S denotes a Carter subgroup of H, then, as HA/A is nilpotent, $H = (A \cap H)S$ by Corollary 3.13 and $G = AS$. As $A' = 1$, $Z(S) \cap A \leq Z(G) \cap A \leq Q \cap A = 1$, and $S \cap A = 1$ as S is nilpotent. Hence $[N_A(S), S] \leq A \cap S = 1$, and $N_A(S) \leq Z(G) \cap A = 1$. Therefore $N(S) = N_A(S)S = S$ and S is a Carter subgroup of G. Thus $x \in S \leq H$, a contradiction showing $A \leq H$, as claimed. Now $H \cap Q$ is a maximal proper definable connected subgroup of Q, and thus it contains $\Phi(Q)$. This is a contradiction as $x \in \Phi(Q)$ and $x \notin H$. □

Corollary 6.6 *Let G be a connected solvable group of finite Morley rank such that $\Phi(G) = 1$. If Q is a Carter subgroup of G, then Q is abelian, $G = G' \rtimes Q$, and $F(G) = G' \times Z(G)$.*

Proof. By Theorem 1.13 and Propositions 3.18 and 6.5, $Q' \leq F(G) \cap \Phi(Q) \leq \Phi(G) = 1$. By Proposition 3.19, G has a definable connected abelian characteristic subgroup A such that $G = A \rtimes Q$. Since Q is abelian, $G' \leq A$. But $G = G'Q$ by Corollary 3.13 and Theorem 1.13, so $A = G'$ and $G = G' \rtimes Q$. Now $F(G) = G' \times (F(G) \cap Q)$. But $Z(G) \leq F(G) \cap N(Q)$, i.e. $Z(G) \leq F(G) \cap Q$ by Theorem 3.11, and as $G = F^\circ(G)Q$ and $Q' = 1$ and $F(G)' = 1$ by Proposition 3.18, $F(G) \cap Q \leq Z(G)$ also. □

6.4 Generalized centralizers

There are other important, and not mentioned yet, subgroups related to Carter subgroups in connected solvable groups of finite Morley rank. If G is a group and $\alpha \in G$, then $E_G(\alpha)$ denotes the set of elements $x \in G$ such that $[x,_n \alpha] = 1$ for some $n \in \mathbb{N}$, where $[x,_0 \alpha] = x$ and $[x,_{n+1} \alpha] = [[x,_n \alpha], \alpha]$. If X is a subset of G, the *generalized centralizer* $E_G(X)$ of X in G is the intersection of $E_G(x)$ for all $x \in X$. This notion was introduced by T. A. Peng in [Pen69a]. As the following example shows, a generalized centralizer is in general not a subgroup of the ambient group, even in an abelian-by-nilpotent finite group.

Example 6.7 [Pen69b] Let $A = \langle a, b \mid a^3 = b^3 = 1,\ ab = ba \rangle$ and D the dihedral group of order 8 $\langle c, d \mid c^2 = d^2 = 1,\ (cd)^4 = 1 \rangle$. Let $G = A \rtimes D$ where D acts on A as follows: $a^c = a^{-1}$, $b^c = b$, and $a^d = b$. Then G is an abelian-by-nilpotent group and $E_G(c)$ is not a subgroup of G.

However, we have the following result in connected solvable groups of finite Morley rank.

Proposition 6.8 [Fré00b, Corollaire 5.17] *Let G be a connected solvable group of finite Morley rank and H a nilpotent subgroup of G. Then $E_G(H)$ is a definable connected subgroup of G and $H \leq F(E_G(H))$.*

In this context there is also a preservation by quotient and an abnormality result.

Proposition 6.9 [Fré00b, §7.2] *Let G be a connected solvable group of finite Morley rank and H a nilpotent subgroup of G.*

a. $E_{G/N}(HN/N) = E_G(H)N/N$ for every (non necessarily definable) normal subgroup N of G.
b. $E_G(H)$ is abnormal in G.

Notice that Proposition 6.9 (*b*) is shown in [Fré00b, §7.2] as a corollary of Proposition 6.9 (*a*). Contrarily to Example 3.17 concerning the set of all abnormal subgroups, the main result here is that there are only finitely many conjugacy classes of such generalized centralizers.

Theorem 6.10 [Fré00b, Théorème 7.9] *Let G be a connected solvable group of finite Morley rank. Then there are at most $2^{rk(G)}$ conjugacy classes of generalized centralizers of nilpotent subgroups.*

By Theorem 4.8 a connected solvable group of finite Morley rank has only finitely many conjugacy classes of maximal definable connected nilpotent subgroups. Theorem 6.10 allows one to strengthen this as follows.

Corollary 6.11 [Fré00b, Corollaire 7.10] *In any connected solvable group of finite Morley rank, there are only finitely many conjugacy classes of maximal nilpotent subgroups.*

6.5 About nonconnected solvable groups

After the preceding extensive study of Carter subgroups in connected solvable groups, we turn our attention in this subsection to nonnecessarily connected solvable groups. Ideally, we would still like Carter subgroups to be characterized as in Corollary 3.12 and in finite solvable groups by being merely nilpotent and selfnormalizing. Unfortunately, the following example prevents from this.

Example 6.12 The group $\mathbb{C}^\times \rtimes \langle \omega \rangle$, where ω is an involution inverting $\mathbb{C}^\times$, has no nilpotent selfnormalizing subgroup.

However, the group of Example 6.12 has a unique conjugacy class of *locally nilpotent* selfnormalizing subgroups, its maximal 2-subgroups. Hence, for nonconnected solvable groups, we rather tend to focus on locally nilpotent selfnormalizing subgroups. This fits well with the connected case, thanks to the following property.

Proposition 6.13 [Fré00b, Corolaire 5.17] *In any connected solvable group of finite Morley rank, locally nilpotent subgroups are nilpotent.*

Besides, it is noteworthy that Proposition 6.13 holds for any, i.e. not necessarily connected, locally nilpotent subgroup. Some versions of the same result are known for *connected* locally nilpotent subgroups in much larger contexts [DW97, Corollary 21].

Locally nilpotent selfnormalizing subgroups are the good candidates for generalizing Carter subgroups in nonconnected solvable groups of finite Morley rank, but there is still a technical problem with them: they are not definable, as Example 6.12 shows. For this reason, we introduce *locally closed* subgroups as subgroups of groups of finite Morley rank which contain the definable closures of all their finite subsets. This notion allows one to study locally nilpotent selfnormalizing subgroups thanks to the following lemma.

Lemma 6.14 [Fré04, Corollaire 3.15] *In any group of finite Morley rank, locally nilpotent selfnormalizing subgroups are locally closed.*

With these definability issues now aside, we have existence and conjugacy.

Theorem 6.15 [Fré04, Théorème 4.6] *If H is a locally closed subgroup of a solvable group of finite Morley rank, then locally nilpotent selfnormalizing subgroups of H exist and are conjugate in H. In particular, in* *any solvable group of finite Morley rank, locally nilpotent selfnormalizing subgroups exist and are conjugate.*

In this context there is also an analog of Corollary 3.13, that the reader can find in [Fré04, Théorème 4.6]. Finally, Theorem 6.15 cannot be generalized to any subgroup of solvable groups of finite Morley rank, since the following example provides a subgroup of a connected solvable group of finite Morley rank, with nonisomorphic nilpotent selfnormalizing subgroups.

Example 6.16 In the standard Borel subgroup of $GL_2(\mathbb{C})$, consider

$$h = \begin{pmatrix} e^{\frac{2i\pi}{3}} & 0 \\ 0 & 1 \end{pmatrix},\ k = \begin{pmatrix} -1 & 1 \\ 0 & 1 \end{pmatrix},\ \text{and } U = \langle h, k \rangle.$$

Then $\langle h \rangle$ and $\langle k \rangle$ are two abelian selfnormalizing subgroups of U of different orders.

7 *Le mot de la fin:* applications beyond solvable groups

The preceding unipotence theory has been developed for some needs in the study of simple groups of finite Morley rank, ultimately motivated by the Algebricity Conjecture. To conclude this paper we merely wish to present a bunch of applications of this theory in nonsolvable groups, together with a few historical remarks.

The most significant theorem concerning simple groups of finite Morley rank, depending on the structure of their Sylow 2-subgroups, is most probably the following at present.

Theorem 7.1 [ABC06] *Let G be a simple group of finite Morley rank with a nontrivial $\tilde{2}$-subgroup. Then G is isomorphic to an algebraic group over an algebraically closed field of characteristic* 2.

Burdges' unipotence theory was not so much needed for these groups, also called groups of *even type.* However the theory of good tori, later on generalized to decent tori, was developed in this context, thanks to the presence of fields of characteristic 2 and what was noticed in Example 2.1 (Wagner's Theorem 1.14 arrived at a very timely moment for that!). More precisely, there was a strong need in [AC05] to transfer arguments from [Jal01b] relying on solvable groups to more general contexts. In particular, the version of Theorem 6.4 restricted to good tori has been proved exactly for that purpose in [Che05].

While the characteristic p unipotence theory was rather well experimented in the context of groups of even type, with Proposition 2.4 often used at least for the prime $p = 2$, the characteristic 0 unipotence theory of §2.3 has originally been motivated by the so-called signalizer functor theory in groups of *odd type*, i.e. groups in which the connected component of a Sylow 2-subgroup is a 2-torus. Aside from Burdges' Thesis, the first application can be found in [Bur04b]. The analogy between this characteristic 0 unipotence theory and Sylow theory in finite groups has then rapidly been developed in [Bur06b].

Subsequently, it became natural to start local analysis of groups of finite Morley rank using this new Sylow theory. It turned out to be particularly useful in the study of *minimal connected simple* groups, i.e. infinite simple groups of finite Morley rank in which each proper definable connected subgroup is solvable. These groups are the core of many preoccupations on simple groups of finite Morley rank, exactly as their finite analogs were for finite groups. In groups of finite Morley rank there is unfortunately no analog of Thompson's complete classification of finite minimal simple groups. However, the paper [Bur06a] provides an encyclopedia of results concerning intersections of Borel subgroups in this context. As a mere example, we mention the following.

Theorem 7.2 [Bur06a] *Let G be a minimal connected simple group of finite Morley rank. Then any nonabelian definable $\tilde{p}$-subgroup of G is contained in a unique Borel subgroup of G.*

This analysis of intersections of Borel subgroups has been used intensively in [BCJ05] and [Del06] for the study of minimal connected simple groups of finite Morley rank and of odd type.

By Theorem 3.2, a nontrivial Sylow $\tilde{p}$-subgroup of minimal unipotence degree is always contained in a Carter subgroup of the ambient group, and in particular a nontrivial definable decent torus is always contained in a Carter subgroup. In case of the presence of such a nontrivial divisible abelian torsion subgroup, the question of the generosity of Carter subgroups is well in hand in minimal connected simple groups, by Theorems 2.3 and 3.11 and Lemma 3.6.

Theorem 7.3 *Let G be a minimal connected simple group of finite Morley rank with a nontrivial divisible abelian torsion subgroup T, and Q a Carter subgroup of G containing T. Then Q is generous in G. In particular $T \leq \tilde{T} \leq Q$ for some maximal divisible abelian torsion subgroup $\tilde{T}$ of G.*

Without divisible torsion, there is in general no way to reach Conjecture 3.9.

To conclude, we mention that Conjecture 2.6 can be proved under very specific circumstances. A group of finite Morley rank is *semisimple* if it has no nontrivial (definable) normal abelian subgroup.

Borovik-Poizat Fusion Argument. *Let G be a connected semisimple group of finite Morley rank and $\tilde{p}$ a unipotence parameter. Assume that each proper definable connected subgroup of G satisfies the conjugacy of Sylow $\tilde{p}$-subgroups. Then any two Sylow $\tilde{p}$-subgroups S and T of G such that $S^G \cap T^G \neq 1$ are conjugate.*

This can be proved by the *fusion* argument used in [BP90] for the conjugacy of Sylow 2-subgroups of Theorem 1.5, invoking here Proposition 2.8 as the appropriate version of the normalizer condition. When S and T are nontrivial 2-subgroups the condition $S^G \cap T^G \neq 1$ can be obtained in an inductive process by the specific properties of elements of order 2. For Sylow $\tilde{p}$-subgroups there is in general no way to obtain such a condition without a generosity property in the style of Conjecture 3.9. That's why the Borovik-Poizat Fusion Argument in the context of $\tilde{p}$-subgroups has not yet found applications without a generosity property.

References

[ABC97] T. Altınel, A. Borovik, and G. Cherlin, Groups of mixed type, *J. Algebra* 192 (1997) no. 2, 524–571.

[ABC06] T. Altınel, A. Borovik, and G. Cherlin, *Simple groups of finite Morley rank*, Book in preparation, 2006.

[AC99] T. Altinel and G. Cherlin, On central extensions of algebraic groups, *J. Symbolic Logic* 64 (1999), no. 1, 68–74.

[AC04] T. Altınel and G. Cherlin, Simple L^*-groups of even type with strongly embedded subgroups, *J. Algebra* 272 (2004) no. 1, 95–127.

[AC05] T. Altınel and G. Cherlin, Limoncello, *J. Algebra* 291 (2005) no. 2, 373–415.

[ACCN98] T. Altinel, G. Cherlin, L. J. Corredor, and A. Nesin, A Hall theorem for ω-stable groups, *J. London Math. Soc. (2)* 57 (1998) no. 2, 385–397.

[Bal73] J. T. Baldwin, α_T is finite for $\aleph_1$-categorical T, *Trans. Amer. Math. Soc.* 181 (1973), 37–51.

[BC02] J. Burdges and G. Cherlin, Borovik-Poizat rank and stability, *J. Symbolic Logic* 67 (2002) no. 4, 1570–1578.

[BCJ05] J Burdges, G. Cherlin, and E. Jaligot, Minimal connected simple groups of finite Morley rank with strongly embedded subgroups, *J. Algebra* 314 (2007) 581-612.

[BHMPW06] A. Baudisch, M. Hils, A. Martin-Pizarro, and F.O. Wagner, Die böse farbe, submitted, 2006.

[BMPZ06] A. Baudisch, A. Martin-Pizarro, and M. Ziegler, Red fields, *J. Symbolic Logic* 72 (2007) no. 1, 207-225.

[BN92] A. V. Borovik and A. Nesin, On the Schur-Zassenhaus theorem for groups of finite Morley rank, *J. Symbolic Logic* 57 (1992) no. 4, 1469–1477.

[BN94a] A. Borovik and A. Nesin, *Groups of finite Morley rank*, The Clarendon Press Oxford University Press, New York, 1994, Oxford Science Publications.

[BN94b] A. V. Borovik and A. Nesin, Schur-Zassenhaus theorem revisited, *J. Symbolic Logic* 59 (1994) no. 1, 283–291.

[BP90] A. V. Borovik and B. P. Poizat, Tores et *p*-groupes, *J. Symbolic Logic* 55 (1990) no. 2, 478–491.

[Bur04a] J. Burdges, *Odd and degenerate types groups of Morley rank*, Doctoral Dissertation, Rutgers University, 2004.

[Bur04b] J. Burdges, A signalizer functor theorem for groups of finite Morley rank, *J. Algebra* 274 (2004), no. 1, 215–229.

[Bur06a] J. Burdges, The Bender method in groups of finite Morley rank, *J. Algebra* 312 (2007), no. 1, 33–55.

[Bur06b] J. Burdges, Sylow theory for $p = 0$ in solvable groups of finite Morley rank, *J. Group Theory* 9 (2006), no. 4, 467–481.

[Car61] R. W. Carter, Nilpotent self-normalizing subgroups of soluble groups, *Math. Z.* 75 (1960/1961), 136–139.

[Che79] G. Cherlin, Groups of small Morley rank, *Ann. Math. Logic* 17 (1979), no. 1-2, 1–28.

[Che05] G. Cherlin, Good tori in groups of finite Morley rank, *J. Group Theory* 8 (2005), no. 5, 613–622.

[Del06] A. Deloro, Groupes simples connexes minimaux algébriques de type impair, to appear in *J. Algebra.*

[DVLT98] F. Dalla Volta, A. Lucchini, and M. C. Tamburini, On the conjugacy problem for Carter subgroups, *Comm. Algebra* 26 (1998) no. 2, 395–399.

[DW97] J. Derakhshan and F. O. Wagner, Nilpotency in groups with chain conditions, *Quart. J. Math. Oxford Ser. (2)* 48 (1997), no. 192, 453–466.

[FJ05] O. Frécon and E. Jaligot, The existence of Carter subgroups in groups of finite Morley rank, *J. Group Theory* 8 (2005), no. 5, 623–644.

[Fré00a] O. Frécon, *Etude des groupes résolubles de rang de Morley fini*, Thèse de Doctorat, Université de Lyon I, 2000.

[Fré00b] O. Frécon, Sous-groupes anormaux dans les groupes de rang de Morley fini résolubles, *J. Algebra* 229 (2000), no. 1, 118–152.

[Fré00c] O. Frécon, Sous-groupes de Hall généralisés dans les groupes résolubles de rang de Morley fini, *J. Algebra* 233 (2000), no. 1, 253–286.

[Fré04] O. Frécon, Sous-groupes de Carter dans les groupes de rang de Morley fini, *J. Symbolic Logic* 69 (2004) no. 1, 23–33.

[Fré06a] O. Frécon, Around unipotence in groups of finite Morley rank, *J. Group Theory* 9 (2006), no. 3, 341–359.

[Fré06b] O. Frécon, Carter subgroups in tame groups of finite Morley rank, *J. Group Theory* 9 (2006), no. 3, 361–367.

[Gas63] W. Gaschütz, Zur Theorie der endlichen auflösbaren Gruppen, *Math. Z.* 80 (1962/1963), 300–305.

[Hum75] J. E. Humphreys, *Linear algebraic groups*, Springer-Verlag, New York, 1975, Graduate Texts in Mathematics, No. 21.

[Jal01a] E. Jaligot, Full Frobenius groups of finite Morley rank and the Feit-Thompson theorem, *Bull. Symbolic Logic* 7(2001) no. 3, 315–328.

[Jal01b] E. Jaligot, Groupes de rang de Morley fini de type pair avec un sous-groupe faiblement inclus, *J. Algebra* 240 (2001) no. 2, 413–444.

[Jal06] E. Jaligot, Generix never gives up, *J. Symbolic Logic* 71 (2006) no. 2, 599–610.

[Mac71] A. Macintyre, On ω_1-categorical theories of fields, *Fund. Math.* 71 (1971), no. 1, 1–25. (errata insert).

[Mor65] M. Morley, Categoricity in power, *Trans. Amer. Math. Soc.* 114 (1965), 514–538.

[Mus04] Y. Mustafin, Structure des groupes linéaires définissables dans un corps de rang de Morley fini, *J. Algebra* 281 (2004) no. 2, 753–773.

[Nes90] A. Nesin, On solvable groups of finite Morley rank, *Trans. Amer. Math. Soc.* 321 (1990) no. 2, 659–690.

[Nes91] A. Nesin, Generalized Fitting subgroup of a group of finite Morley rank, *J. Symbolic Logic* 56(1991) no. 4, 1391–1399.

[New91] L. Newelski, On type definable subgroups of a stable group, *Notre Dame J. Formal Logic* 32(1991) no. 2, 173–187.

[Pen69a] T. A. Peng, Finite soluble groups with an Engel condition, *J. Algebra* 11 (1969), 319–330.

[Pen69b] T. A. Peng, On groups with nilpotent derived groups, *Arch. Math. (Basel)* 20 (1969), 251–253.

[Poi87] B. Poizat, *Groupes stables*, Nur al-Mantiq wal-Ma'rifah, Bruno Poizat, Lyon, 1987.

[Poi01] B. Poizat, L'égalité au cube, *J. Symbolic Logic* 66 (2001) no. 4, 1647–1676.

[PTV04] A. Previtali, M. C. Tamburini, and E. P. Vdovin, The Carter subgroups of some classical groups, *Bull. London Math. Soc.* 36 (2004) no. 2, 145–155.

[TV02] M. C. Tamburini and E. P. Vdovin, Carter subgroups in finite groups, *J. Algebra* 255 (2002) no. 1, 148–163.

[Wag90] F. Wagner, Subgroups of stable groups, *J. Symbolic Logic* 55 (1990) no. 1, 151–156.

[Wag94] F. O. Wagner, Nilpotent complements and Carter subgroups in stable $\mathcal{R}$-groups, *Arch. Math. Logic* 33 (1994) no. 1, 23–34.

[Wag97] F. O. Wagner, *Stable groups*, Cambridge University Press, Cambridge, 1997.

[Wag01] F. Wagner, Fields of finite Morley rank, *J. Symbolic Logic*, 66 (2001) no. 2, 703–706.

[Wag03] F. O. Wagner, Bad fields in positive characteristic, *Bull. London Math. Soc.* 35(2003) no. 4, 499–502.

[Zil77] B. I. Zil'ber, Groups and rings whose theory is categorical, *Fund. Math.* 95(1977) no. 3, 173–188.

Permutation groups of finite Morley rank

Alexandre Borovik
The University of Manchester

Gregory Cherlin
Rutgers University

Introduction

Groups of finite Morley rank made their first appearance in model theory as *binding groups*, which are the key ingredient in Zilber's ladder theorem and in Poizat's explanation of the Picard-Vessiot theory. These are not just groups, but in fact permutation groups acting on important definable sets. When they are finite, they are connected with the model theoretic notion of algebraic closure. But the more interesting ones tend to be infinite, and connected.

Many problems in finite permutation group theory became tractable only after the classification of the finite simple groups. The theory of permutation groups of finite Morley rank is not very highly developed, and while we do not have anything like a full classification of the simple groups of finite Morley rank in hand, as a result of recent progress we do have some useful classification results as well as some useful structural information that can be obtained without going through an explicit classification. So it seems like a good time to review the situation in the theory of permutation groups of finite Morley rank and to lay out some natural problems and their possible connections with the body of research that has grown up around the classification effort.

The study of transitive permutation groups is equivalent to the study of pairs of groups (G, H) with H a subgroup of G, and accordingly one can read much of general group theory as permutation group theory, and vice versa, and, indeed, a lot of what goes on in work on classification

Second author supported by NSF Grant DMS-0100794.
Both authors thank the Newton Institute, Cambridge, for its hospitality during the Model Theory and Algebra program, where the bulk of this work was carried out, as well as CIRM for its hospitality at the September 2004 meeting on *Groups, Geometry and Logic*, where the seed was planted. Thanks to Altınel for continued discussions all along the way.

makes a good deal of sense as permutation group theory—including even the final identification of a group as a Chevalley group, which can go via Tits' theory of buildings, or in other words by recognition of the natural permutation representations of such groups. Many special topics in permutation groups tied up with structural issues were discussed in [7, Chapter 11], with an eye toward applications. See also Part III of [15].

The most important class of permutation groups consists of the *definably primitive* permutation groups, and in finite group theory one has the O'Nan-Scott-Aschbacher classification of these groups into various families, determined mainly by the structure of the socle and the way it meets a point stabilizer. This theorem has been adapted to the context of finite Morley rank by Macpherson and Pillay [14], and is the one really general piece of work in the area to date. We will refer to this fundamental result (or set of results) as MPOSA. Also noteworthy is the classification by Hrushovski of groups acting faithfully and definably on strongly minimal sets [17, Th. 3.27], found here as Proposition 2.5, and the study by Gropp [11] of the rank two case.

It turns out that basic notions of permutation group theory such as primitivity and multiple transitivity have more than one useful analog in the context of groups of finite Morley rank, for two reasons: (a) we are interested particularly in connected groups (and, by implication, sets of Morley degree 1); (b) we are interested in generic behavior. Of course we also impose definability constraints. So we have definable primitivity and some analogs involving connectivity, and we have generic n-transitivity, which is far more common than ordinary n-transitivity. Indeed, sharp 4-transitivity cannot occur on an infinite set [12], while $\mathrm{AGL}(V)$ acts generically sharply $(n+1)$-transitively on V if V has dimension n, with $\mathrm{PGL}(V)$ generically sharply $(n+1)$-transitive on projective space, with similar, though less extreme, statements for other classical groups acting naturally.

In our first section we will explore some of the fundamental definitions and their natural variations. After that we will focus on the following problem, which can be taken up from a number of points of view:

Problem 1 Bound the rank of a definably primitive permutation group of finite Morley rank in terms of the rank of the set on which it acts.

We will show, using soft methods, that there is some such bound. Here we combine MPOSA with some ideas that have come recently out of the classification project. There are two points to the analysis. One aims to drive the stabilizer of a sufficiently long sequence of generic and independent elements to the identity, bounding the length of the sequence. This divides into two parts: (1) getting started: first bound the possible degree of generic multiple transitivity, which is the length of time one waits before anything happens; (2) moving along: once the chain of point stabilizers begins to decrease, argue that the process runs out in bounded time. In the first stage we are not very precise in our estimates, but one may expect that very good bounds should hold in this part of the process, as generically highly transitive groups should be rare outside of known examples. This is a problem which makes sense and is interesting for groups of finite Morley rank in general, for simple algebraic groups acting definably, and even for simple algebraic groups acting algebraically. In the latter case it has been solved in characteristic 0 by Popov [19], using some results of Kimura et al. on rational representations with an open orbit.

The following result, controlling what one might reasonably call "Lie rank", plays an important role in our "soft" analysis and could also be of use in more concrete approaches.

Lemma 3.11 *Let (G,Ω) be a definably primitive permutation group of finite Morley rank, T a definable divisible abelian subgroup of G, T_0 its torsion subgroup, and $O(T)$ the largest definable torsion free subgroup of T. Then* $\mathrm{rk}(T/O(T)) \leq \mathrm{rk}(\Omega)$.

The present paper has an improvisational character. In the light of recent progress (including some directly attributable to the month devoted to this topic at the Newton Institute) it occurred to us that this could be a good time to take up the topic of permutation groups afresh. We thought that it would in particular provide an interesting setting for a review of some of the neglected but interesting areas (representation theory, cohomology) where the theory in the algebraic case offers considerable food for thought. No doubt this is the case, but that is not the paper that has emerged in this round. Rather, the rank-bounding problem described above wound up giving us a sharper but somewhat narrower focus, in which the existing theory plays a substantial role, and problems of linearization are particularly highlighted. While as a result some of our favorite problems are not represented here, we found some others, some entirely new, along with some new reasons for repeating old questions. We hope some of our readers will answer some of these questions quickly, and find better versions of some of the others. Also, we hope that the optimistic tone with which various approaches are described will not give the impression that they do not require proper proofs. The line between the proven and the unproven is certainly fuzzy here and in the long run caution is the best policy (in spite of Keynes).

We also hope that we have dealt with the foundations in a satisfactory way, or, if not, that someone will take up the matter further.

We would like to take this opportunity to thank the staff at the Newton Institute (especially the gentleman who repaired the espresso machine) and the organizers, with extra thanks to Zoé Chatzidakis.

Contents

1 Foundations

We review some basic model theoretic and permutation group theoretic notions in the context of definable group actions, for the most part under the assumption of finite Morley rank, though it may well be worth taking the extra care necessary to work in the stable category systematically. The main notions are transitivity, primitivity, and multiple transitivity and a number of related variants, along with definability, genericity, and connectivity. Most of this is already present in one form or another in [14] but we think it is still worth while to consider the foundations at leisure, and separately from more technical matters. Indeed, we have the feeling that there is still something to be done here, at least at the level of collecting illuminating examples.

The term "permutation group" is generally taken here to refer to a group equipped with a faithful action. But in dealing with intransitive actions, we may consider the restriction of the group to an individual orbit, and there may be a kernel in this case. Sometimes this actually matters. We will be casual about this below; strictly speaking one should insert the term "faithful" almost everywhere below, taking pains to omit it occasionally.

1.1 Transitivity, genericity, connectivity

We work in the definable category, that is with groups and definable actions. We write actions on the right ("α^g"). The following is completely elementary but fundamental.

Lemma 1.1 *Let (G, Ω) be a transitive permutation group, and G_α a point stabilizer (with $\alpha \in \Omega$). Then the action of G on Ω is equivalent to the action of G on the coset space $G_\alpha \backslash G$. In particular the action is interpretable in G if and only if Ω and G_α are interpretable in G, in which case the two actions are definably equivalent.*

One often treats permutation groups (G, Ω) as structures with underlying set Ω and some inherited relations. The most satisfying choice of relations for our present purposes, where the group G is of interest in its own right, are those definable without parameters in the structure (G, Ω).

Lemma 1.2 *Let (G, Ω) be a transitive permutation group and view Ω as a structure equipped with all relations 0-definable in the two-sorted structure (G, Ω), equipped with the action. View G as a structure equipped with its group operation and a distinguished subgroup G_α for some $\alpha \in \Omega$. If the structure on G is stable then G is interpretable in Ω. In particular, if G is stable and Ω has finite Morley rank, then G has finite Morley rank.*

Proof The point stabilizers form a uniformly definable family of subgroups. If G is stable it follows that arbitrary intersections of point stabilizers are in fact point stabilizers of finite sets. In particular the identity subgroup is the stabilizer of some finite subset $A \subseteq \Omega$. Hence the map $g \mapsto A^g$ is $1 - 1$. Therefore the elements of G and the group structure on G are encoded in Ω. □

Problem 2 If (G, Ω) is a permutation group and Ω is stable in the induced language, does it follow that G is stable? Does this hold at least when Ω has finite Morley rank?

Note that in the present paper we will treat the structure (G, Ω) as given, and not just the group G. So in that context the action of interest is always definable, and whether it is definable in some previously given language on G is a question that rarely arises for us here, though when G is, for example, an algebraic group, it may be an important issue.

Genericity is a fundamental notion in groups, and passes to transitive permutation groups.

Lemma 1.3 *Let (G, Ω) be a stable transitive permutation group, $\alpha \in \Omega$, and X a definable subset of Ω. Then the following conditions are equivalent.*

(1) *$\{g \in G : \alpha^g \in X\}$ is generic in G*
(2) *Finitely many G-translates of X cover Ω.*

If Ω has finite Morley rank then an equivalent condition is

(3) $\mathrm{rk}(X) = \mathrm{rk}(\Omega)$

Such a set is called *generic* in Ω.

We note that after identifying Ω with $G_\alpha \backslash G$, the set defined in point (1) is $\bigcup X$.

We now look at connected components.

Lemma 1.4 *Let (G, Ω) be a transitive permutation group of finite Morley rank. Let Ω_0 be an orbit for the connected component G°, Then* $\mathrm{rk}(\Omega_0) = \mathrm{rk}(\Omega)$, $\deg(\Omega_0) = 1$, *and the orbits of G_0 are conjugate under the action of G.*

Proof Since $G^\circ \lhd G$, the orbits of G° are conjugated by G, and as the action of G is transitive it follows that these orbits are conjugate. As there are finitely many such orbits they have the same rank as Ω. Finally, there is a definable bijection between Ω_0 and a coset space for G°, so the Morley degree of Ω_0 is 1. □

We note that the setwise stabilizer $G_0 = G_{\{\Omega_0\}}$ of Ω_0 may be larger than G° and that the former is really the group induced "by G" on Ω_0.

It is useful to extend the notion of genericity to permutation groups which are not necessarily transitive, at least in the context of groups of finite Morley rank. This we do using the rank directly.

Definition 1.5 Let X be a definable set in a structure of finite Morley rank. A definable subset Y of X is *strongly generic* in X if $\mathrm{rk}(Y \setminus X) < \mathrm{rk}(X)$, and *weakly generic* in X if $\mathrm{rk}(Y) = \mathrm{rk}(X)$. When X has Morley degree 1, the two notions are equivalent, and are referred to as genericity. Otherwise, it is prudent to specify which version is meant.

Lemma 1.6 *Let (G, Ω) be a transitive permutation group of finite Morley rank and g an element of G which fixes a strongly generic subset of Ω pointwise. Then $g = 1$.*

Proof Let

$$H = \{g \in G : g \text{ fixes a strongly generic subset of } \Omega \text{ pointwise}\}.$$

Then H is a definable normal subgroup of G.

Take $h \in H^\circ$ generic, and $\alpha \in \Omega$ generic over h. Then h and α are independent, so α is fixed by a generic subset of H°, and hence by H°. As H° is normal in G and G acts transitively, the group H° acts trivially on Ω. That is, $H^\circ = 1$ and H is finite.

Let X be the fixed point set for H in Ω. As H is finite, the set X is generic, hence nonempty. Since X is also G-invariant, we have $X = \Omega$ and thus $H = 1$. □

Definition 1.7 Let (G, Ω) be a permutation group.

(1) G is *generically transitive* on Ω if G has a strongly generic orbit.
(2) G is *generically n-transitive* on Ω if the induced action on Ω^n is generically transitive.
(3) G is *generically sharply n-transitive* on Ω if the induced action on Ω^n has a strongly generic orbit on which G acts regularly.

We will not actually take up the generically sharply transitive case per se, but it is worth mentioning as the subject of [11], particularly since such groups certainly exist.

Example 1 The natural representation of $\mathrm{AGL}(n)$ affords a generically sharply $(n+1)$-transitive action, and the natural projective representation of $\mathrm{PGL}(n)$ affords a generically sharply $(n+1)$-transitive action.

Problem 3 Find all the generically sharply n-transitive actions of algebraic groups over algebraically closed fields, for $n \geq 2$.

Note that by a theorem of Hall [12] there cannot be a sharply n-transitive permutation group on an infinite set for $n \geq 4$, which stands in sharp contrast to the above.

We would suggest that it is reasonable and interesting, though certainly challenging, to aim eventually at an identification of all the generically highly transitive groups of finite Morley rank.

While there is no close connection in general between generic n-transitivity and n-transitivity, it is reasonable to work with transitive actions throughout (unless the set Ω carries some useful structure supported on several orbits, as is the case in most natural representations). For $n = 1$ one just restricts to the strongly generic orbit (possibly picking up a kernel). For $n > 1$ one makes use of the following.

Lemma 1.8 *Let (G, Ω) be a generically n-transitive permutation group of finite Morley rank.*

(1) *(G, Ω) is generically m-transitive for $m \leq n$.*
(2) *If X is the strongly generic orbit for G, then G is generically n-transitive on X*
(3) *If $n > 1$ then Ω has Morley degree 1.*

Proof The first two points are clear. For the last point, we may suppose that G is transitive on Ω. If Ω has Morley degree greater than 1, then Ω contains distinct G°-orbits Ω_0 and Ω_1. But then no element of $\Omega_0 \times \Omega_0$ is conjugate to any element of $\Omega_0 \times \Omega_1$ under the action of G, and we have a contradiction. □

The last argument is a very weak analog, but the best we have, for the statement that a doubly transitive group is primitive.

Lemma 1.9 *Let (G, Ω) be a generically n-transitive permutation group, and let Ω_0 be a G°-orbit. Then (G°, Ω_0) is generically n-transitive.*

Proof If $n = 1$ then this is clear. If $n > 1$ then Ω has Morley degree 1, so Ω^n also has Morley degree 1. If O is the strongly generic orbit for G in Ω^n, and O_0 is a G°-orbit inside O, then $\mathrm{rk}(O_0) = \mathrm{rk}(O) = \mathrm{rk}(\Omega^n)$ and as the Morley degree is 1 it follows that O_0 is strongly generic in Ω^n. □

1.2 Notions of Primitivity

Let us propose a number of notions of primitivity.

Definition 1.10 Let (G, Ω) be a permutation group.

(1) The action is *primitive* if there is no nontrivial G-invariant equivalence relation.

(2) The action is *definably primitive* if there is no nontrivial definable G-invariant equivalence relation.
(3) The action is *virtually definably primitive* if any G-invariant definable equivalence relation has either finite classes or finitely many classes.

We are not much interested in primitivity per se, as this is too not natural in the definable category. On the other hand it happens to be the case that almost all definably primitive permutation groups are primitive ([14], see Lemma 1.17).

It is less clear which of the definable versions of primitivity is to be preferred, and we have a third possibility to offer in a moment (this in turn raises the question as to what the O'Nan-Scott-Aschbacher Theorem should be about in our category). Certainly one desirable criterion is that any permutation representation that may interest us should have a nontrivial primitive quotient. These will be definable, and frequently of Morley degree 1 (or, indeed, with the acting group G connected).

We intend to lay out the relationships among these notions in detail. First we give their translations into "internal" group theoretic terms. For this we will need to restrict attention to transitive group actions, so we begin with this point, which is completely straightforward.

Lemma 1.11 *Let (G, Ω) be a permutation group.*

(1) *If the action is definably primitive, then it is transitive.*
(2) *If the action is virtually definably primitive then either it has finitely many orbits, and is virtually definably primitive on each orbit, or else it has finite orbits, and the set of orbits carries no definable equivalence relation with infinitely many infinite classes.*

Proof Possibly the context for the second point needs to be elucidated. The notion of definability is relative to the structure (G, Ω) containing the action and whatever additional structure it may carry (on G, on Ω, or on both together). We consider the equivalence relation given by the orbits themselves, which is definable in this context. If the orbits are finite then this is not very interesting, but it may meet our definition. □

Remark 1.12 Let (G, Ω) be a stable permutation group with all orbits finite. Then G is finite, since the kernel (which is assumed trivial) is the stabilizer of a finite set, and therefore G acts faithfully on a finite union of orbits.

This remark is actually of some use, when one comes across permutation groups carrying a definable invariant equivalence relation with finite classes. The lemma states that the kernel of the action on the set of classes is finite.

It should now be reasonably clear that nothing of any significance would be lost by including transitivity in the definition of virtual definable primitivity.

Lemma 1.13 *Let (G, Ω) be a transitive permutation group, $\alpha \in \Omega$ fixed, and G_α the point stabilizer.*

(1) *The action is primitive if and only if G_α is a maximal proper subgroup of G.*
(2) *The action is definably primitive if and only if G_α is a maximal definable subgroup of G.*
(3) *The action is virtually definably primitive if and only if for any definable subgroup H of G containing G_α, either $[H : G_\alpha]$ or $[G : H]$ is finite.*

Proof The G-invariant equivalence relations on Ω are classified by the subgroups of G containing G_α. More explicitly, if C is a set containing α, and $G_{\alpha:C} = \{g \in G : \alpha^g \in C\}$, then the following are equivalent

(1) $G_{\alpha:C}$ is a subgroup of G;
(2) $G_{\alpha:C}$ is the setwise stabilizer $G_{\{C\}}$ of C in G;
(3) $\{C^g : g \in G\}$ is a partition of Ω.

Furthermore, when this holds we have $|C| = [G_{\{C\}} : G_\alpha]$ and $|\{C^g : g \in G\}| = [G : G_{\{C\}}]$. □

There is a rule of thumb that says one is always interested in the "connected" versions of classical notions. So we propose a notion of *c-primitivity* intended to be the connected version of primitivity, in two variants.

Definition 1.14 Let (G, Ω) be a transitive permutation group of finite Morley rank, $\alpha \in \Omega$, and G_α the point stabilizer.

(1) The action is *c-primitive* if G_α is a maximal proper definable connected subgroup of G°.

(2) The action is *virtually c-primitive* if G_α° is a maximal proper definable connected subgroup of G°.

One may wish to translate this back into permutation group theoretic terms. Here one needs the notion of a *finite cover* $(G, \hat{\Omega})$ of a permutation group (G, Ω), which is given by a surjective G-invariant map $\pi : \hat{\Omega} \to \Omega$ with finite fibers (really the map $(1, \pi) : (G, \hat{\Omega}) \to (G, \hat{\Omega})$ is the morphism).

Lemma 1.15 *Let (G, Ω) be a transitive permutation group of finite Morley rank.*

(1) *The action is virtually c-primitive if and only if every definable finite cover is virtually definably primitive.*
(2) *The action is c-primitive if and only if every finite cover is an isomorphism, and the action is virtually definably primitive.*

Proof Only the first point requires unwinding, and this is simply a matter of working out the content of the cumbersome criterion given, which can be read as follows: for every definable subgroup G_0 of G_α of finite index, and for every definable subgroup H of G containing G_0, either $[H : G_0] < \infty$ or $[G : H] < \infty$. This can be decoded further to: either $H^\circ = G_\alpha^\circ$ or $H^\circ = G^\circ$—at which point we have virtual c-primitivity. □

One of the goals of permutation group theory is to provide a convenient language for saying useful things about maximal subgroups, and in a finite Morley rank context one thing that one could reasonably ask of the parallel theory is a useful way of looking at maximal connected subgroups. But this is not really the point of view that has been taken to date.

We now take note of reductions to Morley degree 1.

Lemma 1.16 *Let (G, Ω) be a transitive permutation group of finite Morley rank, and Ω_0 an orbit for G° in Ω. Let $G_0 = G_{\{\Omega_0\}}$.*

(1) *If (G, Ω) is definably primitive then Ω has Morley degree 1 and $\Omega_0 = \Omega$, $G_0 = G$.*
(2) *If (G, Ω) is virtually definably primitive then any G_0-invariant definable relation on Ω_0 has finite classes.*
(3) *If (G, Ω) is c-primitive or virtually c-primitive, then (G_0, Ω_0) has the same property.*

On the other hand, this does not mean that we can reduce the group itself to its connected component.

Example 2 Let (H, X) be a definably primitive permutation group of finite Morley rank with H connected. Let K be a finite permutation group acting transitively on a set I. Then the wreath product $H \wr K$ acting on X^I is a definably primitive permutation group of finite Morley rank whose connected component H^I leaves invariant the equivalence relations E_i $(i \in I)$ defined by

$$E_i(a, b) \iff a_i = b_i.$$

Thus (H^I, X^I) is not even virtually definably primitive.

One may still find it profitable in practice to pay particular attention to connected permutation groups, but there is no general reduction of the full theory to that case, and this leads to practical difficulties in situations where one would like to make inductive arguments.

The foregoing example is typical in the sense that the class of definably primitive groups is much richer than the set of connected definably primitive groups precisely because the finite sections can play such an important role. The general MPOSA classification includes several cases that cannot arise as actions of connected groups.

1.3 Primitivity vs. Definable Primitivity

Example 3 A definably primitive action which is not primitive.

Take a large torsion free divisible abelian group in its natural language (or as a $\mathbb{Q}$-vector space) and consider the regular action. This is definably primitive and c-primitive but not primitive.

One may be a bit suspicious of this example since the point stabilizer is trivial, and with good reason. If a regular action is definably primitive then the group G contains no nontrivial proper definable subgroups and hence is abelian and (if infinite) torsion free. On the other hand according to [14, Prop. 2.7] a definably primitive group is primitive unless the point stabilizer is finite. We elaborate slightly on this point.

Lemma 1.17 ([14]) *Let (G, Ω) be a permutation group of finite Morley rank which is definably primitive but not primitive, and let G_α be a point stabilizer. Then G_α is finite, and if $G_\alpha > 1$ then G° is either abelian or quasisimple.*

Proof Suppose that $G_\alpha < H < G$. Let H_0 be the normal closure in H of G°_α. Then by a lemma of Zilber H_0 is definable and connected. Since $G_\alpha \le H_0 G_\alpha \le H$ it follows by definable primitivity that $H_0 = G^\circ_\alpha$, that is $G^\circ_\alpha \triangleleft H$. As $G_\alpha < N(G^\circ_\alpha) \le G$ it follows that $G^\circ_\alpha \triangleleft G$. Now G acts transitively on Ω so it follows that G°_α acts trivially on Ω, hence $G^\circ_\alpha = 1$ and G_α is finite.

Now if K is a definable connected normal subgroup of G then $KG_\alpha = G$ and thus $K = G^\circ$. So if G° is nonabelian then G° is quasisimple, that is perfect with finite center and simple factor group. □

This result has a converse: if the point stabilizer is finite it cannot be maximal, at least if we pass to an uncountable model.

Examples of such groups with G° abelian coming from irreducible representations of finite groups are mentioned in [14]. Another type of example is the following (responding to a question raised in the last paragraph of §2 of [14]).

Example 4 Let $G = \mathrm{PSL}_2(\mathbb{C})$ and consider the action on cosets of $H = \mathrm{Alt}(5)$. Then H is maximal among Zariski-closed proper subgroups, but not among (e.g., countable) proper subgroups. So in the algebraic category this provides a definably primitive but not primitive action.

It can be shown with some additional effort that this action will remain definably primitive in any enrichment of the language for which the group has finite Morley rank.

1.4 Definably Primitive Quotients

One would like to think that the general permutation group of finite Morley rank can be analyzed in terms of transitive and even primitive constituents. The reduction to the transitive case already involves very substantial complications. Even if the action is generically transitive, the action on a generic orbit may have nontrivial kernel, and there may be infinitely many nongeneric orbits.

As far as primitivity is concerned, one would like to find a nontrivial definably primitive quotient of any transitive permutation group of finite Morley rank, but even this is too much to ask.

Example 5 Let G be a Chevalley group over an algebraically closed field of positive characteristic, and let $G(q)$ be the subgroup of $\mathbb{F}_q$-rational points. Typically the proper Zariski closed subgroups of G

containing $G(q)$ are the $G(q')$ for q' a power of q. Thus taking G with its structure as an algebraic group, any proper definable subgroup containing $G(q)$ is finite, and thus if we consider the action of G on $G(q)\backslash G$, this has no definable and definably primitive quotient.

On the other hand, this example is again fairly typical.

Lemma 1.18 *Let (G, Ω) be a transitive permutation group of finite Morley rank. Then the following hold.*

(1) *There is a nontrivial definable quotient of (G, Ω) which is virtually definably primitive.*
(2) *If (G, Ω) is virtually definably primitive then either (G, Ω) is a finite cover of a definably primitive permutation group, or the point stabilizer is finite.*

Proof Let G_α be a point stabilizer.

The first point is immediate: extend G_α to a proper definable subgroup of maximal rank.

For the second point, if $N(G_\alpha^\circ) < G$ then $[N(G_\alpha^\circ) : G_\alpha] < \infty$ by virtual definable primitivity, and we look at $N(G_\alpha^\circ)\backslash G$. On the other hand, if $G_\alpha^\circ \triangleleft G$ then by transitivity G_α° acts trivially, $G_\alpha^\circ = 1$. □

The most useful instance of this is the following.

Corollary 1.19 *Let (G, Ω) be a transitive permutation group of finite Morley rank with G connected simple. Then either (G, Ω) has a nontrivial definably primitive quotient, or the point stabilizer is finite.*

Proof We apply both parts of the previous corollary, getting a nontrivial quotient of (G, Ω) satisfying one of our two conclusions. Note however that if the point stabilizer in the quotient is finite then it was finite in (G, Ω): as G is simple, there is no kernel in this action. □

Problem 4 Suppose that (G, Ω) is a virtually definably primitive permutation group of finite Morley rank with which is not a finite cover of a definably primitive permutation group. Show that G is a Chevalley group of positive characteristic, and the point stabilizer is contained in $G(\mathbb{F}_q)$ for some finite field $\mathbb{F}_q$.

As we have already noticed, c-primitivity is just virtual definable primitivity together with the condition that the point stabilizer be connected.

On the other hand, in general virtual definable primitivity has little to do with c-primitivity, with examples again afforded by actions with finite point stabilizers.

1.5 Generic n-Transitivity Revisited

We have the natural inductive principle: if (G, Ω) is a generically n-transitive permutation group of finite Morley rank, and G_α a point stabilizer with α in the generic orbit, then G_α acts generically $(n-1)$-transitively on Ω.

However, one might also expect generically highly transitive actions to be definably primitive, and this fails badly.

Example 6 Let (H_i, X_i) $(i = 1, 2)$ be generically n-transitive permutation groups of finite Morley rank. Then $(H_1 \times H_2, X_1 \times X_2)$ is generically n-transitive.

One may ask whether there is, nonetheless, a general theory of generically n-transitive groups. There is an initial reduction to the primitive case. Recall that generic n-transitivity passes to connected components.

Lemma 1.20 *Let (G, Ω) be a transitive and generically n-transitive group of finite Morley rank with G connected. Then the following hold.*

(1) $\mathrm{rk}(G_\alpha) \geq (n-1)\,\mathrm{rk}(\Omega)$ *for* $\alpha \in \Omega$.
(2) *Any definable quotient of (G, Ω) is generically n-transitive.*
(3) *If $n > 1$ then there is an infinite, definable, definably primitive quotient* $(\bar{G}, \bar{\Omega})$

Proof The first two points are immediate. For the last point, since G is connected, "infinite" is the same as "nontrivial".

Now Lemma 1.18 tells us that we have an infinite definable and virtually definably primitive quotient $(\bar{G}, \bar{\Omega})$ and that this is either a finite cover of a definably primitive permutation group, or has a finite point stabilizer. But $\mathrm{rk}(\bar{G}_{\bar{\alpha}}) > 0$ so the latter possibility is excluded. □

We now have two good reasons for restricting our attention to the primitive case: (1) in the present article we need to deal with generically n-transitive groups in order to analyze primitive ones; (2) it seems that results on the primitive case may bear strongly on the general case, in view of the foregoing.

Problem 5 Is there an O'Nan-Scott-Aschbacher analysis of generically 2-transitive groups which are not necessarily definably primitive? Are all such groups essentially products of generically n-transitive primitive groups (or generically n'-transitive groups, with n' not much smaller than n)?

This concludes our review of the fundamental definitions. The central notion of [14] was definable primitivity, and this review suggests that one might on occasion prefer to broaden the notion a little, but that the impact of this would be marginal.

2 Bounds on rank

We will organize our discussion around the following result and problem.

Theorem 2.1 *There is a function $\rho : \mathbb{N} \to \mathbb{N}$ such that the following holds. For any virtually definably primitive permutation group (G, Ω) of finite Morley rank we have*

$$\mathrm{rk}(G) \le \rho(\mathrm{rk}(\Omega)).$$

Problem 6 Find good bounds on ρ, where $\rho(r)$ is the maximum rank of a virtually definably primitive permutation group (G, Ω) of finite Morley rank, with $\mathrm{rk}(\Omega) = r$.

Notice the following corollary.

Corollary 2.2 *There is a function $\tau : \mathbb{N} \to \mathbb{N}$ such that the following holds. For any virtually definably primitive permutation group (G, Ω) of finite Morley rank which is generically t-transitive,*

$$t \le \tau(\mathrm{rk}(\Omega)).$$

Problem 7 Find good bounds on τ, where $\tau(r)$ is the maximum degree of generic transitivity associated to a virtually definably primitive permutation group (G, Ω) of finite Morley rank, with $\mathrm{rk}(\Omega) = r$.

We particularly like this last problem, because we have as yet no decent bounds on τ and we imagine there should be very good ones. The gap between τ and ρ is not so large, as we shall show.

Proposition 2.3 *For any primitive permutation group (G, Ω) of finite Morley rank with arbitrary $r = \mathrm{rk}(\Omega)$ we have*

$$r\tau(r) \leq \rho(r) \leq r\tau(r) + \binom{r}{2}.$$

The proof of this Proposition will be given shortly.

Lemma 2.4 *We may take Ω to have Morley degree 1, and we may replace "virtually definably primitive" by "definably primitive," in the definitions of ρ and τ, without altering the values.*

Proof We have seen in Lemma 1.16 that we may restrict attention to Morley degree 1 actions, with all G-invariant relations finite. Furthermore we know that each relevant permutation group (G, Ω) is either a finite cover of a definably primitive one, or has finite point stabilizer. It suffices to observe now that neither the rank nor the degree of generic transitivity can be maximized in the presence of a finite point stabilizer. □

2.1 Examples

Our first order of business is to justify our primitivity hypotheses. After all, any permutation group on a set of rank 0 has rank 0, which is a very good bound. There is also an excellent bound for the rank of a stable group acting transitively on a rank 1 set, due to Hrushovski.

Proposition 2.5 *Let (G, A) be a stable permutation group with A strongly minimal. Then the Morley rank of G is at most 3, and either G° is abelian and regular on A, or G is isomorphic with a Zariski closed subgroup of $\mathrm{PSL}_2(K)$ for some algebraically closed and strongly minimal field K interpretable in G.*

In particular, if the rank of G is $n \leq 3$ then the action of G° is sharply n-transitive on A.

But above rank 1 things become more complicated. Examples of the following type are given in [11], and show that nontransitive permutation groups are not so easily reduced to transitive ones.

Example 7 Let V be a vector space over an algebraically closed field K, L a 1-dimensional vector space over K, $E_V = \mathrm{End}(V)$, $\lambda : V \to L$ a nonzero linear map, and $f : L \to V$ a map from the line L into a space curve in V not contained in a proper subspace of V. We associate with these data the permutation group (E_V, L^2) with E_V viewed as an additive group and with action

$$A.(x,y) = (x, y + \lambda(A.f(x))).$$

(using left-handed notation). Evidently this is faithful, and L^2 has rank two.

The orbits are the rank one sets $L_x = \{x\} \times L$. On L_x, if $v = f(x)$ then the kernel is

$$\{A : A.v \in \ker \lambda\}$$

giving a parametrized family of subspaces of $\mathrm{End}(V)$.

Even in the transitive case there are somewhat similar examples.

Example 8 Let G be an algebraic group with a rational linear representation on the vector space V and let W be a subspace of V which does not contain any nontrivial G-invariant subspace of V. Let $\hat{G} = V \rtimes G$ and consider the transitive permutation group

$$(\hat{G}, W\backslash\hat{G}).$$

This has rank $\mathrm{rk}(G) + \mathrm{rk}(V/W)$, and is not definably primitive since $W < V$. On the other hand it is faithful by the choice of W.

We want $\mathrm{rk}(\hat{G})$ unbounded and $\mathrm{rk}(W\backslash\hat{G})$ bounded, for which we take G fixed and $\dim(V/W)$ bounded. For example G could be simple, V irreducible, and W a hyperplane. Or G could be a torus acting so V so that all weight spaces are 1-dimensional, and W could be a hyperplane avoiding all weight spaces.

In particular G could be a 1-dimensional torus and then $W\backslash\hat{G}$ would be a Morley rank two representation of a group of arbitrarily large rank.

Here a nontrivial invariant equivalence relation is given by the orbits of the normal subgroup V, the quotient has kernel V, and the induced action is the regular action of G on itself. On the other hand the stabilizer of a single equivalence class will be the group V, with kernel a conjugate of W, and this portion of the action resembles the previous example.

These examples are not particularly outlandish, but they do suggest that it may be difficult to get even the coarsest degree of control over imprimitive representations.

On the other hand, we will eventually get a bound on the rank of a *simple* group of finite Morley rank acting definably on a set of specified rank, and from this we will get our general bound in the primitive case.

2.2 Reduction to generic multiple transitivity

We take up the proof of Proposition 2.3. The idea is that if a point stabilizer does not act generically transitively, then the ranks of the orbits of successive point stabilizers (taken along a sequence of independent generic points of Ω) decrease steadily. One difficulty that immediately comes to mind, in view of the last example, is that we have no assurance that we can usefully pass to simple constituents of the induced permutation groups along the way. So we need to apply the initial primitivity hypothesis for the full group at each step.

On the other hand we will see later that by making use of MPOSA one can *sometimes* usefully recover primitivity for such constituents. So in all probability the very soft argument given here can be usefully refined, and the bounds sharpened, using both MPOSA and the structure theory of groups of finite Morley rank. In the present article we will only explore this approach in the opposite case, under the hypothesis of generic multiple transitivity, where our soft bounds are very loose, and one should expect very good bounds by more concrete methods.

Definition 2.6 Let (G, Ω) be a permutation group of finite Morley rank with Ω of Morley degree 1. Let o_k denote the *generic rank* of an orbit for the group G°_α where α is an independent k-tuple of generic elements of Ω. That is, the following set should be generic in Ω:

$$\{\omega \in \Omega : \mathrm{rk}(\omega^{G^\circ_\alpha}) = o_k\}.$$

We would get the same values without taking connected components, but in fact it is the connected components of the stabilizers which interest us.

Lemma 2.7 *Let (G, Ω) be a primitive permutation group of finite Morley rank, $k \geq 1$, and suppose that $0 < o_k < \mathrm{rk}(\Omega)$. Then $o_{k+1} < o_k$.*

Proof Assume the contrary, $o_{k+1} = o_k$. Our idea is then that the orbits of a stabilizer of a sequence of length $k+1$ of independent generic elements do not really depend on the last of the elements, or for that matter any of them, and hence there should be a corresponding G-invariant equivalence relation after eliminating spurious dependencies. This is reminiscent of various standard lines of argument in stability theory and may well be a special case of one of them.

We use the symbol $\approx$ to denote *generic equality* for definable sets: so $A \approx B$ means that $\mathrm{rk}(A \Delta B) < \mathrm{rk}(A), \mathrm{rk}(B)$, and in particular $\mathrm{rk}(A) = \mathrm{rk}(B)$. We also use the *generic quantifier* $\forall^* x$ with the meaning "for a strongly generic set of x". Let $r_0 = \mathrm{rk}(\Omega)$. Our first claim is the following.

(1) For all sequences α, β of either k or $k+1$ independent generic elements of Ω we have
$$\forall^* x \in \Omega \quad [G^\circ_\alpha \cdot x \approx G^\circ_\beta \cdot x].$$

Consider the bipartite graph Γ whose vertices are sequences of length k or $k+1$ of independent generic elements of Ω, with edges corresponding to inclusion or reverse inclusion. As this graph is connected, and the displayed condition in (1) defines an equivalence relation on arbitrary sequences α, β of elements from Ω, it suffices to verify that condition (1) holds for a pair of vertices α, β in Γ joined by an edge. But this is immediate as the orbits in question have Morley degree one and (generically) equal rank. So (1) holds.

Now we define an equivalence relation $E(x, y)$ on Ω as follows:
$$\forall^* \alpha \in \Omega^k \, (G^\circ_\alpha \cdot x = G^\circ_\alpha \cdot y).$$

This is a definable G-invariant equivalence relation on Ω. An equivalent condition is the following, writing $\Omega^{(k)}$ for the set of sequences of k independent generic elements of Ω.
$$\exists \alpha \in \Omega^{(k)} \quad (\alpha \perp x, y) \,\&\, (G^\circ_\alpha \cdot x = G^\circ_\alpha \cdot y).$$

Indeed, as the type over x, y of a sequence of generic elements of Ω independent over x, y is unique, if $G^\circ_\alpha \cdot x = G^\circ_\alpha \cdot y$ for one such sequence α, then the same holds for all such sequences.

We will write E_x for the E-equivalence class of x. The next point is really part of a larger topic (finiteness of weight) but can be dealt with in an *ad hoc* way here. Cf. [8, §5.6]

(2) $$\forall^* x \, \mathrm{rk}(E_x) \geq o_k.$$

Let $\alpha^{(i)}$ be a sequence of $2r_0+1$ independent generic elements of $\Omega^{(k)}$. Then for any pair $x, y \in \Omega$ there is an $i \leq 2r_0+1$ such that $xy \downarrow \alpha^{(i)}$.

Take $x \in \Omega$ generic over $\alpha^{(1)} \ldots \alpha^{2r_0+1}$, and let $C = \bigcap_{i \leq 2r_0+1} G^\circ_{\alpha^{(i)}} \cdot x$. Then by (1) $\mathrm{rk}(C) = o_k$. Choose $y \in C$ with $\mathrm{rk}(y/x) \geq o_k$. Choose i so that $x, y \downarrow \alpha^{(i)}$. Then $E(x,y)$ holds by our second criterion. As E_x is x-definable and contains a point y of rank at least o_k over x, $\mathrm{rk}(E_x) \geq o_k$. This proves (2).

(3) $$\forall^* x \; \mathrm{rk}(E_x) \leq o_k.$$

Take $x \in \Omega$ generic and let $C = E_x$. Choose $y \in C$ generic and independent from x over the parameter C. Then $\mathrm{rk}(C) = \mathrm{rk}(y/\text{"}C\text{"})$ (there is a definable set C on the left side, and a type over the "element" C on the right). Take $\alpha \in \Omega^{(k)}$ with $\alpha \downarrow x, y$. Then $y \in G_\alpha \cdot x$.

As the element "C" is definable from x, and also $x, y \downarrow \alpha$, we have $y \downarrow_{x, \text{"}C\text{"}} \alpha$. But $y \downarrow_{\text{"}C\text{"}} x$, so

$$y \underset{\text{"}C\text{"}}{\downarrow} \alpha, x.$$

In other words, $\mathrm{rk}(y/\alpha x) = \mathrm{rk}(y/\text{"}C\text{"})$. Therefore $\mathrm{rk}(C) = \mathrm{rk}(y/\text{"}C\text{"}) = \mathrm{rk}(y/\alpha x) \leq \mathrm{rk}(G^\circ_\alpha \cdot x) = o_k$. □

Proof of Proposition 2.3. The claim is

$$r\tau(r) \leq \rho(r) \leq r\tau(r) + \binom{r}{2}.$$

Let $t = \tau(r)$ and let (G, Ω) be primitive and generically t-transitive with Ω of rank r. Our claim is that

$$rt \leq \mathrm{rk}(G) \leq rt + \binom{r}{2}.$$

The first inequality is immediate as Ω^t has rank rt.

Now by the definition of τ, (G, Ω) is not generically $(t+1)$-transitive and therefore $o_t < \mathrm{rk}(G)$. Hence by repeated application of the foregoing lemma, $o_{t-1+i} \leq \mathrm{rk}(G) - i$ until o_{t-1+i} becomes 0. In particular $o_{t-1+r} = 0$.

Now let α be a sequence of independent generic elements of Ω, of length $t+r-1$. Then G°_α has finite orbits, generically, hence fixes a generic set of points in Ω, and by Lemma 1.6 we find $G_\alpha = 1$, and $\mathrm{rk}(G_\alpha) = 0$.

Now writing G_i for the connected component of the pointwise stabilizer of $(\alpha_1, \ldots, \alpha_i)$ for $i \le t + r - 1$ (beginning with $G_0 = G^\circ$), we find

$$\mathrm{rk}(G) = \sum_{i<t+r-1} \mathrm{rk}(G_i/G_{i+1}).$$

Furthermore $\mathrm{rk}(G_i/G_{i+1}) = \mathrm{rk}(\alpha_{i+1}^{G_i}) = o_i$, so

$$\mathrm{rk}(G) \le \sum_{i\ge 0} o_i \le rt + \sum_{0\le i\le r-1} (r-1-i) = rt + \binom{r}{2}.$$

□

3 The definable socle

Using some very general results on the definable socle of a definably primitive permutation group [14], we will eventually reduce the proof of Theorem 2.1 to the following Theorem 3.1. The material in [14] is really the beginning of the subject, and if we have addressed other more specialized matters first, it was only to get them out of the way. We will not give a full account of [14] here, but certainly everything there needs to be taken into account and it would also be desirable to continue that analysis further, if possible, at some more substantial level of generality than we envision here. But the MPOSA tool in its present form is more than adequate for applications of the sort we consider here, and here we focus more on the other problems which arise in such cases.

Theorem 3.1 *There is a function σ which bounds the rank of a simple group of finite Morley rank in terms of the rank of any set on which it acts definably and faithfully.*

Because of the simplicity of G, it will suffice to prove Theorem 3.1 in the special case when the group in question is transitive; moreover, in view of Corollary 1.19 we can assume that the action is definably primitive.

Again, this reduces via Proposition 2.3 to the following special case.

Proposition 3.2 *There is a function τ' which bounds the degree of generic multiple transitivity of a simple definably primitive permutation group of finite Morley rank in terms of the rank of the set on which it acts.*

Notice again that the words "definably simple" in the formulation of Proposition 3.2 can be omitted in view of Corollary 1.19.

We will review the necessary information from [14] and then make the reduction of Theorem 2.1 to Theorem 3.1 (and hence to Proposition 3.2).

3.1 The main case division

We now exploit the structure of the socle in a definably primitive permutation group of finite Morley rank. A good deal is known about the possibilities here, and for full details we refer to [14]. Here we emphasize those points that give immediate information relevant to bounds on rank, using comparatively soft methods, though certainly not as soft as those of the previous section. This involves the *definable socle*, which is the subgroup generated by its minimal definable normal subgroups. In any group G of finite Morley rank, it is easy to see that the definable socle is itself definable, and is a finite direct product of finitely many simple groups and an abelian group. If there is no abelian normal subgroup then the socle contains its own centralizer. In particular, if G is infinite, then the definable socle is infinite, and is a direct product of a connected normal subgroup of G (the connected socle) with a finite normal subgroup of G, the latter commuting with G°.

In definably primitive groups the situation is considerably tighter. To begin with we have the following, which is largely the preamble to Theorem 1.1 of [14].

Theorem 3.3 [14, 1.1] *Let (G, Ω) be a definably primitive permutation group of finite Morley rank with Ω infinite. Let B be the definable socle of G. Then one of the following occurs.*

(1) *Affine type: B is abelian, is the unique minimal normal definable subgroup of G, and acts regularly on Ω. G splits as $B \rtimes G_\alpha$ for any $\alpha \in \Omega$ and the action of G on Ω is equivalent to the action of G on B with B acting by translation and G_α by conjugation. Furthermore, B is either torsion free divisible, or an elementary abelian p-group, and there is no nontrivial G_α-invariant definable subgroup in B.*
(2) *B is a finite direct product $T_1 \times \cdots \times T_k$ of isomorphic connected simple groups.*

What is missing in the above summary is the further subdivision of the case 2 into three further subcases, each obtained from wreath products of a corresponding basic type. These types may be referred to suggestively as *regular simple type*, *nonregular almost simple type*, and *simple diagonal action*, and are listed as types 2, 3, and 4(a) in [14]; here type 1 is the affine type, and type 4(b) consists of wreath products of types 2, 3, or 4(a) as well as certain intermediate permutation groups lying between the specified socle and the full wreath product. The further results (1.2-1.4) of [14] then cast additional light on the individual types.

We will in fact need a little more from [14] but we will first see how far Theorem 3.3 takes us.

We make a few comments on the proof of the statement as we have given it. Note first that by definable primitivity, any nontrivial definable normal subgroup acts transitively on Ω. It follows easily that if G has an abelian normal subgroup then any minimal abelian normal subgroup A acts regularly and the group splits as described in case (1); furthermore the definable socle centralizes A and hence also acts regularly, forcing the socle to reduce to A. So one may suppose that all minimal normal subgroups are nonabelian. Furthermore, on general principles each nonabelian minimal definable normal subgroup of G has the structure indicated in (2), so we need only consider the case in which there is more than one minimal definable normal subgroup of G, and they are all nonabelian. In particular, as they commute and act transitively, it follows that each acts regularly. Then for any two minimal normal subgroups H_1, H_2 of G it is easy to see that the point stabilizer $(H_1H_2)_\alpha$ is the graph of an isomorphism of H_1 with H_2 (and this is the starting point for a more careful analysis of B_α).

3.2 The torsion-free divisible case

We give a first indication that this case division is helpful for our present concerns.

Lemma 3.4 *Let (G, Ω) be a definably primitive permutation group whose socle A is torsion free divisible, and let $r = \mathrm{rk}(\Omega)$. Then*

$$\mathrm{rk}(G) \leq r^2 + r.$$

Proof This is very much like saying $\dim(\mathrm{AGL}(V)) \leq \dim(V)^2 + \dim(V)$ and really has the same proof.

By Theorem 3.3 we can identify Ω with A and G with $A \rtimes G_0$ (note that G_0 is actually the point stabilizer of 0 under this identification).

Choose a sequence $(a_1, a_2, \ldots)$ as long as possible with

$$a_{i+1} \notin d(a_1, \ldots, a_i)$$

(in particular $a_1 \neq 0$). Let $A_i = d(a_1, \ldots, a_{i-1})$. As the A_i are also torsion free and divisible, $\mathrm{rk}(A_i) \geq i$ and therefore the sequence $(a_1, \ldots)$ is finite of length $n \leq r$.

Now as $A_n = A$, the point stabilizer $G_{a_1,\ldots,a_n}$ is trivial. Hence writing G_i for the point stabilizer $G_{a_1,\ldots,a_i}$ we have

$$\mathrm{rk}(G_0) = \sum_1^n \mathrm{rk}(G_i/G_{i-1}) \leq nr \leq r^2$$

and our claim follows. □

More to the point is the question as to why this doesn't work in other cases. In the case of elementary abelian socles what is missing is a linear structure to bound the "dimension" (in the sense implicit above) in terms of the rank. In the case of simple socles this argument actually works quite well, as we will see next, but only bounds the rank of G in terms of the rank of the socle, which must still be related to the rank of Ω, since the socle usually does not act regularly.

3.3 Simple socles

Let us extract what we can from the last argument for the case of simple socles, and more generally for nonabelian socles.

Lemma 3.5 *Let S be an $\aleph_0$-saturated simple group of finite Morley rank and rank n. Then there are elements $a_0, \ldots, a_n$ which* definably generate *S in the sense that*

$$S = d(a_0, \ldots, a_n).$$

Notice that we need the saturation hypothesis even in the algebraic case: in positive characteristic there are many locally finite simple algebraic groups.

Proof We claim that we can choose a sequence $(a_0, a_1, \ldots)$ so that setting $S_i = d(a_0, a_1, \ldots, a_i)$ we have $\mathrm{rk}(S_i) \geq i$ until $S_i = S$. This will certainly prove the lemma, padding the sequence if necessary.

We begin with the case $i = 1$, which requires us to choose the two elements a_0, a_1. If G does not have bounded exponent then by the saturation hypothesis we can choose $a_0 = a_1$ to be an element of infinite order. If G does have bounded exponent than for the purposes of this step we may work in a minimal connected simple definable section $\bar{S}$ and lift back to S afterward.

Now as $\bar{S}$ is minimal simple and of bounded exponent it follows easily that its Borel subgroups are nilpotent, and thus $\bar{S}$ is a so-called "bad group". For these there is some useful structural information and in particular any nontrivial finite subgroup lies in a unique Borel subgroup [7]. So if we take a_0, a_1 to be nontrivial elements in two distinct Borel subgroups then $d(a_0, a_1)$ is infinite, as required.

With the base of the induction out of the way, we continue as follows. We suppose that S_i has rank at least i, with $i \geq 1$, and we wish to choose $a_{i+1} = a$ in S so that

$$\mathrm{rk}(d(S_i, a)) > \mathrm{rk}(S_i).$$

If we cannot do this, then we find

$$d(S_i, a)^\circ = S_i^\circ$$

for all $a \in S$, and hence $S_i^\circ \lhd S$, forcing $S_i^\circ = S$, and we are already done. □

And now as before one has the following conclusion.

Lemma 3.6 *Let G be a group of finite Morley rank and suppose that G has a simple definable socle S of rank n. Then $\mathrm{rk}(G) \leq n^2 + n$.*

Proof We can assume without loss of generality that G and S are $\aleph_0$-saturated. Consider the action of G on S by conjugation, which is faithful. Let the sequence $(a_0, a_1, \ldots, a_n)$ be chosen in accordance with the previous lemma. Then the point stabilizer of the sequence fixes S pointwise and is therefore trivial, so we may repeat the computation of Lemma 3.4. □

We have not yet considered the case of a general nonabelian socle, and we will return to this below.

3.4 Elementary abelian socles

Now we take up the case of an elementary abelian socle. For the present subsection, let us fix the notation as follows.

Notation 3.7 (H, A) is a pair consisting of an infinite group H acting definably, definably irreducibly, and faithfully on an elementary abelian p-group A, and the pair has finite Morley rank.

We now apply the version of Zilber's field theorem given in [7, Theorem 9.5].

Theorem 3.8 *Let $V \rtimes G$ be a connected group of finite Morley rank with V, G definable, V abelian and G-minimal, and $C_G(V) = 1$. Suppose that G has an infinite definable abelian normal subgroup K. Then $C_V(G) = 1$, K is central in G, and there is an interpretable algebraically closed field over which V becomes a finite dimensional vector space, K becomes a group of scalars generating the field, and G acts linearly.*

Applying this in the present instance we get the following.

Lemma 3.9 *Suppose that H° has a nontrivial definable connected abelian normal subgroup and let K be a minimal such subgroup. Then there is a definable H°-invariant subgroup A_0 of A, and a finite subgroup K_0 of K, such that K/K_0 acts freely on A_0 and A is a direct sum of H-conjugates of A_0. Furthermore K is central in H°.*

Proof Let $A_0 \leq A$ be a minimal definable connected H°-invariant subgroup. Then A is the sum of the conjugates of A_0 under H. In particular at least one of the conjugates of A_0 is not centralized by K, and we may take A_0 to have this property. Then $K_0 = C_K(A_0)$ is finite by the minimality of K.

Let $\bar{H}^\circ = H^\circ/C_{H^\circ}(A_0)$. Now apply the previous theorem to $A_0 \rtimes H^\circ/C_{H^\circ}(A_0)$, with the subgroup $\bar{K} \cong K/C_K(A_0)$. Then $[K, H^\circ] \leq K_0$ and by connectedness K is central in H°.

Furthermore there is a vector space structure on A_0 with respect to which $\bar{K}$ acts by scalars and $H^\circ/C_{H^\circ}(A_0)$ acts linearly. □

This setup produces the natural bound on the rank of H when H° has a nontrivial connected abelian normal socle.

Lemma 3.10 *Suppose H° has a nontrivial definable connected abelian normal subgroup. Then $\text{rk}(H^\circ) \leq \text{rk}(A)^2$.*

Proof Let K be a minimal definable connected abelian normal subgroup of H° and let A_0 be a minimal definable connected H°-invariant subgroup of A.

Let $r = \mathrm{rk}(A)$, $r_0 = \mathrm{rk}(A_0)$, and let A be the sum of k H-conjugates of A_0 and no fewer. By the minimality of A_0, A modulo a finite subgroup is a direct sum of k subgroups of rank r_0 and thus $r = kr_0$.

The group $\bar{H}^\circ = H^\circ/C_{H^\circ}(A_0)$ acts as a linear group on A_0. If A_0 has dimension d then $\mathrm{rk}(\bar{H}^\circ) \leq d^2(r_0/d) = r_0 d$. Now H° embeds into the direct product of k groups definably isomorphic to $\bar{H}^\circ$ and hence $\mathrm{rk}(H) = \mathrm{rk}(H^\circ) \leq kr_0 d = rd \leq r^2$. □

We still have to deal with the very reasonable possibility that the definable socle of H° is semisimple. In this case what is needed, initially, is a bound on the number of simple factors. For this we need the following preparation, which says, roughly speaking, that tori tend to act generically freely, thereby bounding their ranks. Here the groups playing the role of tori are the definable hulls of divisible abelian torsion subgroups, taken modulo any definable torsion free subgroups.

Lemma 3.11 *Let (G, Ω) be a transitive permutation group of finite Morley rank, T a definable divisible abelian subgroup of G, and $O(T)$ the largest definable torsion free subgroup of T. Then $\mathrm{rk}(T/O(T)) \leq \mathrm{rk}(\Omega)$.*

Proof Let T_0 be the maximal torsion subgroup of T. Note that T_0 is a countable subgroup of T. Therefore in a saturated model we may take a point $\alpha \in \Omega$ generic over T_0. Suppose that a torsion element $t \in T_0$ fixes α. Then t fixes a generic subset of Ω pointwise. Since Ω has Morley degree 1, by Lemma 1.6, we find $t = 1$. In other words, the point stabilizer T_α is torsion free and thus contained in $O(T)$. Hence $\mathrm{rk}(T/O(T)) \leq \mathrm{rk}(T/T_\alpha) \leq \mathrm{rk}(\Omega)$. □

For the next step, we need a result of Wagner on fields of finite Morley rank in positive characteristic.

Theorem 3.12 ([21], cf. [2]) *Let F be a field of finite Morley rank of even type and positive characteristic. Then every definable subgroup of $F^\times$ is the definable hull of its torsion subgroup.*

Accordingly we make the following definition.

Definition 3.13 A divisible abelian group of finite Morley rank is called a *good torus* if every definable subgroup is the definable hull of its torsion subgroup.

The preceding theorem has the following consequence. Recall that a *unipotent* subgroup in a group of finite Morley rank is a solvable definable connected subgroup of bounded exponent. Such groups are in fact nilpotent. There are also "characteristic 0" versions of unipotence but we do not use them here; our unipotence is π-unipotence for some finite set of primes π.

Lemma 3.14 *Let H be a connected solvable group of finite Morley rank acting faithfully on a unipotent group V of bounded exponent. Let $U(H)$ be the maximal unipotent subgroup of H. Then $H/U(H)$ is a good torus.*

Proof Let $1 = V_0 < V_1 < \cdots < V_n = V$ be a chain of definable H-invariant subgroups of V such that successive quotients $A_i = V_i/V_{i-1}$ are H-minimal, that is infinite and without proper definable H-invariant subgroups. It is easy to see that the joint kernel of all the actions of H on the A_i is $U(H)$. Thus we reduce easily to the case in which V is H-minimal and $U(H) = 1$, replacing V by the A_i and H by its various quotients $H/C_H(A_i)$.

Now V is contained in the Fitting subgroup of VH and is self-centralizing in VH, so the connected component $F^\circ(VH)$ of the Fitting subgroup of VH is unipotent, and hence $F^\circ(VH) = V$ since $U(H) = 1$. But by the structure theory for connected solvable groups, $VH/F^\circ(VH)$ is divisible abelian and thus H is divisible abelian. Now Zilber's Field Theorem applies and H acts on V like a subgroup of the multiplicative group of a field. By Wagner's theorem, it is a good torus. □

Lemma 3.15 *Let $K = L_1 \times \cdots \times L_n$ be a product of infinite simple groups acting faithfully on an elementary abelian p-group V, with the pair (V, K) having finite Morley rank. Then $n \le \mathrm{rk}(V)$.*

Proof We argue by induction on the rank of V. If $\mathrm{rk}(V) = 0$, then V is finite and an infinite simple group K cannot act on V faithfully. Therefore we can assume that $\mathrm{rk}(V) > 0$ and $n > 1$. Easily K is faithful on V° so we may suppose V is connected. We may also suppose that $C_V(K) = 1$. If V is not definably K-irreducible and V_0 is a minimal nontrivial K-invariant subgroup, then V_0 is infinite and connected and induction easily yields our claim. So we suppose V is definably K-irreducible. Then $(V \rtimes K, V)$ is a primitive permutation group.

Suppose first that

(1) $L = L_n$ contains a nontrivial unipotent subgroup U.

Then $W = C_V^\circ(U) < V$ is nontrivial.

Let $K_1 = L_1 \times \cdots \times L_{n-1}$. Then K_1 acts on W and if the action is faithful we find $n - 1 \leq \mathrm{rk}(W) < \mathrm{rk}(V)$ and our claim follows.

So we may suppose that L_1 acts trivially on W. Let $V_1 = C_V(L_1) < V$. This contradicts the definable K-irreducibility of V.

So now we consider the alternative, which we may take to be the following.

(2) No factor L_i contains a nontrivial unipotent subgroup.

Now if B is a Borel subgroup of any factor L_i, then $U(B) = 1$ and by the preceding lemma B is a good torus. Since a product of good tori is a good torus [1, Lemma 4.21] we can find a good torus T in K of rank at least n. Then $O(T) = 1$ and thus by Lemma 3.11 $\mathrm{rk}(T) \leq \mathrm{rk}(V)$. So $n \leq \mathrm{rk}(V)$ as claimed. □

Now we use this information to give a bound on the rank of H in terms of the ranks of A and the ranks of the definable simple nonabelian subgroups of H°.

Lemma 3.16 *Suppose that every definable simple nonabelian subgroup of H° has rank at most s, and $r = \mathrm{rk}(A)$. Then the rank of H is at most*

$$\max(r^2, r(s^2 + s)).$$

Proof In view of Lemma 3.10 we will suppose that all minimal normal definable subgroups of H° are simple. Let there be n such. For L a definable normal simple subgroup of H°, let $H_L = H^\circ / C_{H^\circ}(L)$. Then there is an embedding $H^\circ \to \prod_L H_L$ and thus $\mathrm{rk}(H^\circ) \leq n \max_L(\mathrm{rk}(H_L))$

Now identify L with its image in H_L. Then L is a definable simple normal subgroup of H_L, and we claim that L is the definable socle of H_L. Let K be the preimage in H° of $C_{H_L}(L)$. Then $[K, L] \leq C_L(L) = 1$, so $K = C_{H^\circ}(L)$. Thus L is self-centralizing in H_L and is the definable socle.

So by Lemma 3.6 we have $\mathrm{rk}(H_L) \leq s^2 + s$ and therefore $\mathrm{rk}(H) \leq n(s^2 + s) \leq r(s^2 + s)$, taking into account Lemma 3.15. □

Now we go back to the case of nonabelian socles and make a similar reduction.

3.5 Nonabelian socles

Throughout the present subsection (G, Ω) denotes a permutation group of finite Morley rank with nonabelian definable socle. At this point one can usefully examine the subdivision into cases afforded by the full statement of MPOSA, but we continue on with less precise methods. Set $r = \text{rk}(\Omega)$.

Lemma 3.17 *If G has more than one minimal definable normal subgroup, then*

$$\text{rk}(G) \leq r^2 + 2r.$$

Proof Let L be a minimal normal subgroup of G. By our hypothesis $C_G(L)$ is nontrivial and hence by definable primitivity L and $C_G(L)$ both act transitively on Ω. It follows that each acts regularly and in particular $\text{rk}(C_G(L)) = r$, so $\text{rk}(G) \leq r + \text{rk}(G/C_G(L))$.

Now write $L = L_1 \times \cdots \times L_n$ with the L_i isomorphic definable simple normal subgroups of G°. As L acts regularly on Ω we have $\text{rk}(L_i) = r/n$. Then Lemma 3.6 yields

$$\text{rk}(G/C_G(L)) \leq n((r/n)^2 + (r/n)) = r^2/n + r.$$

and our claim follows. □

Now suppose that G contains a unique minimal definable normal subgroup L, which is nonabelian. We should note that if G were connected then that subgroup would in fact be simple, as G acts transitively on the simple factors of a nonabelian minimal normal subgroup, and then the situation would be quite transparent. But what this suggests, correctly, is that primitive groups will have a strong tendency to be disconnected, in general.

In this case we finally need to use the more precise information from [14], describing the point stabilizer L_α, which is a characteristic feature of the MPOSA point of view. There are the following two possibilities:

(a) L is a product of simple factors L_i and $L_\alpha = \prod_i (L_i)_\alpha$ (possibly $L_\alpha = 1$ here);

(b) L is a product of ℓ groups L_i, each of which is a product of k simple factors L_{ij}, with $k \geq 2$, and $(L_i)_\alpha$ is a diagonal subgroup of L_i, and in particular $\text{rk}((L_i)_\alpha)$ is definably isomorphic to L_{ij} for each j.

Lemma 3.18 *If G has a unique minimal normal subgroup L, and its simple factors have rank s, then corresponding to cases (a, b) above we have the following estimates.*

(a) $\mathrm{rk}(G) \le r(s+s^2)$;
(b) $\mathrm{rk}(G) \le 2(r^2+r)$.

Proof Let L be the product of n simple factors, each of rank s. Then our estimate for $\mathrm{rk}(G)$ is

$$n(s^2+s).$$

In case (a) we have $n \le r$. So we consider case (b).

If L acts regularly on Ω then $s = r/n \le r$.

If the point stabilizer L_α is a product of ℓ diagonal subgroups then $n = k\ell$ for some k and $r = \mathrm{rk}(L) - \mathrm{rk}(L_\alpha) = ns - \ell s = (k-1)\ell s$, $s = r/(k-1)\ell = r/(n-\ell) \le (2/n)r$. So $n(s^2+s) \le (4/n)r^2 + 2r \le 2(r^2+r)$ since $n \ge 2$. □

The upshot of all of this is that we need an estimate for the rank of a simple group acting on Ω in terms of the rank of Ω, or, in view of Proposition 2.3, a bound on the degree of generic multiple transitivity for such an action.

Of course, this reduction of a general problem on primitive permutation groups to the simple case is a typical application of the MPOSA point of view.

4 Actions of finite groups on connected solvable groups

The present section has a preparatory character. In the next section we use that fact that generically n-transitive groups have the symmetric group $\mathrm{Sym}(n)$ as a section. Here we examine definable actions of $\mathrm{Sym}(n)$, and related groups, on connected solvable groups of finite Morley rank, looking for lower bounds on the rank. This is an issue which has not arisen in the past and has some affinities with linear representation theory, though an action on a connected abelian group is not necessarily very closely connected with a linear representation, as far as we know. Getting sharp bounds for this sort of problem seems challenging.

There are three natural variations, all of which come into play.

Problem 8 Let Σ be a finite group. Find lower bounds for each of the following.

(1) The minimal rank of a connected solvable group of finite Morley rank which affords a faithful representation of Σ.
(2) The minimal rank of a connected solvable group of finite Morley rank which affords a faithful representation of a central extension of Σ.
(3) The minimal rank of a connected solvable group of finite Morley rank which affords a faithful representation of a group $\hat{\Sigma}$ which covers Σ, i.e. maps homomorphically onto Σ.

We will work with abelian groups rather than solvable groups, but in most cases we can reduce to Σ-minimal groups, in which case there is no difference. But the question makes sense also more generally, without even the hypothesis of solvability.

Actually, as we will see we are interested particularly in actions of finite groups on divisible abelian groups. This brings us a little closer to the characteristic zero linear theory.

The estimate we will need is the following. This is off the main line of our discussion, and any estimate of the kind will suffice for our subsequent purposes.

Proposition 4.1 *Let A be a connected abelian group of finite Morley rank, and Σ a finite group acting definably and faithfully on A. Suppose that Σ maps surjectively onto* $\mathrm{Sym}(n)$. *Then* $\mathrm{rk}(A) \geq \lfloor n/4 \rfloor$

We return to the main line in the next section.

4.1 Generalities

We dispose of some formal points before taking up anything concrete. Recall that the *Frattini* subgroup $\Phi(\Sigma)$ of a finite group Σ is the intersection of its maximal subgroups, and is nilpotent.

Definition 4.2 Let $\Sigma \to \bar{\Sigma}$ be a surjection between finite groups. The map, or by abuse Σ itself, is called a *Frattini cover* of $\bar{\Sigma}$ if the kernel of the map is contained in the Frattini subgroup of Σ.

An equivalent condition is this: no proper subgroup of Σ covers $\bar{\Sigma}$. So we have a trivial but useful starting point.

Lemma 4.3 *Let Σ_0 be a finite group, Σ a finite group mapping homomorphically onto Σ_0. Then Σ contains a subgroup Σ_1 which is a Frattini cover of Σ_0.*

For the proof just take Σ_1 minimal. Notice that if we also have a faithful representation of Σ on a connected abelian group of finite Morley rank, then we have in particular a faithful representation of a Frattini cover. This has the advantage that the kernel of a Frattini cover is nilpotent, so we are getting much more control over the group. What we would prefer is to reduce even further, to central extensions.

Now we take a preliminary look at the minimal modules for our finite groups.

Lemma 4.4 *Let Σ be an almost simple group (i.e., the socle is simple). Let A be a connected abelian group of minimal Morley rank such that Σ acts faithfully and definably on A. Then Σ acts faithfully on some Σ-minimal definable section of A.*

Proof We argue by induction on $\mathrm{rk}(A)$. Let $A_0 \leq A$ be Σ-minimal. If Σ acts faithfully on A_0 or A/A_0 we conclude directly or by induction, respectively. In the alternative case the socle Σ_0 of Σ acts trivially on both factors and therefore each element $a \in A$ gives rise via commutation to a homomorphism $\alpha : \Sigma_0 \to A_0$, from a simple group to an abelian group. So in this case Σ_0 acts trivially on A, a contradiction. □

Next we prepare some general estimates for actions of elementary abelian groups.

Lemma 4.5 *Let E be an elementary abelian p-group and $m = m_p(A)$. If E acts faithfully on the connected p-divisible abelian group A of finite Morley rank, then $\mathrm{rk}(A) \geq m$.*

Proof Let $V \leq A$ be E-minimal. Let E_0 be the kernel of the action of E on V, a subspace of E of codimension 1. If E_0 acts faithfully on A/V then we conclude by induction.

So let E_1 be the kernel of the action of E_0 on A/V. Then E_1 acts trivially on the factors of the chain $1 < V < A$. So for $e \in E_1^{\#}$ commutation with e gives a homomorphism $\epsilon : A/V \to A$ with image a connected elementary abelian p-group, contradicting our hypothesis. □

4.2 Actions of symmetric groups

We aim here at lower bounds for the rank of a connected abelian group on which $\mathrm{Sym}(n)$ acts faithfully, particularly when the abelian group in question is the definable closure of a p-torus for some p. Of course, we will really need to deal with groups covering $\mathrm{Sym}(n)$ and these do not necessarily behave in the same way, so the reader may prefer to pass on to the next subsection which returns to the more general problem, with worse estimates.

Here it would be very useful to have a result of the following kind: if T is the definable closure of a p-torus and is F-minimal under the action of a finite group F, then the corresponding Tate module is F-irreducible. Unfortunately this is nonsensical, as one sees already by taking $F = 1$, but perhaps something can be rescued in this direction.

Before stating our next lemma we note that the symmetric group $\mathrm{Sym}(6)$ is isomorphic to the symplectic group $\mathrm{Sp}(4,2)$ and therefore has a faithful representation on a connected abelian group of Morley rank 4, namely the corresponding vector space over the algebraic closure of $\mathbb{F}_2$.

Lemma 4.6 *Let A be an abelian group on which* $\mathrm{Sym}(n)$ *acts definably and faithfully, with A either a finite elementary abelian p-group or a connected abelian group. Let $\delta(A)$ be $m_p(A)$ in the first case and* $\mathrm{rk}(A)$ *in the second case. Then $\delta(A) \geq n-1$ unless $n = 6$, $d = 4$, and A is an elementary abelian 2-group (finite or connected).*

Proof We proceed by induction on n and then by induction on $\delta(A)$. If $n = 2$ our claim is vacuous. If $n \geq 3$ and $d = 1$ then the transpositions must act by inversion on G, and the action cannot be faithful. So we may suppose throughout that $d \geq 2$ and $n \geq 4$. Write Σ for $\mathrm{Sym}(n)$.

We treat separately the case $n = 4$. Let V be the Klein 4-group in Σ. Let $A_0 < A$ be V-minimal. Then V has a kernel on A_0 and this kernel has three conjugates in Σ. It follows easily that $d \geq 3$. So from now on we suppose

$$n \geq 5.$$

In particular Σ is almost simple and A may be taken to be Σ-minimal in the connected case, irreducible in the finite case. In particular A is either an elementary abelian 2-group, or is 2-divisible.

We deal first with the case of an elementary abelian 2-group. We will use the connected component notation below with an eye on the case in which A is connected. This operator should be interpreted as vacuous ($X^\circ = X$, not $X^\circ = 1$) when A itself is finite.

Fix a transposition $\tau \in \Sigma$. Then $1 \leq [\tau, A] \leq C_A^\circ(\tau) \leq A$. If the three sections $A_1 = [\tau, A]$, $A_2 = C_A^\circ(\tau)/[\tau, A]$, and $A_3 = A/C_A^\circ(\tau)$ are all nontrivial, consider the action of the group $\Sigma_0 = C_\Sigma(\tau) \cong \mathrm{Sym}(n-2)$. If $n \geq 7$ so that $n-2 \geq 5$, then Σ_0 acts faithfully on at least one of the sections A_i and thus we find $\delta(A_i) \geq n-3$, $\delta(A) \geq (n-3)+2 = n-1$ as required. If $n = 6$ then the socle of Sym(4) acts faithfully on at least one of these sections, which therefore has rank at least 2. So we get our lower bound $\delta(A) \geq 4$ in this case. If $n = 5$ we have only to deal with the case in which $\delta(A_i) = 1$ for $i = 1, 2, 3$. Then Sym(3) is generated by involutions and each of them acts trivially on each section, so Sym(3) acts trivially on each section and therefore an element of order 3 acts trivially on A, a contradiction.

This disposes of all cases in which the elementary abelian 2-group A has the three sections A_1, A_2, A_3 all nontrivial. So at least one of these sections is trivial. But τ acts faithfully on A, so we find $C_A^\circ(\tau) = [\tau, A]$ or in other words $[\ker(1-\tau)]^\circ = \mathrm{im}(1-\tau)$. So $\delta(A_1) = \delta(A_3)$. If $n \geq 7$ then again Σ_0 acts faithfully on one of these sections and thus $\delta(A) \geq 2(n-3) \geq n-1$.

We now have to treat the small cases with $n = 5, 6$.

Suppose first $\delta(A_1) = \delta(A_3) = 1$. Let V be a 4-group in Σ. Then $C_A^\circ(V)$ is nontrivial and hence $C_A^\circ(V) = C_A^\circ(i)$ for $i \in V^\#$ or in other words $C_A^\circ(i) = C_A^\circ(j)$ for commuting involutions i, j, and as $n \geq 5$ we conclude that $C_A^\circ(i) = C_A^\circ(\Sigma)$ which may be supposed trivial. Thus we have a contradiction.

If $\delta(A_1) = \delta(A_3) = 2$ and $\delta(A) = 4$ then the only remaining case is $n = 6$, which we allow.

So now we pass to the main case.

A is 2-divisible.

Recall that A is either connected, or is a finite elementary abelian p-group with p odd, now.

Consider an involution τ and let $\Sigma_0 = C_\Sigma(\tau) \cong \mathrm{Sym}(n-2)$. We have

$$A = A_\tau^+ + A_\tau^-$$

with A_τ^+, A_τ^- the subgroups centralized by or inverted by τ respectively. If A is infinite (connected) then the intersection $A_\tau^+ \cap A_\tau^-$ is finite and these groups are connected.

We have actions of Σ_0 on A_τ^+ and A_τ^-, and at least one of these actions is faithful. So $\delta(A) \geq (n-3)+1 = n-2$ with equality only if one of the factors has rank 1 and the other has rank $n-3$.

Suppose $\delta(A_\tau^+) = 1$. The same holds for any transposition as they are conjugate. So if τ, τ' are transpositions whose product σ has order 3, then τ and τ' invert subgroups of corank 1 in A, and hence σ inverts a subgroup of corank at most 2. So in this case $\delta(A) = 2$ and thus by the above $n - 3 = 1$, $n = 4$. Then consider a pair of commuting transpositions τ_1, τ_2. We have $A^{\pm}_{\tau_1} = A^{\mp}_{\tau_2}$ in some order and hence $\tau_1\tau_2$ acts either trivially or by inversion on A. Then the action of $\tau_1\tau_2$ commutes with the action of Σ on A and therefore this action is not faithful. We conclude

$$\delta(A^-_\tau) = 1, \delta(A^+_\tau) = n - 3 > 1$$

for any transposition $\tau \in \Sigma$. In particular $n \geq 5$.

For any pair of commuting transpositions τ, τ', the element τ' centralizes A^-_τ, as otherwise we would have $A^-_\tau = A^-_{\tau'}$ for all such commuting pairs and then A^-_τ is independent of τ, since $n \geq 5$.

For any transposition τ, the action of τ on A/A^-_τ is trivial and in particular any subgroup containing A^-_τ is τ-invariant.

Now we need to work out the action in more detail. Fix two noncommuting transpositions τ_1, τ_2 and let $A_i = A^-_{\tau_i}$ for $i = 1, 2$. Take $a_1 \in A_1$ and choose $a_2 \in A_2$ so that

$$a_1^{\tau_2} = a_1 + a_2.$$

Let $\sigma = \tau_2\tau_1$. Then $a_1^\sigma = -a_1 + a_2^{\tau_1}$ and σ carries A_1 to A_2, so $a_2^{\tau_1} \in a_1 + A_2$ and therefore $a_2^{\tau_1} = a_1 + a_2$. Thus $a_2 = a_1^\sigma$. Similarly $a_2^{\tau_1} = a_2^{\sigma^{-1}} + a_2 = a_1 + a_2$.

Now let τ_i be the elementary transposition $(i, i+1)$ and let $A_i = A^-_{\tau_i}$. Assuming $\delta(A) < n - 1$, let k be minimal so that $\delta(A_1 + \cdots + A_k) < k$. Then $A_k \leq A_1 + \cdots + A_{k-1}$ and therefore $A_1 + \cdots + A_k$ is Σ-invariant, so $k = n - 1$ and $A = A_1 + \cdots + A_{n-2}$ is an almost direct sum.

Consider the cycle $\sigma = (12 \ldots n)$. For any $i < n - 1$ if $\sigma_i = \tau_{i+1}\tau_i$ then $\sigma\sigma_i^{-1}$ commutes with τ_i and thus σ acts on A_i as σ_i does. This holds also for A_{n-1} with a notational variation. So for $a \in A_i$ we have $a^{\tau_{i+1}} = a + a^\sigma$, and similarly $a^{\tau_{i-1}} = a + a^{\sigma^{-1}}$.

Now consider any relation

$$a_1 + \cdots + a_{n-1} = 0$$

with $a_i \in A_i$. Applying τ_i we get

$$a_{i-1}^{\tau_i} - a_i + a_{i+1}^{\tau_i} = a_{i-1} + a_i + a_{i+1}$$

or

$$(*) \qquad a_{i-1}^\sigma + a_{i+1}^{\sigma^{-1}} = 2a_i.$$

From this we easily derive

$$ia_{i+1} = (i+1)a_i^\sigma$$

for all $i < n-1$. For $i = 1$ the equation $(*)$ reduces to the desired form and after that induction applies. But there is also a final equation for $i = n-1$ which reduces to

$$0 = na_{n-1}^\sigma$$

or $na_{n-1} = 0$. As A is divisible it follows that there are only finitely many possible values for a_{n-1}, which contradicts our assumption. □

4.3 Groups covering Sym(n)

Now we look at actions on connected abelian groups of groups Σ mapping homomorphically onto Sym(n).

We first mention some comparatively low dimensional representations of double covers of Sym(n), and the like. The Schur multiplier of Sym(n) is $\mathbb{Z}/2\mathbb{Z}$, but there are two nonisomorphic double covers.

(1) 2.Sym(4) = GL(2, 3) has a 2 dimensional representation.
(2) Both of the double covers of Sym(7) have 4-dimensional representations in characteristic 7.
(3) Alt(8) = SL(4, 2) has 4-dimensional representations in characteristic 2.

Now we give our main estimate for groups covering Sym(n).

Proof of Proposition 4.1. We may suppose that Σ is a Frattini cover of Sym(n).

Suppose first that rk(A) = 1. In this case any abelian subgroup of Σ is cyclic and in particular the Sylow 2-subgroup T of Σ contains a unique involution. So this involution is in the kernel of the map $\Sigma \to$ Sym(n), and T itself is either cyclic or a generalized quaternion subgroup. So the image of T in Sym(n) is abelian, contradicting the structure of Sym(n).

So we may suppose rk(A) ≥ 2. That being the case, we may also suppose $n \geq 8$, but as one might reasonably aim at a sharper lower bound, e.g. $(n-2)/2$, we will only assume here that

$$n \geq 5.$$

Let Γ be the kernel of the given homomorphism $\Sigma \to$ Sym(n). Since $\Sigma \to$ Sym(n) is a Frattini cover, Γ is nilpotent. Let Γ_2 be the Sylow

2-subgroup. If $\Gamma_2 = 1$ then one expects the same estimates as in the preceding subsection, but in any case Σ contains an elementary abelian 2-subgroup E with $m(E) = \lfloor n/2 \rfloor$ and this suffices for our estimate. So we suppose

$$\Gamma_2 > 1.$$

Let $A_0 \leq A$ be Σ-minimal. Let Γ_0 be the kernel of the action of Σ on A_0. If Γ_0 is contained in Γ then we can look at the action of Σ/Γ_0 on A_0 and conclude by induction unless $A = A_0$. On the other hand, if Γ_0 is not contained in Γ then Γ_0 covers $\mathrm{Alt}(n)$ and hence Σ is generated by Γ_0 and a 2-element, and $\Sigma' \leq \Gamma_0$. If $A_0 < A$ we can consider the action of Σ on A/A_0 similarly and conclude by induction unless Σ' acts trivially on A/A_0. So finally Σ' acts trivially on A_0 and A/A_0 and hence the subgroup of Σ' generated by its $2'$-elements acts trivially on A, a contradiction as this covers $\mathrm{Alt}(n)$. So we may suppose that A is Σ-minimal.

Now let Σ_1 be any subgroup of Σ covering $\mathrm{Alt}(n)$. We claim that A is Σ_1-minimal. We may replace Σ_1 by a minimal subgroup covering $\mathrm{Alt}(n)$ and then $\Sigma_1 = \Sigma_1'$ and thus $\Sigma_1 = \Sigma^{(\infty)}$ is normal in Σ. Now suppose A is not Σ_1-minimal. Let $A_0 < A$ be Σ_1-minimal. Then A is the almost direct sum of two conjugates of A_0. Thus $\mathrm{rk}(A) = 2\,\mathrm{rk}(A_0)$. Now there is a subgroup of Σ_1 covering a copy of $\mathrm{Sym}(n-2)$ inside $\mathrm{Alt}(n)$, so we may suppose by induction that $\mathrm{rk}(A_0) \geq \lfloor (n-2)/4 \rfloor$ and thus $\mathrm{rk}(A) \geq \lfloor n/4 \rfloor$ and we conclude. So we may suppose that A is Σ_1-minimal for all such Σ_1.

Now we return to the structure of the nilpotent group Γ and its Sylow 2-subgroup Γ_2. Suppose that Γ_2 is noncyclic and take k minimal so that $Z_k(\Gamma_2)$ is noncyclic. Let $C = Z_{k+1}(\Gamma_2)$ and let Q be the preimage in Γ_2 of $\Omega_1(Z_k(\Gamma_2)/C)$. Let $Q_0 = C_Q(C)$.

Now Γ_2 acts trivially on Q/C and thus there is an action of $\mathrm{Sym}(n)$ on Q/C. If this action is faithful then $m_2(Q/C) \geq 4$ and hence $Q_0 > C$. In this case we can consider the action of $\mathrm{Sym}(n)$ on Q_0/C, and if this is faithful then $m_2(Q_0/C) \geq n-1$ unless $n = 6$ and $m_2(Q_0/C) = 4$.

Suppose $m_2(Q_0/C) \geq n-1$. If Q_0 is elementary abelian then we get a strong estimate. Otherwise, $Q_0' \leq C$ is the subgroup of order 2 and the commutator map $Q_0/C \times Q_0/C \to Q_0'$ gives a symplectic structure on Q_0/C, possibly degenerate. We may take a totally isotropic subspace of Q_0/C of dimension $(n-1)/2$, which lifts back to an abelian subgroup $V \leq Q_0$ for which $m_2(V/C) = (n-1)/2$. Thus V contains an elementary abelian subgroup of rank $(n-1)/2$ and we get the desired estimate.

Now suppose $m_2(Q_0/C) = 4$ and $n = 6$, with $\mathrm{Sym}(n)$ acting faithfully and with $Q_0' > 1$. Let Q_1 be the preimage in Q_0 of $\Omega_1(Q_0/Q_0')$. Then Q_1/Q_0' is elementary abelian of rank 4. As $\mathrm{Sym}(n)$ respects the symplectic structure it coincides with $\mathrm{Sp}(4,2)$ and in particular acts transitively on Q_1/Q_0'. Then as $Q_1 \setminus C$ contains involutions, Q_1 has exponent two and thus is elementary abelian of rank 5. So $\mathrm{rk}(A) \geq 5$.

So we are left with the case in which $\mathrm{Alt}(n)$ centralizes Q/C and hence a subgroup of Σ covering $\mathrm{Alt}(n)$ centralizes Q. If Q contains a noncyclic elementary abelian subgroup E then let $A_0 \leq A$ be E-minimal and let E_0 be the kernel of the action of E on A_0. Then $C_A^\circ(E_0)$ is $C_\Sigma(Q)$-invariant and thus $C_A^\circ(E_0) = A$, implying $E_0 = 1$ and E is cyclic after all.

So Q contains a unique involution and is therefore either cyclic or generalized quaternion. But by our current assumption Q is not cyclic. On the other hand by construction $Q/Z(Q)$ is elementary abelian, and so Q is a quaternion group of order 8, and is centralized by a subgroup of Σ covering $\mathrm{Alt}(n)$.

Taking a 2-element $t \in \Sigma$ which represents a transposition in $\mathrm{Sym}(n)$, we may suppose after adjustment that t centralizes an element a of order 4 in Q_0. Then $C(a)$ covers $\mathrm{Sym}(n)$ so by minimality $a \in Z(\Sigma)$ and in particular $a \in Z(Q)$, a contradiction.

All of this shows that Γ_2 is cyclic. Now one can show in the same way that Γ is cyclic and centralized by $\mathrm{Alt}(n)$ and pursue the matter further, but we will stop here with a simple estimate. We have Γ_2 centralized by a subgroup of Σ covering $\mathrm{Alt}(n)$. Taking an elementary abelian subgroup of $\mathrm{Alt}(n)$ of rank at least $2\lfloor n/4 \rfloor$, and lifting to a 2-subgroup of Σ, we get an elementary abelian subgroup of rank at least $\lfloor n/4 \rfloor$ in Σ, and conclude. □

5 Simple permutation groups

In the present section we will bring to bear a good deal of recent work on simple groups of finite Morley rank to the problem of bounding their rank or degree of multiple transitivity in terms of the rank of a set acted upon faithfully. We begin by recalling some of the relevant information.

5.1 The four types

The 2-Sylow theory for groups of finite Morley rank gives a broad division of all such groups into four types, according to the structure of their 2-Sylow° subgroups (maximal connected 2-subgroups). In general these have the form

$$U * T$$

a central product with finite intersection of a 2-unipotent group U (a connected, definable 2-group of bounded exponent) and a 2-torus T (a divisible abelian 2-group, not definable). The group G is said to be of even, odd, mixed, or degenerate type according as $S = U$, $S = T$, $U, T > 1$, or $S = 1$. A delicate analysis depending on a body of technology developed in the finite case, and some other ingredients, shows the following.

Theorem 5.1 *A simple group of finite Morley rank of even type is algebraic.*

For a group G of finite Morley rank, let $T_2(G)$ be the subgroup generated by all definable hulls of 2-tori in G, and $U_2(G)$ the subgroup generated by all unipotent 2-subgroups.

Theorem 5.2 *Let G be a group of finite Morley rank. Then normal subgroups $T_2(G)$ and $U_2(G)$ have odd and even types respectively. Furthermore, $T_2(G)$ and $U_2(G)$ commute, with $G/[T_2(G) * U_2(G)]$ of degenerate type.*

Theorem 5.2 depends on Theorem 5.1. In fact what one actually shows is that Theorem 5.2 holds for groups whose definable simple sections of even type are all algebraic.

These theorems have the following consequence, which can however be proved much more directly [6].

Theorem 5.3 *Let G be a simple group of finite Morley rank containing no nontrivial divisible torsion subgroup. Then G is of degenerate type.*

Another result of a general character with a comparatively short proof is the following [5].

Theorem 5.4 *Let G be a connected group of finite Morley rank of degenerate type. Then G contains no involutions.*

Actually this can be extended to a similar result holding for any prime p. In that form it reads as follows.

Theorem 5.5 ([5]) *Let G be a connected group of finite Morley rank containing no infinite abelian p-subgroup. Then G contains no elements of order p.*

The classification of simple groups of even type has various structural consequences which are probably not accessible without that classification. For example we have the following [1].

Proposition 5.6 *Let G be a connected group of finite Morley rank of even type and suppose that G contains no nontrivial normal 2-unipotent subgroup. Then G is a central product of definable subgroups*

$$E(G) * \hat{O}(G)$$

where $E(G)$ is a product of quasisimple algebraic groups in characteristic two, and $\hat{O}(G)$ has degenerate type.

These notions apply in our context as follows.

Lemma 5.7 *Let G be a definably primitive permutation group of finite Morley rank. Then G° is not of mixed type.*

Proof Supposing the contrary, $U_2(G)$ is nontrivial by Theorem 5.2. If G contains a nontrivial normal 2-unipotent subgroup then, in view of definable primitivity of G, the definable socle of G is an elementary abelian 2-group. But then the definable socle of G is disjoint from $T_2(G)$ and thus $T_2(G)$ is trivial.

Similarly, if G contains no nontrivial normal 2-unipotent subgroup, then the definable socle of $U_2(G)$ is a product of simple groups of even type and thus the definable socle of G is also a product of simple groups of even type, and in fact all of the factors are isomorphic. Again, this forces the socle to miss $T_2(G)$ and thus $T_2(G) = 1$. □

We can refine this a little when we have generic 2-transitivity as well.

Lemma 5.8 *Let (G, Ω) be a definably primitive and generically 2-transitive permutation group of finite Morley rank. Then G° is either of odd or of even type.*

Proof We have already eliminated mixed type. On the other hand, as G is generically 2-transitive, also G° is generically 2-transitive, and hence there is an involution in G° swapping two generic and independent elements of Ω. Since G° is connected and contains an involution, it is not of degenerate type. □

This really does require primitivity in addition to generic 2-transitivity.

Example 9 Let $G = G_0 \times G_1$, where $G_i = \mathrm{GL}(V_i)$ and V_i are vector spaces over fields K_i, with K_0 of characteristic two and K_1 not of characteristic two. Then G has mixed type and acts generically n-transitively on $V_1 \times V_2$, where n is the minimal dimension involved.

5.2 Even type groups

In the simple case, we can use the classification in even type to reduce the analysis to the specific case of Chevalley groups. The theory in odd type is not sufficiently advanced to allow this kind of sweeping reduction to known groups, though there is a large body of work also in this direction.

Lemma 5.9 *Let (G, Ω) be a transitive group of finite Morley rank, where the group G is isomorphic as an abstract group to a Chevalley group over an algebraically closed field. Then the algebraic dimension of a maximal torus T of G is at most $\mathrm{rk}(\Omega)$, and if the characteristic of the base field is positive then $\mathrm{rk}(T) \leq \mathrm{rk}(\Omega)$.*

Proof By Lemma 3.11, $\mathrm{rk}(T/O(T)) \leq \mathrm{rk}(\Omega)$. Now T is definably a direct product of 1-dimensional tori T_i and $O(T) = \prod O(T_i)$, so $T/O(T)$ is also a product $\prod T_i/O(T_i)$ and therefore $\mathrm{rk}(T/O(T))$ is at least the number of factors.

Now in the case of positive characteristic we also know that T is a good torus by Wagner's results, and thus $O(T) = 1$. □

Proposition 5.10 *Let (G, Ω) be a simple permutation group of finite Morley rank and even type. Then the rank of G is bounded by a function of the rank of Ω (e.g., $4\,\mathrm{rk}(\Omega)^2$).*

Proof Due to the simplicity of G, we can assume without loss of generality that G is transitive on Ω. If d is the algebraic dimension of G and ℓ the Lie rank, then for the classical groups of types $A_\ell, B_\ell, C_\ell, D_\ell$ we have $d = \ell(\ell+2)$ or $\ell(2\ell \pm 1)$, so at most $2\ell^2 + \ell$. This leaves aside only the exceptional groups of types $E_6 - E_8, F_4, G_2$, for which the Lie ranks and dimensions are: $(6, 78)$; $(7, 133)$; $(8, 248)$; $(4, 52)$; $(2, 14)$ respectively.

Let T be a maximal torus in G. Now if f is the rank of the base field (visible in various ways in the group, notably via the root groups), then $\mathrm{rk}(G) = df$ and $\mathrm{rk}(T) = \ell f$. So it suffices to bound d/ℓ in terms of $\mathrm{rk}(\Omega)$, and since $\ell \leq \mathrm{rk}(T) \leq \mathrm{rk}(\Omega)$ by Lemma 5.9, this is certainly possible. □

So in view of Lemma 5.8 the problem now is to get some control over actions of simple groups of odd type without an explicit classification (and even for Chevalley groups in characteristic zero there is something still to analyze).

5.3 Odd type groups

We now need some general structural properties of odd type groups, and we have to elaborate a little on the existing theory. Our starting point is the following.

Definition 5.11 A *decent torus* is a divisible abelian group of finite Morley rank which is the definable hull of its torsion subgroup.

Theorem 5.12 ([9]) *Let T be maximal definable decent torus in a connected group G of finite Morley rank. Then the generic element of G belongs to a unique conjugate of $C^\circ(T)$. Furthermore, any two maximal decent tori of G are conjugate.*

The following inessential variant can be proved the same way, or simply deduced from the foregoing.

Proposition 5.13 *Let G be a connected group of finite Morley rank and T a maximal p-torus of G. Then the generic element of G lies in one and only one conjugate of $C^\circ(T)$. Furthermore, any two maximal p-tori of G are conjugate in G.*

Proof Since any maximal p-tori extend to maximal decent tori, the conjugacy statement here follows from the preceding.

Now let $\hat{T}$ be a maximal decent torus containing the given p-torus T. Then for a generic set of elements $a \in G$, a lies in one and only one conjugate of $C^\circ(\hat{T})$, and in particular a centralizes a conjugate of T. We may suppose a centralizes $\hat{T}$.

If a centralizes the maximal p-torus T_1 of G, and $\hat{T}_1$ is a maximal decent torus in $C^\circ(a)$ containing T_1, then T and T_1 are conjugate in

$C^\circ(a)$ and hence $\hat{T}_1$ is also a maximal decent torus of G. Thus $\hat{T}_1 = \hat{T}$ (by our choice of a) and $T_1 \leq \hat{T}$. So by maximality of T and T_1, we have $T = T_1$. □

In a similar vein the following will be useful.

Lemma 5.14 *Let G be a connected group of finite Morley rank containing no nontrivial unipotent p-subgroup. Then for a generic element a of G, the group $d(a)$ is p-divisible.*

Proof Let T be a maximal p-torus of G. Let $X \subseteq G$ be the set of elements belonging to a unique conjugate of $C^\circ(T)$. This set is generic in G and disjointly covered by conjugates of $X \cap C^\circ(T)$. So it suffices to look at generic elements of $C^\circ(T)$. In other words, we may suppose that the maximal p-torus T is central in G.

We claim then that all elements of order p in G belong to T. Let $\hat{T} = d(T)$ be the definable hull of T, a divisible abelian group whose p-torsion lies in T. Our claim is that $G/\hat{T}$ contains no p-torsion. Otherwise, $\bar{G} = G/\hat{T}$ contains a nontrivial connected definable abelian p-subgroup $\bar{K}_0$, whose definable hull is the image of a connected nilpotent subgroup K of G. Now K is a connected nilpotent group containing p-torsion not in T, and as T is a maximal p-torus in K there must be a nontrivial p-unipotent subgroup in K, contradicting our hypothesis.

Now consider $a \in G$ generic over T (which is a countable set). Then $d(a)$ has the form $A \times T_0$ with A p-divisible and T_0 a finite cyclic p-group. In particular $T_0 \leq T$.

Now $a = bt$ with $b \in A$ and $t \in T_0$. So $AT_0 = d(a) \leq d(b) \times \langle t \rangle$ and thus $A = d(b)$, $T_0 = \langle t \rangle$. But $b = at^{-1}$ is also generic over T, and $d(b)$ is p-divisible. As there is a unique generic type in G, we must have $d(g)$ p-divisible for any realization g of this type. □

Unlike most results of this general character, the lemma does not yield a generic *definable set* of elements g for which the definable hull $d(g)$ is p-divisible. For example, working in an ordinary algebraic torus over an algebraically closed field, any generic set will contain p-elements for all primes p different from the characteristic, and for such elements the definable hull is the finite cyclic group generated by them.

The following specialized result will be quite useful for the analysis of groups of odd type as permutation groups.

Proposition 5.15 *Let G be a connected group of finite Morley rank and* *odd type, and let T be a maximal 2-torus of G. Then T contains all the involutions in $C(T)$.*

Proof It is convenient to generalize this a little, allowing G also to be of degenerate type (i.e. $T = 1$), in which case we arrive at the nontrivial but known result that degenerate type groups contain no involutions.

We suppose toward a contradiction that G is a minimal counterexample, and that $i \in G$ is an involution centralizing a maximal torus T of G, with $i \notin T$. We make a number of reductions. Note that i cannot belong to any proper definable connected subgroup containing T. Furthermore G is nonabelian.

$$(1) \qquad Z^\circ(G) = 1.$$

The connected abelian subgroup $Z^\circ(G)d(T)$ does not contain i. Therefore in the quotient $\bar{G} = G/Z^\circ(G)$, the subgroup $\bar{T}$ is a maximal 2-torus centralizes by an involution $\bar{i}$ which lies outside $\bar{T}$. If $Z^\circ(G) > 1$ then we contradict the minimality of G (or $T \leq Z^\circ(G)$ and then we reduce to the degenerate case). So $Z^\circ(G) = 1$.

$$(2) \qquad \text{Without loss of generality, } Z(G) = 1.$$

What we actually have is $Z(G)$ finite, and we wish to factor out the center. For this it suffices to check first that $i \notin Z(G)T$. Assuming the contrary, we may even take $i \in Z(G)$.

Now a generic element of G lies in a conjugate of $C^\circ(T)$. So for a generic set of elements $g \in G$, both g and gi lie in subgroups of the form $C^\circ(T_1)$, $C^\circ(T_2)$ for some conjugates T_1, T_2 of T. In particular $T_1, T_2 \leq C^\circ(g)$. So T_1 and T_2 are conjugate in $C^\circ(g)$. Since $g \in C^\circ(T_1)$, we find $g \in C^\circ(T_2)$ as well. Since $gi \in C^\circ(T_2)$ we find $i \in C^\circ(T_2)$. So conjugating in G, we find $i \in C^\circ(T)$ as well. Now $C^\circ(T)/d(T)$ has degenerate type and hence contains no involutions. That is, $i \in d(T)$ and thus $i \in T$, a contradiction. Replacing G by $\bar{G} = G/Z(G)$, we may suppose $Z(G) = 1$.

As $C^\circ(i) < G$, minimality of G yields

$$(3) \qquad i \notin C^\circ(i).$$

This is the sort of pathological situation in which a genericity argument becomes available.

We consider the coset $iC^\circ(i)$. A generic element of this coset has the form ia with a a generic element of $C^\circ(i)$ (treating i as a fixed parameter), and in particular there is one and only one conjugate T_a of T in $C^\circ(i)$ centralizing a.

Now $d(a)$ is 2-divisible by Lemma 5.14, and $[d(a) : d(a^2)] \leq 2$, so $d(a) = d(a^2)$ and thus $a \in d(ai)$ and $d(ai) = d(a, i)$. Hence $C(ai) = C(a) \cap C(i)$ and thus T_a is the is also the unique conjugate of T in G commuting with ai.

Let $X = \{aC^\circ(i) : a \text{ is generic in } C^\circ(i) \text{ over } i\}$. Suppose that $x \in X \cap X^g$ for some $g \in G$, so that

$$x = ai = (bi)^g$$

with a and b generic over i in $C^\circ(i)$. Then $T_a = T_{ai}$ and $T_b = T_{bi}$. For notational simplicity suppose that $T_a = T$. Then $T_a = T_b^g$. There is also some $h \in C^\circ(i)$ with $T^h = T_b$. Then $T^{hg} = T$ and $hg \in N(T)$.

Furthermore we have $i^g = i^{hg} \in C(T) \cap iC^\circ(i)$, since $i^g \in d(bi)^g = d(ai)$. Writing $i^g = ij$ with $j \in C^\circ(i)$, we see that $j \in C^\circ(i)$ commutes with T, so by induction $j \in T$. Thus $i \in i\Omega_1(T)$. This is a finite set.

Now suppose the element x lies in $X^{g_1} \cap X^{g_2}$ and the conjugates i^{g_1}, i^{g_2} are equal. Then $g_1 g_2^{-1} \in C(i)$, and therefore $X^{g_1} = X^{g_2}$. So all of this shows that x belongs to only finitely many distinct conjugates of X. Let the number of such conjugates be k.

The type X (over i) does not have $k + 1$ distinct conjugates which intersect. So the same applies, by compactness, to some i-definable subset $\hat{X}$ of $iC^\circ(i)$ containing X. We claim that $\bigcup \hat{X}$ is generic in G.

We have $N(\hat{X}) \leq N(C^\circ(i)\langle i\rangle)$ since $\hat{X}$ generates $C^\circ(i)\langle i\rangle$. So by a Frattini argument we have $N(\hat{X}) \leq C(i)N(T\langle i\rangle)$. Now $N^\circ(T\langle i\rangle) = C^\circ(T\langle i\rangle) \leq C(i)$, so $N^\circ(\hat{X}) = C^\circ(i)$. Now a standard rank computation (first used in the theory of bad groups, and much used recently, cf. [13]), shows that the union $\bigcup \hat{X}^G$ is generic in G.

However for all $x \in \hat{X}$, the group $d(x)$ is not 2-divisible, while for generic $g \in G$, the group $d(g)$ is 2-divisible, and we arrive at a contradiction. □

We tidy this up a little more.

Corollary 5.16 *Let G be a connected group of finite Morley rank and odd type, and let T be a maximal 2-torus of G. Then T contains all the 2-elements in $C(T)$.*

Proof Supposing on the contrary that t is a 2-element in $C(T)$ outside

T, we may take t to have order 2 modulo T. Then taking $t_0 \in T$ with $t_0^2 = t^2$ we have $tt_0^{-1} \in T$ by the previous proposition. □

5.4 Primitive simple groups of odd type

Now we come back to permutation groups. We are interested in working with the action of a maximal 2-torus in a definably primitive simple permutation group of odd type.

Lemma 5.17 *Let (G, Ω) be a definably primitive permutation group of finite Morley rank with G of odd type, and let $r = \mathrm{rk}(\Omega)$. Then for any k there is an $N \leq kr + 1$ such that for any sequence*

$$x_1, \ldots, x_N, \ldots, x_{N+k}$$

of independent generic elements of Ω, the point stabilizer $G_{N+k} = G_{x_1,\ldots,x_{N+k}}$ contains a maximal 2-torus of $G_N = G_{x_1,\ldots,x_N}$.

Proof Take an infinite sequence $x_1, x_2, \ldots$ of independent generic elements of Ω. Let G_i be the point stabilizer $G_{x_1,\ldots,x_i}$. Let T_i be the definable hull of a maximal 2-torus of G_i, and let $r_i = \mathrm{rk}(T_i/O(T_i))$.

The sequence r_i is monotonically decreasing and takes on at most $r+1$ values in view of Lemma 3.11. Thus over the interval $[1, \ldots, k(r+1)+1]$ there must be an interval $[N, N+k]$ of length k over which the function is constant. Here $N \leq kr + 1$.

Suppose now that T_1 is the definable hull of a maximal 2-torus of G_{N+k} and that T_2 is the definable hull of a maximal 2-torus of G_N with $T_1 \leq T_2$. As $r_N = r_{N+k}$ we have

$$\mathrm{rk}(T_1/O(T_1)) = \mathrm{rk}(T_2/O(T_2)).$$

Thus $T_2 = T_1 O(T_2)$ and hence the maximal 2-torus of T_2 is contained in T_1, as claimed. □

Proposition 5.18 *Let (G, Ω) be a definably primitive permutation group of finite Morley rank with G connected and of odd type. Then the degree of generic transitivity of G is at most $4r^2 + 5r + 1$ for $r = \mathrm{rk}(\Omega)$.*

Proof Take an infinite sequence $x_1, x_2, \ldots$ of independent generic elements of Ω. Let $k = 4r + 4$. Suppose that G is generically $[k(r+1)+1]$-transitive. Apply the preceding lemma to get a corresponding value of

N. Let H be the point stabilizer $G_{x_1,\dots,x_N}$ and Let F be the set

$$\{x_{N+1},\dots,x_{N+k}\}.$$

Then H_F is the point stabilizer $H_{x_{N+1},\dots,x_{N+k}}$. As G is generically $(N+k)$-transitive, the setwise stabilizer $H_{\{F\}}$ induces the full symmetric group $\mathrm{Sym}(k)$ on F.

Fix a maximal 2-torus T_0 of H_F and let T be its definable hull. By the Frattini argument there is a subgroup of $H_{\{F\}}$ normalizing T and covering $\mathrm{Sym}(k)$. Modulo the kernel of the action on $T/O(T)$, the group still covers $\mathrm{Sym}(k)$, as its 2-elements are forced to act nontrivially on T_0 by Corollary 5.16.

This then yields $\mathrm{rk}(T/O(T)) \geq \lfloor k/4 \rfloor > r$ by Proposition 4.1, violating our estimate in Lemma 3.11. □

Note that for $r = 2$ this gives a bound on generic transitivity of at most 27 for a definably primitive action. Everything depends on the bound in Proposition 4.1 and there is considerable room for improvement here, particularly when r is low.

The bound in [11] for *sharp* generic transitivity without assuming definable primitivity is 6. In the actual example of $\mathrm{SL}(n)$, with generic sharp $(n+1)$-transitivity, a maximal torus of rank $n-1$ appears in the stabilizer of n independent points with a group $\mathrm{Sym}(n)$ acting on it, so the various estimates do not give up too much up to this point. But after fixing another generic independent point one loses the whole torus in this case. It would also be helpful to understand whether double covers of $\mathrm{Sym}(n)$ can occur in the normalizer of a maximal 2-torus.

5.5 Tying up the loose ends

Finally, we have to assemble a proof of Theorem 2.1 from several pieces developed in the paper.

Notice, first of all, that results of Section 3 reduce Theorem 2.1 to its special case for simple groups G, Theorem 3.1.

To get a bound for $\mathrm{rk}(G)$ in Theorem 3.1, it will suffice, in view of Proposition 2.3, to bound the generic transitivity degree of a simple group G in terms of the rank of the set on which it acts; this intermediate result is formulated as Proposition 3.2.

In a proof of Proposition 3.2, we may assume that G has sufficiently high generic transitivity degree; in particular, G is generically 2-transitive, hence contains involutions and is therefore a group of odd or even type (Lemma 5.8). The bound in the even case is provided by Proposition 5.10; the odd case is dealt with in Proposition 5.18.

This completes the proof of Theorem 2.1.

6 Generic multiple transitivity

6.1 An extremal case

The following conjecture has the potential to involve a good deal of the existing theory of groups of finite Morley rank.

Problem 9 Let G be a connected group of finite Morley rank acting faithfully, definably, transitively and generically $(n+2)$-transitively on a set Ω of Morley rank n. Then the pair (G, Ω) is equivalent to the projective linear group $\mathrm{PGL}_{n+1}(F)$ acting on the projective space $\mathbb{P}^n(F)$ for some algebraically closed field F.

Of course the transitivity hypothesis is superfluous as one may restrict to the relevant orbit in any case. It is here just to keep the notation clean. Note that we aim to show in particular that the action of G is primitive.

This conjecture would give natural bounds for generic multiple transitivity of permutation groups of finite Morley rank. But the main point is that it provides a convenient sandbox for trying out recently developed methods, in particular those used in the ongoing analysis of groups of odd type. A major part of this conjecture is the "affine group" case which might come in to the analysis of the point stabilizer of the group from Problem 9; see Problem 13 below.

One reason that induction may be useful is the following.

Lemma 6.1 *Let (G, Ω) be a transitive and generically n-transitive permutation group of finite Morley rank. Let E be a definable G-invariant equivalence relation on Ω. Then $(G, \Omega/E)$ is generically n-transitive (though not necessarily faithful).*

Proof As G acts transitively on Ω/E, the equivalence classes have constant rank. So generic sets in Ω and Ω/E correspond. □

Taking an inductive approach to Problem 9, then under the stated hypotheses it follows that any proper G-invariant definable equivalence relation E will have finite classes. One can then show that there is a maximal such relation E, definable by $E(\alpha, \beta) \iff G^\circ_\alpha = G^\circ_\beta$. There is a potential issue with the uniform definability of G°_α but that is not a problem for transitive actions. Modulo this maximal E, we have a definably primitive action. So if we have the desired result for definably primitive actions, then the pair $(G/K, \Omega/E)$ is identified, with K the kernel of the action on the quotient. Now K leaves each class of E invariant, so K° acts trivially on Ω, hence K is finite and G is a central extension of G/K. In particular K is contained in a maximal torus T of G. Now T leaves some class C of E invariant, and being connected fixes C pointwise. Hence K has fixed points, and being central acts trivially on Ω. So $K = 1$. So without loss of generality one may suppose (G, Ω) is definably primitive in Problem 9.

Thus we may bring [14] to bear again, and now the group G is connected. This is a very special case, and the classification then reduces to the following possibilities.

(1) The affine case: $G = A \rtimes H$ with H a point stabilizer acting generically $(n+1)$-transitively on an elementary abelian group of rank n.
(2) The socle is a simple group L acting regularly on Ω and $G = L \rtimes H$ with H a point stabilizer acting generically $(n+1)$-transitively on the simple group L of rank n as a group of automorphisms.
(3) The socle is a simple group L acting nonregularly and G lies between L and $\mathrm{Aut}(L)$.
(4) Diagonal action: the socle is the product of two isomorphic simple groups $L_1 \times L_2$. $L_1 \times L_2 \le G \le \mathrm{Aut}(L_1) \times \mathrm{Aut}(L_2)$ and furthermore $G/(L_1 \times L_2) \le \Delta(\mathrm{Out}(L))$ where $L \cong L_1 \cong L_2$ and $\Delta(\mathrm{Out}(L))$ is the subgroup of $\mathrm{Out}(L_1) \times \mathrm{Out}(L_2)$ corresponding to the diagonal subgroup via an isomorphism of L_1 with L_2. In other words, if $\hat{G}$ is the subgroup of $\mathrm{Aut}(L_1) \times \mathrm{Aut}(L_2)$ consisting of pairs (α, β) with $\alpha \equiv \beta \mod \mathrm{Inn}(L)$, we want

$$L_1 \times L_2 \le G \le \hat{G}.$$

The set Ω is defined by specifying the point stabilizer, which is the diagonal subgroup $\Delta(\mathrm{Aut}(L))$ (or its intersection with G).

In the third and fourth cases the condition of finiteness of Morley rank is extremely restrictive. If L is a Chevalley group then any connected subgroup of $\mathrm{Aut}(L)$ containing L will coincide with L itself, while $\mathrm{Aut}(L)$ includes everything induced by a field automorphism.

In Problem 9 our target is type (3) and therefore one of the things one wants to do, presumably, is to remove the other possibilities. However another idea is to use induction and focus on the point stabilizer. Probably one should do both, but the second seems like the main avenue of attack.

Let us examine the basic case of type (4), where $G = L \times L$ and $L_\alpha = \Delta(L)$. This is equivalent to the natural action of $L^{\mathrm{op}} \times L$ on L on the left and right, which as an action of $L \times L$ becomes $(g, h).a = g^{-1}ah$. Thus $G_1 = \Delta(L)$ and the action of G_1 on L is equivalent to the action of L on L by conjugation.

The general case of (4) is similar. The set Ω can be identified with $\Delta(L)\backslash L \times L$ which has as representatives either copy of L, and after this identification G_1 becomes a subgroup of $\mathrm{Aut}(L)$ containing L and acting on Ω as it acts on L.

Now in both cases (2) and (4) there is a definable simple normal subgroup L acting regularly on Ω, on which the point stabilizer G_α acts as a group of automorphisms. Thus in these cases $\mathrm{rk}(L) = n$ and G_α acts generically $(n+1)$-transitively on L. By the proof of Lemma 3.5 if $a_1, \ldots, a_n$ are n independent generic elements of L, then $d(a_1, \ldots, a_n) = L$ unless $\mathrm{rk}(d(a_1, \ldots, a_\ell)) \leq \ell - 1$ for all $\ell \leq n$. In particular the generic element of L has finite order.

Problem 10 Let L be a connected $\aleph_0$-saturated group of finite Morley rank such that for two independent generic elements a, b we have $\mathrm{rk}(d(a, b)) \leq 1$. Show that L is nilpotent of bounded exponent.

This, or something similar, would get rid of these two cases for our present problem.

Alternatively, as we will see later, in many such cases one can force the socle to act generically doubly transitively and in particular get an involution into L, in which case it is definitely of unbounded exponent, as it is not of degenerate type.

One may also use Lemma 3.11 to study the point stabilizer $M = G_\alpha$ on an appropriate definable quotient Ω/E. Here one hopes to arrive at a configuration where modulo the kernel K of the action, M/K contains a subgroup $\mathrm{PGL}_n(F)$, while K has Prüfer rank at most 1. This configuration deserves close study. In particular, one should try to deal with the following.

Problem 11 If a simple algebraic group L acts definably on a group K of degenerate type, then $[K, L]$ is nilpotent.

There is a similar conjecture for actions of the definable hull of a 2-torus on a degenerate type group, which would give the solution to this problem as a special case. But perhaps this version is more accessible.

6.2 The affine case

For the affine case, we want the following.

Problem 12 Let G be a connected group of finite Morley rank acting faithfully, definably, and generically t-transitively on an abelian group V of Morley rank n. Then $t \leq n$.

Later, with an eye on the point stabilizer, we may also want the extremal affine case treated fully. Whether or not it enters directly into the treatment of Problem 9, it is of independent interest.

Problem 13 Let G be a connected group of finite Morley rank acting faithfully, definably, and generically n-transitively on a connected abelian group V of Morley rank n. Then V has a structure of a n-dimensional vector space over an algebraically closed field F of Morley rank 1, and G is $\mathrm{GL}_n(F)$ in its natural action on F^n.

This problem connects closely to the structure theory for groups of finite Morley rank. Note that Problem 12 is a special case of Problem 13.

In either of these problems A should be G-minimal, if we are in an inductive setting. We have only to consider the possibility of a nontrivial G-invariant finite subgroup A_0, in which case G fixes A_0 pointwise. Passing to A/A_0 we either have a contradiction or an identification of G/K acting on A/A_0 with $\mathrm{GL}_n(F)$ in its natural action, where $K = C_G(A_0)$. As usual K° acts trivially A and K is finite. As G is connected we find $K = 1$. So now we have

$$1 \to A_0 \to A \to A/A_0 \to 1$$

a sequence of G-modules with A_0 finite and trivial, and A/A_0 natural. The associated 2-cocycle with values in A_0 is "generically constant" in each variable, the extension so should split definably, another point to be checked. As A is connected we have finally $A_0 = 1$.

In particular A is either torsion free or an elementary abelian p-group for some p. In the context of Problem 12 only the elementary abelian case is possible, and in the context of Problem 13 one may have the torsion free case, with each subgroup of the form $d(a)$ of rank 1. But in the latter case one has also that the definable closure of k independent generic elements has rank k and that its setwise stabilizer acts generically k-transitively. This seems like enough to complete the analysis of the torsion free case in the full generality of Problem 13, by direct induction (perhaps by characterizing the underlying projective geometry, or by more group theoretic methods). Some particular arguments of a more concrete character may also be needed in low rank cases.

Now return to Problem 12 in the case in which A is an elementary abelian p-group.

Consider any nontrivial definable G-invariant equivalence relation E such that a generic element of A belongs to an infinite class. Then G acts generically t-transitively on the quotient of the generic orbit modulo E and if $t > n$ this forces the ranks of the generic classes to be 1, by induction on n, and then the action on the quotient and the induced group G/K can be identified with $(\mathrm{PGL}(n-1, F), \mathbb{P}^{n-1}(F))$. In particular the setwise stabilizer of a single generic E-class C is its pointwise stabilizer in this representation, and acts on the class C, which however has rank 1. It follows that the whole class C has a nontrivial pointwise stabilizer G_C in G. Let $\hat{C}$ be the fixed point set of G_C. This is a proper subgroup of A. One can recover $\hat{C}^\circ$ from its generic element. So this gives rise to another G-invariant equivalence relation on the generic orbit in A, again with classes of rank 1, but now each class is generic in the group it generates. So we cover A by a collection of G-invariant subgroups with finite intersections. Actually the intersection of each subgroup with the union of its distinct conjugates is finite and G-invariant so by G-minimality these intersections are trivial.

Now it seems the kernel K of the quotient action is acting as an abelian subgroup on each class C (group $C \cup \{0\}$) and hence is itself abelian. And also a good torus. So central in G. Hence semi-regular. So if this is nontrivial then we have just put a linear structure on A and that is the end of generic $(n+1)$-transitivity. So K is finite and G is a covering group of $\mathrm{PGL}(n-1, F)$. But this acts generically sharply n-transitively on the quotient so after fixing n generic independent points in A we are out of group elements.

All of which says that as one expects any G-invariant equivalence relation has generically finite classes. And for such a class C and $a \in C$, C is contained in the fixed point set for G_a°, which is a finite subgroup of A $(a \neq 0)$. So we have again a primitive quotient of (G, A) or rather (G, O) with O the generic orbit, and there is at worst a finite kernel. So

G is itself subject to MPOSA. An infinite abelian normal subgroup in G would linearize A again and lead to a contradiction, so the socle of G is again simple or a product of two simple groups; actually quasisimple allowing for the finite kernel.

Two simple groups will lead again to linearization and a contradiction so the socle is a single simple subgroup L. As L acts transitively on the primitive quotient of the generic orbit it also acts transitively on that orbit.

Suppose A contains a nontrivial infinite definable proper subgroup. Then fixing a minimal one A_0, A is more or less a direct sum of conjugates of A_0 by finitely many elements of L. Then the generic element of A is a sum of independent generic elements in these conjugates (which can all be identified with A_0). So the generic $(n+1)$-transitivity of G on A gives the same thing on A_0, in lower rank, and a contradiction. So A is minimal in this strong sense.

Now G contains involutions since it is generically doubly transitive. An involution has finite centralizer on A so inverts A. But we have a 2-element interchanging generic independent elements so an involution swaps them or fixes them, a contradiction.

So it seems that in Problem 9 our original group G must have a simple socle L acting nonregularly.

If this socle is of even type then it is a Chevalley group, and $G = L$, and one is in the algebraic case, more or less. Actually as the language is enriched there are potentially more permutation representations than in the algebraic category. For the truly algebraic case generically multiply transitive actions of simple groups (also reductive groups) have been studied by Popov [19] by a method which is limited to characteristic zero for technical reasons. One does not get beyond generic 4-transitivity except in the case of the series A_ℓ, as anticipated.

Problem 14 Extend Popov's work to all characteristics and to the finite Morley rank permutation group category in which the groups are Chevalley groups, or products of Chevalley groups and tori.

In our category the base field can vary from group to group in a direct product, and it is not clear what a "torus" actually is, but good tori would be a natural candidate for the role. In any case it is the base case of Chevalley groups which is critical here.

This approach also leaves low values of n to deal with. Passing through this kind of classification is cumbersome for low values of n and one should look for more intrinsic methods.

The basis of the induction is discussed again, more generally, in Problems 15 and 16 formulated below.

If the even type case is eliminated then while G is of odd type, the socle of G may, in principle, be of either degenerate or odd type. It is not clear how to limit things further on an a priori basis, so it would seem to be high time to take up the point stabilizer, which also has a generically highly transitive action.

Note incidentally that we may assume

The point stabilizer is connected.

This amounts to replacing Ω by a finite cover where the point stabilizer is replaced by its connected component.

If one succeeds in identifying the point stabilizer (and the ambient group) after this change, then all that remains is to check that it is self-normalizing in the larger group.

6.3 The point stabilizer

In the context of Problem 9 it would be nice to show that the point stabilizer has a faithful primitive permutation representation, and to argue that it falls in the affine class, then identify the affine action explicitly as the action of $\mathrm{GL}(n)$ on its natural module. After which one would still have to identify the original group, and the whole of the point stabilizer. However it is not so clear how to extract an appropriate representation from the data.

Let us consider the action of the point stabilizer G_α on Ω. This may be definably imprimitive. In that case one gets equivalence classes of rank 1 and a quotient of G_α to which induction applies. The connected component of the kernel operates on a collection of rank 1 sets and is therefore either solvable or has factors of type $\mathrm{PSL}(2)$. So one has, in this case, a great deal of information about G_α and one may hope to disentangle the situation from this point.

On the other hand the action of G_α on its generic orbit may be definably primitive, and while this again restricts the structure of G_α, it is not so strong, and needs to be driven toward an eventual contradiction. In particular G_α may have simple socle, in which case we are forced again to deal with simple groups having an unclear structure.

However, if L is the simple socle of G, then L_α contains the socle of G_α and by primitivity acts transitively on the generic orbit of G_α. So in this case at least L acts generically doubly transitively and in particular contains involutions, so is of odd type.

6.4 Small affine groups

Returning now to the affine case, we comment on the case of low rank, which would seem to require separate treatment.

Problem 15 Let G be a connected group of finite Morley rank acting faithfully and definably on an abelian group V of Morley rank 2. Then either G is solvable, or V has a structure of a 2-dimensional vector space over an algebraically closed field F and G is one of the groups $\mathrm{SL}_2(F)$ and $\mathrm{GL}_2(F)$ in their natural action on $V = F^2$.

Problem 15 is believed to be solved, though all such claims must be checked.

Problem 16 Let G be a connected group of finite Morley rank acting faithfully, definably and irreducibly on an abelian group V of Morley rank 3. Then one of the following holds:

- V is of finite exponent $p > 2$ and G is a simple p-group.
- V has a structure of a 3-dimensional vector space over an algebraically closed field F and $G = ZL$ where Z is a subgroup of the group of scalar matrices and L is one of the groups $\mathrm{PSL}_2(F)$ (in its irreducible representation as a 3-dimensional orthogonal group) or $\mathrm{SL}_3(F)$ (in its natural action on F^3).

For the treatment of Problem 16, one would have to collate all known information on "small" groups of odd type, including Altseimer's work on characterization of PSL_3 by centralizers of involutions. When G happens to be $\mathrm{PSL}_2(F)$, turning V into a vector space over F could be a problem (unless this case is already covered by Meierfrankenfeld [16]—but we have not checked the details).

The $\mathrm{PSL}_2(F)$ case of Problem 16 is likely to come out of the configuration which appears to be covered by the recent theorem by Adrien Deloro [10] on minimal simple groups of Prüfer rank 1.

Fact 6.2 Let G be a minimal simple group of odd type and Prüfer rank 1. Assume that the centralizer of a toric involution is not a Borel subgroup in G. Then $G \simeq \mathrm{PSL}_2(F)$ for an algebraically closed field F.

Eric Jaligot has informed us that the proof of Deloro's theorem as such uses the solvability assumption only for "local" subgroups in G: that, subgroups of the form $N_G^\circ(A)$ for an abelian subgroup $A < G$. Exactly this condition, solvability of local subgroups, will naturally appear in the treatment of Problem 16. It would be nice if someone would undertake the task of checking that all papers on minimal simple groups actually use only this weaker assumption; this would make Fact 6.2 immediately applicable in the analysis of Problem 16.

The following fact could be useful in proofs of Problems 15 and 16. (Again, the proof of this "fact" is being written up.)

Fact 6.3 Let G be a simple group of finite Morley rank acting faithfully and definably on an elementary abelian p-group V. If G contains no infinite elementary abelian p-subgroups then the following statements hold.

- Every connected solvable subgroup of G is abelian.
- If C is a Carter subgroup of G then $C_G(C)$ is generous in G.
- Carter subgroups of G are conjugate.
- G contains no involutions.

In the course of the proof of Fact 6.3 the following observation arises; one would expect it to be very useful in the proof of Problem 16.

Fact 6.4 Let H be a connected group of finite Morley rank acting faithfully and definably on an elementary abelian p-group V. Assume that $H = LT$ where $L \lhd H$ and T is an divisible abelian group equal to the definable closure of its 2-torsion. If L contains no infinite elementary abelian p-subgroups then $[L, T] = 1$.

6.5 An inductive step for Problem 13

The inductive step in the treatment of Problem 13 also appears to be natural: we take a generic point $v \in V$ and consider

$$U = C_V(C_G(v));$$

the aim is to show that $\mathrm{rk}(U) = 1$ and that therefore $H = N_G^\circ(U)/U$ acts generically $(n-1)$-transitively on V/U, which would allow us to apply the inductive assumption and conclude that $\bar{H} = H/C_H(V/U)$ is $\mathrm{GL}_{n-1}(F)$.

The next step is to glue together a Curtis-Tits system in a big subgroup G^* of G from a Curtis-Tits system in $[\bar{H}, \bar{H}] \simeq \mathrm{SL}_{n-1}(F)$ and from a Curtis-Tits system in a similar section of an appropriate conjugate of H. We discuss Curtis-Tits systems in the next subsection.

Finally, if a Curtis-Tits system in G^* is constructed and G^* is identified with $\mathrm{GL}_n(F)$ for an algebraically closed field F, the structure of an F-vector space on V is introduced by Meierfrankenfeld's characterization of natural modules for classical groups [16].

6.6 The Curtis-Tits Theorem

We start by quoting a very general form of the Curtis-Tits Theorem in the formulation due to Timmesfeld [20].

Fact 6.5 ([20]) Let Φ be an irreducible spherical root system of Tits rank at least 3, with fundamental system Π and Dynkin diagram Δ. Let G be any group generated by rank one groups $X_r = \langle A_r, A_{-r}\rangle$ for $r \in \Pi$, with unipotent subgroups A_r, A_{-r} satisfying the condition

$$N_{X_r}(A_r) \cap N_{X_r}(A_{-r}) \leq N(X_s)$$

for all $r, s \in \Pi$. Set $X_{rs} = \langle X_r, X_s\rangle$ for $r, s \in \Pi$ distinct, and assume the following all hold.

(1) X_r, X_s commute for r, s not connected in Δ.

(2) If r, s are connected in Δ, then there is a group $\bar{X} = \bar{X}_{rs}$ of Lie type with root system Φ_{rs} (the span of r, s in Φ), which is generated by subgroups $\bar{A}_\alpha$ for $\alpha \in \Phi_{r,s}$, and there is a surjective homomorphism $\phi_{rs} : X_{rs} \to \bar{X}_{rs}$, such that:

   (a) $\phi_{rs}[A_\alpha] = \bar{A}_\alpha$ for $\alpha \in \Phi_{rs}$;

   (b) $\ker \phi_{rs} \leq Z(X_{rs})$;

   (c) If $\bar{X}_{rs}$ is defined over a field of order 2 or 3, or is of the form $\mathrm{PSL}_3(4)$, then $\ker \phi_{rs}$ is a $2'$-group or a $3'$-group respectively.

Then there is a group $\bar{G}$ of Lie type $\mathcal{B}$, with root system Φ and with fundamental system Π, and there is a surjective homomorphism $\sigma : G \to \bar{G}$ mapping the groups $A_{\pm r}$ for $r \in \Pi$ onto the corresponding fundamental root groups and their opposites in $\bar{G}$. Furthermore, $\ker \sigma \leq$

$Z(G) \cap H$, where H is the subgroup generated by the groups $H_r = N_{X_r}(A_r) \cap N_{X_r}(A_{-r})$ for $r \in \Pi$.

The following case is the one which concerns us here.

Proposition 6.6 *Let Φ be an irreducible root system (of spherical type) and rank at least 3, and let Π be a system of fundamental roots for Φ. Let X a group generated by subgroups X_r for $r \in \Pi$, Set $X_{rs} = \langle X_r, X_s \rangle$. Suppose that X_{rs} is a group of Lie type Φ_{rs} over an infinite field, with X_r and X_s corresponding root SL_2-subgroups with respect to some maximal torus of X_{rs}. Then $X/Z(X)$ is isomorphic to a group of Lie type via a map carrying the subgroups X_r to root SL_2-subgroups.*

Since the only simple algebraic groups which we expect to appear in Problems 9 and 13 are groups PSL_n and SL_n over an algebraically closed field, we can specialize the Curtis-Tits Theorem even further.

The following fact is essentially Fact 5.2 of [4].

Proposition 6.7 *Let G be a group of finite Morley rank generated by a family of subgroups K_i, $i = 1, \ldots, n-1$, $n \geqslant 3$. Assume that the following conditions hold:*

- *All K_i are isomorphic to $\mathrm{SL}_2(F)$ for some algebraically closed field F of characteristic $\neq 2$.*
- *$[Z(K_i), Z(K_j)] = 1$ for all $i, j = 1, \ldots, n-1$.*
- *$[K_i, K_j] = 1$ if $|i-j| > 1$.*
- *$\langle K_i, K_j \rangle \simeq \mathrm{SL}_3(F)$ if $|i-j| = 1$.*

Then G is isomorphic to a factor group of the group $\mathrm{SL}_n(F)$ by a (finite) subgroup from the center.

Problem 16, which describes "3-dimensional" groups, could be very useful in the control of groups generated by two "root" SL_2-subgroups.

6.7 Pseudoreflection groups

Finally, Problem 13 may reduce to yet another Problem, also concerned with very familiar objects.

Problem 17 Let G be a connected group acting definably, faithfully and irreducible on a abelian group V (written additively). Assume that G contains a *pseudoreflection subgroup*, that is, an abelian subgroup R such that

(1) $V = [V, R] \oplus C_V(R)$

(2) R acts transitively on the set of non-trivial elements in $[V, R]$.

Then V has a structure of a vector space over an algebraically closed field F such that $[V, R]$ is one-dimensional subspace and R acts on $[V, R]$ as the multiplicative group of F, and $G = \mathrm{GL}(V)$.

It would be interesting to see whether the theory of pseudoreflection groups can be transferred from the case of groups of even type, where it features prominently in the classification theory of simple groups of finite Morley rank and even type, to the context of groups of odd type.

Notice that in the special case when we have $\mathrm{rk}([V, R]) = 1$, which is the only case needed for treating Problem 13, Problem 17 should follow easily from the results of Problems 15 and 16. However, at this point another theme comes into the plot: *classical involutions.*

Indeed, for any conjugate R^g of R, the group $L = L_g = \langle R, R^g \rangle$ centralizes $U = C_V(R) \cap C_V(R^g)$ and acts on V/U. But $\mathrm{rk}(V/U) \leqslant 2$, and it will follow from Problem 15 that $\bar{L} = L/C_L(V/U)$ is isomorphic either to an abelian group $R * R^g$ (central product), or to a soluble group $F^+ \rtimes F^*$, or to $\mathrm{GL}_2(F)$ for an algebraically closed field F, with R and R^g being one-dimensional algebraic tori.

The case in which all of the subgroups $\bar{L}_g$ are solvable, $g \in G$, is very peculiar and should lead to a contradiction with the assumption that G acts on V irreducibly.

Otherwise we choose L such that $\bar{L} \simeq \mathrm{GL}_2(F)$. It should be easy to show that $C_L(V/U) = 1$; after that, if we denote by J the derived subgroup of $L \simeq \mathrm{GL}_2(F)$, we get a remarkable subgroup:

- $J \simeq \mathrm{SL}_2(F)$;
- if z is the involution from $Z(J)$ then $J \lhd C_G(z)$;
- Let J' be a conjugate of J and z' be the involution in $Z(J')$. If z and z' commute then

$$\langle J, J' \rangle \simeq J \times J', J * J' \text{ or } \mathrm{SL}_3(F)$$

(this is where Problem 16 would be useful).

The classical involution analysis in the spirit of Aschbacher and Berkman [4] should hopefully lead to the configuration of the Curtis-Tits Theorem (in the version of Proposition 6.7), and ultimately yield a subgroup $G^* \simeq \mathrm{SL}_n(F)$.

We can repeat a comment made in §6.5: if a Curtis-Tits system in G^* is constructed and G^* is identified with $\mathrm{SL}_n(F)$ for an algebraically closed field F, the structure of a F-vector space on V is introduced by Meierfrankenfeld's characterization of natural modules for classical groups [16]. After that the identification of G with $\mathrm{GL}(V)$ is likely to be very straightforward.

Notice also that Problem 17 could possibly provide a very efficient way to develop the Inductive Step in Problem 13: if, in the notation of §6.5, we can indeed conclude that $H/C_H(V/U)$ is $\mathrm{GL}_{n-1}(F)$, then a pseudoreflection subgroup from $\mathrm{GL}_{n-1}(F)$ in its action on V/U is likely to induce a pseudoreflection subgroup action on V.

6.8 One more problem

The following arises in conjunction with attempts to analyze intersections of Carter subgroups in groups of degenerate type, which is a major focus of interest in the general theory, and may not be entirely out of place here.

Problem 18 Let A be a connected abelian group of Morley rank n, and let $(A_i : i \in I)$ be a uniformly definable family of pairwise distinct connected definable subgroups of rank k in A. Show that $\mathrm{rk}(I) \le \binom{n}{k}$.

The bound is achieved when A is a vector space over a field of rank 1. The case $n = 2$ and $k = 1$ is of particular interest.

7 Problem list

We list here the problems that turned up along the way.

Problem 1 (Introduction) Bound (tightly, if possible) the rank of a definably primitive permutation group of finite Morley rank in terms of the rank of the set on which it acts.

Problem 2 (§1) If (G, Ω) is a permutation group and Ω is stable in the induced language, does it follow that G is stable? Does this hold at least when Ω has finite Morley rank?

Problem 3 (§1) Find all the generically sharply n-transitive actions of algebraic groups over algebraically closed fields, for $n \ge 2$.

Problem 4 (§1) Suppose that (G, Ω) is a virtually definably primitive permutation group of finite Morley rank with which is not a finite cover of a definably primitive permutation group. Show that G is a Chevalley group of positive characteristic, and the point stabilizer is contained in $G(\mathbb{F}_q)$ for some finite field $\mathbb{F}_q$.

Problem 5 (§1) Is there an O'Nan-Scott-Aschbacher analysis of generically 2-transitive groups which are not necessarily definably primitive? Are all such groups essentially products of generically n-transitive primitive groups (or generically n'-transitive groups, with n' not much smaller than n)?

Problem 6 (§2) Find good bounds on ρ, where $\rho(r)$ is the maximum rank of a virtually definably primitive permutation group (G, Ω) of finite Morley rank, with $\mathrm{rk}(\Omega) = r$.

Problem 7 (§2) Find good bounds on τ, where $\tau(r)$ is the maximum degree of generic transitivity associated to a virtually definably primitive permutation group (G, Ω) of finite Morley rank, with $\mathrm{rk}(\Omega) = r$.

Problem 8 (§5) Let Σ be a finite group. Find lower bounds for each of the following.

(1) The minimal rank of a connected solvable group of finite Morley rank which affords a faithful representation of Σ.
(2) The minimal rank of a connected solvable group of finite Morley rank which affords a faithful representation of a central extension of Σ.
(3) The minimal rank of a connected solvable group of finite Morley rank which affords a faithful representation of a group $\hat{\Sigma}$ which covers Σ, i.e. maps homomorphically onto Σ.

Problem 9 (§6) Let G be a connected group of finite Morley rank acting faithfully, definably, transitively and generically $(n+2)$-transitively on a set Ω of Morley rank n. Then the pair (G, Ω) is equivalent to the projective linear group $\mathrm{PGL}_{n+1}(F)$ acting on the projective space $\mathbb{P}^n(F)$ for some algebraically closed field F.

Problem 10 (§6) Let L be a connected $\aleph_0$-saturated group of finite Morley rank such that for two independent generic elements a, b we have $\mathrm{rk}(d(a, b)) \leq 1$. Show that L is nilpotent of bounded exponent.

Problem 11 (§6) If a simple algebraic group L acts definably on a group K of degenerate type, then $[K, L]$ is nilpotent.

Problem 12 (§6) Let G be a connected group of finite Morley rank acting faithfully, definably, and generically t-transitively on an abelian group V of Morley rank n. Then $t \leq n$.

Problem 13 (§6) Let G be a connected group of finite Morley rank acting faithfully, definably, and generically n-transitively on a connected abelian group V of Morley rank n. Then V has a structure of a n-dimensional vector space over an algebraically closed field F of Morley rank 1, and G is $\mathrm{GL}_n(F)$ in its natural action on F^n.

Problem 14 (§6) Extend Popov's work to all characteristics and to the finite Morley rank permutation group category in which the groups are Chevalley groups, or products of Chevalley groups and tori.

Problem 15 (§6) Let G be a connected group of finite Morley rank acting faithfully and definably on an abelian group V of Morley rank 2. Then either G is solvable, or V has a structure of a 2-dimensional vector space over an algebraically closed field F and G is one of the groups $\mathrm{SL}_2(F)$ and $\mathrm{GL}_2(F)$ in their natural representations.

Problem 16 (§6) Let G be a connected group of finite Morley rank acting faithfully, definably and irreducibly on an abelian group V of Morley rank 3. Then one of the following holds:

- V is of finite exponent $p > 2$ and G is a simple p-group.
- V has a structure of a 3-dimensional vector space over an algebraically closed field F and $G = ZL$ where Z is a subgroup of the group of scalar matrices and L is one of the groups $\mathrm{PSL}_2(F)$ (in its irreducible representation as a 3-dimensional orthogonal group) or $\mathrm{SL}_3(F)$ (in its natural action on F^3).

Problem 17 (§6) Let G be a connected group acting definably, faithfully and irreducible on a abelian group V (written additively). Assume that G contains a *pseudoreflection subgroup*, that is, an abelian subgroup R such that $V = [V, R] \oplus C_V(R)$ and R acts transitively on the set of non-trivial elements in $[V, R]$. Then V has a structure of a vector space over an algebraically closed field F such that $[V, R]$ is one-dimensional subspace and R acts on $[V, R]$ as the multiplicative group of F, and $G = \mathrm{GL}(V)$.

Problem 18 (§6) Let A be a connected abelian group of Morley rank n, and let $(A_i : i \in I)$ be a uniformly definable family of pairwise distinct connected definable subgroups of rank k in A. Show that $\mathrm{rk}(I) \leq \binom{n}{k}$.

References

[1] T. Altınel, A. Borovik, and G. Cherlin, *Simple Groups of Finite Morley Rank*, Amer. Math. Soc. xvi+554 pp, to appear.
[2] T. Altınel and G. Cherlin, On groups of finite Morley rank of even type, *J. Algebra*, 264 (2003), 155-185.
[3] T. Altınel and G. Cherlin, Simple L^*-groups of even type with strongly embedded subgroups, *J. Algebra* 272 (2004), 95–127.
[4] A. Berkman, The Classical Involution Theorem for groups of finite Morley rank, *J. Algebra* 243 (2001), 361–384.
[5] A. Borovik, J. Burdges, and G. Cherlin, Involutions in groups of finite Morley rank of degenerate type, *Selecta Mathematica* 13 (2007), 1–22.
[6] A. Borovik, J. Burdges, and G. Cherlin, Simple groups of finite Morley rank of unipotent type, in *Algebra, Logic, Set Theory. Festschrift für Ulrich Felgner zum 65. Geburtstag* (B. Löwe, ed.), College Publications, 2007, pp. 47–61.
[7] A. Borovik and A. Nesin, *Groups of Finite Morley Rank*, Oxford Science Publications, The Clarendon Press, Oxford University Press, New York, 1994.
[8] S. Buechler, *Essential Stability Theory*, Perspectives in Mathematical Logic, Springer, 1996.
[9] G. Cherlin, Good tori in groups of finite Morley rank, *J. Group Theory*, 8 (2005), 613-621.
[10] A. Deloro, Groupes simples minimaux algébriques de type impair, to appear in *J. of Algebra*.
[11] U. Gropp, There is no sharp transitivity on q^6 when q is a type of Morley rank 2, *J. Symbolic Logic* 57 (1992), 1198–1212.
[12] M. Hall, Jr., On a theorem of Jordan, *Pacific J. Mathematics* 4 (1954), 219-226.
[13] E. Jaligot, Generix never gives up, *J. Symbolic Logic* 71 (2006), 599–610.
[14] H. D. Macpherson and A. Pillay, Primitive permutation groups of finite Morley rank, *Proc. London Math. Soc.* 70 (1995), 481–504.
[15] R. Kaye and H. D. Macpherson, *Automorphisms of First-Order Structures*, Oxford Science Publications, The Clarendon Press, Oxford University, 1994.
[16] U. Meierfrankenfeld, A characterization of the natural module for classical groups, Preprint.
[17] B. Poizat, *Groupes Stables*, Nur Al-Mantiq Wal-Ma'rifah, Villeurbanne, France, 1987; English translation 2001, AMS.
[18] B. Poizat, Modestes remarques à propos d'une conséquence inattendue d'un résultat surprenant de Monsieur Frank Olaf Wagner, *J. Symbolic Logic* 66 (2001), 1637–1646.
[19] V. Popov, Generically multiply transitive algebraic group actions, Preprint, 2005.

[20] F. Timmesfeld, *Abstract Root Subgroups and Simple Groups of Lie-Type*, Monographs in Mathematics 95, Birkhäuser, Basel, 2001, xiii+389 pp.

[21] F. Wagner, Fields of finite Morley rank, *J. Symbolic Logic*, 66 (2001), 703–706.

A survey of asymptotic classes and measurable structures

Richard Elwes
University of Leeds

Dugald Macpherson
University of Leeds

1 Introduction

In this article we survey a body of results about classes of finite first order structures in which definable sets have a rather uniform asymptotic behaviour. Non-principal ultraproducts of such classes may have unstable theory, but will be supersimple of finite rank. In addition, they inherit a definable *measure* on definable sets.

The starting point for this work was the following result of [11] on definability in finite fields.

Theorem 1.1 ([11]) *Let $\varphi(\bar{x}, \bar{y})$ be a formula in the language L_{rings} for rings, with $\bar{x} = (x_1, \ldots, x_n)$ and $\bar{y} = (y_1, \ldots, y_m)$. Then there is a positive constant C, and a finite set D of pairs (d, μ) with $d \in \{0, \ldots, n\}$ and μ a non-negative rational number, such that for each finite field $\mathbb{F}_q$ and $\bar{a} \in \mathbb{F}_q^m$,*

$$\left||\varphi(\mathbb{F}_q^n, \bar{a})| - \mu q^d\right| \leq C q^{d-(1/2)} \qquad (*)$$

for some $(d, \mu) \in D$.

Furthermore, for each $(d, \mu) \in D$, there is a formula $\varphi_{(d,\mu)}(\bar{x})$ which defines in each finite field $\mathbb{F}_q$ the set of tuples $\bar{a}$ such that $(*)$ holds.

This result rests on the Lang-Weil estimates for the number of $\mathbb{F}$-rational points of an absolutely irreducible variety defined over the finite field $\mathbb{F}$. The proof uses partial quantifier elimination for pseudofinite fields, derivable from the paper of Ax [2] which introduced pseudofinite fields: any formula $\varphi(\bar{x})$ is a boolean combination of formulas $\exists y(g(\bar{x}, y) = 0)$, where $g \in \mathbb{Z}[\bar{X}, Y]$. (In fact, by adding a set C of constants to the language, the authors arrange that φ is equivalent to a *conjunction* of similar formulas, rather than an arbitrary boolean combination; there is an alternative effective proof in [20], based on Galois stratification.) The reason why μ may be a fraction (rather than an integer as in Lang-Weil), is that the existential quantifier ranges over a finite set. Oversimplifying considerably, the set defined by $\exists y(g(\bar{x}, y) = 0)$ is, for some k, the image of a k-to-1 projection map of the variety defined by $g(\bar{x}, y)$; thus, its measure is $\frac{1}{k}$ times that of $\{(\bar{x}, y) : g(\bar{x}, y) = 0\}$.

As a short application of Theorem 1.1, noted in [11], it follows that the finite field $\mathbb{F}_q$ is not uniformly definable in $\mathbb{F}_{q^2}$. Theorem 1.1 also yields that pseudofinite fields are supersimple of rank 1, with an associated definable and finitely additive notion of measure on the definable sets.

The present article is a survey of recent work stimulated by Theorem 1.1. The idea, initiated in [43], is to consider classes of finite structures in which definable sets satisfy the same kind of asymptotic behaviour as for finite fields. In the initial work of [43], only a 1-dimensional version was considered, and the error terms were as in Theorem 1.1. More recently, Elwes [18] has developed a theory of higher dimensional asymptotic classes where the error terms are weaker – see Definition 2.1 below.

There is also a notion of *measurable structure*: a supersimple structure of finite SU-rank equipped with the kind of measure function on definable sets which exists, by virtue of Theorem 1.1, in pseudofinite fields. Any ultraproduct of members of an asymptotic class will be measurable (Theorem 3.9 below) but there are also measurable structures which do not even have the finite model property, so cannot arise from asymptotic classes; see Theorem 3.12. Familiar supersimple structures such as the random graph, smoothly approximable structures, and pseudofinite fields, are all measurable.

One of the main results in this area is a theorem of Ryten [51], stemming from earlier work of Ryten and Tomašić [52], that for any natural number d, the collection of all finite simple groups of Lie rank at most d forms an asymptotic class (Theorem 6.1 below). This opens the possibility of variations of the Algebraicity Conjecture of Cherlin and Zilber that any simple group of finite Morley rank is an algebraic group. Namely, the following seems reasonable, and, unlike the Algebraicity Conjecture, incorporates the classes of *twisted* finite simple groups.

Conjecture 1.2 If G is a simple group with measurable theory, then G is a Chevalley group (possibly of twisted type) over a pseudofinite field.

In Section 2 below we give the current definition of *asymptotic class*, and some examples. Measure is introduced in Section 3, again with examples and discussion of the connection with asymptotic classes, and with Hrushovski's notion of *unimodular* theory. Smoothly approximable structures provide an important class of examples, and we describe in Section 4 how these fit into the framework. In Section 5, we survey results from [51] and [52] on measure in ACFA, and an extension of Theorem 1.1 to finite difference fields. This yields the above result on finite simple groups of fixed Lie rank, sketched in Section 6. We turn in Section 7 to measurable groups, and asymptotic classes of groups, of low dimension. The main results here, due to Elwes and Ryten, are that any 2-dimensional asymptotic class of groups consists of groups with a uniformly definable soluble subgroup of bounded index (Theorem 7.5), and an analogue of Hrushovski's result on groups of finite Morley rank in a stable theory which act transitively on a strongly minimal set. The paper concludes with a section on open questions.

If $\mathcal{C}$ is a class of finite structures in a language L, then the *asymptotic theory* of $\mathcal{C}$ is the collection of all sentences which hold in all but finitely many members of $\mathcal{C}$. Equivalently, it consists of those sentences which hold of any non-principal ultraproduct of members of $\mathcal{C}$. Notation is introduced locally, but we try to stick to the convention that for a formula $\varphi(\bar{x}, \bar{y})$, $\bar{x} = (x_1, \ldots, x_n)$, $y = (y_1, \ldots, y_m)$, with y the parameter variables. The algebraic closure of a field K is denoted $\tilde{K}$. We write $A \mathop{\smile\!\!\!\!\!\!|}_C B$ to mean that the sets A and B are independent over C (in the sense of model-theoretic non-forking).

Though this material is close to stability and simplicity, our intention is that little knowledge of simplicity theory is needed to follow this paper; the reader can refer to [48] and [60]. For details of the ranks mentioned (S_1-rank, D-rank, SU-rank), see [35]. For background on the model theory of finite and pseudofinite fields, we refer to the original paper [2], the survey [14], or to [21]. For background on ACFA, see [12]. For the structure of the finite simple groups of Lie type, see [10].

Acknowledgement: We thank the referee for a meticulous and helpful report.

2 Asymptotic classes

Definition 2.1 (Elwes, [18]) Let $N \in \mathbb{N}$, and let $\mathcal{C}$ be a class of finite L-structures, where L is a finite language. Then we say that $\mathcal{C}$ is an **N-dimensional asymptotic class** if the following hold.

(i) For every L-formula $\varphi(\bar{x},\bar{y})$ where $l(\bar{x})=n$ and $l(\bar{y})=m$, there is a finite set of pairs $D\subseteq(\{0,\dots,Nn\}\times\mathbb{R}^{>0})\cup\{(0,0)\}$ and for each $(d,\mu)\in D$ a collection $\Phi_{(d,\mu)}$ of pairs of the form $(M,\bar{a})$ where $M\in\mathcal{C}$ and $\bar{a}\in M^m$, so that $\{\Phi_{(d,\mu)}:(d,\mu)\in D\}$ is a partition of $\{(M,\bar{a}): M\in\mathcal{C},\bar{a}\in M^m\}$, and

$$\left||\varphi(M^n,\bar{a})|-\mu|M|^{\frac{d}{N}}\right|=o(|M|^{\frac{d}{N}})$$

as $|M|\longrightarrow\infty$ and $(M,\bar{a})\in\Phi_{(d,\mu)}$.

(ii) Each $\Phi_{(d,\mu)}$ is $\emptyset$-definable, that is, $\{\bar{a}\in M^m:(M,\bar{a})\in\Phi_{(d,\mu)}\}$ is uniformly $\emptyset$-definable across $\mathcal{C}$.

We may write D_φ for D, and will call $\{\Phi_{(d,\mu)}:(d,\mu)\in D\}$ a *(definable) asymptotic partition.* We write $h(\varphi(M^n,\bar{a})):=(\mathrm{Dim}(\varphi(M^n,\bar{a})),$ $\mathrm{Meas}(\varphi(M^n,\bar{a}))):=(d,\mu)$ where $(M,\bar{a})\in\Phi_{(d,\mu)}$, except that if $d=\mu=0$ we work with the convention that $\mathrm{Dim}(\varphi(M^n,\bar{a}))=-1$. We call $\mathcal{C}$ a *weak asymptotic class* when $\mathcal{C}$ satisfies the asymptotic criteria (i) for all φ, but the $\Phi_{(d,\mu)}$ are not assumed to be definable.

Remark 2.2 1. In this context the o-notation in (i) means the following: for every $\varepsilon>0$ there is $Q\in\mathbb{N}$ such that for all $M\in\mathcal{C}$ with $|M|>Q$ and all $\bar{a}\in M^m$ where $(M,\bar{a})\in\Phi_{d,\mu}$, we have

$$\left||\varphi(M^n,\bar{a})|-\mu|M|^{\frac{d}{N}}\right|<\varepsilon|M|^{\frac{d}{N}}.$$

2. For every member of an N-dimensional asymptotic class, the universe of the structure is viewed as being N-dimensional.

3. Unlike the presentation in [43] and [18], the definition covers all formulas $\varphi(\bar{x},\bar{y})$, not just those in which $\bar{x}$ is a singleton. It is a theorem (Theorem 2.3 below) that this is equivalent to the same condition just for formulas $\varphi(x,\bar{y})$.

4. The focus of [43] was on the case when $N=1$ (the case for finite fields).

5. The general theory below works equally well with tighter error terms: in 1 above, replace $=o(|M|^{\frac{d}{N}})$ by $<C|M|^{\frac{d}{N}-\frac{1}{2N}}$, where C is a constant depending on φ. Many of the known examples of asymptotic classes satisfy this constraint. However it is not clear that envelopes of smoothly approximable structures satisfy it. For example, if p,q are distinct primes, and M is the disjoint union of two infinite-dimensional vector spaces, one over $\mathbb{F}_p$, the other over $\mathbb{F}_q$, is M approximated by an asymptotic class satisfying the tighter error terms?

6. It is easy to create artificial examples of N-dimensional classes (for any N) in which the μ are irrational, or even transcendental. This

can be done just using sets with a unary predicate. We do not know of natural examples with transcendental μ.

7. It is immediate that any reduct of an asymptotic class is a weak asymptotic class. However, the definability clause (ii) can be lost under reducts. Elwes [18, Section 2.2] has shown that if $\mathcal{C}$ is an asymptotic class, and $\mathcal{D}$ is a class of finite structures uniformly interpretable in $\mathcal{C}$, then $\mathcal{D}$ is a weak asymptotic class. If $\mathcal{C}$ and $\mathcal{D}$ are uniformly bi-interpretable (without parameters), then $\mathcal{D}$ will also be an asymptotic class. This is relevant to Section 6 below (bi-interpretations between classes of simple groups and fields or difference fields).

The following theorem gives a more easily recognised criterion for being an asymptotic class. By ensuring that the condition is really one on one-variable definable sets, it gives an analogy to o-minimality, and related minimality notions.

Theorem 2.3 (Lemma 2.1.2 of [18]) *Suppose that $\mathcal{C}$ is a class of finite structures which satisfies Definition 2.1 (clauses (i) and (ii)) for $n = 1$, i.e. for definable sets in 1 variable. Then $\mathcal{C}$ is an N-dimensional asymptotic class.*

Sketch of the proof. The proof is inductive. For a definable subset X of M^{n+1} consider the projection $\pi : M^{n+1} \to M$ to the first variable, and apply the assumption to $\pi(X)$, and the inductive hypothesis to the fibres X_a ($a \in \pi(X)$). The definability clause (ii) is essential.

We draw attention to the following *non-example*. The class of all finite total orders is not an asymptotic class of any dimension; for, as a varies through a finite ordering, the formula $x < a$ defines an arbitrary proportion of the domain. In fact, by Proposition 3.9 (and Corollary 3.7) below, any nonprincipal ultraproduct of an asymptotic class is supersimple, so cannot interpret a partial order with an infinite chain (i.e. cannot have the *strict order property*).

Example 2.4 1. By Theorem 1.1, the collection of all finite fields forms a 1-dimensional asymptotic class. This class, when restricted to fixed characteristic, has an important expansion. For any fixed p, and positive integers m, n with $m \geq 1$, $n > 1$, and $(m, n) = 1$, there is a 1-dimensional asymptotic class $\mathcal{C}_{(m,n,p)}$ of difference fields, namely $\mathcal{C}_{(m,n,p)} = \{(\mathbb{F}_{p^{kn+m}}, \mathrm{Frob}^k) : k > 0\}$ – see Theorem 5.8. The automorphism here is not uniformly definable in the field. The classes $\mathcal{C}_{(m,n,p)}$

are significant for finite simple groups: $\mathcal{C}_{1,2,2}$ is uniformly parameter bi-interpretable with the classes of Suzuki groups ${}^2B_2(2^{2k+1})$ and the Ree groups ${}^2F_4(2^{2k+1})$, and $\mathcal{C}_{(1,2,3)}$ with the class of Ree groups ${}^2G_2(3^{2k+1})$. See Section 6.

2. It is shown in [43, Theorem 3.14] that the collection of all finite cyclic groups is a 1-dimensional asymptotic class. This follows from the partial quantifier elimination ('near model completeness') of Szmielew [55] for abelian groups. The multiplicative groups of finite fields form a subclass.

3. For any odd prime p, the class $\mathcal{C}_p$ of finite extraspecial p-groups of exponent p forms a 1-dimensional asymptotic class (see [43, Proposition 3.11]). Here, a group G is *extraspecial* if $G' = Z(G) = \Phi(G)$ (the Frattini subgroup), and G' is isomorphic to the cyclic group $\mathbb{Z}/p\mathbb{Z}$. An extraspecial group of exponent p is a central product of several copies of the unique non-Abelian group of order p^3 and exponent p, and a finite one will have order p^{2t+1} for some t. The groups in $\mathcal{C}_p$ are bounded-by-abelian (in fact, $(\mathbb{Z}/p\mathbb{Z})$-by-abelian), but not abelian-by-bounded, that is, they do not have an abelian normal subgroup of bounded index. The quotient $G/Z(G)$ carries definably the structure of a vector space over the field $\mathbb{F}_p$, equipped with an alternating bilinear form which comes from the commutator map. An infinite ultraproduct of groups in $\mathcal{C}_p$ will be ω-categorical, smoothly approximable (see Section 4), and supersimple of rank 1, but not stable. See also Theorem 7.3, and the remark after Lemma 7.4, for partial converses.

4. If q is a prime power with $q \equiv 1 \pmod 4$, then there is a graph P_q (known as a *Paley graph*) with vertex set the finite field $\mathbb{F}_q$, with vertices a, b joined if $a - b$ is a square. By [43], the class $\mathcal{C}$ of all Paley graphs is a 1-dimensional asymptotic class. Essentially, the reason is that by a theorem of Bollobás and Thomason [7] (see also [8, Ch.XIII.2]), if U, W are disjoint sets of vertices of P_q with $m := |U \cup W|$, and $v(U, W)$ is the number of vertices of P_q joined to everything in U and to nothing in W, then

$$|v(U,W) - 2^{-m}q| \leq \frac{1}{2}(m - 2 + 2^{-m+1})q^{\frac{1}{2}} + m/2.$$

It follows that an infinite ultraproduct of Paley graphs is elementarily equivalent to the random graph, so has quantifier elimination. This persists in sufficiently large finite Paley graphs, so corresponding asymptotic estimates hold for *all* formulas in one variable, and hence, by Theorem 2.3, for all formulas. The definability clause of Definition 2.1 is easily verified.

There are analogous classes, interpretable in finite fields, associated with other homogeneous structures. For example, if one considers primes $q \equiv 3 \pmod 4$, and defines $a \rightarrow b$ (on $\mathbb{F}_q$) whenever $a - b$ is a square, one obtains the *Paley tournaments*. These approximate the random tournament as above; the analogue of the Bollobás-Thomason result was proved in [22]. For more on this, see [56], and the results of Tomašić [58] mentioned at the end of Section 3.

There is an analogue of the Paley graphs for arity 3 (so a class of 3-hypergraphs), considered in [43, Example 3.6]. For $q \equiv 1 \pmod 4$ one considers a hypergraph on $\mathbb{F}_q$ where $\{a, b, c\}$ is an edge if a, b, c are distinct and $(a - b)(b - c)(a - c)$ is a square. The asymptotic theory of such hypergraphs is that of the generic homogeneous *two-graph* (a 3-hypergraph which is a reduct of the random graph), not the generic 3-hypergraph.

5. By Proposition 3.3.2 of [18], if $\mathcal{M}$ is a smoothly approximable structure, then $\mathcal{M}$ is the union of a chain of 'envelopes' which form an asymptotic class. A basic example is an infinite vector space over a finite field, obtained as a union of infinitely many finite dimensional vector spaces. This example is 1-dimensional but in general such classes will be of higher dimension. See Section 4. As a special case, consider the smoothly approximable structure consisting of a set equipped with an equivalence relation with infinitely many infinite classes. It is smooothly approximated by a chain of structures each of size t^2 (with t varying) equipped with an equivalence relation with t classes all of size t. The latter class of structures is an asymptotic class of dimention 2.

6. Suppose that $\mathcal{C}$ is a class of finite structures such that every infinite ultraproduct of members of $\mathcal{C}$ is strongly minimal. Then $\mathcal{C}$ is a 1-dimensional asymptotic class (Lemma 2.5 of [43]). This is an easy consequence of Theorem 2.3.

In particular, let d be a positive integer, and let $\mathcal{C}_d$ be the collection of all finite graphs of valency d whose automorphism group is transitive on the vertex set. Then $\mathcal{C}_d$ is a 1-dimensional asymptotic class. The reason is that any infinite ultraproduct M of members of $\mathcal{C}_d$ is itself vertex transitive: take the ultraproduct of 2-sorted structures, with a second sort for the automorphism group. It follows that M is strongly minimal, so the last paragraph applies.

In the next section, it will be shown that if $\mathcal{C}$ is an N-dimensional asymptotic class, then any infinite ultraproduct of members of $\mathcal{C}$ is supersimple of rank at most N (and, furthermore, *measurable*). A number of other model-theoretic properties of ultraproducts can be recognised directly from asymptotics. One such is the following. The main issue is to show, by an elementary argument with indiscernibles, that if the asymptotic conditions hold then $\varphi(x, \bar{y})$ is unstable in some ultraproduct.

Proposition 2.5 ([43]) *Let $\mathcal{C}$ be a 1-dimensional asymptotic class. Then some ultraproduct of members is unstable if and only if there is a formula $\varphi(x, \bar{y})$, and for each $k \in \mathbb{N}$ some $M \in \mathcal{C}$ and $\bar{a}_1, \ldots, \bar{a}_k \in M^{\ell(\bar{y})}$ with*

(a) $|\varphi(M, \bar{a}_i)| \geq k$ *for each* $i = 1, \ldots, k$, *and*

(b) $|\varphi(M, \bar{a}_i) \Delta \varphi(M, \bar{a}_j)| \geq k$ *for all distinct* $i, j \in \{1, \ldots, k\}$.

In Chapter 5 of [18], finitary criteria are given for ultraproducts to be ω-categorical, or all to be 1-based.

3 Measurable structures

It was already noted in [11, 4.10, 4.11] that, because of Theorem 1.1, pseudofinite fields are supersimple of S_1-rank 1, and that there is a definable 'measure' on the definable sets. Below, following Section 5 of [43], we generalise this.

Definition 3.1 An infinite L-structure M is *measurable* if there is a function $h : \mathrm{Def}(M) \to \mathbb{N} \times \mathbb{R} \cup \{(0,0)\}$ (we also write $h(X)$ as $(\mathrm{Dim}(X), \mathrm{Meas}(X))$ or $(\mathrm{Dim}, \mathrm{Meas})(X)$) such that the following hold.

(1) For each L-formula $\varphi(\bar{x}, \bar{y})$ there is a finite set $D \subset \mathbb{N} \times \mathbb{R}^{>0} \cup \{(0,0)\}$, so that for all $\bar{a} \in M^m$ we have $h(\varphi(M^n, \bar{a})) \in D$.

(2) If $\varphi(M^n, \bar{a})$ is finite then $h(\varphi(M^n, \bar{a})) = (0, |\varphi(M^n, \bar{a})|)$.

(3) For every L-formula $\varphi(\bar{x}, \bar{y})$ and all $(d, \mu) \in D_\varphi$, the set $\{\bar{a} \in M^m : h(\varphi(M^n, \bar{a})) = (d, \mu)\}$ is $\emptyset$-definable.

(4) (Fubini) Let $X, Y \in \mathrm{Def}(M)$ and $f : X \to Y$ be a definable surjection. Then there are $r \in \omega$ and $(d_1, \mu_1), \ldots, (d_r, \mu_r) \in (\mathbb{N} \times \mathbb{R}^{>0}) \cup \{(0,0)\}$ so that if $Y_i := \{\bar{y} \in Y : h(f^{-1}(\bar{y})) = (d_i, \mu_i)\}$, then $Y = Y_1 \cup \ldots \cup Y_r$ is a partition of Y into non-empty disjoint definable sets. Let $h(Y_i) = (e_i, \nu_i)$ for $i \in \{1, \ldots, r\}$. Also let $c := \mathrm{Max}\{d_1 + e_1, \ldots, d_r + e_r\}$, and suppose (without loss) that this maximum is attained by $d_1 + e_1, \ldots, d_s + e_s$. Then $h(X) = (c, \mu_1\nu_1 + \ldots + \mu_s\nu_s)$.

If $X \in \mathrm{Def}(M)$ and $h(X) = (d, \mu)$, we call d the *dimension* of X and μ the *measure* of X, and h the *measuring function.* We say that a complete theory T is *measurable* if it has a measurable model (see Remark 3.8 (1)).

Example 3.2 The basic motivating example of a measurable structure is a pseudofinite field. Essentially, this was shown in [11]. If $X \subset F^n$ is definable (F a pseudofinite field) then $\mathrm{Dim}(X)$ (which equals its S_1-rank or D-rank) is just the algebraic-geometric dimension of the Zariski closure of X in $\tilde{F}^n$. The measure of any absolutely irreducible variety in F will be 1. Measurability of any pseudofinite field was used in [28], in a new proof of the well-known fact that any almost simple algebraic group has a universal cover. The main point was that, by measure considerations, if F is a pseudofinite field and G_1, G are connected algebraic groups defined over F, and $f : G_1 \to G$ is an isogeny defined over F, then $|\,\mathrm{Ker}(f) \cap G_1(F)| = |G(F) : f(G_1(F))|$.

A key question is whether, conversely, every measurable field is pseudofinite (see Section 8). An algebraically closed field cannot be measurable – see the remark after Proposition 3.15.

Generalising the case of pseudofinite fields, Proposition 3.9 below shows that asymptotic classes yield measurable structures.

Definition 3.1 is slightly different from that in [43], since we do not specify in the definition that M has a supersimple theory. However, it follows from the next few lemmas, in particular Corollary 3.6, that indeed if M is measurable, then $\mathrm{Th}(M)$ is supersimple of finite rank. Similar arguments were communicated to the authors by Ryten, and can also be found in [32]. The computations below are with D-rank, but by [35, Section 6], in a supersimple theory in which D-rank is finite, D-rank, S_1-rank, and SU-rank all agree for formulas. See [35] for definitions of these ranks. In 3.3–3.6 below the sets are taken in an ambient structure M, which is assumed to be measurable.

Lemma 3.3 *Let $n \in \omega$, and $A_1, \ldots, A_n$ be definable sets, where* $\mathrm{Dim}(A_i) = d$ *for each i.*

(i) $\mathrm{Dim}(\bigcup_{i=1}^n A_i) = d$.
(ii) If, in addition we have $\mathrm{Dim}(A_{i_1} \cap A_{i_2}) < d$ *for each distinct i_1, i_2, then*

$$(\mathrm{Dim}, \mathrm{Meas})(\bigcup_{i=1}^{n} A_i) = (d, \sum_{i=1}^{n} \mathrm{Meas}(A_i)).$$

Proof. We proceed by induction on n. For $n = 2$, pick any distinct $a_1, a_2, a_3 \in M$, and define $f : A_1 \cup A_2 \to \{a_1, a_2, a_3\}$ by $f(\bar{x}) := a_1$ if $\bar{x} \in A_1 \backslash A_2$; $f(\bar{x}) := a_2$ if $\bar{x} \in A_2 \backslash A_1$; and $f(\bar{x}) = a_3$ if $\bar{x} \in A_1 \cap A_2$. Then (i) and (ii) are immediate from the Fubini condition.

Suppose now that both statements hold for $n = k - 1$. Then given $A_1, \ldots A_k$, we may first apply the inductive hypothesis to $A_1, \ldots A_{k-1}$, and derive statement (i) by applying the case $n = 2$ to $\bigcup_{i=1}^{k-1} A_i$ and A_k.

Now for (ii), we know by the inductive hypothesis that

$$(\mathrm{Dim}, \mathrm{Meas})(\bigcup_{i=1}^{k-1} A_i) = (d, \sum_{i=1}^{k-1} \mathrm{Meas}(A_i)).$$

Now $A_k \cap \bigcup_{i=1}^{k-1} A_i = \bigcup_{i=1}^{k-1} (A_k \cap A_i)$, and so by (i) we know that $\mathrm{Dim}(A_k \cap \bigcup_{i=1}^{k-1} A_i) < d$. Therefore we may apply the case $n = 2$ to find that $\mathrm{Meas}(\bigcup_{i=1}^{k} A_i) = \mathrm{Meas}(\bigcup_{i=1}^{k-1} A_i) + \mathrm{Meas}(A_k) = \sum_{i=1}^{k-1} \mathrm{Meas}(A_i) + \mathrm{Meas}(A_k) = \sum_{i=1}^{k} \mathrm{Meas}(A_i)$, as required. □

Corollary 3.4 (Dim, Meas) *is monotonic, that is, whenever* $A \subseteq B$ *are definable, then* $(\mathrm{Dim}, \mathrm{Meas})(A) \leq (\mathrm{Dim}, \mathrm{Meas})(B)$ *(under the lexicographic ordering).*

For convenience we assume that all the following occurs in the home sort.

Lemma 3.5 *Let* X *be a definable set,* $\varphi(\bar{x}, \bar{y})$ *an* L*-formula, and* $(\bar{b}_i : i \in \omega)$ *an indiscernible sequence where for each* $i \in \omega$ *we have* $\varphi(M^n, \bar{b}_i) \subseteq X$. *Suppose that* $\{\varphi(M^n, \bar{b}_i) : i \in \omega\}$ *is inconsistent. Then* $\mathrm{Dim}(X) > \mathrm{Dim}(\varphi(M^n, \bar{b}_i))$.

Proof. Suppose not. Suppose $\mathrm{Dim}(X) = \mathrm{Dim}(\varphi(M^n, b_i)) = d$. Then, as $\{\varphi(M^n, \bar{b}_i) : i \in \omega\}$ is inconsistent, by compactness there exists some minimal k such that $\mathrm{Dim}(\varphi(M^n, \bar{b}_1) \cap \ldots \cap \varphi(M^n, \bar{b}_{k+1}) \cap \varphi(M^n, \bar{b}_{k+2})) < d$.

For $i \geq 1$ define $A_i := \varphi(M^n, \bar{b}_1) \cap \ldots \cap \varphi(M^n, \bar{b}_k) \cap \varphi(M^n, \bar{b}_{k+i})$. Notice that by indiscernibility and the minimality of k, we have $\mathrm{Dim}(A_i) = d$, and for $i_1 \neq i_2$ also $\mathrm{Dim}(A_{i_1} \cap A_{i_2}) < d$. Say $\mathrm{Meas}(A_i) = \mu$. Thus by Lemma 3.3, for any $t \geq 1$ we have $(\mathrm{Dim}, \mathrm{Meas})(\bigcup_{i=1}^{t} A_i) = (d, t\mu)$. But then by Corollary 3.4 $\mathrm{Meas}(X) \geq t\mu$ for all t, which is clearly impossible. □

Corollary 3.6 *For any definable set* X, *we have* $D(X) \leq \mathrm{Dim}(X)$.

Proof. We proceed by induction on r, showing $(*)$: if $\mathrm{Dim}(X) \leq r$ then $D(X) \leq r$. The case $r = 0$ is automatic from clause (2) of Definition 3.1.

For the inductive step, assume $(*)$ below r, and suppose for a contradiction that $\mathrm{Dim}(X) = r$, and $D(X) \geq r + 1$.

By definition of D-rank, there is an indiscernible sequence $(\bar{b}_i : i \in \omega)$ and an L-formula $\varphi(\bar{x}, \bar{y})$ where $\{\varphi(\bar{x}, \bar{b}_i) : i \in \omega\}$ is inconsistent, and for each $i \in \omega$ we have $D(\varphi(M^n, \bar{b}_i)) \geq r$ and $\varphi(M^n, \bar{b}_i) \subseteq X$. By Lemma 3.5, we have $\mathrm{Dim}(\varphi(M^n, \bar{b}_i)) < r$ for each i. It follows by the inductive hypothesis that $D(\varphi(M^n, \bar{b}_i)) < r$ for each i, which is a contradiction. □

Corollary 3.7 *If M is measurable, then M has a supersimple theory.*

Remark 3.8 1. It is immediate from the definition, and noted in [43], that measurability is a property of a *theory*; that is, if M is measurable and $M \equiv N$, then N is measurable.

2. Less obviously, if M is measurable, and we adjoin to M finitely many sorts from M^{eq}, then the resulting structure is measurable – see [43, Proposition 5.10]. In fact, the class of measurable theories is closed under bi-interpretability.

3. A measurable structure may have many different measuring functions. For example, for the random graph, there is a measure corresponding to any edge probability p, where $0 < p < 1$: we let p be the measure of the set of neighbours of a vertex [43, 5.12]. This is generalised in Theorem 3.10 below, and the remarks after it. If a theory is measurable, then, rather as with Theorem 2.3, the measuring function is determined by its restriction to definable sets in one variable.

4. It can happen that in a measurable structure M there are definable sets X_1 and X_2 with $\mathrm{Dim}(X_1) = \mathrm{Dim}(X_2)$ but $D(X_1) \neq D(X_2)$. For example, consider a structure (M, P, E) where P is a unary predicate picking out an infinite subset with infinite complement, and E is an equivalence relation partitioning $P(M)$ into infinitely many infinite classes, but with $\neg P(M)$ a single E-class. Then $D(P(M)) = 2$ and $D(\neg P(M)) = 1$, but we can artificially choose a dimension and measure such that $\mathrm{Dim}(P(M)) = \mathrm{Dim}(\neg P(M)) = 2$.

One can modify this example to produce a measurable structure in which D-rank is not definable. Indeed, consider the structure

$$(M, E, F_i)_{i \in \omega},$$

where E is an equivalence relation whose classes $\{X_i : i \in \mathbb{Z}\}$ are all infinite, and for each $i \in \omega$, F_i is an equivalence relation which agrees with E except on X_i, which it partitions into infinitely many infinite classes. There is a measuring function on M under which each X_i has dimension 2. But $D(X_i) = 1$ for $i < 0$ and $D(X_i) = 2$ for $i \geq 0$, and the latter are not uniformly definable.

5. If M is measurable with measuring function (Dim, Meas), and $S \subset M^n$ is definable, there is an induced finitely additive probability measure μ_S on the σ-algebra generated by the definable subsets of S: for definable $X \subset S$, put

$$\mu_S(X) = \begin{cases} \frac{\text{Meas}(X)}{\text{Meas}(S)} & \text{if } \text{Dim}(X) = \text{Dim}(S), \\ 0 & \text{if } \text{Dim}(X) < \text{Dim}(S). \end{cases}$$

If S is defined by the formula $\psi(\bar{x})$, we sometimes write it as μ_ψ.

6. Ben-Yaacov [3] defines, in a measurable theory, the relation $\bar{a} \mathop{\smile\!\!\!\!\!\mid}^{d}_{C} \bar{b}$, to mean $\text{Dim}(\bar{a}/C) = \text{Dim}(\bar{a}/C \cup \{\bar{b}\})$, where, if $\bar{a} = (a_1, \ldots, a_n)$,

$$\text{Dim}(\bar{a}/C) := \min\{\text{Dim}(\varphi(M^n, \bar{c})) : \bar{c} \text{ in } C \text{ and } M \models \varphi(\bar{a}, \bar{c})\}.$$

He shows that this relation coincides with non-dividing, which, by simplicity, agrees with non-forking.

Proposition 3.9 ([18]) *Let $\mathcal{C} = \{M_i : i \in \omega\}$ be an N-dimensional asymptotic class, and $M = \prod_{i \in I} M_i/\mathcal{U}$ be an infinite ultraproduct of members of $\mathcal{C}$. Then M is measurable, with* (Dim, Meas)$(M) = (N, 1)$.

Proof. We use the definability clause (ii) of Definition 2.1 to define (Dim, Meas)$(\varphi(\bar{x}, \bar{a}))$ for any $\bar{a} \in M^m$, assigning the appropriate d, μ. Easy asymptotic arguments show that this is indeed a measure. □

It follows that all the classes of examples listed in Examples 2.4 yield corresponding examples of measurable structures. In particular, the random graph is measurable, as it is an ultraproduct of Paley graphs; and any infinite vertex transitive graph of finite valency is measurable.

We mention some further examples and constructions. The first comes from the generic predicate construction of [13].

Theorem 3.10 (5.11 of [43]) *Let T be a complete measurable theory over a language L with quantifier elimination, eliminating the quantifier $\exists^\infty$, such that for all $M \models T$ and $A \subset M$,* $\text{acl}(A) = \text{dcl}(A)$. *Let P be a unary predicate not in L, let $L' := L \cup \{P\}$, let S be a sort of T, and let $T_{P,S}$ be the model companion of the theory of L'-structures satisfying T,* *with P interpreted by a subset of S. Let T' be any completion of $T_{P,S}$. Then T' is measurable.*

In the proof, one uses the measuring function μ_T for T, fixes $p \in \mathbb{R}$ with $0 < p < 1$, and for $M \models T_{P,S}$, assigns to $P(M)$ the measure $p\mu_T(S)$. The partial quantifier elimination of [13, 2.6] makes this measure extendable to *all* definable sets. We expect that the assumption $\mathrm{acl}(A) = \mathrm{dcl}(A)$ is not needed, but it was assumed in [43] to avoid complications.

As noted in [43], it follows from Theorem 3.10 that for any $k \geq 2$, the universal homogeneous k-uniform hypergraph Γ_k (that is, a universal homogeneous structure with a single symmetric irreflexive k-ary relation) is measurable. It is not obvious that for $k \geq 2$ this structure is an ultraproduct of an asymptotic class, as there is no obvious analogue for the Paley graphs of Example 2.4 (4). However, Beyarslan in [6] has shown that for any $k \geq 2$, Γ_k is interpretable in a pseudofinite field, so it must at least be an ultraproduct of a *weak* asymptotic class.

The motivating example of a measurable structure is a pseudofinite field. Any pseudofinite field arises as the fixed field of the automorphism in a generic difference field, that is, in a model of ACFA. This, and the last theorem, suggests the following construction technique for measurable structures, given by Hrushovski in Proposition 11.1 of [30]. The definability of measure was not explicitly stated in [30], but uniqueness was, and as noted in [18, 3.4.2], definability follows from uniqueness by Beth's Theorem.

Theorem 3.11 ([30]) *Let D be any strongly minimal set over a language L, assume that* $\mathrm{Th}(D)$ *has the definable multiplicity property (DMP) and elimination of imaginaries, let σ be a generic automorphism of a sufficiently saturated model of* $\mathrm{Th}(D)$, *and let $K = \mathrm{Fix}(\sigma)$, an L-substructure. Then K is measurable, of dimension 1.*

We remark that under assumption (DMP), by [13, 3.11(2)], there is a generic automorphism (i.e. a model companion of the theory of expansions of D by automorphisms).

Recall Hrushovski's fusion construction [31], which, given two strongly minimal sets M_1, M_2 with (DMP) in disjoint languages L_1, L_2 respectively, yields a new strongly minimal set M in $L_1 \cup L_2$ whose reduct to each L_i is M_i. It is shown in [31] that this construction preserves the (DMP). Thus, there is a generic automorphism σ of M, and $\mathrm{Fix}(\sigma)$

will be measurable. As shown in Section 3.4 of [18], this yields the following, when applied to the fusion of two algebraically closed fields of characteristics p_1 and p_2.

Theorem 3.12 ([18]) *Let L_1, L_2 be disjoint languages for rings, and $L := L_1 \cup L_2$, and p_1, p_2 be distinct primes. Then there is a measurable structure M whose reduct to each language L_i is a pseudofinite field of characteristic p_i.*

Note that M is not elementarily equivalent to an ultraproduct of an asymptotic class, since there do not exist positive integers a_1, a_2 with $p_1^{a_1} = p_2^{a_2}$.

Example 3.13 ([43]) For any field F, any infinite vector space over F, in the language of F-modules, is a measurable structure. In the particular case when F is infinite, such a structure cannot be an ultraproduct of an asymptotic class. The point here, essentially, is that if M is *any* strongly minimal set with (DMP) and definable Skolem functions, then M is measurable, with dimension and measure equal to Morley rank and degree. In vector spaces over an infinite field, this is applied after naming a non-zero vector by a constant symbol.

The following definition is given in [26].

Definition 3.14 Let T be a complete theory. We say that T is *unimodular* if for any $M \models T$, definable sets X, Y in M^{eq}, and definable surjections $f_i : X \to Y$ such that f_i is k_i-to-1 (for $i = 1, 2$, and with k_i a positive integer), we have $k_1 = k_2$.

We say that M is unimodular if $\mathrm{Th}(M)$ is.

Clearly any theory with the finite model property (e.g. an ultraproduct of an asymptotic class) is unimodular. In fact, the following is almost immediate from the Fubini condition.

Proposition 3.15 *Let M be measurable. Then M is unimodular.*

In the main applications of measure seen by the authors so far, only unimodularity (plus finite rank supersimplicity with definability of some dimension or rank) is used.

It follows immediately from Proposition 3.15 that an algebraically closed field cannot be measurable. For example, in characteristic not equal to 2, if F is an algebraically closed field, then the identity map is

1-to-1, but the map $x \mapsto x^2$ is 2-to-1 on $F\setminus\{0\}$. More generally, we have the following theorem. Recall that a sufficiently saturated structure M with supersimple theory is said to be *1-based* if, for any subsets A, B of M^{eq}, $A \mathop{\smile\!\!\!\!\!\!\mid}_{\mathrm{acl}^{\mathrm{eq}}(A)\cap\mathrm{acl}^{\mathrm{eq}}(B)} B$.

Theorem 3.16 (Theorem 4.2.6 of [18]) *Any unimodular stable theory of finite U-rank (and hence any measurable stable theory) is 1-based.*

The main point in the proof is that by [26] (see also 2.4.15 and 5.3.2 of [48]), every minimal type of the theory will be 1-based. By the coordinatisation of such structures by minimal types, this is sufficient (see 2.5.8 of [48]).

We have the following easy observation.

Proposition 3.17 *Let M be ω-categorical. Then M is unimodular.*

Proof. Suppose that M is ω-categorical, and X, Y are definable sets in M^{eq} and $f_i : X \to Y$ are definable surjections with f_i k_i-to-1 (for $i = 1, 2$, and with $k_i \in \mathbb{N}$). By adding finitely many sorts to M, naming finitely many parameters, and adding dummy variables, we may suppose that X, Y are disjoint subsets of M, and X, Y and the f_i are $\emptyset$-definable. Let $a \in X$, and $D = \mathrm{acl}(a)\cap(X\cup Y)$. Then D is finite, by ω-categoricity, and is closed under the f_i and f_i^{-1}. Thus, f_i induces a k_i-to-1 map from $D \cap X$ onto $D \cap Y$ and it follows by counting that $k_1 = k_2$. □

Itay Ben-Yaacov [3] has investigated the connections between measurable theories and continuous model theory, and in particular, with measure algebras in an unbounded continuous logic. His main result is a measure-theoretic version of the independence theorem [35] for simple theories. A version was proved earlier by very different means, for pseudofinite fields, by Tomašić [59].

We mention also some work of Tomašić [58] which, in the context of pseudofinite fields, makes connections between measure and exponential sums, and ω-categoricity of reducts. It generalises the use of characters in the proof of Example 2.4 (4) for Paley graphs. See also the survey article by Szönyi [56] for combinatoiral applications of related results.

First, consider a measurable structure M with measuring function (d, μ), and a definable set S in M, with induced probability measure μ_S on the definable subsets of S, yielding a measure space $(S, \mathcal{M}_S, \mu_S)$, where $\mathcal{M}_S$ is the σ-algebra of subsets of S generated by the definable subsets. Then if $f : S \to \mathbb{C}$ is measurable with respect to this measure space (with $\mathbb{C}$ equipped with the Euclidean topology), we may form the integral $\int_S f d\mu_S$. In practice, Tomašić is only concerned with functions of the form $f = \Sigma_{i=1}^t \alpha_i \chi_{A_i}$, where the A_i are definable sets partitioning S and $\alpha_i \in \mathbb{C}$, and here, $\int_S f d\mu_S = \Sigma_{i=1}^t \alpha_i \mu_S(A_i)$. Tomašić investigates additive and multiplicative characters on a pseudofinite field F, and proves the following. As elsewhere in this paper, $\tilde{F}$ denotes the algebraic closure of the field F.

Theorem 3.18 ([58]) *Let X be an absolutely irreducible variety over a pseudofinite field F and let f be a rational function on X. Suppose either*

(i) χ is a multiplicative character of F of order $k > 1$ and f is not a kth power of a rational function on $\tilde{X}$ (the corresponding variety over $\tilde{F}$), or

(ii) χ is a nontrivial additive character of F of the form $\chi_{a,1}$ (notation from [58]) and f is not of the form $g^p - g$ for any $g \in \tilde{F}(\tilde{X})$.

Then $\int_{X(F)} \chi \circ f = 0$.

Tomašić then considers reducts of a pseudofinite field F of the following form. Let X be an absolutely irreducible variety over F, let f be a regular function on $X \times X$, and let $\chi : X \to \mathbb{C}$ be a multiplicative character of order $k > 1$ (with the assumption of Theorem 3.18(i)). Given $x, y \in X(F)$, define the binary relation $R_j(x, y)$ to hold if $\chi(f(x, y)) = e^{2\pi i j/k}$. Then there is a binary structure $(X(F), R_0, \ldots, R_{k-1})$ interpretable in F. Ultraproducts of Paley graphs are really a special case, with $X(F) = F$, $f(x, y) = x - y$, and χ the quadratic character. Using Theorem 3.18, Tomašić shows that under certain conditions such reducts are ω-categorical, and their measures, and the corresponding structures in finite fields, share the 'equidistribution' properties of the theorem of Bollobás and Thomason in Example 2.4 (4). In particular, if $X(F) = F$, f is a symmetric polynomial defining a conic, and χ is a quadratic character, then (F, R_0) is ω-categorical.

We mention also work of E. Kowalski [36] on estimates for exponential sums over definable subsets of finite fields. It generalises the corresponding estimates for varieties (Weil, Deligne, and others) and the main theorem of [11].

4 Smoothly approximable structures

The class of *smoothly approximable* structures is a class of ω-categorical supersimple structures of finite rank which properly contains the class of ω-categorical ω-stable structures (so in particular the totally categorical structures). A deep structure theory is developed in [15], which includes, for example, proofs of the equivalence of smooth approximation, Lie coordinatisability, and other notions, such as 'strong 4-quasifiniteness'. In [15] there is also a proof of a version of quasi-finite axiomatisability, a Lachlan-style shrinking and stretching theory, results on definable groups (they must be finite-by-abelian-by-finite) and much else.

Following Definition 2.1.1 of [15], we say that a finite substructure N of a structure M is a *k-homogeneous substructure* of M if all $\emptyset$-definable relations on M induce $\emptyset$-definable relations of N, and for any pair $\bar{a}, \bar{b}$ of k-tuples from N, they have the same type in N if and only if they have the same type in M. An ω-categorical structure M is *smoothly approximated* if it is a union of a chain $(M_i : i \in \omega)$ of finite substructures, where for each i, M_i is an $|M_i|$-homogeneous substructure of M.

Smoothly approximated structures with primitive automorphism groups were classified in [33]. Based on this, a list of rank 1 *Lie geometries* is identified in [15, Section 2.1.2]. Typical examples are vector spaces (or their projective and affine versions) over finite fields, possibly equipped with bilinear forms, but there is a rather more mysterious example, the quadratic geometry. The sorts and languages to handle these geometries are chosen with care in [15] to ensure flexibility in handling slightly different automorphism groups (e.g. semilinear automorphisms, which involve field automorphisms), and quantifier elimination and weak elimination of imaginaries in certain cases. The authors define a *Lie coordinatised structure* to be one built by covering constructions from Lie geometries in a way indexed by a tree of finite height, and then prove that a countable structure is smoothly approximated if and only if it is Lie coordinatisable (i.e. bi-interpretable with a Lie cooordinatised structure).

It is easily seen that the Lie geometries arise as direct limits of 1-dimensional asymptotic classes. For example, given a finite field $\mathbb{F}_q$, let $\mathcal{C}$ be the collection of all finite dimensional vector spaces V over $\mathbb{F}_q$, equipped with a non-degenerate alternating bilinear form $\beta : V \times V \to \mathbb{F}_q$. We may code β by introducing a binary relation symbol R_a for each $a \in F_q$, with $R_a(x, y)$ whenever $\beta(x, y) = a$. Witt's Lemma, which says that partial isometries extend to total isometries, gives quantifier elimination, so we can reduce to considering the cardinalities of sets of the form $\{x : \beta(x, v_1) = a_1 \wedge \ldots \wedge \beta(x, v_r) = a_r\}$, where $a_1, \ldots, a_r \in \mathbb{F}_q$, and $v_1, \ldots, v_r \in V$ are linearly independent. Such a set has size $\frac{1}{q^r}|V|$. Thus, there is some resemblance with the way, in the random graph, 1-types over a finite set $\{x_1, \ldots, x_r\}$ are determined by edges/non-edges to each x_i, with all sets of adjacencies equiprobable (assuming edge probability $\frac{1}{2}$); but unlike in the Paley graphs, we get precise results on sizes of definable sets. The random graph is not smoothly approximable.

Theorem 4.1 (Section 3.3.2 of [18]) *(i) Let M be a smoothly approximable structure. Then M is the union of a sequence of substructures $(M_i : i < \omega)$ such that each M_i is a $|M_i|$-homogeneous substructure of M, and $\{M_i : i \in \omega\}$ is an asymptotic class.*
(ii) Every smoothly approximable structure is measurable.

The M_i are also called *envelopes* in [15]. Theorem 4.1(i) rests on rather precise information in [15, Proposition 5.2.2] on the sizes of definable sets in envelopes; these cardinalities are given by polynomials in certain dimensions. As a result, for smoothly approximated structures for which only one canonical projective geometry is involved in the coordinatisation, the asymptotics for the envelopes are much better than that required in Definition 2.1 – see [18, Proposition 3.3.5].

Part (ii) of Theorem 4.1 follows immediately from (i) and Proposition 3.9, since any smoothly approximable structure satisfies the asymptotic theory of any approximating chain $(M_i : i \in \omega)$ of envelopes.

One question throughout this paper concerns the extent to which the examples of asymptotic classes and measurable structures go beyond finite and pseudofinite fields, and structures interpretable in them. It is conceivable that any smoothly approximable structure is interpretable in a product of pseudofinite fields, in which case Theorem 4.1 does not provide new examples. Certainly, we have

Proposition 4.2 ([43]) *Let M be a smoothly approximable Lie geometry. Then M is interpretable in a pseudofinite field.*

As an example, consider an $\aleph_0$-dimensional vector space V over a finite field $\mathbb{F}_p$, equipped with a non-degenerate symmetric bilinear form $\beta : V \times V \to \mathbb{F}_p$. This is approximated by the family $(V_n : n \geq 1)$, where V_n is an n-dimensional vector space. We may identify V_n with $\mathbb{F}_{p^n}$, viewed as a vector space over $\mathbb{F}_p$. There is a trace map $\mathrm{Tr} : \mathbb{F}_{p^n} \to \mathbb{F}_p$, and we may put $\beta(x, y) = \mathrm{Tr}(xy)$ for any $x, y \in \mathbb{F}_{p^n}$. The trace map is uniformly definable (so definable in the limit). Indeed, its kernel has index p in $(\mathbb{F}_{p^n}, +)$, and is uniformly defined as $\{x^p - x : x \in \mathbb{F}_p^n\}$ (Hilbert's Theorem 90). To define the trace, we just specify its value on each of the p cosets of the kernel.

5 Measure and difference fields

The families of Suzuki and Ree finite simple groups are not uniformly interpretable in finite fields. However, as is clear from their constructions, they are uniformly interpretable in certain finite difference fields. In this and the next section, we describe work of Ryten, based on joint work of Ryten and Tomašić, which ensures that *all* families of finite simple groups of fixed Lie rank form asympotic classes.

Ryten's starting point is the following theorem of Hrushovski.

Theorem 5.1 ([31]) *Let $\tilde{K}$ be an algebraically closed field of characteristic p, let $t \in \mathbb{N}$, and $q = p^t$. Suppose $V(\bar{x})$ is an algebraic variety over $\tilde{K}$, and $W(\bar{x}\bar{z}) \subset V(\bar{x}) \times V^q(\bar{z})$ is an irreducible subvariety, where the polynomials defining V^q are the images under the automorphism $y \mapsto y^q$ of those defining V. Assume $\mathrm{Dim}(W) = \mathrm{Dim}(V) = d$, and that the projection $\pi_1 : W \to V$ is dominant of degree δ and $\pi_2 : W \to V^q$ is quasi-finite of purely inseparable degree δ'. Then there is a constant C depending on the total degree of W such that*

$$\left|\left|\{\bar{x}\bar{z} \in W(\tilde{K}) : \bar{z} = \bar{x}^q\}\right| - \frac{\delta}{\delta'}q^d\right| \leq Cq^{d-\frac{1}{2}}.$$

From this, Hrushovski proves

Theorem 5.2 ([31]) *Any non-principal ultraproduct of difference fields $(\tilde{\mathbb{F}}_p, \mathrm{Frob}^k)$ is a model of ACFA.*

Using these results, Ryten and Tomašić prove the following.

Theorem 5.3 ([52]) *Let $\theta(\bar{x}, \bar{y})$ be a formula in the language of difference rings. Then there is a constant $C \in \mathbb{R}^+$ and a finite set D of pairs (d, μ) with $d \in \mathbb{Z} \cup \{\infty\}$ and $\mu \in \mathbb{Q}^+ \cup \{\infty\}$ such that in each difference field $(\tilde{\mathbb{F}}_p, \mathrm{Frob}^k)$, and for any $\bar{a} \in \tilde{\mathbb{F}}_p^m$,*

$$\left||\theta(\bar{x}, \bar{a})| - \mu p^{kd}\right| \leq Cp^{k(d-\frac{1}{2})}$$

holds for some $(d, \mu) \in D$.

Moreover, for each $(d, \mu) \in D$ there is a formula $\theta_{d,\mu}(\bar{y})$ such that for *each* $(\tilde{\mathbb{F}}_p, \mathrm{Frob}^k)$*, the above estimate holds for* $\theta(\tilde{\mathbb{F}}_p, \bar{a})$ *with* (d,μ) *if and only if* $(\tilde{\mathbb{F}}_p, \mathrm{Frob}^k) \models \theta_{d,\mu}(\bar{a})$.

The proof has a somewhat different presentation in Chapter 2 of [51]. The theorem is derived from Theorem 5.1 rather as Theorem 1.1 is derived from the Lang-Weil estimates. Formulas with quantifiers are handled using Theorem 5.2, together with the following partial quantifier elimination for ACFA.

Proposition 5.4 (1.5 and 1.6 of [12]) *If* $\theta(\bar{x},\bar{y})$ *is a formula in the language of difference rings, then*

$$ACFA \models \theta(\bar{x},\bar{y}) \Leftrightarrow \bigvee_{i-1}^{k} \exists t \theta_i(\bar{x},\bar{y},t),$$

where $\theta_i = \theta_i(\bar{x},\sigma(\bar{x}),\ldots,\sigma^\ell(\bar{x}),\bar{y},\sigma(\bar{y}),\ldots,\sigma^\ell(\bar{y}),t,\sigma(t),\ldots,\sigma^\ell(t))$ *is a quantifier free formula in the language of rings, and for any* $(M,\sigma) \models ACFA$ *and* $(\bar{x}_0,\bar{y}_0,t_0) \in \theta_i(M)$*,* t_0 *is algebraic (in the sense of fields) over* $\{\sigma^i(\bar{x}_0),\sigma^i(\bar{y}_0) : 0 \le i \le \ell\}$.

Ryten and Tomašić also note that, by Theorem 5.3, in a model of ACFA, the family of *finite dimensional* sets is measurable, under a natural variant of Definition 3.1.

In [51], Ryten uses Theorem 5.3 to investigate an important class of finite difference fields. Fix a prime p, and integers m,n with $m \ge 1$, $n > 1$, and $(m,n) = 1$. Let $\mathcal{C}_{m,n,p}$ be the collection $\{(\mathbb{F}_{p^{kn+m}}, \mathrm{Frob}^k) : k > 0\}$ of finite difference fields. By Frob^k we understand its restriction to $\mathbb{F}_{p^{kn+m}}$, and notice that on this domain it is a solution in σ to the equation $\mathrm{Frob}^m \circ \sigma^n = \mathrm{id}$.

Consider the following conditions on a difference field (K,σ). Below, with m,n as above, if (K,σ) is a difference field of characteristic p with $\mathrm{Frob}^m \circ \sigma^n = \mathrm{id}$, we say that the extension of difference fields $(K,\sigma) \subseteq (L,\sigma')$ is *generic* if $\mathrm{Fix}(\mathrm{Frob}^m \circ \sigma'^n) = K$.

(1) K is a pseudofinite field of characteristic p.

(2) σ is an automorphism of K which satisfies $\mathrm{Frob}^m \circ \sigma^n = \mathrm{id}$.

(3) Suppose $U \subset \mathbb{A}^{nN}$ is an absolutely irreducible variety defined over K and let $\sigma(U)$ be the variety obtained from U by applying σ to the coefficients of the defining polynomials; let the variables of U be $(x_{11}\ldots x_{n1}\ldots x_{1N}\ldots x_{nN})$, those of $\sigma(U)$ be $(y_{11}\cdots y_{n1}\cdots y_{1N}\cdots y_{nN})$. Suppose $V \subset U \times \sigma(U)$ is an absolutely irreducible variety defined over K, whose definition includes the equations $y_{ij} = x_{i+1,j}$ and $y_{nj}^{p^m} = x_{1j}$

for $i = 1, \ldots, n-1$ and $j = 1, \ldots, N$. Suppose that V projects generically onto U and $\sigma(U)$, and that W is a K-algebraic set properly contained in V. Then there is a point $x \in V(K) \setminus W(K)$ such that $x = (a, b)$, where $a = (a_{ij:1\leq i\leq n, 1\leq j\leq N})$, $b = (b_{ij} : 1 \leq i \leq n, 1 \leq j \leq N)$, $a \in U$, $b \in \sigma(U)$, and $b_{ij} = \sigma(a_{ij})$ for each i, j.

(4) Let $K \subseteq L \subseteq H$ be a tower of finite field extensions. Suppose that $(K, \sigma) \subseteq (L, \sigma')$ is a generic extension of difference fields. Then there is an extension of difference fields $(L, \sigma') \subseteq (H, \sigma'')$ such that (H, σ'') is a generic extension of (K, σ).

As suggested by the referee, it may be more elegant to replace (4) by the statement:

(4') any lifting of σ to $\tilde{K}$ commutes with the action of $\mathrm{Aut}(\tilde{K}/K)$.

It can be shown that (1)–(4) (and (4')) are first order expressible. Furthermore, basic facts about ACFA yield the following.

Proposition 5.5 ([51]) *Let (M, τ) be a model of ACFA of characteristic p, let $K := \mathrm{Fix}(\mathrm{Frob}^m \circ \tau^n)$, and put $\sigma := \tau|_K$. Then $(K, \sigma) \models (1) - (4)$.*

By Proposition 5.5, conditions (1)–(4) axiomatise a theory, denoted $\mathrm{PSF}_{(m,n,p)}$. Ryten proves for $\mathrm{PSF}_{(m,n,p)}$ some results similar to those known for ACFA and PSF. The completions (and types) are described through the following theorem.

Theorem 5.6 ([51]) *Let (F, σ) and (E, τ) be models of $\mathrm{PSF}_{(m,n,p)}$ with a common substructure K (so $\sigma|_K = \tau|_K$). Then $(F, \sigma) \equiv_K (E, \tau)$ if and only if $(F \cap \tilde{K}, \sigma|_{F\cap\tilde{K}}) \cong_K (E \cap \tilde{K}, \tau|_{E\cap\tilde{K}})$.*

From this, near model completeness is proved: that is, for every formula $\varphi(\bar{x})$ there is a formula $\theta(\bar{x})$ which is a boolean combination of existential formulas (in fact, formulas of a rather specific type), such that $\mathrm{PSF}_{(m,n,p)} \vdash \varphi(\bar{x}) \leftrightarrow \theta(\bar{x})$.

Using Theorem 5.2, Ryten proves the following theorem.

Theorem 5.7 ([51]) $\mathrm{PSF}_{(m,n,p)}$ *is the asymptotic theory of the class $\mathcal{C}_{(m,n,p)}$ of finite difference fields.*

We show part of this, that every non-principal ultraproduct

$$(N, \sigma) = \prod_{i\in\mathbb{N}} (\mathbb{F}_{p^{nk_i+m}}, \mathrm{Frob}^{k_i})/\mathcal{U}$$

satisfies $\mathrm{PSF}_{(m,n,p)}$. Indeed, if $(M, \tau) = \prod_{i\in\mathbb{N}}(\tilde{\mathbb{F}}_p, \mathrm{Frob}^{k_i})/\mathcal{U}$, then

$(M,\tau) \models$ ACFA by Theorem 5.2. Also, $N = \mathrm{Fix}(\mathrm{Frob}^m \circ \sigma^n)$ and $\sigma = \tau|_N$, so $(N,\sigma) \models \mathrm{PSF}_{(m,n,p)}$ by Proposition 5.5.

Theorem 5.8 ([51]) *$\mathcal{C}_{(m,n,p)}$ is a 1-dimensional asymptotic class.*

To see this, let $\mathcal{D}_p := \{(\tilde{\mathbb{F}}_p, \mathrm{Frob}^k) : k \in \mathbb{N}\}$, a class of difference fields. There is a bijection $\mathcal{F} : \mathcal{D}_p \to \mathcal{C}_{(m,n,p)}$, with $\mathcal{F}((\tilde{\mathbb{F}}_p, \mathrm{Frob}^k)) = (\mathbb{F}_{p^{kn+m}}, \mathrm{Frob}^k)$. There is a corresponding function ^{Fix} defined on the set of formulas of the language L_{diff} of difference rings, defined inductively. If $(K,\sigma) \in \mathcal{D}_p$ and $\mathcal{F}((K,\sigma)) = (M,\sigma) \in \mathcal{C}_{(m,n,p)}$, then for any L_{diff} formula $\varphi(y)$ and tuple $\bar{a}$ from M,

$$(M,\sigma) \models \varphi(\bar{a}) \Leftrightarrow (K,\sigma) \models \varphi^{\mathrm{Fix}}(\bar{a}).$$

From this and Theorem 5.2, the uniform asymptotic estimates for formulas in $\mathcal{C}_{(m,n,p)}$ are rapidly derived. The definability clause (Definition 2.1(ii)) requires a little more work.

Finally, Ryten shows that in a *pure* pseudofinite field (in fact, in any pure bounded PAC field) the only definable field automorphisms are powers of the Frobenius. Thus, $\mathrm{PSF}_{(m,n,p)}$ is a proper expansion of the theory of pseudofinite fields of characteristic p, and $\mathcal{C}_{(m,n,p)}$ is not interpretable in any class of pure finite fields.

In [51, Chapter 3], Ryten proves a number of other results about $\mathrm{PSF}_{(m,n,p)}$, indicating that its complexity is somewhere between that of PSF and ACFA. As for pseudofinite fields in [11], it is possible to add constants to the language to ensure, in the partial quantifier elimination, that only *positive* boolean combinations of existential formulas are used, so model-completeness in the expanded language is obtained. Model-theoretic independence is characterised algebraically. Elimination of imaginaries (over a language with constants for an elementary submodel) is proved.

6 Asymptotic classes of simple groups

There is a natural notion of *uniform parameter bi-interpretability* between two classes $\mathcal{C}$ and $\mathcal{D}$ of finite structures. We do not give it formally, but it requires a matching between $\mathcal{C}$ and $\mathcal{D}$, and that each element of $\mathcal{C}$ is bi-interpretable (not just mutually interpretable) with the corresponding element of $\mathcal{D}$, possibly using parameters, but with uniformity of the interpretation across the families.

The non-abelian finite simple groups, excluding the sporadics, are the alternating groups $\mathrm{Alt}(n)$, the classical groups of Lie rank n over the finite field $\mathbb{F}_q$, and certain exceptional groups, namely $E_6(q)$, $E_7(q)$, $E_8(q)$, $F_4(g)$, $G_2(q)$, and the twisted groups ${}^3D_4(q)$, ${}^2E_6(q)$, the Ree groups ${}^2F_4(2^{2k+1})$, the Suzuki groups ${}^2B_2(2^{2k+1})$, and the Ree groups ${}^2G_2(2^{3k+1})$. We may regard $\mathrm{Alt}(n)$ as having Lie rank n. We emphasise that the twisted groups (including the classical unitary groups) do not arise over algebraically closed fields, since their definition depends on finite field extensions. They do not have finite Morley rank analogues.

Ryten shows that, for every family of non-abelian finite simple groups of fixed Lie rank, other than those of Suzuki and Ree groups, the family is uniformly parameter bi-interpretable (infact, *bi-definable*) with a class of finite fields. With some additional work, it follows that such a family is an asymptotic class. Care is needed here with the definability clause in Definition 2.1, because of the role of parameters in the bi-interpretation. For a given formula $\varphi(\bar{x},\bar{y})$, the corresponding $\Phi_{(d,\mu)}$ should be definable *without parameters*.

For the Suzuki and Ree groups, the situation is more complicated. In fact, the classes of groups $\{{}^2F_4(2^{2k+1}) : k \in \mathbb{N}\}$, $\{{}^2B_2(2^{2k+1}) : k \in \mathbb{N}\}$, are uniformly parameter bi-interpretable with $\mathcal{C}_{(1,2,2)}$, and the class $\{{}^2G_2(3^{2k+1}) : k \in \mathbb{N}\}$ is uniformly parameter bi-interpretable with $\mathcal{C}_{(1,2,3)}$. A word of explanation is necessary here. The twisted group ${}^2E_6(q)$, for example, is the fixed point set inside $E_6(q^2)$ of a certain automorphism σ which is a product of a graph automorphism (an automorphism arising from the symmetry of the E_6 Dynkin diagram), and the Frobenius. All this data is definable in the field $\mathbb{F}_{q^2}$, with which ${}^2E_6(q)$ is bi-definable, uniformly in q. However, in the case ${}^2G_2(3^{2k+1})$, for example, the automorphism σ is a product of a graph automorphism and the automorphism $x \mapsto x^{3^k}$, a proper (unbounded) power of the Frobenius. So to define ${}^2G_2(3^{2k+1})$ uniformly, it is necessary to be able to define the automorphism Frob^k of $\mathbb{F}_{3^{2k+1}}$, and hence to work in $\mathcal{C}_{(1,2,3)}$.

It is hard to identify the right sources for the model-theoretic relationship between finite simple groups and fields. In work in his PhD thesis, not subsequently published, Thomas showed definability of the corresponding (pure) field in each finite simple group, uniformly across each class of groups. There are related results in, for example, [39]. The interpretation of the groups in the (difference) fields is pretty clear. For the Suzuki and Ree groups it seems to have been known to Hrushovski for a long time. Some consequences (e.g. decidability) are mentioned in the introduction of [31].

The above results, together with Theorem 5.8, yield the following theorem of Ryten.

Theorem 6.1 ([51]) *Any family of non-abelian finite simple groups of fixed Lie rank is an asymptotic class.*

We mention two further results, beyond the measurable/asymptotic class context. A group is *pseudofinite* if it is an infinite model of the theory of finite groups.

Theorem 6.2 ([44]) *Let G be a stable pseudofinite group. Then G has a definable soluble subgroup of finite index.*

The key point is that a non-principal ultraproduct of finite simple groups of fixed Lie rank, though supersimple, is not stable, as a pseudofinite field is interpretable in it. The proof of the theorem below rests on this, together with slightly delicate arguments with chain conditions, and basic facts about finite nilpotent groups. It is not possible here to strengthen 'soluble' to nilpotent: as noted independently by Khélif and Zilber, and not published, there is a metabelian pseudofinite stable group which is not nilpotent-by-finite.

Theorem 6.3 (Wilson [61]) *Every pseudofinite simple group is elementarily equivalent to a Chevalley group (possibly twisted) over a pseudofinite field.*

This builds on earlier work of Felgner. By Ryten's work, 'elementarily equivalent' can be strengthened to 'isomorphic' in Theorem 6.3. In [50], Point proves that an ultraproduct of simple Chevalley groups of fixed type (possibly twisted) is *isomorphic* to a Chevalley group over the ultraproduct of the fields, and is simple.

7 Groups of low dimension

Theorem 6.1 above yields immediately the converse to Conjecture 1.2: any Chevalley group (possibly twisted) over a pseudofinite field is measurable. In this section we sketch beginnings of a general structure theory for measurable groups, with a view to Conjecture 1.2. Many of the results are jointly due to Elwes and Ryten. They can be found in [18], and a joint paper [19] is in preparation.

The central definability result for groups of finite Morley rank is the

Zilber Indecomposability Theorem, which generalises the Indecomposability Theorem for algebraic groups. In the supersimple case, we have the following, proved first by Hrushovski [30] for S_1-theories, that is, supersimple theories of finite S_1-rank in which the S_1-rank is definable.

Theorem 7.1 (Wagner[60] Theorem 5.4.5) *Let G be a definable group in a supersimple theory of finite rank, and let X_i $(i \in I)$ be definable subsets of G. Then there is a definable subgroup H of G, and $n \in \omega$, and $i_1, \ldots, i_n \in I$, where*
(i) $H \leq X_{i_1}^{\pm 1} \cdot X_{i_2}^{\pm 1} \cdot \ldots \cdot X_{i_n}^{\pm 1}$
(ii) X_i/H is finite for each $i \in I$.

Proof. This follows from 5.4.5 and 5.5.4 of [60], by compactness. □

This is used repeatedly in arguments described below. It also yields, for example, that any non-abelian definably simple measurable group is simple, and that in a measurable theory, the derived subgroup of a definable group is always definable.

A second useful tool, already heavily used by Wagner for groups in supersimple theories, is Schlichting's Theorem [54], proved independently by Bergman and Lenstra [5]. A family of subgroups $\mathcal{H}$ is *uniformly commensurable* if there is $n \in \mathbb{N}$ bounding $|H : H \cap K|$ for all $H, K \in \mathcal{H}$. The version below is Theorem 4.2.4 of [60].

Theorem 7.2 ([54], [5]) *Let G be a group and $\mathcal{H}$ a uniformly commensurable family of subgroups. Then there is a subgroup N which is uniformly commensurable to $\mathcal{H}$ and is invariant under all automorphisms of G which stabilise $\mathcal{H}$ setwise. In particular, if $\mathcal{H}$ consists of all conjugates of some $H \leq G$, then N is normal in G.*

It is well-known that every rank 1 superstable group is abelian-by-finite. In the supersimple case, one expects 'abelian-by-finite' to be replaced by 'finite-by-abelian-by-finite', in view of the extraspecial groups considered in Example 2.4(3). However, this is an open question. The following result of Elwes and Ryten is proved in [18, 6.0.11]; it was proved under the stronger assumption of measurability in [43].

Theorem 7.3 ([18]) *If G is unimodular supersimple of rank 1, then G has a definable normal subgroup H of finite index, such that H has a finite central subgroup Z, with H/Z abelian.*

The proof is a counting argument with unimodularity. It uses the fact that if G is a BFC group (a group with a finite bound on the size of its conjugacy classes) then the derived subgroup G' is finite; see Theorem 3.1 of [47]. The key to the counting argument is the following lemma, proved in [43] under measurability; the proof under unimodularity is easily extracted from that of [18, 6.0.11]. It eliminates the possibility of measurable groups with finitely many non-identity conjugacy classes, all of full dimension.

Lemma 7.4 ([18], [19]) *Let G be a group defined in a supersimple unimodular theory of finite rank. Then there is $g \in G \setminus \{1\}$ such that $C_G(g)$ is infinite.*

In [43], a cruder argument is given to prove the asymptotic class analogue of Theorem 7.3. It states that any 1-dimensional asymptotic class of *finite* groups consists of groups which are bounded-by-abelian-by-bounded. The argument applies to any class of finite groups all of whose ultraproducts are supersimple of rank 1.

These results, and the analogue for superstable groups, suggest that any measurable 2-dimensional group should be soluble-by-finite. This question is open, but some progress in this direction was made in [19]. Using Theorems 7.3 and 7.1 (to replace any 1-dimensional normal subgroup by a definable one), Elwes and Ryten show that any counterexample interprets a simple group G of dimension and S_1-rank 2. Further information is obtained: for example, the group G must have conjugacy classes (in fact, infinitely many) of dimension 1, and it must also have at least 1, but at most finitely many, conjugacy classes of dimension 2. Such examples could not arise as ultraproducts of an asymptotic class, by the classification of finite simple groups. Thus, they prove the following; a proof can be found in [18] or [19].

Theorem 7.5 *Let $\mathcal{C}$ be a 2-dimensional asymptotic class of groups. Then there is $d \in \mathbb{N}$ such that each group G in $\mathcal{C}$ has a subgroup of index at most d which is soluble of derived length at most 4 (and uniformly definable in the class).*

Elwes and Ryten have also investigated measurable structures (G, X), where G is a group with a definable faithful action on X. The intended analogy is with [42], on the structure of primitive permutation groups of finite Morley rank (see also [9] in this volume). More specifically, they generalise the theorem of Hrushovski [25], that in a stable theory, a connected group of finite Morley rank acting definably and transitively on a strongly minimal set must have Morley rank at most 3, and either acts regularly (so is a strongly minimal group) or is of form $\mathrm{AGL}(1, F)$ or $\mathrm{PSL}(2, F)$ in the natural action (F an algebraically closed field).

A permutation group G on a set X is *primitive* if there is no proper non-trivial G-invariant equivalence relation on X, or equivalently if each stabiliser G_x ($x \in X$) is a maximal subgroup of G. If G, X, and the G-action on X are definable in some structure, we say that (G, X) is *definably primitive* if there is no *definable* proper non-trivial G-invariant equivalence relation on X, or equivalently if point stabilisers are definably maximal. Definable primitivity implies transitivity, as the orbit equivalence relation is definable. We shall say that (G, X) is a *measurable group action* if G, X, and an action of G on X are all definable in a measurable structure. It is an *asymptotic group action* if it is elementarily equivalent to an infinite ultraproduct of an asymptotic class of group actions.

Proposition 7.6 ([19]) *Let (G, X) be a measurable faithful group action, with G infinite, and suppose that G acts faithfully and definably primitively on X. If* $\mathrm{Dim}(X) < \mathrm{Dim}(G)$, *then G is primitive on X.*

The proof uses Theorems 7.2 and 7.1, and in fact 'measurable' can be weakened to 'supersimple, finite rank, and eliminates $\exists^\infty$'. To start the proof (and illustrate a useful argument), define $\sim$ on X, putting

$$x \sim y \Leftrightarrow |G_x : G_x \cap G_y| < \infty,$$

(where G_x denotes the stabiliser of $x \in X$). Then $\sim$ is a G-invariant equivalence relation on X, so is definable, so either there is a single $\sim$-class, or $\sim$-classes are singletons. The first case is easily eliminated. For if there is a single $\sim$-class, then (because measurable theories eliminate the quantifier $\exists^\infty$), there is a fixed upper bound on the indices $|G_x : G_x \cap G_y|$ over all $x, y \in X$. Thus, by Theorem 7.2, there is $N \triangleleft G$ uniformly commensurable to all the G_x, and the proof of 7.2 yields that N is definable. Since N is normal and non-trivial, it is transitive on X. However, as N is commensurable with G_x, the orbit of x under N is finite, a contradiction as X is infinite.

Theorem 7.7 ([19]) *Let (G, X) be an asymptotic group action which is faithful and definably primitive, with* $\mathrm{Dim}(X) = 1$. *Then* $\mathrm{Dim}(G) \leq 3$, *and one of the following holds.*

(i) $\mathrm{Dim}(G) = 1$, *and G has a definable subgroup B of finite index which is torsion-free divisible abelian and acts regularly on X. (In this situation, for the description we only require measurability, not an asymptotic group action.)*

(ii) $\mathrm{Dim}(G) = 2$. *In this case, there is a definable pseudofinite field K, and a definable infinite subgroup T of the multiplicative subgroup* K^*, *such that G is isomorphic to* $K^+ \rtimes T$ *(a subgroup of* $\mathrm{AGL}(1, K)$*) in its natural action on* $(K, +)$*).*

(iii) $\mathrm{Dim}(G) = 3$. *Here G has a unique minimal normal subgroup T, which is definable and isomorphic to* $\mathrm{PSL}_2(K)$ *(K a pseudofinite field) and its action on X is its natural action on the projective line.*

A non-regular example in (i) would be $K^+ \rtimes \{+1, -1\}$, where K is a pseudofinite field in characteristic 0. For (ii), one might take K to be an ultraproduct of fields $\mathbb{F}_p$ ($p \equiv 1 \pmod 4$) and T to be the squares of K.

Sketch of the Proof. First, if $\mathrm{Dim}(G) = 1$, then G is finite-by-abelian-by-finite by Theorem 7.3, and in particular G has a subgroup B of finite index defined as the union of the finite conjugacy classes of G. As $B \lhd G$, B is transitive (the orbits of B would yield a definable G-invariant partition of X). Also the derived subgroup B' is finite, so trivial by definable primitivity, so B is abelian, and hence acts regularly on X. The remaining analysis in this case is easy.

If $\mathrm{Dim}(G) = 2$, then by Theorem 7.5, G is soluble-by-finite. A minimal normal subgroup A of G will be abelian, and transitive on X (by primitivity); as A is abelian it acts regularly on X. It follows that we may identify A with X and G with a semidirect product $A \rtimes G_1$, where G_1 is the stabiliser of the identity of A (in its action by conjugation). For the full description in (ii), a version of the Zilber Field Theorem is used.

Finally, suppose that $\mathrm{Dim}(G) \geq 3$. In this case one may apply the O'Nan-Scott Theorem, the standard reduction theorem for finite primitive permutation groups; see [40] for a careful account of it. This, and the fact that $\mathrm{Dim}(X) = 1$, reduces us to a situation where G has a unique minimal definable normal subgroup T, and T is simple and $T \leq G \leq \mathrm{Aut}(T)$. A further analysis, using extensive information about finite simple groups, reduces to the case when $T = \mathrm{PSL}(2, K)$ (K a pseudofinite field) in its action on the projective line. Tools used include Aschbacher's description of maximal subgroups of finite classical groups [1].

As with Theorem 7.5, it would be very nice to have a version of Theorem 7.7 for groups with measurable theories, but the above proof makes heavy use of finite group theory.

We mention one further result which can be proved under the assumption of measurability, but is open under just supersimplicity.

Theorem 7.8 ([43]) *Let G be an ω-saturated measurable group. Then G has an infinite abelian subgroup.*

In the proof, one looks at a counterexample G of minimal dimension. The exponent must be finite (as otherwise there are infinite cyclic subgroups), and by the minimality assumption, every definable subgroup is finite or of finite index. Let $N := \{g \in G : g^G \text{ is finite}\}$. Then N is finite or of finite index. But $N \neq \{1\}$, for by Lemma 7.4 there is $g \in G \setminus \{1\}$ with $C_G(g)$ infinite; then $|G : C_G(g)|$ is finite, so $g \in N$. If N is infinite, then we may assume that $N = G$, so $|G : C_G(g)|$ is finite for all $g \in G$, and it is then easy to construct an infinite abelian subgroup. Finally, if N is finite but non-trivial, apply similar arguments to G/N.

Wagner has noted that a similar proof yields that any measurable group has an infinite *definable* finite-by-abelian subgroup.

8 Further questions

If M is a 1-dimensional measurable structure, then it has SU-rank 1, and so algebraic closure gives a pregeometry. It is natural to ask whether this pregeometry satisfies the Zilber Trichotomy: trivial, locally modular (so 1-based) non-trivial, and field-like. A trivial geometry arises for the random graph or any vertex transitive graph of finite valency. Smoothly approximable Lie geometries are locally modular, and, except for a pure set, non-trivial. And pseudofinite fields (as well as ultraproducts of classes $C_{m,n,p}$ and some examples arising from Theorem 3.12) are non locally modular.

It should be possible to construct a non locally modular 1-dimensional measurable structure which does not interpret an infinite group. Let M be the saturated, strongly minimal set constructed in [27], which is known to have (DMP), is not locally modular, and has no infinite interpretable group. Then M admits a generic automorphism σ. It seems that a variant of Theorem 3.12 for finite Morley rank, mentioned in Remark (viii) at the end of [30], should apply in M^{eq} (to ensure elimination of imaginaries), and yield that if $N = \text{Fix}(\sigma)$ then the induced structure on N is measurable of dimension 1. We can ensure that N is not 1-based: for algebraic closure in N is the same as that induced from M, and (M, id) embeds in some model (M', σ') of the theory of generic automorphisms of models of $\mathrm{Th}(M)$, so that $N := \mathrm{Fix}(\sigma')$ contains M, so contains a witness to non 1-basedness. Also, an infinite definable group in N would yield, by a group configuration argument, an infinite definable group in M, which is impossible.

However, we ask

Question 1. If M is a 1-dimensional measurable structure which is a non-principal ultraproduct of an asymptotic class, and M is not 1-based, must M interpret an infinite field?

The question can also be asked for structures interpretable in PSF. In the same vein, we ask whether measurability can be used to sharpen the group configuration results of [4].

Question 2. Is every ω-categorical measurable structure 1-based?

For this, a key example is Hrushovski's construction in [29] of an ω-categorical non-locally modular supersimple rank 1 structure. It is unimodular, by Proposition 3.17, but we do not know whether it is measurable. There is more chance of a positive answer to Question 2 for structures interpretable in PSF. On Question 2, Elwes [18, 5.2.3] has obtained some partial results analogous to Proposition 8 of [26].

Question 3. Is every measurable field pseudofinite?

By the results of Ax [2], a field F is pseudofinite if and only if it is perfect, quasifinite (i.e. has absolute Galois group $\hat{\mathbb{Z}}$), and satisfies the PAC condition; the latter asserts that any absolutely irreducible variety defined over F has an F-rational point. It is straightforward that any supersimple field of finite rank is perfect. By an argument of Scanlon ([43, Theorem 5.18 and Appendix], see also [53]), any measurable field is quasifinite, that is, has a unique extension of degree n for each n. We do not know whether every measurable field satisfies the PAC condition. It was shown in [49] that any supersimple division ring is commutative and has absolute Brauer group, so the norm from any finite extension to the field is surjective. By [45], any generic elliptic or hyperelliptic curve defined over a supersimple field F has an F-rational point. If F is supersimple and has a unique quadratic extension (e.g. if F is measurable), then by [46] *any* elliptic curve defined over F has an F-rational point.

Halupczok [23] has investigated a weakening of measure, where, roughly speaking, the Fubini property is dropped but measure is assumed to be invariant under definable bijections. He has shown that for perfect PAC fields with procyclic absolute Galois group (i.e. with at most one extension of each finite degree) there is such a measure, but that in general there is no such measure.

Question 4. Is every measurable simple group a (possibly twisted) Chevalley group over a pseudofinite field?

Of course, Question 4 requires a positive answer to Question 3. Question 4 looks difficult, but it has a positive answer, in unpublished work of Hrushovski, for groups definable in Ryten's theories $\mathrm{PSF}_{(m,n,p)}$ of measurable pseudofinite difference fields. In fact, Hrushovski classifies infinite definable simple groups in ACFA, without use of the classification of finite simple groups. Dello Stritto [personal communication] has partial results which, modulo a positive answer to Question 3 and an analogue for measurable difference fields, are likely to answer Question 4 positively for groups with a BN pair of rank at least 2.

It would also be interesting to obtain, without use of the classification, structural results on asymptotic classes of finite simple groups.

Hrushovski (Appendix to [30]) has suggested that one might use results on measurable groups, in conjunction with a version of Theorem 3.11 for structures of finite Morley rank, to study groups of finite Morley rank.

Question 5. Is every 2-dimensional measurable group soluble-by-finite? Equivalently, is every non-abelian infinite measurable simple group of dimension at least 3?

Following Proposition 4.2, we ask:

Question 6. Is every smoothly approximable structure interpretable in a pseudofinite field?

It would be helpful to compare measurability with some similar notions in the literature. One such is unimodularity, and Proposition 3.17 suggests the following.

Question 7. Is every ω-categorical supersimple (finite rank) theory measurable?

An example to try would again be Hrushovski's construction from [29].

Question 8. (Macintyre) Is there a natural example of a measurable structure in which some measures are transcendental numbers?

We finish with some comments on other counting and measure principles.

There is a notion of measure investigated by Keisler in [34], and more recently in [32]. The authors of the latter consider a sort X in a sufficiently saturated model $\bar{M}$, and define a Keisler measure on X to be a finitely additive probability measure on $\mathrm{Def}_1(X)$, the collection of parameter-definable subsets of X. A type, for example, is just a zero-one measure. Here we are not requiring a measure on subsets of all powers of M, so there is no Fubini-like assumption. If M is measurable, then for any sort S, there is a definable Keisler measure on S: the two measures will agree for definable subsets of S of dimension $\mathrm{Dim}(S)$, and sets of lower dimension will have Keisler measure 0.

Keisler measure behaves particularly well for theories with the NIP (that is, theories without the independence property) so is orthogonal to the context of the present paper. For example, the authors show that if T has the NIP, and μ is a Keisler measure on X, then there are boundedly many $\sim_\mu$-classes of definable subsets of X, where $Y \sim_\mu Z$ whenever $\mu(Y \triangle Z) = 0$. They investigate existence and uniqueness of Keisler measures on groups.

Next, model-theoretic ideas of Euler characteristic were developed initially for o-minimal theories (see e.g. [17]) but more generally by Krajiček [37] and later Krajiček and Scanlon [38]. The latter define a *strong ordered Euler characteristic* on a structure M to be a function $\chi : \mathrm{Def}(X) \to R$, where R is a partially ordreed ring, the image of χ takes values amone the non-negative elements of R, and we have

(a) $\chi(X) = \chi(Y)$ if X are in definable bijection,
(b) $\chi(X \times Y) = \chi(X).\chi(Y)$,
(c) $\chi(X \cup Y) = \chi(X) + \chi(Y)$ if $X \cap Y = \emptyset$, and
(d) $\chi(E) = c\chi(B)$ if $s : E \to B$ is a definable function and $c = \chi(f^{-1}(b))$ for each $b \in B$.

The Euler characteristic is *non-trivial* if $0 < 1$ in R and the image of χ is not just $\{0\}$.

Unlike for our notion of measure, there is no associated dimension, and indeed, the analogue of (c) for measure only holds if $\mathrm{Dim}(X) = \mathrm{Dim}(Y)$. However, under this definition very similar counting arguments are available. For example, Scanlon (see the Appendix of [43]) has shown that any field with strong ordered Euler characteristic is perfect and quasifinite, and these conclusions follow by the same argument for measurable fields.

Other questions about measure, and variations on measure are suggested by Hrushovski at the end of [30]. For example, there is a suggestion to consider finitely additive measures into $\mathbb{Q}[T]$, which could incorporate the current notion of dimension (exponent of leading term) and measure (coefficient of leading term).

References

[1] M. Aschbacher, On the maximal subgroups of the finite classical groups, *Invent. Math.* *76* (1984), 469–514.

[2] J. Ax, The elementary theory of finite fields, *Ann. Math.* 88 (1968), 239–271.

[3] I. Ben-Yaacov, On measurable first order theories and the continuous theory of measure algebras, preprint.

[4] I. Ben-Yaacov, I. Tomašić, F. O. Wagner, Constructing an almost hyperdefinable group, *J. Math. Logic*, 4 (2004), 181–212.

[5] G. Bergman, H.W. Lenstra, Subgroups close to normal subgroups, *J. Algebra* 127 (1989), 80–97.

[6] O. Beyarslan Interpreting random hypergraphs in pseudofinite fields, preprint `arXiv:math/0701029`.

[7] B. Bollobás, A. Thomason, Graphs which contain all small graphs, *Europ. J. Comb.* *2* (1981), 13–15.

[8] B. Bollobás, *Random Graphs*, Academic Press, New York, 1985.

[9] A. Borovik, G. Cherlin, Permutation groups of finite Morley rank, this volume.

[10] R. Carter, *Simple groups of Lie type*, vol. XXVIII of *Pure and Applied Mathematics*, Wiley, 1972.

[11] Z. Chatzidakis, L. van den Dries, A.J. Macintyre, Definable sets over finite fields, *J. Reine Angew. Math.* 427 (1992), 107–135.

[12] Z. Chatzidakis, E. Hrushovski, The model theory of difference fields, *Trans. Amer. Math. Soc.* 351 (1999), 2997–3071.

[13] Z. Chatzidakis, A. Pillay, Generic structures and simple theories, *Ann. Pure Appl. Logic* 95 (1998), 71–92.

[14] Z. Chatzidakis, Model theory of finite and pseudo-finite fields, *Ann. Pure Apl. Logic* 88 (1997), 95–108.

[15] G. Cherlin, E. Hrushovski, *Finite structures with few types*, Annals of Mathematics Studies No. 152, Princeton University Press, Princeton, 2003.

[16] A. Chowdhury, B. Hart, Z. Sokolović, Affine covers of Lie geometries and the amalgamation property, *Proc. London Math. Soc. (3)* 85 (2002), 513–563.

[17] L. van den Dries, *Tame topology and o-minimal structures*, London Math. Soc Lecture Notes 248, Cambridge University Press, 1998.

[18] R. Elwes, *Dimension and measure in finite first order structures*, PhD thesis, University of Leeds, 2005.

[19] R. Elwes, M. Ryten, Measurable groups of low dimension, in preparation.

[20] M. Fried, D. Haran, M. Jarden, Effective counting of the points of definable sets over finite fields, *Isr. J. Math.* 85 (1994), 103–133.

[21] M. Fried, M. Jarden, *Field arithmetic*, 2nd Ed., Springer, Berlin, 2005.

[22] R. L. Graham, J. H. Spencer, A constructive solution to a tournament problem, *Canad. Math. Bull.* 14 (1971), 45–48.

[23] I. Halupczok, A measure for perfect PAC fields with pro-cyclic Galois group, *J. Algebra 310* (2007), 371–395.
[24] W. Hodges, *Model Theory*, Cambridge University Press, Cambridge, 1993.
[25] E. Hrushovski, Almost orthogonal regular types, *Ann. Pure Appl. Logic* 45 (1989), 139–155.
[26] E. Hrushovski, Unimodular minimal structures, *J. London Math. Soc. (2)* 46 (1992), 385–396.
[27] E. Hrushovski, A new strongly minimal set, *Ann. Pure Appl. Logic* 62 (1993), 147–166.
[28] E. Hrushovski, A. Pillay, Definable subgroups of algebraic groups over finite fields, *J. Reine Angew. Math.* 462 (1995), 69–91.
[29] E. Hrushovski, Simplicity and the Lascar group, unpublished notes, 1997.
[30] E. Hrushovski, Pseudofinite fields and related structures, in *Model theory and applications* (Eds. L. Bélair, Z. Chatzidakis, P. D'Aquino, D. Marker, M. Otero, F. Point, A. Wilkie), Quaderni di Matematica, vol. 11, Caserta, 2005, 151–212.
[31] E. Hrushovski, The first order theory of the Frobenius, preprint `arXiv:math/0406514` .
[32] E. Hrushovski, Y. Peterzil, A. Pillay, Groups, measures, and the NIP, preprint `arXiv:math/0607442`.
[33] W.M. Kantor, M.W. Liebeck, H.D. Macpherson, $\aleph_0$-categorical structures smoothly approximated by finite substructures, *Proc. London Math. Soc. (3)* 59 (1989), 439–463.
[34] H.J. Keisler, Measures and forking, *Ann. Pure Appl. Logic 43* (1987), 119–169.
[35] B. Kim, A. Pillay, Simple theories, *Ann. Pure Appl. Logic* 88 (1997), 149–164.
[36] E. Kowalski, Exponential sums over definable subsets of finite fields, preprint `arXiv:math/0504316v2`, to appear in *Israel J. Math.*
[37] J. Krajiček, Uniform families of polynomial equations over a finite field and structures admitting an Euler characteristic of definable sets, *Proc. London Math. Soc.* (3) 81 (2000), 257–284.
[38] J. Krajiček, T. Scanlon, Combinatorics with definable sets: Grothendieck rings are Euler characteristics, *Bull. Symb. Logic* 6 (2000), 311-330.
[39] L. Kramer, G. Röhrle, K. Tent, Defining k in $G(k)$, *J. Alg.* 216 (1999), 77–85.
[40] M.W. Liebeck, C.E. Praeger, J. Saxl, On the O'Nan-Scott theorem for finite primitive permutation groups, *J. Austral. Math. Soc.* 44 (1988), 389–396.
[41] I.D. Macdonald, Some explicit bounds in groups with finite derived subgroups, *Proc. London Math. Soc. (3)* 11 (1961), 23–56.
[42] H.D. Macpherson, A. Pillay, Primitive permutation groups of finite Morley rank, *Proc. London Math. Soc. (3)* 70 (1995), 481–504.
[43] H.D. Macpherson, C. Steinhorn, One-dimensional asymptotic classes of finite structures, *Trans. Amer. Math. Soc.*, to appear, available at `http://www.amsta.leeds.ac.uk/pure/staff/macpherson/macpherson.html`.
[44] H.D. Macpherson, K. Tent, Stable pseudofinite groups, *J. Algebra* 312

(2007), 550–561.

[45] A. Martin-Pizarro, A. Pillay, Elliptic and hyperelliptic curves over supersimple fields, *J. London Math. Soc (2)* 69 (2004), 1–13.

[46] A. Martin-Pizarro, F.O. Wagner, Supersimplicity and quadratic extensions, preprint, (available at `http://www.newton.cam.ac.uk/preprints/NI05035.pdf`)

[47] B.H. Neumann, Groups covered by permutable subsets', *J. London Math. Soc.* 29 (1954), 236–248.

[48] A. Pillay, *Geometric stability theory*, Oxford University Press, Oxford Logic Guides 32, 1996.

[49] A. Pillay, T. Scanlon, F.O. Wagner, Supersimple fields and division rings, *Math. Research Letters* 5 (1998), 473–483.

[50] F. Point, Ultraproducts and Chevalley groups, *Arch. Math. Logic* 38 (1999), 355–372.

[51] M. Ryten, *Results around asymptotic and measurable groups*, PhD thesis, University of Leeds, 2007.

[52] M. Ryten, I. Tomašić, ACFA and measurability, *Selecta Mathematica* 11 (2005), 523–537.

[53] T. Scanlon, Fields admitting nontrivial strong ordered Euler characteristics are quasifinite, unpublished manuscript, available at `http://math.berkeley.edu/~scanlon/papers/papers.html`.

[54] G. Schlichting, Operationen mit periodischen Stabilisatoren, *Arch. Math.* 34 (1980), 97–99.

[55] W. Szmielew, Elementary properties of abelian groups, *Fund. Math.* 41 (1955), 203–271.

[56] T. Szönyi, Some applications of algebraic curves in finite geometry and combinatorics, in *Surveys in Combinatorics* (Ed. R. Bailey), London Math. Society Lecture Notes 241, Cambridge University Press, Cambridge, 1997, pps. 197–236.

[57] S. Thomas, *Classification theory of simple locally finite groups*, PhD thesis, University of London, 1983.

[58] I. Tomašić, Exponential sums in pseudofinite fields and applications, *Illinois J. Math.* 48 (2004), 1235–1257.

[59] I. Tomašić, Independence, measure and pseudofinite fields, *Selecta Math.* 12 (2006), 271–306.

[60] F.O. Wagner, *Simple theories*, Kluwer, Dordrecht, 2000.

[61] J.S. Wilson, On pseudofinite simple groups, *J. London Math. Soc. (2)* 51 (1995), 471–490.

Counting and dimensions

Ehud Hrushovski[†]
The Hebrew University of Jerusalem

Frank Wagner[‡]
Université de Lyon, Université Lyon 1 and CNRS

Summary

We prove a theorem comparing a well-behaved dimension notion to a second, more rudimentary dimension. Specialising to a non-standard counting measure, this generalizes a theorem of Larsen and Pink on an asymptotic upper bound for the intersection of a variety with a general finite subgroup of an algebraic group. As a second application we apply this to bad fields of positive characteristic, to give an asymptotic estimate for the number of $\mathbb{F}_q$-rational points of a definable multiplicative subgroup similar to the Lang-Weil estimate for curves over finite fields.

Introduction

In [1] Larsen and Pink show that if H is a "sufficiently general" finite subgroup of a connected almost simple algebraic group G, then for any subvariety X of G

$$|H \cap X| \leq c \cdot |H|^{\dim(X)/\dim(G)},$$

where the constant c depends only on the form of G and X, but not on H (in other words, G and X are allowed to vary in a constructible family). This theorem was recast (in somewhat greater generality) in model-theoretic form by the first author of the present paper, and rediscovered by the second author in the context of bad fields. In the general form it allows to give an upper bound, for suitable minimal structures

Work done during the semester on Model Theory and Applications to Algebra and Analysis at the Isaac Newton Institute Cambridge, whose hospitality is gratefully acknowledged.
† Supported by Israel Science Foundation grant no. 244/03
‡ Membre junior de l'Institut universitaire de France

with a well-behaved dimension d, of a rudimentary dimension δ (which may for instance be derived from counting measure in a quasi-finite subset) in terms of the original dimension d, typically giving Larsen-Pink like estimates for increasing families of finite subsets. We offer two proofs of the theorem: a more rapid one using types, and a more explicit construction using definable sets. The latter proof could in principle be used to get effective estimates on the constant c.

1 The Main theorem

Definition 1.1 Let $\mathfrak{M}$ be an uncountably saturated structure. A *dimension theory* on $\mathfrak{M}$ is an automorphism-invariant map d from the class of definable sets into $\mathbb{N}$, together with a formal element $-\infty$, satisfying

(1) $d(\emptyset) = -\infty$ and $d(\{x\}) = 0$ for any point x.
(2) $d(X \cup Y) = \max\{d(X), d(Y)\}$.
(3) Let $f : X \to Y$ be a definable map.
  (a) If $d(f^{-1}(y)) = n$ for all $y \in Y$, then $d(X) = d(Y) + n$, for all $n \in \mathbb{N} \cup \{-\infty\}$.
  (b) $\{y \in Y : d(f^{-1}(y)) = n\}$ is definable for all $n \in \mathbb{N} \cup \{-\infty\}$.

It follows that $d(X \times Y) = d(X) + d(Y)$, and $d(X) = d(Y)$ if X and Y are definably isomorphic. By uncountable saturation, $d(f^{-1}(y))$ takes only finitely many values for $y \in Y$. Note that the trivial dimension $d(X) = 0$ for non-empty X is allowed.

Below, A will denote a countable subset of M.

Definition 1.2 For a partial type π let $d(\pi) := \min\{d(X) : X \in \pi\}$; note that the minimum is necessarily attained. If $p = \text{tp}(x/A)$, put $d(x/A) := d(p)$.

For two partial types π, π' over A let

$$\pi \otimes_A \pi' := (\pi \times \pi') \cup \{\neg X : X \ A\text{-definable}, d(X) < d(\pi) + d(\pi')\}.$$

Definition 1.3 Let $\mathfrak{M}$ be a structure with dimension d. A definable subset $F \subset M^3$ is a *correspondence* on $\mathfrak{M}$ if the projection to the first two coordinates is surjective with 0-dimensional fibres. We put

$$F(X) := \{y \in M : \models \exists (x, x') \in X \ F(x, x', y)\}, \text{ and}$$
$$F^{-1}(y) := \{(x, x') \in M^2 : \models F(x, x', y)\}.$$

If $\mathcal{F}$ is a set of correspondences, $\mathfrak{M}$ is *$\mathcal{F}$-minimal* if for any A and partial 1-types π, π' over A with $0 < d(\pi) \leq d(\pi') < d(M)$ and a partial type ρ over A extending $\pi \otimes_A \pi'$, there is $F \in \mathcal{F}$ with $d(F(\rho)) > d(\pi')$.

Roughly speaking, a structure is $\mathcal{F}$-minimal if it is generated from any definable subset by repeated applications of the correspondences in $\mathcal{F}$.

Lemma 1.4 *The following are equivalent:*

(1) *$\mathfrak{M}$ is $\mathcal{F}$-minimal.*
(2) *For any $x, x' \in M$ and set A with $0 < d(x/A) \leq d(x'/A) < d(M)$ and $d(xx'/A) = d(x/A) + d(x'/A)$ there is $F \in \mathcal{F}$ and $y \in F(xx')$ with $d(F^{-1}(y) \cap \mathrm{tp}(xx'/A)) < d(x/A)$.*
(3) *For any A and A-definable X, X' with $0 < d(X) \leq d(X') < d(M)$ and $(x, x') \in X \times X'$ there is A-definable $W \subseteq X \times X'$ with $(x, x') \in W$ such that either $d(W) < d(X \times X')$ or $d(F^{-1}(y) \cap W) < d(X)$ for some $F \in \mathcal{F}$ and $y \in F(xx')$.*

Proof Suppose $\mathfrak{M}$ is $\mathcal{F}$-minimal, and consider x, x', A as in (2). Put $\pi := \mathrm{tp}(x/A)$, $\pi' := \mathrm{tp}(x'/A)$ and $\rho := \mathrm{tp}(xx'/A)$. Since $d(xx'/A) = d(x/A) + d(x'/A)$ we have $\rho \supseteq \pi \otimes_A \pi'$, so there is $F \in \mathcal{F}$ with $d(F(\rho)) > d(\pi')$. In particular there is $y \in F(xx')$ with $d(y/A) > d(x'/A)$. Let $k = d(F^{-1}(y) \cap \mathrm{tp}(xx'/A))$, and choose A-definable $W \in \mathrm{tp}(xx'/A)$ with $d(W) = d(xx'/A)$ and $d(F^{-1}(y) \cap W) = k$, and A-definable $Y \in \mathrm{tp}(y/A)$ with $d(F^{-1}(y') \cap W) = k$ for all $y' \in Y$. Then

$$\begin{aligned} d(x/A) + d(x'/A) = d(xx'/A) &= d(W) \geq d(F \cap (W \times Y)) \\ &= d(Y) + k \geq d(y/A) + k > d(x'/A) + k \end{aligned}$$

(the first inequality holds, since the projection of $F \cap (W \times Y)$ to W has fibres of dimension 0), whence $d(x/A) > k = d(F^{-1}(y) \cap \mathrm{tp}(xx'/A))$.

For the converse, consider partial types π, π' and ρ over A as in the definition of $\mathcal{F}$-minimality, and take $xx' \models \rho$. Since $\rho \supseteq \pi \otimes_A \pi'$ we have $d(\pi) = d(x/A)$, $d(\pi') = d(x'/A)$, $d(\rho) = d(xx'/A)$ and $d(x/A) + d(x'/A) = d(xx'/A)$. By (2) there is $F \in \mathcal{F}$ and $y \in F(xx')$ with $d(F^{-1}(y) \cap \mathrm{tp}(xx'/A)) < d(x/A)$. Choose A-definable $W \in \mathrm{tp}(xx'/A)$ with $k = d(F^{-1}(y) \cap \mathrm{tp}(xx'/A)) = d(F^{-1}(y) \cap W)$, and A-definable $Y \in \mathrm{tp}(y/A)$ with $d(Y) = d(y/A)$ and $d(F^{-1}(y') \cap W) = k$ for all

$y' \in Y$. Then

$$\begin{aligned} d(x/A) + d(x'/A) &= d(xx'/A) = d(\rho) \leq d(F \cap (W \times Y)) \\ &= d(Y) + k = d(y/A) + k < d(y/A) + d(x/A) \end{aligned}$$

(for all $uu' \models \rho$ there is v with $uu'v \in F \cap (W \times Y)$, whence the first inequality), whence $d(F(\rho)) \geq d(y/A) > d(x'/A) = d(\pi')$.

The equivalence (2) $\Leftrightarrow$ (3) follows from the fact that for any partial type π there is $X \in \pi$ with $d(\pi) = d(X)$. □

Example 1 A field of finite Morley rank (possibly with additional structure) is $\{+, \times\}$-minimal.

Proof Suppose $0 < \mathrm{RM}(x/A) \leq \mathrm{RM}(x'/A)$ and $x \mathop{\smile\!\!\!\!\!\!|}_A x'$. If both $\mathrm{RM}(x, x'/x + x', A) \geq \mathrm{RM}(x/A)$ and $\mathrm{RM}(x, x'/xx', A) \geq \mathrm{RM}(x/A)$, then $x \mathop{\smile\!\!\!\!\!\!|}_A x + x'$ and $x \mathop{\smile\!\!\!\!\!\!|}_A xx'$. Let x_0, x_1 be independent realizations of $\mathrm{stp}(x/A, x')$. Since $x_0 + x'$ and $x_1 + x'$ realize the same strong type over A, they realize the same non-forking extension to A, x_0, x_1; a strong automorphism over A, x_0, x_1 mapping $x_0 + x'$ to $x_1 + x'$ will map $x_0 - x_1 + x'$ to x', whence $x_0 - x_1 + x' \models \mathrm{stp}(x'/A)$. As $x' \mathop{\smile\!\!\!\!\!\!|}_A x_0 - x_1$, we get $x_0 - x_1 \in \mathrm{stab}^+(x'/A)$; similarly $x_0 x_1^{-1} \in \mathrm{stab}^\times(x'/A)$. As x_0, x_1 are independent non-algebraic, both stabilizers are infinite; note that obviously $\mathrm{stab}^+(x'/A)$ is $\mathrm{stab}^\times(x'/A)$-invariant. However, in a field K of finite Morley rank the only definable additive subgroup A invariant under an infinite multiplicative subgroup is K itself (otherwise $\{c \in K : cA \leq A\}$ would define an infinite subring, and hence an infinite subfield, a contradiction). Thus $\mathrm{stab}^+(x'/A) = K$, and $\mathrm{RM}(x'/A) = \mathrm{RM}(K)$. □

Example 2 Let G be a simple algebraic group (or more generally, a simple group of finite Morley rank, possibly with additional structure). Let $\mathcal{F}$ be the collection of maps $F_c(x, y) = cx^{-1}c^{-1}y$, where c runs over a countable Zariski-dense subgroup Γ (respectively, subgroup Γ not contained in any proper definable subgroup of G). Then G is $\mathcal{F}$-minimal.

Proof In any group of finite Morley rank, $d = \mathrm{RM}$ is additive and definable. So consider $A \supseteq \Gamma$ and $x \mathop{\smile\!\!\!\!\!\!|}_A x'$ with $0 < \mathrm{RM}(x/A) \leq \mathrm{RM}(x'/A)$, and suppose $\mathrm{RM}(x, x'/cx^{-1}c^{-1}x', A) \geq \mathrm{RM}(x/A)$ for all $c \in \Gamma$. Then $x \mathop{\smile\!\!\!\!\!\!|}_A cx^{-1}c^{-1}x'$, whence $x^{-c^{-1}} \mathop{\smile\!\!\!\!\!\!|}_A x^{-c^{-1}} x'$ for all $c \in \Gamma$. So for any two independent realizations x_0, x_1 of $\mathrm{stp}(x/A, x')$ both $x_0^{-c^{-1}} x'$ and

$x_1^{-c^{-1}}x'$ satisfy the unique non-forking extension of $\text{stp}(x^{-c^{-1}}x'/A)$ to A, x_0, x_1, and $(x_0x_1^{-1})^{c^{-1}}x' \models \text{stp}(x'/A)$. Since $x_0, x_1 \mathop{\smile\!\!\!\!\!\!\,|}_A x'$ this means that $(x_0x_1^{-1})^{c^{-1}} \in \text{stab}(x'/A)$ for any two independent realisations x_0, x_1 of $\text{stp}(x/A)$, and any $c \in \Gamma$. So this stabilizer is an infinite definable subgroup, as is the intersection H of its Γ-conjugates. But then the normalizer of H contains Γ, whence G by our choice of Γ; since H is infinite and G is simple, we get $H = G = \text{stab}(x'/A)$. Therefore $\text{tp}(x'/A)$ is generic, and $\text{RM}(x'/A) = \text{RM}(G)$. □

Definition 1.5 Let $\mathfrak{M}$ be any structure. A *quasi-dimension* on $\mathfrak{M}$ is a map δ from the class of definable sets into an ordered abelian group G, together with a formal element $-\infty$, satisfying

(1) $\delta(\emptyset) = -\infty$, and $\delta(X) > -\infty$ implies $\delta(X) \geq 0$.
(2) $\delta(X \cup Y) = \max\{\delta(X), \delta(Y)\}$, and $\delta(X \times Y) = \delta(X) + \delta(Y)$.
(3) For any definable $X \subseteq M^k$ and projection π to some of the coordinates, if $\delta(\pi^{-1}(\bar{x})) \leq g$ for all $\bar{x} \in \pi(X)$, then $\delta(X) \leq \delta(\pi(X))+g$, for all $g \in G \cup \{-\infty\}$.

We can now state the main theorem.

Theorem 1.6 *Let $\mathfrak{M}$ be an $\mathcal{F}$-minimal structure, where $\mathcal{F}$ is a set of $\emptyset$-definable correspondences for some dimension d. Let δ be a quasi-dimension on $\mathfrak{M}$ such that*

(0) *$d(X) = 0$ implies $\delta(X) \leq 0$ for all definable X.*
(4) *For any $F \in \mathcal{F}$ and definable $X \subseteq M^2$, $Y \subseteq M$ we have $\delta(F \cap (X \times Y)) \geq \delta(X)$, provided for all $xx' \in X$ there is $y \in Y$ with $F(xx'y)$.*

Then $d(M)\delta(X) \leq d(X)\delta(M)$ for any definable set $X \subseteq M$.

Remark 1.7 (1) $\delta(F \cap (X \times Y)) \leq \delta(X)$ follows from axiom (3) and the fact that the fibres of the projection $F \cap (X \times Y) \to X$ have d-dimension zero, and hence δ-dimension zero.
(2) Requirement (4) holds in particular if $\mathcal{F}$ consists of definable functions, and δ is invariant under definable bijections.

The idea of the proof will be that given a set X, by $\mathcal{F}$-minimality there is a sequence $(F_1, \ldots, F_n)$ of correspondences such that for $Y_1 = X$

and $Y_{i+1} = F_i(X, Y_i)$, we get $Y_n = \mathfrak{M}$, and the kernels of the maps $X \times Y_i \to F(X, Y_i)$ all have smaller dimension than X. By inductive hypothesis the kernels have small δ; since $\delta(M)$ is $n\,\delta(X)$ minus δ of the kernels, we get the desired upper bound for $\delta(X)$.

Proof Clearly we may assume $d(M) > 0$. We use induction on $d(X)$. For $d(X) = 0$ the assertion follows from condition (0). So suppose the assertion holds for dimension less than k, and $d(X) = k$. Put $\alpha = \delta(M)/d(M)$ and suppose $\delta(X) \geq \alpha k$.

Lemma 1.8 *Let $X, Y \subseteq M$ be B-definable with $0 < d(X) \leq d(Y)$. Then there is a B-definable finite partition $X \times Y = W_0 \cup \cdots \cup W_n$, correspondences $F_i \in \mathcal{F}$ and sets $Z_i \subseteq F(W_i)$ for $i = 1, \ldots, n$, such that*

(1) *$d(W_i) = d(X) + d(Y)$ for $i > 0$, and $d(W_0) < d(X) + d(Y)$.*
(2) *for all $i > 0$ we have $d(Z_i) > d(Y)$, and $d(F^{-1}(z) \cap W_i) = d(X) + d(Y) - d(Z_i)$ for all $z \in Z_i$.*

Proof For $F \in \mathcal{F}$ and B-definable $W \subseteq X \times Y$ put

$$W_F := \{(x, y) \in W : \exists z \in F(xy)\; d(F^{-1}(z) \cap W) < d(X)\}, \text{ and}$$
$$Z_F := \{z \in F(W_F) : d(F^{-1}(z) \cap W) < d(X)\}.$$

By Lemma 1.4(3) the B-definable sets

$$\{V \subset X \times Y : d(V) < d(X) + d(Y)\} \cup \{W_F : F \in \mathcal{F}, W\ B\text{-definable}\}$$

cover $X \times Y$. By compactness a finite subset covers $X \times Y$; shrinking the sets if necessary, we may assume that the sets form a partition of $X \times Y$. For $i = d(X) - 1, d(X) - 2, \ldots, 0$ partition every Z_F involved into parts

$$Z_F^i := \{z \in Z_F : d(F^{-1}(z) \cap (W_F \setminus \bigcup_{j>i} W_F^j)) = i\},$$

and put $W_F^i = F^{-1}(Z_F^i) \cap (W_F \setminus \bigcup_{j>i} W_F^j)$. Let W_0 be the union of those sets of dimension strictly less than $d(X)+d(Y)$, and enumerate the others as $W_1, \ldots, W_n$ and $Z_1, \ldots, Z_n$, respectively, with correspondences $F_1, \ldots, F_n$. This satisfies the conditions. □

We inductively choose a tree of subsets of M with $Y_\emptyset := X$ and $d(Y_{\eta'}) < d(Y_\eta)$ whenever $\eta' < \eta$ is a proper initial segment. Suppose we have found Y_η. If $d(Y_\eta) = d(M)$ this branch stops. Otherwise put $Y = Y_\eta$ in Lemma 1.8 and let $Y_{\eta i} := Z_i$ for $i > 0$. Put $F_{\eta i} := F_i$, $W_{\eta i} := W_i$, and $n_{\eta i} := n_i = d(X) + d(Y_\eta) - d(Y_{\eta i})$. As $d(Y_{\eta i}) > d(Y_\eta)$ for all η, the tree is finite. Let m be the maximal length of a branch, and put $m_\eta = m - |\eta|$, where $0 \le |\eta| \le m$ is the length of η.

Lemma 1.9 *If $W \subset X^{m_{\eta i}} \times Y_{\eta i}$ with $d(W) < d(X^{m_{\eta i}} \times Y_{\eta i})$, then $d((id_{X^{m_\eta - 1}} \times F_{\eta i})^{-1}(W) \cap (X^{m_\eta - 1} \times W_{\eta i})) < d(X^{m_\eta} \times Y_\eta)$.*

Proof Since the fibres have constant dimension $n_{\eta i}$, we have

$$\begin{aligned} d((id_{X^{m_\eta - 1}} \times F_{\eta i})^{-1}(W) \cap (X^{m_\eta - 1} \times W_{\eta i})) &= d(W) + n_{\eta i} \\ &< d(X^{m_{\eta i}} \times Y_{\eta i}) + d(X) + d(Y_\eta) - d(Y_{\eta i}) \\ &= d(X^{m_\eta} \times Y_\eta). \end{aligned}$$

□

If $d(Y_\eta) = d(M)$ put $V_m = \emptyset$, and if V_{η_i} has been defined for all $i > 0$ put

$$V_\eta := (X^{m_\eta - 1} \times W_{\eta 0}) \cup \bigcup_{i>0} [(id_{X^{m_\eta - 1}} \times F_{\eta i})^{-1}(V_{\eta i}) \cap (X^{m_\eta - 1} \times W_{\eta i})].$$

Then inductively $d(V_\eta) < d(X^{m_\eta} \times Y_\eta)$. In particular $d(V_\emptyset) < d(X^{m+1})$.

Lemma 1.10 *If $W \subset X^n$ with $d(W) < n\, d(X)$, then $\delta(W) < n\, \delta(X)$.*

Proof We use induction on n, the assertion being trivial for $n = 0, 1$. So assume it holds for n, and consider $W \subseteq X^{n+1}$. Let π be the projection of W to the first n coordinates, and put $W_i = \{\bar{x} \in \pi(W) : d(\pi^{-1}(\bar{x})) = i\}$ for $i \le k$. Since $d(W) < d(X^{n+1})$, we have $d(W_k) < d(X^n)$. So by inductive hypothesis

$$\begin{aligned} \delta(\pi^{-1}(W_k)) \le \delta(W_k \times X) &= \delta(W_k) + \delta(X) \\ &< \delta(X^n) + \delta(X) = (n+1)\, \delta(X). \end{aligned}$$

On the other hand, for $\bar{x} \in W_i$ with $i < k$ we have

$$\delta(\pi^{-1}(\bar{x})) \le \alpha\, d(\pi^{-1}(\bar{x})) = \alpha\, i$$

by our global inductive hypothesis. Hence by requirement (3)

$$\delta(\pi^{-1}(W_i)) \le \delta(W_i) + \alpha\, i \le \delta(X^n) + \alpha\, (k-1) < (n+1)\, \delta(X)$$

since we assume $\delta(X) \ge \alpha\, k$. Thus

$$\delta(W) = \max_{i \le k} \delta(\pi^{-1}(W_i)) < (n+1)\, \delta(X).$$

□

It follows that $\delta(V_\emptyset) < \delta(X^{m+1})$, and

$$(m+1)\,\delta(X) = \delta(X^{m+1}) = \delta((X^{m_\emptyset} \times Y_\emptyset) \setminus V_\emptyset).$$

For $\bar{y} \in (X^{m_{\eta i}} \times Y_{\eta i}) \setminus V_{\eta i}$

$$d((id_{X^{m_\eta - 1}} \times F_{\eta i})^{-1}(\bar{y}) \cap [(X^{m_\eta - 1} \times W_{\eta i}) \setminus V_\eta]) \leq n_{\eta i} < k,$$

so by inductive hypothesis

$$\delta((id_{X^{m_\eta - 1}} \times F_{\eta i})^{-1}(\bar{y}) \cap [(X^{m_\eta - 1} \times W_{\eta i}) \setminus V_\eta]) \leq \alpha\, n_{\eta i}.$$

Hence

$$\begin{aligned}\delta((id_{X^{m_\eta - 1}} \times F_{\eta i}) \cap ([(X^{m_\eta - 1} \times W_{\eta i}) \setminus V_\eta] \times [(X^{m_{\eta i}} \times Y_{\eta i}) \setminus V_{\eta i}]))\\ \leq \delta((X^{m_{\eta i}} \times Y_{\eta i}) \setminus V_{\eta i}) + \alpha\, n_{\eta i}\end{aligned}$$

by assumption (3), and

$$\begin{aligned}\delta((id_{X^{m_\eta - 1}} \times F_{\eta i}) \cap ([(X^{m_\eta - 1} \times W_{\eta i}) \setminus V_\eta] \times [(X^{m_{\eta i}} \times Y_{\eta i}) \setminus V_{\eta i}]))\\ \geq \delta((X^{m_\eta - 1} \times W_{\eta i}) \setminus V_\eta])\end{aligned}$$

by assumption (4). Since $(X^{m_\eta} \times Y_\eta) \setminus V_\eta) = \bigcup_{i>0} (X^{m_\eta - 1} \times W_{\eta i}) \setminus V_\eta)$,

$$\begin{aligned}\delta((X^{m_\eta} \times Y_\eta) \setminus V_\eta) &= \max_{i>0} \delta((X^{m_\eta - 1} \times W_{\eta i}) \setminus V_\eta)\\ &\leq \max_{i>0} \delta((X^{m_{\eta i}} \times Y_{\eta i}) \setminus V_{\eta i}) + \alpha\, n_{\eta i}.\end{aligned}$$

On the other hand, $d(X) + d(Y_\eta) = d(Y_{\eta i}) + n_{\eta i}$ for all η and $i > 0$. Let η be the branch which corresponds always to the maximum of the δ-dimensions. Summing over the initial segments of η we obtain

$$\begin{aligned}(m+1)\,\delta(X) = \delta((X^m \times Y_\emptyset) \setminus V_\emptyset) &\leq \delta(X^{m_\eta} \times Y_\eta) + \alpha \sum_{\emptyset < \eta' \leq \eta} n_{\eta'}\\ &= m_\eta\, \delta(X) + \delta(Y_\eta) + \alpha \sum_{\emptyset < \eta' \leq \eta} n_{\eta'}\\ &\leq (m - |\eta|)\, \delta(X) + \delta(M) + \alpha \sum_{\emptyset < \eta' \leq \eta} n_{\eta'},\end{aligned}$$

whereas

$$(|\eta| + 1)\, d(X) = d(Y_\eta) + \sum_{\emptyset < \eta' \leq \eta} n_{\eta'} = d(M) + \sum_{\emptyset < \eta' \leq \eta} n_{\eta'}.$$

Therefore

$$(|\eta| + 1)\, \delta(X) \leq \alpha\, (d(M) + \sum_{\emptyset < \eta' \leq \eta} n_{\eta'}) = \alpha\, (|\eta| + 1)\, d(X),$$

and $\delta(X) \leq \alpha\, d(X)$. This proves the theorem. □

We shall now give a second, type-based proof for Theorem 1.6.

Proof We use induction on $d(X) =: k$, the assertion following from condition (0) if $k = 0$. For partial types $(\pi_i : i < m)$ and $(\pi'_j : j < n)$ and rationals α_i and α'_j we put

$$\sum_{i<m} \alpha_i\, \delta(\pi_i) \leq \sum_{j<n} \alpha'_j\, \delta(\pi'_j)$$

if for every choice of $X'_j \in \pi'_j$ there are $X_i \in \pi_i$ with $\sum_{i<m} \alpha_i\, \delta(X_i) \leq \sum j < n\alpha'_j\, \delta(X'_j)$. Note that $\leq$ is transitive.

Claim *It is enough to prove the assertion for complete types.*

Proof of Claim Let X be an A-definable set, and $\mathfrak{X}$ the collection of A-definable $X' \subseteq X$ such that $d(M)\delta(X') \leq d(X')\delta(M)$. Then $\mathfrak{X}$ is closed under finite unions, so either $d(M)\delta(X) \leq d(X)\delta(M)$, or there is a type $p \in S(A)$ completing the partial type $\{X \setminus X' : X' \in \mathfrak{X}\}$. By assumption $d(M)\delta(p) \leq d(p)\delta(M)$. So there are A-definable $X_1, X_2 \in p$ with $d(M)\delta(X_1) \leq d(p)\delta(M)$ and $d(X_2) = d(p)$. But then $X_1 \cap X_2 \in \mathfrak{X}$, a contradiction. □

So let $p \in S_1(A)$ with $d(p) = k$. For ease of notation we assume that the value group G of δ is divisible. Clearly we may also assume that $d(M)\,\delta(p) \geq d(p)\,\delta(M)$.

Claim *If $p' \in S_1(A)$, there is $q \in S_2(A)$ extending $p \otimes_A p'$ with $\delta(p) + \delta(p') \leq \delta(q)$.*

Proof of Claim Suppose not, and consider

$$\mathfrak{X} := \{X \subseteq M^2 \text{ } A\text{-definable} : \delta(p) + \delta(p') \not\leq \delta((p \times p') \cup \{X\})\}.$$

Then $\mathfrak{X}$ is closed under finite unions, and we can put

$$\rho := (p \times p') \cup \{\neg X : X \in \mathfrak{X}\},$$

a consistent partial type. By assumption $d(\rho) < d(p)+d(p')$, as otherwise we could complete ρ to a type q with $d(q) = d(p) + d(p')$, whence $q \supseteq p \otimes_A p'$ and $\delta(p) + \delta(p') \leq \delta(q)$. Hence the projection to the second coordinate has fibres of dimension $i < k$. So there are A-definable sets $X \in p$, $X' \in p'$ and $X \times X' \supset Y \in \rho$ with $d(X) = d(p)$, $d(X') = d(p')$, $d(Y) = d(\rho)$ and $d(Y \cap (X \times \{x'\})) = i$ for all $x' \in X'$. By inductive hypothesis $d(M)\delta(Y \cap (X \times \{x'\})) \leq i\,\delta(M)$ for all $x' \in X'$, so by property (3)

$$\delta(Y) \leq \delta(X') + i\,\delta(M)/d(M) \leq \delta(X') + i\,\delta(p)/d(p)\,;$$

as one can choose Y depending on X' we get

$$\delta(\rho) \leq \delta(p') + \frac{i}{k}\delta(p) < \delta(p') + \delta(p),$$

since $\delta(p)$ is bounded below by $\delta(M)\,d(p)/d(M)$, a contradiction to the definition of ρ. □

By $\mathcal{F}$-minimality there is $n < \omega$, a sequence $p = p_0, p_1, \ldots, p_n$ of complete types over A, a complete A-type $q_i \supseteq p \otimes_A p_i$ with $\delta(p) + \delta(p_i) \leq \delta(q_i)$ for $i < n$, and correspondences $(F_i : i < n)$ in $\mathcal{F}$, such that p_{i+1} is a completion of $F_i(q_i)$ for all $i < n$ with $d(p_i) < d(p_{i+1})$, and $d(p_n) = d(M)$. For $i < n$ put $R_i := F_i \cap (q_i \times p_{i+1})$, and choose A-definable sets $X \in q_i$, $X' \in p_{i+1}$ and $Y \in R_i$ with $d(X) = d(q_i) = d(p) + d(p_i)$, $d(X') = d(p_{i+1})$, $Y \subseteq X \times X'$, and such that the fibres of the projection π of Y to the last coordinate have constant dimension $j_i = d(\pi^{-1}(a))$, where $a \models X'$. Then

$$d(X') + j_i = d(Y) = d(X) = d(p) + d(p_i) < d(p) + d(X')$$

by axiom (3)(a). By inductive hypothesis $\delta(\pi^{-1}(a)) \leq j_i\,\delta(M)/d(M)$ for all $a \in X'$, whence $\delta(Y) \leq \delta(X') + j_i\,\delta(M)/d(M)$. Letting X' converge to p_{i+1} and Y to R_i, we obtain $\delta(R_i) \leq \delta(p_{i+1}) + j_i\,\delta(M)/d(M)$.

Since condition (4) implies $\delta(q_i) \leq \delta(F_i \cap (q_i \times p_{i+1})) = \delta(R_i)$, we get

$$\delta(p) + \delta(p_i) \leq \delta(q_i) \leq \delta(R_i) \leq \delta(p_{i+1}) + j_i\,\delta(M)/d(M).$$

Summing the inequalities for $i < n$, we obtain

$$(n+1)\,\delta(p) \leq \delta(p_n) + \frac{\delta(M)}{d(M)}\sum_{i<n} j_i = \delta(M) + \frac{\delta(M)}{d(M)}\sum_{i<n} j_i.$$

On the other hand,

$$d(M) + \sum_{i<n} j_i = d(p_n) + \sum_{j<n}[d(p) + d(p_i) - d(p_{i+1})] = (n+1)\,d(p),$$

whence

$$(n+1)\,\delta(p) \leq \frac{\delta(M)}{d(M)}[d(M) + \sum_{i<n} j_i] = \frac{\delta(M)}{d(M)}(n+1)\,d(p),$$

which proves the theorem. □

Remark 1.11 The above proof of Theorem 1.6 defined the relation $\delta(\pi) \leq \delta(\pi')$ without actually defining the quantities $\delta(\pi)$. Perhaps for other applications an invariant $\delta(\pi)$ for types may be useful. We sketch now how this may be done.

Definition 1.5 requires δ to be a function into the non-negative elements of a linearly ordered group G that can be assumed divisible. In place of this, let us gain generality by taking $G = (G, +, 0, <)$ to be a divisible linearly ordered commutative semi-group. This means that (1)–(2) below hold; we may as well assume (3); we assume cancellation only in the limited form (4), with respect to a distinguished element $\delta(M)$.

(1) $(G, +, 0)$ is an additive semi-group, with every element uniquely divisible by any positive integer.
(2) $<$ is a linear ordering, and $x \leq y$ implies $x + z \leq y + z$.
(3) For any $x \in G$ there is $k < \omega$ with $0 \leq x \leq k\,\delta(M)$.
(4) $x + \delta(M) > x$ for any x.

It follows that $x + \frac{1}{n}\,\delta(M) > x$ for any x and integer $n > 0$.

These more general assumptions have the advantage that the semi-group G can be completed by means of Dedekind cuts. The assumptions continue to hold; in particular (4) does, since if U is a Dedekind cut invariant under adding $\delta(M)$, then by (3) it must include all of G, but Dedekind cuts are assumed bounded.

Now for any partial type $\pi = \bigwedge_{i \in I} X_i$ we can define $\delta(\pi) = \inf_{i \in I} \delta(X_i)$. The earlier definition of the inequality is now a consequence. Whether the greater generality has any additional use, we do not know.

Corollary 1.12 *Under the same hypotheses as Theorem 1.6, let $X \subset M^n$ be definable. Then $d(M)\,\delta(X) \leq d(X)\,\delta(M)$.*

Proof We use induction on n, the assertion being Theorem 1.6 for $n = 1$. For $X \subseteq M^{n+1}$ let π be the projection to the first n coordinates, and partition $Y := \pi(X)$ into sets

$$Y_i := \{\bar{x} \in Y : d(\pi^{-1}(\bar{x}) \cap X) = i\}.$$

Let $X_i := \pi^{-1}(Y_i) \cap X$, then $(X_i : i \leq d(M))$ partitions X, and

$$d(X) = \max_{i \leq d(M)} d(X_i) = \max_{i \leq d(M)} d(Y_i) + i.$$

For every $i \leq d(M)$ and $\bar{x} \in Y_i$ Theorem 1.6 yields $\delta(\pi^{-1}(\bar{x}) \cap X) \leq \alpha i$, with $\alpha = \delta(M)/d(M)$. By inductive hypothesis $\delta(Y_i) \leq \alpha\, d(Y_i)$, so

$$\begin{aligned}\delta(X) = \max_{i \leq d(M)} \delta(X_i) &\leq \max_{i \leq d(M)} \delta(Y_i) + \alpha i \\ &\leq \alpha \max_{i \leq d(M)} d(Y_i) + i = \alpha\, d(X).\end{aligned}$$

□

Remark 1.13 If $\mathfrak{M}$ is $\mathcal{F}$-minimal, then $\mathfrak{M}^n$ can be shown to be minimal with respect to the induced set of correspondences; this yields an alternative proof of Corollary 1.12.

2 An example that counts

Let $(\mathfrak{M}_n : n < \omega)$ be a family of $\mathcal{L}$-structures for some language $\mathcal{L}$, and Γ_n finite subsets of M_n. For some ultrafilter on ω let $\langle \mathfrak{M}, \Gamma \rangle$ be the ultraproduct of the structures $\langle \mathfrak{M}_n, \Gamma_n \rangle$. The ultraproduct of the counting measures on the Γ_n yields a finitely additive measure μ on the definable subsets of Γ which takes values in some non-standard real closed field $\mathbb{R}^*$. Note that $\langle \mathfrak{M}, \Gamma, \mathbb{R}^*, \mu, \log \rangle$ is $\aleph_0$-saturated (in fact, even $\aleph_1$-saturated).

Let I be the convex hull of $\mathbb{Z}$ in $\mathbb{R}^*$, and $\pi : \mathbb{R}^* \to \mathbb{R}^*/I$ the natural (additive) quotient map. For a definable subset X of Γ define

$$\delta(X) = \pi \log \mu(X),$$

and note that $\delta(X) = 0$ if and only if $\log \mu(X) \in I$, that is $\mu(X) \in I$, in other words $\mu_n(X_n) = O(1)$ in the factors, that is X is finite in the ultraproduct. For a definable subset Y of M we put $\delta(Y) := \delta(Y \cap \Gamma)$.

Lemma 2.1 *Assume that $\mathfrak{M}$ has a dimension d such that $d(X) = 0$ implies X finite, and Γ is closed under the correspondences (i.e. for all $xx' \in \Gamma^2$ and $y \in M$ such that $F(xx'y)$ holds, $y \in \Gamma$ as well). Then δ satisfies conditions (0)–(4) from Theorem 1.6.*

Proof (1) is obvious. For (2) note that

$$\mu(X \cup Y) \leq \mu(X) + \mu(Y) \leq 2 \max\{\mu(X), \mu(Y)\},$$

whence $\log(\mu(X \cup Y)) \leq \log 2 + \max\{\log \mu(X), \log \mu(y)\}$. Since $\log 2 \in I$, we get $\delta(X \cup Y) \leq \max\{\delta(X), \delta(Y)\}$; the other inequality follows from monotonicity.

We claim that for any definable map $f : X \to Y$, if $\delta(f^{-1}(y)) \leq \alpha$ for all $y \in Y$, then there is $r \in \mathbb{R}^*$ with $\pi(r) = \alpha$ and $\log \mu(f^{-1}(y)) \leq r$ for all $y \in Y$. Indeed, pick any $r_0 \in \mathbb{R}^*$ with $\pi(r_0) = \alpha$. Put

$$Y_n := \{y \in Y : \log \mu(f^{-1}(y)) \leq r_0 + n\}.$$

Then $Y_n \subset Y_{n+1}$ for all $n < \omega$, and $Y = \bigcup_{n<\omega} Y_n$; by $\aleph_0$-saturation there is n_0 with $Y = Y_{n_0}$. Then $r := r_0 + n_0$ will do.

This shows (3). Finally, (4) is clear, since the fibres of the projection of any $F \in \mathcal{F}$ to the first two coordinates must have d-dimension zero, hence be finite in the ultraproduct, and thus uniformly finite in the factors; they are non-empty by closedness of Γ under $\mathcal{F}$. □

Unwinding the definitions, for this choice of δ (and suitable dimension d) the inequality $d(M)\delta(X) \leq \delta(M)d(X)$ becomes

$$|X_n \cap \Gamma_n| \leq O(|\Gamma_n|^{d(X)/d(M)}).$$

Possible choices for d include algebraic dimension, Morley rank, Shelah rank, Lascar rank, SU-rank or S_1-rank, whenever it is finite, additive and definable in the pure $\mathcal{L}$-structure $\mathfrak{M}$.

Remark 2.2 Uniformity in parameters of the constant intervening in the O-notation follows automatically from compactness.

Remark 2.3 Note that for any definable map $f : X \to Y$:

(1) If $\delta(f^{-1}(y)) \geq \alpha$ for all $y \in Y$, then $\delta(Y) + \alpha \leq \delta(X)$.
(2) If $\delta(f^{-1}(y)) \leq \alpha$ for all $y \in Y$ and $f(X \cap \Gamma) \subseteq \Gamma$, then $\delta(X) \leq \delta(Y) + \alpha$.

In particular δ is invariant under definable bijections f preserving Γ (i.e. $x \in \Gamma$ if and only if $f(x) \in \Gamma$).

3 An application

We shall now give the model-theoretic formulation of the theorem by Larsen and Pink alluded to in the introduction.

Theorem 3.1 [1] *Let G_n be a simple algebraic group varying in an algebraic family and Γ_n a finite subgroup such that in the ultraproduct G the subgroup Γ is Zariski-dense. Then for any subvariety V of G*

$$|V_n \cap \Gamma_n| \leq O(|\Gamma_n|^{\dim(V)/\dim(G)}).$$

Proof Since G_n varies in an algebraic family, G is a simple algebraic group, and $d = \dim = \mathrm{RM}$ is finite, additive and definable. Let $\mathcal{F}$ be the collection of maps $F_c(x, y) = cx^{-1}c^{-1}y$, where c runs over a countable Zariski-dense subgroup Γ_0 of Γ. Clearly Γ is $\mathcal{F}$-closed; moreover G is $\mathcal{F}$-minimal by Example 2. Theorem 1.6 and Lemma 2.1 yield the result. □

Corollary 3.2 [1] *In the setting of Theorem 3.1 consider $a \in \Gamma$ with $\mathrm{RM}(C_G(a)) > 0$, $\mathrm{RM}(a^G) > 0$ and $\delta(G) > 0$. Then Γ meets both $C_G(a)$ and a^G in infinite sets.*

Proof Using the definable map $x \mapsto a^x$ and translation maps between $C_G(a)$ and its cosets, we see that

$$\mathrm{RM}(C_G(a)) + \mathrm{RM}(a^G) = \mathrm{RM}(G), \text{ and}$$
$$\delta(C_G(a)) + \delta(a^G) = \delta(G).$$

If $\alpha = \delta(G)/\mathrm{RM}(G)$, then $\delta(C_G(a)) \leq \alpha\,\mathrm{RM}(C_G(a))$ and $\delta(a^G) \leq \alpha\,\mathrm{RM}(a^G)$ by Theorem 1.6 and Lemma 2.1, so equality must hold. □

4 Bad fields

A *bad field* [2] is a structure $\langle K, 0, 1, +, -, \cdot, T\rangle$ of finite Morley rank, where T is a predicate for a distinguished infinite proper connected multiplicative subgroup (or even a non-algebraic connected subgroup of $(K^\times)^n$ for some n, but these shall not be considered here). Such an object appears naturally when considering a faithful action of an abelian group M on an M-minimal abelian group A, the whole of finite Morley rank: We obtain that there is an algebraically closed field K such that $A \cong K^+$ and $M \hookrightarrow K^\times$; one knows that the image of M generates K additively, but *a priori* it could be a proper subgroup. In particular, the possible existence of bad fields (and of bad groups) prevents us from proving an analogue of the Feit-Thompson theorem for simple groups of finite Morley rank, namely that they contain an involution (or, indeed, any torsion element at all).

In [3] the second author showed that under the assumption that there are infinitely many prime numbers of the form $(p^n-1)/(p-1)$ (called *p-Mersenne primes*), there is no bad field of characteristic $p>0$. In [4] he obtained an asymptotic estimate for the number of $\mathbb{F}_q$-rational points of a multiplicative subgroup of rank 1; this shows the nonexistence of bad fields with $\mathrm{RM}(T)$ of rank 1 modulo a slightly weaker number-theoretic hypothesis. We can now obtain an analogous asymptotic estimate for multiplicative subgroups of arbitrary rank.

For two functions f and g on $\mathbb{N}$ we put $f \asymp g$ if there are positive constants c, c' with $cf(n) \leq g(n) \leq c'f(n)$ for all $n \in \mathbb{N}$.

Theorem 4.1 *For any definable subset X of a bad field K of positive characteristic and any finite subfield $\mathbb{F}_q \leq K$ we have $|X \cap \mathbb{F}_q| \leq O(q^{\mathrm{RM}(X)/\mathrm{RM}(K)})$. In particular $|T \cap \mathbb{F}_{p^n}| \asymp p^{n\,\mathrm{RM}(T)/\mathrm{RM}(K)}$.*

Proof Let $\langle K, T\rangle$ be a bad field of characteristic $p>0$. We put $\mathfrak{M}_n = \langle K, T\rangle$ for all $n<\omega$, and $\Gamma_n = \mathbb{F}_{p^n}$; our correspondences $\mathcal{F}$ will be addition and multiplication. Clearly Γ is closed under $\mathcal{F}$, and K is $\mathcal{F}$-minimal by Example 1. So Theorem 1.6 and Lemma 2.1 imply the first assertion.

By [3, Theorem 2] there is an $\emptyset$-definable partial function $f: K \to T$ with generic domain and an integer $\ell > 0$ such that $f(ta) = t^\ell f(a)$ for all $a \in \mathrm{dom}(f)$ and all $t \in T$ (in particular $\mathrm{dom}(f)$ is closed under multiplication by T). By connectivity T is ℓ-divisible, so all fibres have the same rank, namely $\mathrm{RM}(K) - \mathrm{RM}(T)$. Hence the number of $\mathbb{F}_q$-points on a fibre is bounded by $O(q^{1-\alpha})$, where $\alpha = \mathrm{RM}(T)/\mathrm{RM}(K)$. Moreover, the complement of the domain has rank at most $\mathrm{RM}(K)-1$, so its number of $\mathbb{F}_q$-points is bounded by $O(q^{1-1/\mathrm{RM}(K)})$. Since $\mathbb{F}_q$ is precisely the set of fixed points of the definable automorphism $x \mapsto x^q$, it is closed under all $\mathbb{F}_q$-definable functions. Hence the number of $\mathbb{F}_q$-points of T is at least $(q - O(q^{1-1/\mathrm{RM}(K)}))/O(q^{1-\alpha}) \geq cq^\alpha$ for some constant c. □

Definition 4.2 Let π be a set of prime numbers. For an integer n the *π-part* n_π is the biggest π-number (with all prime divisors in π) dividing n.

Corollary 4.3 *Suppose $\langle K, T\rangle$ is a bad field of characteristic $p>0$,* *and let π be the set of prime orders of elements in T. Then*

$$(p^n - 1)_\pi \asymp p^{\alpha n},$$

with $\alpha = \mathrm{RM}(T)/\mathrm{RM}(K)$.

Proof Since T is divisible, it is a direct sum of Prüfer groups. Hence if k is the subfield of K with p^n elements and q is a prime dividing $|T \cap k^\times|$, then T contains all of the q-part of $k^\times$. Thus $|T \cap k| = (p^n - 1)_\pi$. □

Definition 4.4 Let $0 < \alpha < 1$. A set π of primes is (p, α)-*balanced* if $\big((p^n - 1)_\pi\big) \asymp p^{\alpha n}$. It is p-balanced if it is (p, α)-balanced for some α with $0 < \alpha < 1$.

Note that if π is (p, α)-balanced, then the complement of π is $(p, 1 - \alpha)$-balanced.

Corollary 4.5 *If there is no p-balanced set, then there is no bad field of characteristic p.*

Proof This follows immediately from Corollary 4.3. □

References

[1] Michael J. Larsen and Richard Pink, Finite subgroups of algebraic groups, Preprint 1998, available at `http://www.math.ethz.ch/~pink/`

[2] Bruno Poizat, *Groupes Stables*, Nur al-Mantiq wal-Marifah, Villeurbanne, 1987. Translated as: *Stable Groups*, AMS, 2002.

[3] Frank O. Wagner, Bad fields of positive characteristic, *Bulletin of the London Mathematical Society*, 35 (2003), 499-502.

[4] Olivier Roche and Frank O. Wagner, Bad fields with a torus of rank 1, *Proceedings of the Euro-conference on Model Theory and Applications, Ravello 2002*, pp. 367–377, Quaderni di Matematica, Seconda Università di Napoli, Caserta, 2005.

A survey on groups definable in o-minimal structures

Margarita Otero†
Universidad Autónoma de Madrid

1 Introduction

Groups definable in o-minimal structures have been studied for the last twenty years. The starting point of all the development is Pillay's theorem that a definable group is a definable group manifold (see Section 2). This implies that when the group has the order type of the reals, we have a real Lie group. The main lines of research in the subject so far have been the following:

(1) Interpretability, motivated by an o-minimal version of Cherlin's conjecture on groups of finite Morley rank (see Sections 4 and 3).

(2) The study of the Euler characteristic and the torsion, motivated by a question of Y. Peterzil and C. Steinhorn and results of A. Strzebonski (see Sections 6 and 5).

(3) Pillay's conjectures (see Sections 8 and 7).

On interpretability, we have a clear view, with final results in Theorems 4.1 and 4.3 below. Lines of research (2) and (3) can be seen as a way of comparing definable groups with real Lie groups (see Section 2). The best results on the Euler characteristic are those of Theorems 6.3 and 6.5. The study of the torsion begins the study of the algebraic properties of definable groups, and the best result about the algebraic structure of the torsion subgroups is Theorem 5.9. On the other hand, the cases in which Pillay's conjectures are proved are stated in Theorems 8.3 and 8.8. In my opinion, what is most beautiful about the proofs of the conjectures is that they make an essential use of most of the results obtained for definable groups; they also bring into the game properties of o-minimal structures, such as NIP and properties of Keisler measures,

† Partially supported by the Newton Institute, MODNET FP6-MRTN-CT-2004-512234 and GEOR MTM2005-02568

which have not been used so far, for the study of definable groups; and they introduce the notion of compact domination which opens a new line of research.

For the rest of the paper, I fix an o-minimal structure $\mathcal{M}$ and 'definable' means 'definable in $\mathcal{M}$'. Any other assumption on $\mathcal{M}$ is stated either at the beginning of a section or within the statements of the results. (See Section 2 for more conventions.)

A group G is *definable* if both the set and the graph of the group operation are sets definable (with parameters) in $\mathcal{M}$. Here are some examples of definable groups when $\mathcal{M}$ is an o-minimal expansion of a real closed field: the additive group of M; the multiplicative group of M; the algebraic subgroups of $\mathrm{GL}(n, M)$ and $H(M)$, the M-rational points of an (abstract) algebraic group H defined over the field M. Since every compact Lie group is isomorphic, as a Lie group, to a (linear) real algebraic group (see [9]), compact Lie groups are also examples of groups definable over the ordered real field. In [47] A. Strzebonski gives a good list of examples of semialgebraic groups. More examples can be found in [40] (see comments after Theorem 5.2 below). In [35], Peterzil, Pillay and Starchenko give examples of definable subgroups of $\mathrm{GL}(n, M)$ – for certain o-minimal expansions $\mathcal{M}$ of the real field – which are not definably isomorphic to semialgebraic groups (compare with Theorem 4.6 below).

When the order type of $\mathcal{M}$ is $(\mathbb{R}, <)$, Theorem 2.1 (see below) implies that a definable group G is a Lie group. It is because of this last fact that Lie groups have been a good source for the study of definable groups. Much research in the field has been done comparing both situations: Lie and definable. I refer the reader to [8] and [9] for results about Lie groups. Properties of definable groups which correspond to analogous properties of Lie groups can be found below (e.g., Theorems 2.1, 4.8, 5.9 and 6.7). Sometimes definable groups behave better than Lie groups (e.g., definable subgroups are closed), but some others existence results in Lie groups (*e.g.*, the existence of subgroups or isomorphisms) cannot be ensured in the definable context. Other sources for the study of definable groups have been groups of finite Morley rank or groups definable in a strongly minimal structure. We can find good examples of such results in Sections 3 and 4.

If $\mathcal{M}$ is the ordered real field then a definable group G is a Nash group. That is, G is equipped with a semialgebraic manifold structure in which the maps involved in the manifold structure are Nash (real-analytic and semialgebraic), and multiplication and inversion are also Nash maps. E. Hrushovski and A. Pillay study in [19] groups definable in local fields and they prove that if G is a (Nash) group definable in the real field, then there is an algebraic group H defined over $\mathbb{R}$, and a Nash isomorphism between neighbourhoods of the identity of G and $H(\mathbb{R})$ (see Theorem A in [19]). This local isomorphism is the best possible because it is not always possible to lift a local Nash isomorphism to a global one. In affine groups they avoid this obstacle (note that the embedding obtained by Robson's theorem – see Section 2 – does not need to be a Nash map). More precisely, they prove that if G is a connected Nash group over $\mathbb{R}$, for which there is a Nash embedding of G into some $\mathbb{R}^l$, then there is a Nash surjective homomorphism with finite kernel between G and the real connected component of the set of real points of an algebraic group defined over $\mathbb{R}$ (see Theorem B in [19]). The proofs of these two results make use of the study of geometric structures over the reals, also developed in [19].

I have written this paper as a collection of known results on definable groups with hopefully precise references, and there are also some open problems scattered through the text. I have tried to give the taste of the proofs or at least to give some information about what one can expect to find if going to the original papers. For basic properties of o-minimal structures I refer the reader to [10] and [45].

I started working on this paper during my visit to Cambridge within the Programme Model Theory and Applications to Algebra and Analysis in the spring 2005. I would like to thank both the Newton Institute for their hospitality, and Sergei Starchenko for many instructive comments on some of the results stated here. I also thank Alessandro Berarducci for helpful conversations.

2 The manifold structure and basic properties

I begin by stating the following essential result on definable groups due to A. Pillay.

Theorem 2.1 *Let G be a definable group. Then G can be equipped with a definable manifold structure making G a topological group.*

This is Proposition 2.5 in [42]. For a detailed description of the manifold structure see Section 1.1 in [33]. The ingredients of the proof are the following. Firstly, Pillay proves that the (model theoretic) dimension of a definable set coincides with its geometrical dimension (via cells) – Lemma 1.4 in [42]. Then, he proves the following key lemma.

Lemma 2.2 *Let G be a definable group. Let X be a large subset of G. Then finitely many translates of X cover G.*

This is Lemma 2.4 in [42], where X *large* means that $\dim(G \setminus X) < \dim G$. For the rest of the proof of the theorem, A. Pillay adapts to the o-minimal context a proof by E. Hrushovski of Weil's theorem that an algebraic group over an algebraic closed field can be defined from *birational data.*

It is easy to see that the topology of our definable group $G \subset M^n$, say, obtained in Theorem 2.1 does not coincide in general with the topology induced by the ambient space M^n: take for instance $[0, 1) \subset \mathbb{R}$ and let the operation be addition *mod* 1 (with the definable manifold topology $1-x$ tends to 0 when x tends to 0). However, as G is a topological group, it is a (Hausdorff) regular manifold (regular as a topological space) and when $\mathcal{M}$ is an expansion of a real closed field, G can be embedded (as a manifold) in some M^l; this is due to Robson's theorem (see Theorem 1.8 in Section 10 in [10]). Therefore, the following convention can be used for the rest of the paper:

When $\mathcal{M}$ is an expansion of a real closed field all the topological concepts about definable groups refer to the topology induced by the ambient space. Otherwise, the concepts mentioned refer to the manifold topology.

In particular, the notion of *definable compactness*, introduced by Y. Peterzil and C. Steinhorn – Definition 1.1 in [40] – is equivalent to that of being closed and bounded, when we work over an expansion of a real closed field. This is because in Theorem 2.1 in [40], they prove that both concepts coincide when the topology is the one induced by the ambient space. Note also that when $\mathcal{M}$ is an expansion of a real closed field the usual definition of derivative makes sense and we can speak of *C^p–maps* for $p \geq 0$. In this case, the same proof of Theorem 2.1 (which is the case $p = 0$) gives that a definable group carries on a definable C^p–manifold structure. See Section 1.1 and 1.2 in [33] for details. Also in [33], Y. Peterzil, A. Pillay and S. Starchenko adapt the proof of Theorem 2.1 above to prove – Theorem 2.12 there – the following result.

Theorem 2.3 *Let $p \geq 0$ and let $\mathcal{M}$ be an expansion of a real closed field if $p > 0$. Let G be a definable group, A a definable set and α a definable transitive action of G on A. Then, A can be equipped with a definable C^p-manifold structure making α a definable C^p-action.*

Note that when $\mathcal{M}$ is an expansion of a real closed field and N is a definable normal subgroup of a definable group G, we have three topologies on G/N: the topology as a definable group (from Theorem 2.1), the quotient topology (where G gets the topology from Theorem 2.1) and the topology obtained via Theorem 2.3 (for the action of G on G/N by left multiplication); A. Berarducci proves in Theorem 4.3 in [2] – using the trivialization theorem – that the three topologies indeed coincide.

Next I state some corollaries of Theorem 2.1.

Corollary 2.4 *Let G be a definable group, and let H be a definable subgroup of G. Then,*

(i) H is closed;

(ii) if G is infinite then it has an infinite definable abelian sugbroup;

(iii) the following three are equivalent: H open, H has finite index, and $\dim H = \dim G$;

(iv) the definably connected component of the identity in G, G^0, is the smallest subgroup of finite index and moreover it is normal in G, and

(v) G has the descending chain condition on definable subgroups.

Once we have an explicitly defined topology on G, (i) and (ii) are proved by A. Pillay in [41] (Propositions 2.7 and 5.6 respectively, noting that the group topology makes G topologically totally transcendental); (iii) is Lemma 2.11 in [42] and (iv) and (v) can be easily obtained from (iii). S. Strzebonski gave a different proof of (iv) and (v) – Theorem 2.6 in [47] – provided $\mathcal{M}$ is an expansion of a group. For more on descending chain conditions see Theorem 8.3 below.

Corollary 2.5 *Let G be a definable group, $\mathbb{G} = (G, \cdot)$ and A an abelian subgroup of G (definable or not). Then, the centre of the centralizer $C_{\mathbb{G}}(A)$ is a $\mathbb{G}$-definable abelian subgroup containing A, which is normal if A is normal.*

This follows from (v) above (see e.g. Cor. 1.17 in [33]). It is because of this last corollary (and thinking of linear algebraic groups) that we say that a definable group is *semisimple* if it has no infinite abelian normal definable subgroup.

In several papers on definable groups in (general) o-minimal structures the extra assumption of having definable choice is added just to be able to speak about *definable* quotient groups or having a *definable* set of representatives of the cosets. The following result, due to M. Edmundo – Theorem 7.2 in [13] – allows us to eliminate this extra assumption.

Theorem 2.6 *Let G be a definable group and let $\{T(x) : x \in X\}$ be a definable family of non-empty definable subsets of G. Then, there is a definable function $t\colon X \to G$ such that for $x, y \in X$ we have $t(x) \in T(X)$ and if $T(X) = T(Y)$ then $t(x) = t(y)$.*

3 Small dimension

In this section I consider definable groups of dimension less than or equal to three. I begin with dimension one and the work of V. Razenj in [46].

Proposition 3.1 *Let G be a definably connected 1-dimensional group. Then,*

(i) G is abelian;

(ii) G does not have bounded exponent, and

(iii) either, G is definably compact and the torsion subgroups $G[m]$ are isomorphic to $\mathbb{Z}/m\mathbb{Z}$ (for each $m > 0$), or G is torsion–free.

Corollary 2.4(ii) above gives (i). For (ii) see Proposition 1, and for (iii) Propositions 3 and 4, all of them in [46].

Razenj makes use of Theorem 2.1 and adapts the classical classification of 1-dimensional topological Hausdorff manifolds to obtain that $G\setminus\{pt.\}$ has either one definably connected component ($\mathbb{S}^1$-type), or two components ($\mathbb{R}$-type). Then, using the Reineke classification of strongly minimal groups (note that G as a pure group is strongly minimal, see Remark 6.5 in [41]), he obtains the following (see Theorem p.272 in [46]).

Theorem 3.2 *Let G be a definably connected 1-dimensional group. Then (as an abstract group) G is isomorphic to either $\bigoplus_p \mathbb{Z}_{p^\infty} \oplus \bigoplus_\delta \mathbb{Q}$ or $\bigoplus_\delta \mathbb{Q}$, $\delta \geq 0$.*

For more on 1-dimensional groups definable in an arbitrary o-minimal structure see Theorem 8.1 below. When $\mathcal{M}$ is the real field our definable group is a Nash group (see Section 1), and J. Madden and C. Stanton classify in [24] (see also [25]) the 1-dimensional Nash groups, up to Nash isomorphism. When $\mathcal{M}$ is an expansion of a real closed field, Strzebonski proves in Theorem p.204 in [48] that a definably connected 1-dimensional group is definably isomorphic to either an abelian group on either the set (0,1) or the set $[0, 1)$. On the other hand, when $\mathcal{M}$ is an o-minimal expansion of a real closed field, there are two natural torsion-free 1-dimensional definably connected groups: M_a (the additive group of $\mathcal{M}$) and M_m (the multiplicative group of the positive elements of $\mathcal{M}$). C. Miller and S. Starchenko prove – Theorem C in [26] – the following result as a consequence of their Growth Dichotomy Theorem (Theorem A in [26]) and a result in [29].

Theorem 3.3 *Let $\mathcal{M}$ be a polynomially bounded expansion of a real closed field. Up to definable isomorphism, there are exactly two definably connected torsion-free 1-dimensional groups: M_a and M_m.*

Miller and Starchenko posed in [26] the following question.

Question 3.4 *Let $\mathcal{M}$ be an expansion of a real closed field. Is every definably connected torsion free definable 1-dimensional group definably isomorphic to M_a or M_m?*

Definable groups of dimension two and three are analysed by A. Nesin, A. Pillay and V. Razenj in [28]. Their work is a continuation of that of V. Razenj above.

Theorem 3.5 *Let G be a definably connected 2-dimensional group. Then G is solvable. Moreover, either G is abelian, or $Z(G) = 1$ and G is definably isomorphic to a semi-direct product of R_a and R_m, where R is a definable real closed field.*

This is Theorem 2.6 in [28], noting that firstly, by Theorem 2.6 we can eliminate the assumption on strong elimination of imaginaries, and secondly, by J. Johns Theorem in [20], the open mapping theorem holds in any o-minimal structure.

The situation is not so clear for abelian definable 2-dimensional groups. Peterzil and Steinhorn give in [40] examples of definably compact 2-dimensional abelian groups definable in an o-minimal expansion of a real closed field which are not a direct sum of definable 1-dimensional subgroups (see the comments after Theorem 5.2 below). On the other hand, by Corollary 6.6 below, a torsion-free group G definable in an o-minimal expansion of a real closed field cannot be definably compact, so by Theorem 5.2 G has a definable torsion-free 1-dimensional subgroup H which must be divisible. If G is commutative, H abstractly has a complement, but we do not know what happens in the definable context. The following question was posed in [40] and in [38].

Question 3.6 *Let $\mathcal{M}$ be an expansion of a real closed field. Is every abelian torsion-free 2-dimensional group a direct sum of definable 1-dimensional subgroups?*

I end this section with a result of Nesin, Pillay and Razenj in [28].

Theorem 3.7 *Let G be a definably connected non-solvable 3-dimensional group. Then $G/Z(G)$ is definably isomorphic to either* $\mathrm{PSL}_2(R)$ *or* $\mathrm{SO}_3(R)$*, for some definable real closed field R.*

The existence of two types depends on whether a 2-dimensional subgroup of G exists ($\mathrm{PSL}_2(R)$), or not (see Propositions 3.5 and 3.1 in [28], noting that again as above we can remove the extra assumptions). The proof in the first case is done in the spirit of both Cherlin's analysis of groups of Morley rank 3 and [27]. This result also proves an o-minimal version of Cherlin's conjecture for non-solvable 3-dimensional groups (see Section 4 below).

4 Interpretability and linear groups

Classical results tell us that in a compact simple Lie group, the Lie structure is implicitly defined from the abstract group structure, that is, any group automorphism is an (analytic) homeomorphism; see Corollary 3.7 in [43] for a model-theoretic version of this fact. A. Nesin and A. Pillay prove in [27] that this Lie structure is actually (explicitly) definable from the abstract group structure. Namely, if $\mathbb{G} = (G, \cdot)$ is a simple (centreless) compact Lie group (definable in the real field) then, there is an isomorphic copy K of the real field interpretable in $\mathbb{G}$, and there is also a $\mathbb{G}$-definable isomorphism between G and a Nash group over K (see Theorem 0.1 in [27]). This last result and Theorems 3.5 and 3.7 above, give us examples of definable groups in which a field is interpretable. Y. Peterzil, A. Pillay and S. Starchenko prove in [33] and [34] (see also [32] for a survey), that an infinite definably simple group $\mathbb{G} = (G, \cdot)$ interprets a field R and G is $\mathbb{G}$–definably isomorphic to a linear group over R. The full strength of their analysis of definably simple groups is the following result, which is also a strong version of an o-minimal analogue to Cherlin's conjecture. I recall that an algebraic group defined over a field k is said to be *k-simple* if it has no nontrivial normal algebraic subgroups defined over k.

Theorem 4.1 *Let* $\mathbb{G} = (G, \cdot)$ *be an infinite definably simple (centreless) group. Then, there is a real closed field* k *such that one and only one of the following holds:*

(1) $\mathbb{G}$ *and the field* $k[\sqrt{-1}]$ *are bi-interpretable, and* G *is* $\mathbb{G}$*-definably isomorphic to* $H(k[\sqrt{-1}])$*, where* H *is a linear algebraic group defined over* $k[\sqrt{-1}]$.

(2) $\mathbb{G}$ *and the field* k *are bi-interpretable, and* G *is* $\mathbb{G}$*-definably isomorphic to the semialgebraic connected component of a group* $H(k)$*, where* H *is a* k*-simple algebraic group defined over* k.

This is Theorem 1.1 in [34]. Note that it also proves Cherlin's conjecture for groups of finite Morley rank which happen to be definable in an o-minimal structure. The proof is divided in two main parts. Firstly, given G as in the hypothesis of the theorem they find a real closed field R *definable in* $\mathcal{M}$, such that G is definably isomorphic to a semialgebraic subgroup of some $\mathrm{GL}(n, R)$ (this is carried out in [33]). To find a field they make use of the Trichotomy theorem (proved by Y. Peterzil and S. Starchenko in [36]) to prove that a definably connected centreless group G is a direct product of (unidimensional) groups, each one defined over a definable real closed field (actually, they get that each one of the groups – in the direct product – is a linear group over the corresponding field). See Theorems 3.1 and 3.2 in [33]. Once we have a field, we may suppose that our o-minimal structure is an expansion of a real closed field, and in this situation any definable centreless group is definably isomorphic to a subgroup of $\mathrm{GL}(n, M)$ (see below in this Section). If moreover the group is semisimple we have the following result.

Theorem 4.2 *Let* $\mathcal{M}$ *be an expansion of a real closed field. Let* G *be a definably connected centreless semisimple group. Then,* G *is definably isomorphic to a semialgebraic linear group over* M.

This is Theorem 2.37 in [33]. For its proof, Peterzil, Pillay and Starchenko develop Lie algebra machinery, and they study the Lie algebra of G and the adjoint representation. A key result there is that the dimension of a semisimple Lie algebra coincides with the dimension of its automorphism group. The latter is obtained by transfer from the reals.

Then, for the rest (and second part) of the proof of Theorem 4.1 above, we may suppose that our G is a semialgebraic subgroup of some $\mathrm{GL}(n, R)$ where R is a definable (in $\mathcal{M}$) real closed field (this second part is carried out in [34]). The first goal is to find a field K interpretable in $\mathbb{G}$. The latter is first proved in the solvable not nilpotent case (see Theorem 2.12 in [34]) through an o-minimal version of a theorem by Zil'ber (see Theorem 2.6 in [34]). Now, going back to G in our general hypothesis, the last part of the proof is based on the study of the geometric structures partly developed in [19].

Y. Peterzil and S. Starchenko make use of the above results to give the following characterization of definable groups which interpret a field.

Theorem 4.3 *Let $\mathbb{G} = (G, \cdot)$ be a definable infinite group. Then, $\mathbb{G}$ interprets an infinite field if and only if G is not abelian-by-finite.*

This is Corollary 5.1 in [37], noting that by Theorem 2.6 the quotient groups are definable (see also Corollary 6.3 in [13]). Firstly, note that to avoid local modularity, for G to interpret a field, we need G not to be abelian-by-finite. They prove that this condition is also sufficient. By the results in [33] and [34], stated above, they reduce to the case in which G is not semisimple. Then, they obtain an infinite definable family of homomorphisms between two abelian definable groups (either subgroups or quotients of subgroups of G). In this situation they have already proved – Theorem 4.4 in [37] – that a field is interpretable in $\mathbb{G}$ (the field is actually defined in a quotient of subgroups of G). In turn, the proof of Theorem 4.4 in [37] is based on the study of $\bigvee$-definable groups. (For more on $\bigvee$-definable groups see section 2.1 in [34] and also [15] by M. Edmundo.)

The following theorem on definable permutation groups is proved by H.D. Macpherson, A. Mosley and K. Tent (Theorem 1.1 in [23]), and in particular gives one more example of how to find a field via a definable group. They consider definable groups G equipped with a definable faithful transitive action on a set Ω. Such a permutation group (G, Ω) is *definably primitive* if there is no proper nontrivial definable G-invariant equivalence relation on Ω. A transitive action on Ω is said to be *regular* if the stabilizer of a point is the identity.

Theorem 4.4 *Let (G, Ω) be a definably primitive permutation group. In the case when G has a non-trivial abelian normal subgroup, assume also that G is definably connected and not regular. Then, (G, Ω) is definably isomorphic to a semialgebraic permutation group over a definable real closed field.*

Firstly, note that the extra assumption when G has a nontrivial abelian normal subgroup is needed. The proof of Theorem 4.4 is based on a fine structure theorem for definably primitive permutation groups (Theorem 1.2 in [23]) which is an analogue of the O'Nan-Scott theorem for *finite* primitive permutation groups. Macpherson, Mosley and Tent apply Theorem 4.4 to give a description of definable permutation groups (G, Ω) where $\dim \Omega = 1$ – Theorem 1.5 in [23]. They propose as an application of their results the following problem (for more problems on definable permutation groups see p.669 in [23]).

Problem 4.5 *Classify definably primitive permutation groups (G, Ω) such that G has finitely many orbits on Ω^2.*

In [49] K. Tent classifies sharply n-transitive infinite definable permutation groups for $n = 2, 3$ (there are no infinite 4-transitive groups definable in an o-minimal structure), where *sharply n-transitive* means that the stabilizer of n distinct points is trivial. Note that Problem 4.5 is a generalization of this last result.

By the above results on (semi)simple groups, the study of definable linear groups can be seen as a continuation of them. On the other hand, any definably connected centreless group defined over an expansion of a real closed field is definably isomorphic to a linear group (see Corollary 3.3 in [29], proved by Y .Peterzil, A. Pillay and myself). Y. Peterzil, A. Pillay and S. Starchenko study in [35] linear groups defined over an expansion of a real closed field. They obtain the following results.

Theorem 4.6 *Let $\mathcal{M}$ an expansion of a real closed field. Let G be a definably connected subgroup of some $\mathrm{GL}(n, M)$. Then G is semialgebraic, provided it is either nilpotent, semisimple or definably compact. If G is not semialgebraic then there are definable functions $f_1, \ldots, f_k$ such that G is defined over $(M, +, \cdot, f_1, \ldots, f_k)$ and either $k = 1$ and f_1 is some exponential function or the f_i's are power functions.*

The nilpotent case is Proposition 3.10 in [35]. The rest of the cases are treated in Theorems 4.3 and 4.6 and Corollary 4.2 in [35]. Their proofs are based on Lie algebra machinery, together with Theorem 4.7 below. The latter – Theorem 4.1 in [35] – also makes use of results on Lie algebras, especially of the classical fact that the commutator of a Lie subalgebra of the general Lie algebra (over an algebraic closed field) is an *algebraic* Lie algebra.

Theorem 4.7 *Let $\mathcal{M}$ an expansion of a real closed field. Let G be a definably connected subgroup of some $\mathrm{GL}(n, M)$. Then, there are semi-algebraic subgroups G_1, G_2 of $\mathrm{GL}(n, M)$ such that $G_2 < G < G_1$ with G_2 normal in G_1 and G_1/G_2 abelian (i.e. G is an extension of a definable subgroup of a abelian semialgebraic group by a semialgebraic group). Moreover, there are abelian definably connected subgroups $A_1, \ldots, A_k$ of G such that $G = G_2 \cdot A_1 \cdots A_k$.*

Peterzil, Pillay and Starchenko prove – Theorem 5.1 in [35] – the following result. They make use of their results in this section – in particular of Theorem 4.1 – and the classification of the simple Lie algebras over $\mathbb{R}$.

Theorem 4.8 *Let G be a definably simple (centreless) group. Then,*

(i) $(G, \cdot)$ is elementary equivalent to $(H, \cdot)$ for some simple Lie group H, and

(ii) G is definably isomorphic to some semialgebraic linear group, defined over $\mathbb{R}_{alg}$ (= the real algebraic numbers).

5 Algebraic aspects

In this section I collect various properties of different nature which give some information about the algebraic structure of definable groups (see also Section 6 below). However, most of the time the hypotheses on the definable group are not algebraic.

Theorem 5.1 *Let G be an infinite definable group. Then,*

(i) G does not have bounded exponent, and

(ii) if G is abelian then the torsion subgroup $G[m]$ is finite, for each $m > 0$.

In dimension one we get the result by Proposition 3.1. The general case is proved by A. Strzebonski in Proposition 6.1 in [47] (once we remove the extra assumption via Theorem 2.6). Clearly (ii) is obtained from (i), and (i) is reduced to the abelian case via Corollary 2.4 (2). For (i) in the abelian case Strzebonski makes use of his results on Euler characteristic (see Section 6).

Theorem 5.2 *Let G be a definable not definably compact group. Then, there is a 1-dimensional torsion-free definable subgroup of G.*

This result is proved by Y. Peterzil and C. Steinhorn (Theorem 1.2 in [40]). The assumption on G ensures the existence of a non-completable curve in G. This allows them to equip G with an equivalence relation based on *infinitesimal* neighbourhoods and tangency at infinity (inspired by a concept of tangency due to Zil'ber). See Definition 3.2 in [40]. They prove that the equivalence class of the identity element of G is the required subgroup. Peterzil and Steinhorn also give examples of both *(a)* a definably compact abelian group with no definable infinite subgroups, and *(b)* a definable group which is not definably compact and which has no definably compact subgroup; both are over the real field (see Examples 5.2 and 5.6 in [40], based on some examples in [47]). Example (a) indicates that the assumption on G in Theorem 5.2 cannot be avoided. On the other hand, any connected Lie group is homeomorphic to $K \times \mathbb{R}^l$, where K is a maximal compact subgroup. Example (b) tell us that we cannot hope for this in the definable context.

The analysis of definably (semi)simple groups and linear groups in Section 4 gives the following results.

Theorem 5.3 *Let G be a definably connected semisimple centreless group. Then G is definably isomorphic to a direct product of definably connected definably simple groups.*

This is Theorem 4.1 in [33], observing (as is done in Remark 4.2 there) that by the results on linear groups (Theorem 5.1 in [35]), the factor groups are definably connected and hence definably simple.

Theorem 5.4 *Let $\mathcal{M}$ be an expansion of a real closed field. Let G be a definably connected subgroup of some $\mathrm{GL}(n, M)$. Then G is an almost semi-direct product of a normal solvable definable subgroup N and a semialgebraic semisimple subgroup H (i.e., $G = NH$ where $N \cap H$ is finite).*

This is Theorem 4.5 in [35]. The ingredients of the proof are: the analysis of the associated Lie algebra and Levi's decomposition theorem of real Lie algebras.

Question 5.5 *Can we eliminate the assumption: "$G < \mathrm{GL}(n, M)$" in Theorem 5.4?*

Theorem 5.6 *Let G be a definably compact group. Then, either G is abelian-by-finite or $G/Z(G)$ is semisimple. In particular, if G is solvable then it is abelian-by-finite.*

This is Corollary 5.4 in [37], observing, as usual, that we can remove the extra hypothesis via Theorem 2.6. For the proof, Peterzil and Starchenko consider two cases depending on whether $Z(G)$ is finite or not. The proof follows the lines of the proof of Theorem 4.3 noting (as observed in [40]) that an infinite field cannot be definably compact.

In [13] M. Edmundo studies group extensions and solvable groups. He proves the following results.

Theorem 5.7 *Let G be a definable group and let N be a definable normal subgroup of G. Then there is a definable extension*

$$1 \to N \to G \to H \to 1,$$

with definable section $s\colon H \to G$.

Theorem 5.8 *Let G be a definably connected solvable group. Then, there is a definable normal subgroup N of G such that G/N is definably compact and N is $K \times H$, where K is the definably connected definably compact maximal dimensional subgroup of G, and H is a direct product of subgroups, each of which is defined over either a definable semibounded o-minimal expansion of a group or a definable real closed field, where 'semibounded' means generated by sets which are bounded together with sets which are linear.*

Theorem 5.7 is Corollary 3.11 in [13] and is a special case of Theorem 2.6. The ingredients of this case are the Trichotomy Theorem from [36] and non-orthogonality as in [33]. The proof of the description of the definable solvable groups in Theorem 5.8 (see Theorem 5.8 in [13]) is based on the study of definable extensions also developed in [13] and also make use of Theorem 4.2 in [12] and Theorem 5.2.

Continuing with the resemblance between definable groups and Lie groups, one could expect that a definably compact definably connected abelian group should resemble a torus of $\mathcal{M}$. Even though we know, by examples in [40] (see example (a), above in this section), that such a group cannot be a direct product of $\mathrm{SO}(2, M)$'s, one could expect that the structure of the torsion is that of a real torus. That this is the case is proved in [16] by M. Edmundo and myself. However the result does not give information of how the torsion points lie within the group (see Question 7.4).

Theorem 5.9 *Let $\mathcal{M}$ be an expansion of a real closed field. Let G be a definably compact definably connected abelian n-dimensional group. Then, the torsion subgroups $G[m] \cong (\mathbb{Z}/m\mathbb{Z})^n$, for each $m > 0$.*

This is Theorem 1.1(b) in [16]; the 1-dimensional case is Proposition 3.1(iii). Here are the ingredients of the proof: Firstly, we obtain some properties of definable covering maps and making use of Corollary 2.10 in [4] (proved by A. Berarducci and myself), we prove that if H is a definably connected abelian group then, there is an $s \geq 0$ such that both the o-minimal fundamental group of H is $\mathbb{Z}^s$, and the torsion subgroups $H[m]$ are $(\mathbb{Z}/m\mathbb{Z})^s$, for each $m > 0$. Hence it remains to prove that if H is definably compact (as is G) then $s = \dim H$. This is done by developing some o-minimal cohomology machinery – Theorem 6.7 below – based on o-minimal homology (developed by A. Woerheide in [50]), and making use of Theorem 5.2 in [5], the latter proved by A. Berarducci and myself. Along the way, we prove that both the o-minimal fundamental group and the o-minimal cohomology $\mathbb{Q}$-algebra of G are isomorphic to the (classical) fundamental group and cohomology $\mathbb{Q}$-algebra of a real torus with Lie dimension n.

We have the following corollary to the proof of Theorem 5.9.

Corollary 5.10 *Let $\mathcal{M}$ be an expansion of a real closed field. Let G be a definably connected abelian group. Then,*

(i) G is torsion-free if and only if G is definably simply-connected, and

(ii) if there is a $k > 1$ for which there is a $g \in G$ of order k, then for every $k > 1$ there is $g \in G$ of order k.

I finish this section with two recent results on algebraic properties of definably compact groups which again are o-minimal versions of results on Lie groups.

Theorem 5.11 *Let G be a definably compact definably connected group. Then,*

(i) there is a definably connected abelian subgroup T of G such that $G = \bigcup_{g \in G} T^g$, and

(ii) G is divisible (i.e. the map $x \mapsto x^m$ on G is surjective for each $m > 0$).

Firstly, note that (ii) follows from (i) and the fact that any definably connected abelian group is divisible – divisibility is obtained by Theorem 5.1(ii) (see *e.g.* Proof of Theorem 2.1 p.170 in [16]). When $\mathcal{M}$ is an expansion of a real closed field, (i) is proved by A. Berarducci in Theorem 6.12 in [2] making use of Theorem 3.3 in [5]. The general case is proved by M. Edmundo in Proposition 1.2 in [14]. His proof is based on Theorem 5.6 and transfer from the reals together with Theorem 4.2. Note that the definable compactness assumption is needed because of examples such as $SL_2(\mathbb{R})$. Both proofs of the divisibility – via (i) – simplify an unpublished proof by myself.

6 Euler characteristic and torsion

In an o-minimal structure, besides the dimension we have another definable invariant: the o-minimal Euler characteristic (E). If X is definable $E(X) := \sum_{C \in \mathcal{D}} (-1)^{\dim C}$, where $\mathcal{D}$ is a cell decomposition of X. The map E is well defined, invariant under definable bijections and $E(X \times Y) = E(X)E(Y)$, for X and Y definable; see Chapter 4.2 in [10]. When $\mathcal{M}$ is an expansion of a real closed field, we can triangulate X and, if X is definably compact, $E(X) = \chi(X(\mathbb{R}))$ where χ is the classical Euler characteristic and $X(\mathbb{R})$ denotes the realization over $\mathbb{R}$ of a (closed) simplicial complex determined by X; the definable compactness assumption is essential because, even over the real field, $E((0,1)) = -1 \neq 1 = \chi((0,1))$, where $(0,1)$ denotes the unit interval.

A. Strzebonski in [47] makes use of the o-minimal Euler characteristic to study groups definable in o-minimal structures. (He has the extra assumption of $\mathcal{M}$ having definable choice. However, he uses this to have definable quotients – in this case we can avoid it via Theorem 2.6 – and to give another proof of the descending chain condition on definable subgroups.) Strzebonski develops the nice idea of making the Euler characteristic play the role (in definable groups) that cardinality plays in finite groups. Of course, $E(G) = card(G)$ if G is finite, so his results are generalizations of (basic) results on finite groups. For p a prime *or zero* and G a definable group, he defines G to be a p-group if for any proper definable subgroup H of G we have $E(G/H) \equiv 0\,(mod\,p)$ (with equality if p is zero), where G/H denotes a definable choice of representatives for left cosets of H in G. Note that, in general, both $E(G/H)$ and $E(H)$ divide $E(G)$, for any definable subgroup H of G. Strzebonski proves the three Sylow theorems, with p-Sylow subgroups as maximal definable p-subgroups. As a basis case, he has the following generalization of Cauchy's theorem on finite groups, which links the o-minimal Euler characteristic with the torsion of a definable group.

Theorem 6.1 *Let G be a definable group and p a prime number. If p divides $E(G)$ then G has an element of order p. In particular, if $E(G) = 0$ then G has nontrivial torsion for each prime p.*

This is Lemma 2.5 in [47]. He considers the usual action of $\mathbb{Z}/p\mathbb{Z}$ on the set $\{(g_1, \ldots, g_p) \in G \times \cdots \times G \mid g_1 \cdots g_p = 1\}$ and works with the Euler characteristic instead of cardinality.

For the proof of the following result see Propositions 4.1 and 4.2, both in [47].

Proposition 6.2 *Let G be a definable group. Then,*
(i) if $E(G) = \pm 1$, then G is uniquely divisible (i.e., the map $x \mapsto x^m$ is a bijection for each $m > 0$), and
(ii) if G is abelian and definably connected then $E(G) = 0, \pm 1$.

The most interesting cases of p-groups are the 0-groups. Note that if G is a p-group and $E(G) \neq 0$ then G is finite (see Remark 2.15 in [47]). On the other hand, Strzebonski proves – Corollary 5.17 in [47] – that 0-groups are abelian. (See [2], by A. Berarducci, for more on 0-groups.)

A. Strzebonski conjectured – first section in [47] – that we do not need G to be abelian in Proposition 6.2 (ii). In [33] (see the end of the introduction therein) it is observed that the results in [33] prove this conjecture.

Theorem 6.3 *Let G be a definably connected group. Then $E(G) = 0, \pm 1$. Moreover, if G is semisimple then $E(G) = 0$.*

I thank Sergei Starchenko for giving the following details to me. We work by induction on the dimension of G. Either there is an abelian normal definable subgroup N of G with $\dim(G/N) < \dim G$, or G is semisimple. In the first case, since $E(G) = E(G/N)E(N)$, the result follows by induction and Proposition 6.2(ii). For the semisimple case, we will show that $E(G) = 0$. We may clearly suppose that $Z(G) = 1$. Now, by Theorem 3.1 in [33] (see the comments after Theorem 4.1, above) we may also suppose G is unidimensional and hence, by Theorem 3.2 in [33], that $\mathcal{M}$ is an expansion of a real closed field. Then, our semisimple centreless definably connected group is the definably connected component of $Aut(g)$, the automorphism group of the Lie algebra g of G (see proof of Theorem 2.37 in [33]). Again, it suffices to prove that $E(Aut(g)) = 0$, but now $Aut(g)$ is an algebraic subgroup of a $\mathrm{GL}(n, M)$. For $M = \mathbb{R}$, the automorphism group of a semisimple Lie algebra has elements of order p for each p, and transferring this to M, we get $E(Aut(g)) = 0$.

As a corollary of Proposition 6.2 and Theorem 6.3 we have the following result.

Corollary 6.4 *Let G be a definably connected group. If G is torsion-free then it is uniquely divisible.*

In connection with Theorem 5.2 above, Y. Peterzil and C. Steinhorn ask in [40] the question if every definably compact group has at least nontrivial torsion (see Question 5.8 there, and a survey by Y. Peterzil on this torsion problem in [30]). By Theorem 6.1 above, one way of giving an affirmative answer to the above question is proving that the Euler characteristic of such a group is zero. In March 2000 M. Edmundo announced the following result.

Theorem 6.5 *Let $\mathcal{M}$ be an expansion of a real closed field. Let G be an infinite definably compact group. Then $E(G) = 0$.*

Corollary 6.6 *Let $\mathcal{M}$ be an expansion of a real closed field. Let G be definable group. Suppose there is a definable normal subgroup H of G such that G/H is an infinite definably compact group (in particular this holds if G is definably compact). Then G has an element of order p, for each prime p.*

Edmundo's nice idea was to consider o-minimal cohomology (dualizing o-minimal homology developed by Woerheide in [50]), and then follow a classical proof. Once we have proved that the relevant cohomology algebra is not trivial, the classical proof is just a bit of linear algebra. I give the ingredients. The basic properties of the o-minimal cohomology are developed in section 3 in [16] by M. Edmundo and myself. In particular we prove the following result.

Theorem 6.7 *Let $\mathcal{M}$ be an expansion of a real closed field. Let G be an infinite definably connected n-dimensional group. Then, the o-minimal cohomology $\mathbb{Q}$-vector space $H^*(G;\mathbb{Q}) = \bigoplus_{m=0}^{n} H^m(G;\mathbb{Q})$ can be equipped with a structure of graded $\mathbb{Q}$-algebra generated by $y_1, \ldots, y_r$ $(r \geq 0)$ such that*
(i) $H^0(G;\mathbb{Q}) \cong \mathbb{Q}$ (1 say, generates $H^0(G;\mathbb{Q})$ as $\mathbb{Q}$-vector space);
(ii) each $y_j \in H^{m_j}(G;\mathbb{Q})$ with $deg\, y_j = m_j$ odd $(1 \leq j \leq r)$, and
(iii) $B = \{y_{j_1} \cdots y_{j_l} : 1 \leq j_1 < \cdots < j_l \leq r\}$ together with 1 (from (i)) form a basis of the $\mathbb{Q}$-vector space $H^*(G;\mathbb{Q})$ (for a monomial $x = y_{j_1} \cdots y_{j_l}$ we say $len\, x = l$).

This is Theorem 3.4 and its corollaries in [16]. The ingredients of the proof are the following. Dualizing the results in [50] we obtain that $H^*(G;\mathbb{Q})$ is a graded $\mathbb{Q}$-vector space, $H^m(G;\mathbb{Q}) = 0$ for each $m > n$ and (i) (the latter because G is definably connected); the $\mathbb{Q}$-algebra structure is obtained via a Hopf-algebra structure and a classical classification of Hopf algebras.

We now go back to the proof of Theorem 6.5. We may suppose G is definably connected. Since G is definably compact, the top homology group is nontrivial (this is proved in [5] Theorem 5.2, and also independently by M. Edmundo); dualizing we get $H^*(G;\mathbb{Q}) \not\cong \mathbb{Q}$ and hence $r > 0$, in Theorem 6.7. On the other hand, by the triangulation theorem and after dualizing, the results in [50] (linking simplicial and singular o-minimal homology) give $E(G) = \sum_{m=0}^{n} (-1)^m \dim H^m(G;\mathbb{Q})$. For $m > 0$, $\dim H^m(G;\mathbb{Q}) = card\{x \in B\colon deg\, x = m\}$, hence $E(G) = 1 + \sum_{x \in B} (-1)^{deg\, x}$. Now, by (ii) in Theorem 6.7 above, the graded $\mathbb{Q}$-algebra product makes $deg\, x \equiv len\, x\, (mod\, 2)$ for any monomial $x \in B$, so we can substitute $deg\, x$ by $len\, x$ in the last equality. Finally, note that there are $\binom{r}{l}$ monomials of length l in B, hence $E(G) = \sum_{l=0}^{r} \binom{r}{l} (-1)^l$, which is 0, since $r > 0$.

There are other proofs of Theorem 6.5. In [5] A. Berarducci and myself proved the result (see the proof of Corollary 3.4 there) via a

Lefschetz fixed point theorem for o-minimal expansions of real closed fields, which in turn is proved by transfer from the reals; the transferring can be done after we prove that the top homology group of a definably compact definably connected group is $\mathbb{Z}$ (see Theorem 5.2 in [5]). Using a differential topology approach, in [3], we define the Lefschetz number of id_G, $\Xi(G)$, as a self intersection number of the diagonal in $G \times G$ (see Definition 9.11, there) and we prove in Theorem 11.4 also there, that $\Xi(G) = 0$; our hope was – as in the classical case and as we conjectured – that $\Xi(G) = E(G)$. This conjecture has recently been proved by Y. Peterzil and S. Starchenko in [39]; they also give there – Corollary 4.6 – another proof of Theorem 6.5. They extend further the results in [3] introducing definable Morse functions and prove that $E(G)$ is the degree of a map, a degree that they have shown is 0. To finish with the different proofs of Theorem 6.5, let me note that we can also get one via the structure of the torsion – Theorem 5.9 – which gives $E(G) = 0$, for G abelian, and then make use of either Theorem 6.3 and Theorem 5.6, or the existence of an infinite abelian subgroup – Corollary 2.4(ii). If we think only of the torsion, the best result is via Theorem 5.9 and Corollary 2.4(ii), which gives an element of order m for each $m > 0$. Another proof of the existence of torsion is in [18] (Remark 2 after Corollary 8.4); this one is based on the facts (see Section 8, below) that a definable group has DCC for type-definable subgroups of bounded index, that a definably compact group has *fsg* (see Definition 8.5 below) and also on the classical fact that a compact Lie group has torsion.

Question 6.8 *Can we eliminate the assumption: "$\mathcal{M}$ is an expansion of a real closed field" in Theorem 6.5 or in Corollary 6.6?*

Y. Peterzil and S. Starchenko have recently been studying (see [38]) torsion-free groups definable in o-minimal expansions of real closed fields. Note that by Corollary 6.6 such a group cannot be definably compact, and by Theorem 6.1, it must have o-minimal Euler characteristic ± 1; moreover, it must also be solvable (see Claim 2.11 in [38]). They develop in [38] definable group extensions and prove in Corollary 5.8 the following result.

Theorem 6.9 *Let $\mathcal{M}$ be an o-minimal expansion of a real closed field. Let G be a torsion-free n-dimensional definable group. Then $E(G) = (-1)^n$. Moreover, G is definably diffeomorphic to M^n.*

When $\mathcal{M}$ is $\mathbb{R}_{an}$ (or any polynomially bounded o-minimal expansion of a real closed field in which any definable function $f: M \to M$ has a definable Puiseux-like expansion at infinity) Peterzil and Starchenko get the following structure theorem (see Theorem 4.14 together with Theorem 4.9 in [38]).

Theorem 6.10 *Let G be a connected abelian torsion-free group definable in $\mathbb{R}_{an}$. Then, G is definably isomorphic to a direct sum of $(\mathbb{R}, +)^k$ and $(R^{>0}, \cdot)^m$, for some $k, m \geq 0$.*

7 Genericity and measure

In this section $\mathcal{M}$ is a sufficiently saturated o-minimal expansion of a real closed field, small *or* bounded *means small with respect to the degree of saturation of $\mathcal{M}$ and* model *means elementary substructure of $\mathcal{M}$.*

The motivation for considering measures and generic sets in the o-minimal context was Pillay's conjecture (see Section 8). A definable subset X of a definable group H is *left* (resp. *right*) *generic* (in H) if finitely many left (resp. right) translates of X cover H. And, X is *generic* (in H) if it is both left and right generic. A first result on generic sets already appeared in Lemma 2.1 in [42], where A. Pillay proves that large sets are generic (see Lemma 2.2 above). Since X generic in H implies $\dim X = \dim H$, generic sets lie between large and having the same dimension. Note also that if p is a complete type in H and X is (left) generic in H, then a (resp. left) translate of X must be in p. If a definable group H is not definably compact, then the complement of a nongeneric set does not need to be generic (take $G = (M, +)$ and $X = (0, +\infty)$). However, this is the only obstruction in the abelian case:

Theorem 7.1 *Let G be an abelian definably compact group. If $X \subset G$ is definable and not generic then $G \setminus X$ is generic.*

This is proved when G is a torus – Proposition 5.6 of [6] by A. Berarducci and myself – using the existence of a measure on definable sets (see below in this section). The general case is proved by Y. Peterzil and A. Pillay – Corollary 3.9 in [31] – as a corollary of the following result of A. Dolich in [11] (see also Appendix in [31]).

Theorem 7.2 *Let X be a definably compact set and $\mathcal{M}_0$ a small model. Then the following are equivalent:*

(a) the set of $\mathcal{M}_0$-conjugates of X is finitely satisfiable, and

(b) X has a point in $\mathcal{M}_0$.

We have the following corollary to Theorem 7.1, see Lemma 3.12 and Corollary 3.10 in [31].

Corollary 7.3 *Let G be an abelian definably compact group. Then,*

(i) the set $\mathcal{I} = \{X \subset G \colon X$ is definable and not generic$\}$ is an ideal of $Def(G)$ (= the Boolean algebra of definable subsets of G);

(ii) for every $X \subset G$ definable and left-generic, the stabilizer

$$Stab_{\mathcal{I}}(X) := \{g \in G \colon gX \triangle X \in \mathcal{I}\}$$

is a type-definable subgroup of G ($\triangle$ = symmetric difference), and

(iii) there is a complete generic type (= every formula in the type defines a generic set) in G.

In Section 8 the commutativity assumption – in both Theorem 7.1 and Corollary 7.3 – will be substituted by *fsg* (see Definition 8.5). An answer to the following question will give us some information about how the torsion points lie in a definably compact group (see the comments just before Theorem 5.9).

Question 7.4 *Let G be a definably connected definably compact group. Let X be a generic subset of G. Does X contain a torsion point of G?*

We do not know the answer even if we assume $\mathcal{M}$ an expansion of a real closed field and hence the positive solution to Pillay's conjectures (see Theorem 8.8 below) or if we assume G commutative. We have the same open questions for X large in G.

A *Keisler measure* μ on a definable set X is a finitely additive probability measure on $Def(X)$; hence $\mu \colon Def(X) \to [0, 1]$, where $[0, 1]$ is the unit real interval. For instance, a type is a 0-1 valued measure on any definable set. Keisler measures were introduced by Keisler in [21]. If μ is a Keisler measure on X, we have the following equivalence relation: $Y \sim_\mu Z$ if $\mu(Y \triangle Z) = 0$, for $Y, Z \subset X$ definable. The following result was obtained in [18] as a corollary of Keisler's work in [21].

Theorem 7.5 *Let X be definable and μ a Keisler measure on X. Then, there are boundedly many $\sim_\mu$-classes. In particular, there is a small model $\mathcal{M}_0$ such that every definable subset Y of X is $\sim_\mu$-equivalent to some M_0-definable subset of X.*

For a proof see Corollary 3.4 in [18] noting that by Corollary 3.10 in [45], the theory of an o-minimal structure has NIP.

Observe that if μ is a left-invariant Keisler measure on a definably compact group H, then any left generic subset of H must have positive measure. Over the reals, there exist (nondefinable) subsets of the two-dimensional torus with positive measure and empty interior (hence, *non-generic*). We will see below that, in the definable context, generics are the only subsets of a definably compact group with positive Keisler measure (see Corollary 8.9). With the results we have so far we can use the existence of a measure to prove the following property of stabilizers (defined in Corollary 7.3 above).

Corollary 7.6 *Let G be a definably compact abelian group. Then, for every $X \subset G$ definable and generic, $Stab_{\mathcal{I}}(X)$ is a type-definable subgroup of G of bounded index.*

This is Proposition 6.3(ii) in [18]. By the fact that every abelian group is amenable (i.e., it has a (left)-invariant finitely additive measure over *all* subsets), G – being abelian – has in particular an invariant Keisler measure μ. By Theorem 7.5 there are bounded many $\sim_\mu$-classes. Since generic subsets of G have positive measure, the type-definable subgroup $Stab_{\mathcal{I}}(X)$ must have bounded index.

8 Pillay's conjectures

The fact that we have a definable manifold topology (Theorem 2.1), the results on definably simple groups (Theorems 4.2 and 4.8) and the structure of the torsion of definably compact abelian groups (Theorem 5.9) induce us to think that a definably compact group must be a non-standard version of a real Lie group; that is, if we quotient out by the right subgroup of infinitesimals (see e.g., Proposition 3.8 in [43]), a definably compact group becomes a Lie group. The definable compactness assumption is needed (see below) to ensure that we do not collapse the group when we take the quotient. More formally, the following conjectures were formulated by A. Pillay around 1998, and formally stated during the Problem Session of the ECMTA in honour of A.J. Macintyre. See Problem 19 p.464 in [1], and Conjecture 1.1 in [44].

Pillay's conjectures. *Let $\mathcal{M}$ be sufficiently saturated and let G be a definably connected group. Then,*

(C1) *G has a smallest type-definable subgroup of bounded index,* G^{00}*;*

(C2) G/G^{00} *is a compact connected Lie group, when equipped with the logic topology;*

(C3) *if moreover, G is definably compact then the Lie dimension of* G/G^{00} *is equal to the (o-minimal) dimension of G, and*

(C4) *if moreover, G is definably compact and abelian, then* G^{00} *is divisible and torsion-free.*

A type-definable equivalence relation Q on a definable set is *bounded* if X/Q is small. The *logic topology* was introduced by D. Lascar and A. Pillay in [22]. With X and Q as above, we say that $C \subset X/Q$ is closed if its preimage under the natural projection is type-definable in X (see Definition 3.1 in [22] or Definition 2.3 in [44]).

In the rest of this section I fix a sufficiently saturated o-minimal structure $\mathcal{M}$ *and 'small' means small with respect to the degree of saturation of* $\mathcal{M}$*.*

If H is any type-definable normal subgroup of bounded index of a definable group G, then G/H with the logic topology is a compact Hausdorff topological group (see Lemma 3.3 in [22] and Lemmas 2.5, 2.6 and 2.10 in [44]). This, in particular also means that the projection map from G to G/H is *definable* over some small model, in the sense of Definition 2.1 in [18].

The rest of this section is dedicated to stating the cases in which the conjectures are proved to be true and to giving the ingredients of the proofs.

Theorem 8.1 *Conjectures* **C1–C4** *are true when G has dimension one.*

This is Proposition 3.5 of [44]. The proof is based on Razenj's analysis of 1-dimensional definable groups (see Section 3). In the definably compact case Pillay makes use of the torsion subgroup of G which is abstractly isomorphic to the torsion subgroup of the circle group of $\mathbb{S}^1$ (see Proposition 3.1(iii)), to define G^{00}, and then he proves that G/G^{00} is $\mathbb{S}^1$. In the non-definably compact case he gets $G^{00} = G$.

Theorem 8.2 *Conjectures* **C1–C4** *are true when G is definably simple (centreless).*

This is Proposition 3.6 in [44] and its proof is obtained using the properties of definably simple groups. By Theorem 4.8 we can assume that

G is a semi-algebraic subgroup of $\mathrm{GL}(n, R)$, where R is a saturated real closed field (containing the reals) and G is defined over $\mathbb{R}$. Then making use of some properties of Lie groups, Pillay proves that the kernel of the standard part map $\mathrm{st} : G \to G(\mathbb{R})$ is G^{00} and the logic topology in $G/\ker(\mathrm{st})$ coincide with the standard topology on $G(\mathbb{R})$.

Theorem 8.3 *Conjectures* **C1** *and* **C2** *are true. Moreover, if G is abelian then G^{00} is divisible.*

See Theorem 1.1 in [7] by A. Berarducci, Y. Peterzil, A. Pillay and myself. Firstly, we make use of Proposition 2.12 in [44] in which Pillay establishes the equivalence between *(a)* **C1** and **C2** are true for G, and *(b)* G has the DCC (descending chain condition on type-definable subgroups of bounded index). With the aim of proving (b) note first that if N is a normal definable subgroup of G, and N and G/N have DCC, then so does G – Lemma 1.10 in [7]; making use of the latter, we can reduce the situation to the abelian and the semisimple cases. The semisimple case reduces to the simple case via Theorem 5.3 – in which we know that **C1** and **C2** are true. It hence remains to prove (b) in the abelian case. Associated to a type-definable set we have a concept of *definably connected* (= intersection of a small directed family of definable connected sets; see Definition 2.1 and Lemma 2.2 in [7]), and any type-definable set is a disjoint union of a small number of maximal definably connected type-definable subsets – Theorem 2.3 in [7]. Now we argue by contradiction. If there were a descending chain of type-definable subgroups of G then this would also happen in a countable language with everything defined over a countable model $\mathcal{M}_0$. Since $\mathcal{M}_0$ is small, there is a smallest M_0-type-definable subgroup of G of bounded index H, say. This H must be normal and definably connected, and also divisible (see Claim 3 in p.311 in [7], a proof which also yields that G^{00} is divisible, once we have proved it exists). This implies that the compact group G/H is connected and locally connected – Lemma 2.6 and Theorem 3.9 in [7] – and has finite m-torsion, for each $m > 0$. These facts together with a classical characterization of abelian compact Lie groups makes G/H a compact Lie group. But now the descending chain of subgroups of G induces a descending chain of closed subgroups in G/H, a contradiction.

Note that conjectures **C1** and **C2** do not ensure $G^{00} \neq G$. In fact, for G not definably compact we have already encountered the case $G^{00} = G$

in dimension one, and in general if G is commutative and torsion-free then $G^{00} = G$ (Corollary 1.2 in [7]). When $\mathcal{M}$ is a saturated model of *any* o-minimal structure the rest of the conjectures are still unproved, and we do not even know if $G^{00} \neq G$. Recently, the rest of the conjectures have been proved by E. Hrushovski, Y. Peterzil and A. Pillay in [18] provided $\mathcal{M}$ is an expansion of a real closed field, and even more recently when $\mathcal{M}$ is an ordered vector space in [17] by P. Eleftheriou and S. Starchenko. I will give the ingredients of the proof in [18]. The key idea in their proof comes with the property *fsg* (see Definition 8.5 below). Hrushovski, Peterzil and Pillay prove the abelian case by nicely putting together some previous results and then reducing to the abelian and semisimple cases using the fact that definably compact groups have *fsg*. The ingredients follow.

Theorem 8.4 *Conjectures* **C3** *and* **C4** *are true if both $\mathcal{M}$ is an expansion of a real closed field, and G is abelian.*

This is in Lemma 8.2 in [18]. By Theorem 8.3 we know that G^{00} exists and is divisible. By the hypothesis on $\mathcal{M}$ we can apply the results from Section 7. By Corollary 7.6, for every definable $X \subset G$ (generic), $Stab_{\mathcal{I}}(X)$ is a type-definable subgroup of G of bounded index, and hence it contains G^{00}. By Proposition 3.13 in [31], for each $m > 0$, there is a definable generic subset X of G such that $Stab_{\mathcal{I}}(X) \cap G[m] = 0_G$. Therefore, the divisible group G^{00} is also torsion-free (and hence, **C4** is true) and so $(G/G^{00})[m] \cong G[m]$, for each $m > 0$. But G/G^{00} is an abelian compact Lie group, hence a torus and so $(G/G^{00})[m] \cong (\mathbb{Z}/m\mathbb{Z})^l$, where l is the Lie dimension of G/G^{00}. On the other hand, again because we are over an expansion of a real closed field, we can apply Theorem 5.9, and get $G[m] \cong (\mathbb{Z}/m\mathbb{Z})^n$ where n is the dimension of G. From these three isomorphisms, we obtain that **C3** is also true.

For the next results (in [18]) we need the following.

Definition 8.5 Let G be a definable group. We say that G has *fsg* (finitely satisfiable generics) if there is some global type $p(x)$ in G and some small model $\mathcal{M}_0$ such that every left translate

$$gp = \{\varphi(x) \colon \varphi(g^{-1}) \in p\}$$

of p with $g \in G$, is finitely satisfiable in $\mathcal{M}_0$.

Proposition 8.6 *Let G have fsg, witnessed by p and $\mathcal{M}_0$, and let X be a definable subset of G. Then,*

(a) if X is left (resp. right) generic then X is generic, and

(b) conditions (i), (ii) and (iii) of Corollary 7.3 hold for G. Moreover, in (iii), the generic type can be taken to be p.

Theorem 8.7 *Let $\mathcal{M}$ be an expansion of a real closed field. Let G be a definable group and N be a definable normal subgroup of G. Then,*

(1) if N and G/N have fsg, then so does G, and

(2) if G is definably compact then G has fsg.

See Proposition 4.2 in [18] for Proposition 8.6. Theorem 8.7 (1) is Proposition 4.5 in [18]; the proof of (2) in the commutative case is based on Theorem 7.5 and Theorem 7.2 (see Lemma 8.2 in [18]); the proof of (2) in the general case is reduced – by (1) and (2) in the commutative case – to the semisimple case, and this in turn is reduced to the simple case via Theorem 5.3; the simple case is solved via Theorem 4.8 in Proposition 4.6 of [31], the latter is based on Theorem 4.6 in [6] (see Theorem 8.1(i) in [18]).

Finally, we can state the following.

Theorem 8.8 *Conjectures* **C1**–**C4** *are true when $\mathcal{M}$ is an expansion of a real closed field.*

This is Theorem 8.1(ii) in [18] (taking into account the comments after Remark 2.3 there). By Theorems 8.3 and 8.4, it remains to prove conjecture **C3**, and this is done by induction on dim $G(>0)$. Either G is semisimple or there is a normal definable abelian subgroup N of G with dim $G/N <$ dim G. The semisimple case is reduced – via Theorem 5.3 – to the definably simple case, for which we know – Theorem 8.2 – the conjecture is true. On the other hand – by Theorem 8.4 – **C3** is true for N, and by induction for G/N. Knowing the latter, we still have to prove **C3** for G, but now we also know – Theorem 8.7(2) – that G has *fsg*. For the rest of the proof one makes use of Proposition 8.6 and the fact that N has finite n-torsion for each $n > 0$.

We have the following result which is partly a corollary to the proof of the conjectures.

Corollary 8.9 *Let $\mathcal{M}$ be an expansion of a real closed field. Let G*

be a definably compact definably connected group. Then, there is a left invariant Keisler measure μ on G. Moreover, for any definable $X \subset G$, X is generic if and only if $\mu(X) > 0$.

This is Proposition 6.2 in [18] (including its proof) noting that by Corollary 3.10 in [45] the theory of an o-minimal structure has NIP, and by the proof of Theorem 8.8, G has *fsg*.

References

[1] L. Bélair, Z. Chatzidakis, P. D'Aquino, D. Marker, M. Otero, F. Point and A.J. Wilkie (eds.), *Model Theory and Applications*, Quad. Mat. 11, Dept. Mat. Sec. Univ. Napoli, Caserta 2002.

[2] A. Berarducci, Zero-groups and maximal tori in *Logic Colloquium'04*, A. Andretta *et al.* (eds.), LNL 29, 33-45, to appear.

[3] A. Berarducci and M. Otero, Intersection theory for o-minimal manifolds, *Ann. Pure Appl. Logic* 107 (2001) 87-119.

[4] A. Berarducci and M. Otero, o-Minimal fundamental group, homology and manifolds, *J. London Math. Soc.* 65 (2002) 257-270.

[5] A. Berarducci and M. Otero, Transfer methods for o-minimal topology, *J. Symbolic Logic* 68 (2003) 785-794.

[6] A. Berarducci and M. Otero, An additive measure in o-minimal expansions of fields, *Quarterly J. Math.* 55 (2004) 411-419.

[7] A. Berarducci, M. Otero, Y. Peterzil and A. Pillay, Descending chain conditions for groups definable in o-minimal structures, Ann. Pure Appl. Logic 134 (2005) 303-313.

[8] R. Carter, G. Segal and I. Macdonald, *Lectures on Lie Groups and Lie Algebras*, Cambridge Univ. Press 1995.

[9] C. Chevalley, *Theory of Lie Groups*, Princeton Univ. Press 1946.

[10] L. van den Dries, *Tame Topology and o-Minimal Structures*, London Math. Soc. Lect. Notes, vol. 248, Cambridge Univ. Press 1998.

[11] A. Dolich, Forking and independence in o-minimal theories, J. Symbolic Logic 69 (2004) 215-240.

[12] M. Edmundo, Structure theorems for o-minimal structures, *Ann. Pure Appl. Logic* 102 (2000) 159-181.

[13] M. Edmundo, Solvable groups definable in o-minimal expansions of groups, *J. Pure Appl. Algebra* 185 (2003) 103-145.

[14] M. Edmundo, A remark on divisibility of definable groups, *Math. Logic Quart.* 51 (6) (2005) 639-641.

[15] M. Edmundo, Covers of groups definable in o-minimal structures, *Illinois J. Math.* 49 (2005) 99-120.

[16] M. Edmundo and M. Otero, Definably compact abelian groups, *J. Math. Logic* 4 (2) (2004) 163-180.

[17] P. Eleftheriou and S. Starchenko, Groups definable in ordered vector spaces over ordered division rings, *J. Symb. Logic*, 31pp. to appear.

[18] E. Hrushovski, Y. Peterzil and A. Pillay, Groups, measures and NIP, *J. Amer. Math. Soc.*, 34pp. to appear.

[19] E. Hrushovski and A. Pillay, Groups definable in local fields and pseudo-finite fields, *Israel J. Math.* 85 (1994) 203-262.

[20] J. Johns, An open mapping theorem for 0-minimal structures, *J. Symbolic Logic* 66 (2001) 1817-1820.

[21] H.J. Keisler, Measures and forking, *Ann. Pure Appl. Logic* 43 (1987) 119-169.

[22] D. Lascar and A. Pillay, Hyperimaginaries and automorphism groups, *J. Symbolic Logic* 66 (2001) 127-143.

[23] D. Macpherson, A. Mosley and K. Tent, Permutation groups in o-minimal structures, *J. London Math. Soc.* 62 (2000) 650-670.

[24] J. Madden and C. Stanton, One-dimensional Nash groups, *Pacific J. Math.* 154 (1992) 331-334.

[25] J. Madden, Errata correction to One-dimensional Nash groups, *Pacific J. Math.* 161 (1993) 393.

[26] C. Miller and S. Starchenko, A growth dichotomy for o-minimal expansions of ordered groups, *Trans. Am. Math. Soc.* 350 (1998) 3505-3521.

[27] A. Nesin and A. Pillay, Some model theory of compact Lie groups, *Trans. Am. Math. Soc.* 326 (1991) 453-463.

[28] A. Nesin, A. Pillay and V. Razenj, Groups of dimension two and three over o-minimal structures, *Ann. Pure Appl. Logic* 53 (1991) 279-296.

[29] M. Otero, Y. Peterzil and A. Pillay, Groups and rings definable in o-minimal expansions of real closed fields, *Bull. London Math. Soc.* 28 (1996) 7-14.

[30] Y. Peterzil, Some topological and differentiable invariants in o-minimal structures in *Model Theory and Applications*, L. Bélair *et al.* (eds.), 309-323, Quad. Mat.11, Dept. Mat. Sec. Univ. Napoli, Caserta 2002.

[31] Y. Peterzil and A. Pillay, Generic sets in definably compact groups, *Fund. Math.* 193 (2007) 153-170.

[32] Y. Peterzil, A. Pillay and S. Starchenko, Simple groups definable in o-minimal structures in *Logic Colloquium'96*, J.M. Larrazabal *et al.* (eds.), 211-218, LNL 12, Springer, Berlin 1998.

[33] Y. Peterzil, A. Pillay and S. Starchenko, Definably simple groups in o-minimal structures, *Trans. Am. Math. Soc.* 352 (2000) 4397-4419.

[34] Y. Peterzil, A. Pillay and S. Starchenko, Simple algebraic and semialgebraic groups over real closed fields, *Trans. Am. Math. Soc.* 352 (2000) 4421-4450.

[35] Y. Peterzil, A. Pillay and S. Starchenko, Linear groups definable in o-minimal structures, *J. Algebra* 247 (2002) 1-23.

[36] Y. Peterzil and S. Starchenko, A trichotomy theorem for o-minimal structures, *Proc. London Math. Soc.* 77 (1998) 481-523.

[37] Y. Peterzil and S. Starchenko, Definable homomorphisms of abelian groups in o-minimal structures, *Ann. Pure Appl. Logic* 101 (2000) 1-27.

[38] Y. Peterzil and S. Starchenko, On torsion-free groups in o-minimal structures, *Illinois J. Math.* 49 (2005) 1299-1321.

[39] Y. Peterzil and S. Starchenko, Computing o-minimal topological invariants using differential topology, *Trans. Am. Math. Soc.* 359 (2007) 1375-1401.

[40] Y. Peterzil and C. Steinhorn, Definable compactness and definable subgroups of o-minimal groups, *J. London Math. Soc.* 59 (1999) 769-786.

[41] A. Pillay, First order topological structures, *J. Symbolic Logic* 52 (1987) 763-778.

[42] A. Pillay, On groups and fields definable in o-minimal structures, *J. Pure Appl. Algebra* 53 (1988) 239-255.

[43] A. Pillay, An application of model theory to real and p-adic groups, *J. Algebra 126* (1989) 134-146.

[44] A. Pillay, Type-definability, compact Lie groups, and o-minimality, *J. Math. Logic* 4 (2004) 147-162.

[45] A. Pillay and C. Steinhorn, Definable sets in ordered structures I, *Trans. Am. Math. Soc.* 295 (1986) 565-592.

[46] V. Razenj, One dimensional groups over an o-minimal structure, *Ann. Pure Appl. Logic* 53 (1991) 269-277.

[47] A. Strzebonski, Euler characteristic in semialgebraic and other o-minimal structures, *J. Pure Appl. Algebra* 96 (1994) 173-201.

[48] A. Strzebonski, One-dimensional groups in o-minimal structures, *J. Pure Appl. Algebra* 96 (1994) 203-214.

[49] K. Tent, Sharply n-transitive groups in o-minimal structures, *Forum Math.* 12 (2000) 65-75.

[50] A. Woerheide, *O-Minimal homology*, PhD Thesis, University of Illinois at Urbana-Champaign, 1996.

Decision problems in Algebra and analogues of Hilbert's tenth problem

A tutorial presented at American Institute of Mathematics and Newton Institute of Mathematical Sciences

Thanases Pheidas
University of Crete

Karim Zahidi
University of Antwerp

Contents

Introduction

One of the first tasks undertaken by Model Theory was to produce elimination results, for example methods of eliminating quantifiers in formulas of certain structures. In almost all cases those methods have been effective and thus provide algorithms for examining the truth of possible theorems. On the other hand, Gödel's Incompleteness Theorem and many subsequent results show that in certain structures, constructive elimination is impossible. The current article is a (very incomplete) effort to survey some results of each kind, with a focus on the decidability of existential theories, and ask some questions at the intersection of Logic and Number Theory. It has been written having in mind a mathematician without prior exposition to Model Theory. Our presentation will consist of four parts.

Part A deals with positive (decidability) results for analogues of Hilbert's tenth problem for substructures of the integers and for certain local rings.

Part B focuses on the 'parametric problem' and the relevance of Hilbert's tenth problem to conjectures of Lang.

Part C deals with the analogue of Hilbert's tenth problem for rings of Analytic and Meromorphic functions.

Part D is an informal discussion on the chances of proving a negative (or could it be positive?) answer to the analogue of Hilbert's tenth problem for the field of rational numbers.

A central undecidability result in our presentation will be Hilbert's tenth problem, which asked:

Give a procedure which, in a finite number of steps, can determine whether a polynomial equation (in several variables) with integer coefficients has or does not have integer solutions.

The answer by Matiyasevich ([43]), following work of Davis, Putnam and J. Robinson, was negative ('no such algorithm can exist').

Analogous questions can be asked for domains other than the ring of integers. In trying to ask questions "similar" to Hilbert's tenth problem (from now on denoted by HTP) in rings other than the rational integers, one has to specify the kind of 'diophantine equations' one considers. For example, say that we want to ask HTP for the ring of polynomials $\mathbb{C}[z]$ in one variable, z, with complex coefficients. Let $\mathcal{D}$ be a class of 'diophantine equations' over $\mathbb{C}[z]$, that is, polynomial equations in many variables, with coefficients in $\mathbb{C}[z]$. The analogue of HTP for this class is "does there exist an algorithm to decide, given any equation in $\mathcal{D}$, whether that equation has or does not have solutions in $\mathbb{C}[z]$?". It is obvious that if one chooses $\mathcal{D}$ to be the set of all 'diophantine equations' over $\mathbb{C}[z]$ then the answer to the question is NO, simply because $\mathcal{D}$ is uncountable (algorithms, in the classical sense, treat only countable problems). On the other hand, one may take $\mathcal{D}$ to be the set of 'diophantine equations' which have coefficients in $\mathbb{Z}[z]$ or in $\mathbb{Z}[i][z]$ ($i = \sqrt{-1}$)

or in $\mathbb{Z}$. Each of these choices gives a different 'analogue of HTP for $\mathbb{C}[z]$'. In this paper we will be specifying $\mathcal{D}$ by specifying the *language* in which we work. Notice that the analogue of HTP for $\mathbb{C}[z]$ for the class $\mathcal{D}$, asked for systems of diophantine equations (rather than single equations), is really the question of decidability of the positive-existential theory of $\mathbb{C}[z]$ in the language L which contains symbols for addition, multiplication, equality, and constant symbols for the coefficients of the equations of $\mathcal{D}$ (or constant symbols whose interpretations generate exactly the set of coefficients of equations of $\mathcal{D}$). For example, the analogue of HTP (for systems of equations) for $\mathbb{C}[z]$ with $\mathcal{D}$ the set of equations with coefficients in $\mathbb{Z}$ is equivalent to the question of decidability of the positive-existential theory of $\mathbb{C}[z]$ in the language $L_r = \{+, \cdot\,; =; 0, 1\}$, while that with $\mathcal{D}$ the set of equations with coefficients in $\mathbb{Z}[z]$ is equivalent to the question of decidability of the positive-existential theory of $\mathbb{C}[z]$ in the language $L_z = \{+, \cdot\,; =; 0, 1, z\}$.

We will consider structures (models) such as the field $\mathbb{C}(z)$. Each structure comes with a language, i.e. a set of symbols for the relations, functions and distinguished elements of the structure. For example we consider $\mathbb{C}(z)$ as an L_z-structure, with symbol for the relations $=$, the functions $+$ (addition) and $\cdot$ (multiplication) and the distinguished elements 0, 1 and z. The *first order sentences* of the language of the structure are the sentences built using the symbols of the language, variables ranging over the universe of the structure, quantifiers ($\exists$ and $\forall$) and logical connectives, by the usual rules. The *existential* (resp. *positive-existential*) sentences are those that start with existential quantifiers which are followed by a quantifier-free formula (resp. by a quantifier-free formula which is a disjunction of conjunctions of relations - negations of relations are not allowed in this case). The *(full) theory* (resp. *existential theory, positive-existential theory*) of the structure is the set of sentences (resp. existential sentences, positive-existential sentences) which are true in the structure.

We say that the theory (resp. existential theory, positive-existential theory) of a structure is *decidable* if there exists an algorithm which determines whether any given sentence (resp. existential sentence, positive-existential sentence) is true or false in the structure - otherwise we say that the theory is *undecidable*.

We present a list of decidability properties of some structures of common use. The first three lines in the table of the next page contain substructures of the ring of rational integers: $(\mathbb{Z}, +, n \mapsto 2^n, 0, 1)$ is the structure of $\mathbb{Z}$ with addition and the partial function $n \mapsto 2^n$ (with

	ex. th. in L_T	ex. th. in L_z	full th.
$(\mathbb{Z},+,0,1)$	Y		Y
$(\mathbb{Z},+,n\mapsto 2^n,0,1)$	Y		Y
$(\mathbb{Z},+,\vert,0,1)$	Y		N
$(\mathcal{O}_K,+,\vert,0,1)$	conj. N		N
$\mathbb{Z}$	N		N
$\mathcal{O}_K$	conj. N		N
$\mathbb{Q}$	?		N
$\mathbb{F}_q[z],\mathbb{R}[z],\mathbb{C}[z]$	N	N	N
$\mathbb{F}_q(z)$	?	N	N
$\mathbb{R}(z)$	?	N	N
$\mathbb{C}(z)$	?	?	?
$\mathcal{H}(\{a\})$	Y	Y	Y
$\mathcal{H}(\mathcal{U})$	Y	?	N
$\mathcal{H}(\mathbb{C})$	?	?	N
$\mathcal{M}(\mathcal{U})$	Y	?	?
$\mathcal{M}(\mathbb{C})$	?	?	?

domain $\mathbb{N}$), and $(\mathbb{Z},+,|,0,1)$ the structure of $\mathbb{Z}$ with addition and divisibility. On the fourth line is the structure of addition and divisibility in a ring $\mathcal{O}_K$ of integers of the number field K - which is assumed not to be imaginary quadratic. The positive-existential theories of those structures have been shown to be of the same hardness as the positive-existential theories of the corresponding ring structures, but it is an open problem whether the latter are undecidable for arbitrary K (see the comment below). In the first four lines the columns 'ex. th.' and 'full th.' show the decidability properties of the existential and the full theory of the structure, respectively. The remaining structures are ring structures: $\mathbb{Q}$ is the field of rational numbers, $\mathbb{R}$ the field of real numbers, $\mathbb{C}$ the field of complex numbers, $\mathbb{F}_q$ is the finite field with q elements, $B[z]$ the ring of polynomials in the variable z with coefficients in the ring B, $B(z)$ the corresponding field of rational functions in z, $\mathcal{H}(\mathcal{D})$ the ring of analytic functions of the variable z as that ranges in an open superset of the subset $\mathcal{D}$ in the complex plane, $\mathcal{M}(\mathcal{D})$ is the corresponding field of meromorphic functions, U is the open unit disk. L_T is the language $\{+,\cdot;=;0,1\}$ which, for rings of functions is augmented by the predicate T which is interpreted as 'x is not a constant function'.

For rings of functions of the variable z the language L_z is as above. The first column shows whether the positive existential theory of the ring in the language L_T is decidable or not ('Y' means decidable, 'N' means undecidable, 'conj. N' means 'conjectured to be undecidable', '?' denotes an open problem), the second column corresponds to the similar properties in the language L_z and the third column to that of the full theory in the language L_T for the rings $\mathbb{Z}$, $\mathcal{O}_K$ and $\mathbb{Q}$ and the language L_z for the remaining rings.

Note that it is known that the theories of many rings $\mathcal{O}_K$ are undecidable (e.g. for abelian K) and it has been conjectured that all of them are, but the question for arbitrary K remains open.

For a fast introduction to applications of Model Theory to Algebra the reader may consult [9] and [66]. The solution of HTP can be found in [15], and is explained very nicely to the non-expert in [14]. Surveys of questions similar to the present paper's are [47], [51], [52] and [59]. Surveys of elimination ('decidability') techniques and results can be found in [60] (and many later more specialized articles, from the Algebraist's point of view).

We are indebted to F. Campana, J. Demeyer and T. Scanlon for various comments towards improvements of this paper.

1 Part A: Decidability results

For a quick introduction to a basic elimination technique, whose origin lies in the early days of Model Theory, solve the following Exercise.

Say that we work in a language L. A theory T (i.e. a subset of the set of formulas) of L *admits elimination of quantifiers* if any L-formula $\phi(x)$ is equivalent in T to an existential formula.

Exercise 1.1 (a) Let T be an L-theory. Assume that any formula of the form $\exists x\ \psi(x, y)$, where x is one variable, y is a tuple of variables and ψ is quantifier-free, is equivalent in T to a quantifier-free formula. Prove that T admits elimination of quantifiers.

[Hint: Recall that any formula is equivalent to one in *prenex normal form* (i.e. a sequence of quantified variables, followed by a quantifier-free formula). Thus it suffices to consider only formulas of that form. You may then use induction on the number of quantifiers. Also note that to prove that $\forall x \phi(x, y)$ (with ϕ quantifier-free) is quantifier-free, it suffices to prove that its negation (which is an existential formula) is quantifier-free.]

(b) Consider the field $\mathbb{C}$ of complex numbers as an L_r-structure. Consider a system of polynomial equations and inequations

$$S(x,y) \ : \ \bigwedge_i f_i(x,y) = 0 \wedge \bigwedge_j g_j(x,y) \neq 0$$

where x is one variable, $y = (y_1, \ldots, y_m)$ and each y_k is a variable, the indices i and j range over some finite sets I and J respectively, and each f_i and g_i is a polynomial in $\mathbb{Z}[x, y_1, \ldots, y_m]$. Prove that the formula $\exists x \ S(x,y)$ is equivalent to a quantifier-free formula,

[Hint: First notice that the existential quantifier distributes over $\vee$ ('or'). In the next few lines a 'polynomial' is an element of $\mathbb{Z}[x,y]$. Note that we may assume that J is empty or a singleton, i.e. there is no inequation or a single inequation in the system. First assume that J is empty and use the theory of *resultants* (see any graduate book in Algebra) to eliminate the existential quantifier (but the resulting equivalent existential formula contains, in general, inequations). In case J is a singleton, show that one can reduce to the previous case, generalizing to many variables the following observations about two variables:
Say that y is one variable. Consider the system

$$T(x,y) : \ f(x,y) = 0 \wedge g(x,y) \neq 0.$$

(i) Show that a system $T_1 : \ T(x,y) \wedge h(x,y) = 0$, with h a nontrivial polynomial in x and y and with degree in x lower than that of f, is equivalent to a system like T (possibly with some relations involving only the variable y) but with degree of f in x lower than the original one. Achieve this by euclidean division of f by h: there is a polynomial a in y and polynomials h_1 and q in (x,y) and with degree of h_1 in x lower than that of h such that $af = qh + h_1$ (a is the highest degree coefficient of h as a polynomial in x); then the system T_1 is equivalent to the disjunction of the systems $T(x,y) \wedge h - ax^r = 0 \wedge a = 0$, where r is the degree in x of h, and $h = 0 \wedge h_1 = 0 \wedge ag \neq 0$. Observe that both the latter systems have sum of degrees in x of the polynomial equations lower than that of T_1. Iterate.

(ii) Using the euclidean algorithm for polynomials, find a polynomial d which is a greatest common divisor in $\mathbb{Q}(y)[x]$ of f and g with respect to x and for which the following holds formally: $bd = uf + vg$ for some polynomials u, v, b, with v having degree in x lower than that of f and with b being a no-zero polynomial of the variable y only. Then T is equivalent to the disjunction of systems $f = 0 \wedge bd \neq 0$ and $f = 0 \wedge v = 0 \wedge g \neq 0$. The second of the latter leads to 'simpler' systems, by the previous paragraph. The first one has degree in x of the polynomial inequation less than the original one.

(iii) Iterating the results of the last paragraph, observe that T is equivalent to a disjunction of systems which have the form $f = 0$ together with some relations involving only the variable y. Then observe that for any value of y, $f(x, y) = 0$ is satisfiable if f is non-trivial as a polynomial in x (since $\mathbb{C}$ is algebraically closed). Hence the satisfiability of each of the latter systems is equivalent to the satisfiability of a disjunction of systems of the variable y only. Hence the existential quantifier $\exists x$ has been eliminated.

Use (a) to conclude that the theory of $\mathbb{C}$ in L_r admits elimination of quantifiers. Observe that the procedure of finding a quantifier-free formula equivalent to a given formula, is effective. Conclude that the theory of $\mathbb{C}$ in L_r is decidable, that is, there is an algorithm to determine whether any given sentence of L_r is true or false in $\mathbb{C}$.

1.1 Presburger arithmetic

A basic, old result is the decidability of Presburger arithmetic, i.e. the theory of the ordered additive group of integers. Presburger showed that **Z** admits elimination of quantifiers in a language which extends the language $L_P = \{+, \geq; 0, 1\}$ with predicates for $a \equiv b \bmod m$ for every integer m (actually Presburger considered the theory of natural numbers, not the integers, and in a language slightly different from L_P but with the same expressive power).

Exercise 1.2 Show that the theory of $\mathbb{Z}$ in the language L_P is 'model-complete' in the following way:

(a) For each $k \in \mathbb{N}$ with $k \geq 2$ define the one-place predicate M_k to mean '$M_k(x) \leftrightarrow x$ is a multiple of k'. Extend the language L_P by the predicates M_k to obtain the language L_P^M.

(b) Observe that each quantifier-free formula of L_P^M is equivalent to a finite disjunction of formulas of the form

$$\bigwedge_i (f_i(\bar{x}) = 0) \wedge \bigwedge_j (M_{k_j}(g_j(\bar{x}))) \wedge \bigwedge_i (h_i(\bar{x}) \geq 0)$$

where each f_i and g_j is a polynomial of degree 1 in the variables of the tuple $\bar{x}$.

(c) Prove that each existential L_P^M-formula, i.e. of the form $\exists \bar{x}\, \phi(\bar{x}, \bar{y})$ where $\phi(\bar{x}, \bar{y})$ is quantifier free, is equivalent to a quantifier-free L_P^M-

formula. You can do this by eliminating the existential quantifiers, one at a time. For example, to eliminate the quantifier $\exists x$ from

$$\exists x[a - x \geq 0 \wedge x \geq b \wedge M_2(x+c)]$$

(a, b and c can be linear polynomials in some variables and x does not occur in any of them) observe that it is equivalent to the disjunction of the formulas that correspond to the cases (A) $a \geq b+1$, (B) $a = b$ and b is even and c is even, and (C) $a = b$ and a is odd and c is odd.

(d) Fix a language and a theory. Assume that each existential formula is equivalent (in the theory) to some quantifier-free formula. Then show that each formula of the language has the same property (i.e. is equivalent to a quantifier-free formula).

One can ask how one can enrich the Presburger language L_P and still retain decidability for the existential theory of $\mathbb{Z}$ in this enriched language. Of course, as soon as multiplication is definable from the functions and predicates in the language, one gets undecidability. The bibliography on this subject is quite extensive. For some information see [10] and the bibliography of [51]. Below we will see some results of this kind, as well.

1.2 Addition and divisibility

The results we present are from [38] (similar results were obtained in [6]), [39] and [40].

A natural extension of L_P consists of adding a binary predicate for the divisibility relation $|$ (that is, $a|b$ if and only if $\exists c : b = ac$). J. Robinson showed in [53] that multiplication can be defined from addition and divisibility by a first-order formula and hence, the first-order theory of $\mathbb{Z}$ in this language is undecidable. In contrast, Lipshitz, in [38] (and, independently, Bel'tyukov in [6]) showed that the existential theory of $\mathbb{Z}$ in the language $L_| = L_P \cup \{|\}$ is decidable. The same is true for any ring of integers of an imaginary quadratic extension of the rationals (in this case the predicate $>$ has to be excluded from the language). This result is optimal in several ways: multiplication is positive-existential in the $L_|$-theory of any ring of algebraic integers in a number field other than the rationals and imaginary quadratic ([39] and [40]); hence the $L_|$-existential-theory of those rings has the same decidability property as the ring theory which has been conjectured – by Denef and Lipshitz, but not yet proved – to be undecidable. These results were later generalized to polynomial rings over fields: the existential theory of addition and divisibility in a polynomial ring $k[t]$ over a field k (in the language $L = L_P \cup \{t, |\}$), is decidable if and only if the existential theory of k is decidable; the positive existential theory of $A[t_1, t_2]$ in the language $L = L_P \cup \{t_1, t_2, |\}$ is undecidable for any commutative domain A.

We give a short account of the proof of [38]:
Let $L_| = L_P \cup \{|\}$ where $|$ will be interpreted by '$a|b \leftrightarrow \exists c[b = ac]$'.

(a) Show that every existential sentence of $L_|$ is equivalent to a disjunction of formulas of the form $\exists x \phi(x)$ where

$$(1) \qquad \phi(x) : \bigwedge_i (f_i(x)|g_i(x)) \bigwedge_j (h_j(x) \geq 0)$$

where each f_i, g_i and h_j is a polynomial of degree at most 1 in the variables of the tuple of variables $x = (x_1, \ldots, x_m)$. To do this one has to

(a1) Eliminate non-divisibilities: observe that $\neg a|b$ is equivalent to 'a greatest common divisor of a and b is different from a and $-a$' which is equivalent to

$$\exists s, x, y\ [s|a \wedge s|b \wedge a|x \wedge b|y \wedge s = x - y \wedge \neg(s + a = 0) \wedge \neg(s - a = 0)].$$

(a2) Eliminate equations, e.g. if the equation $2x - y = 0$ occurs then one can substitute all occurrences of the variable x by $\frac{y}{2}$, clear denominators by multiplying all terms by 2, and add the divisibility $2|y$.

In the rest work with notation as in (1).

(b) Impose all possible relative orderings on the variables of x, e.g. $x_1 \geq \cdots \geq x_m$, and all possible orderings on the terms f_i, g_i and h_i. Each one of these cases gives a sentence (the disjunction of all these is equivalent to the initial sentence). So assume that (1) corresponds to such a sentence.

Draw conclusions that eliminate some variables, for example, if a divisibility $x + f(y)|x + g(y)$ occurs and $x, x + f(y), x + g(y) \geq 0$ and the variable x is $\geq$ to all the variables of the tuple y, then draw the conclusion that $\frac{x+g(y)}{x+f(y)}$ is an integer $\leq M + 1$ where M is the maximum absolute value of the coefficients of $f - g$ (**Exercise**: Why?); Consider the cases and eliminate the variable x (conclude with fewer variables).

(c) Impose certain 'implied' divisibilities, for example, from $f_1|f_2$ and $f_2|f_3$ conclude that $f_1|f_3$ (and add it to the given list of divisibilities).

(d) Iterate (b) and (c). Show that a finite number of iterations concludes with systems which are 'diagonal', in the sense that in each divisibility $f|g$ there is a variable that occurs in g, which is 'bigger' than all variables of f; in addition these systems are closed under the operations mentioned in (c).

(e) Show that diagonal systems satisfy the following 'local-to-global' principle: Let $\phi(x) : \bigwedge_i (f_i(x)|g_i(x)) \bigwedge_j (h_j(x) \geq 0)$ be a diagonal system and let $\phi'(x)$ be the system obtained from ϕ by deleting all inequalities (so ϕ' contains only divisibilities). Then one can effectively find a natural number N, such that $\phi(x)$ has a solution in $\mathbb{Z}$ if and only if for each prime number $p \leq N$ $\phi'(x)$ has a solution x in $\mathbb{Z}_p$ such that $f_i(x) \not\equiv 0 \mod p^N$. Use the deciability of the theory of each $\mathbb{Z}_p$ to conclude.

Remark The local-to-global principle mentioned in (e) fails for systems in general, e.g. $x + 2|2x + 3$ is satisfiable in each $\mathbb{Z}_p$ but the system $x + 2|2x + 3 \wedge x \geq 0$ is not satisfiable in $\mathbb{Z}$.

Exercise 1.3 The following give some of the main ideas in [39].

Consider the ring of integers $\mathcal{O}$ of a real quadratic number field K (e.g. $K = \mathbb{Q}(\sqrt{5})$). Consider known the following fact: There is a unit ϵ_0 which is 'fundamental' i.e. all units of $\mathcal{O}$ are of the form $\pm\epsilon_0^n$ with $n \in \mathbb{Z}$.

(a) Show that if $n|m$ in $\mathbb{Z}$ then $\epsilon_0^n - 1|\epsilon_0^m - 1$ in $\mathcal{O}$.

It can be shown that a sort of converse is also true: There is a $d \in \mathbb{Z}$ such that setting $\epsilon = \epsilon_0^d$ we have that $n|m$ in $\mathbb{Z}$ if and only if $\epsilon^n - 1|\epsilon^m - 1$ in $\mathcal{O}$. For the rest assume that one has in the language a name (constant symbol) for a unit ϵ with these properties and assume that the set $\{\epsilon^n : n \in \mathbb{Z}\}$ is positive existential in the language $L_| \cup \{\epsilon\}$ (all the mentioned facts are true).

(b) Assume that $m, n \neq 0$ and $m \neq n$. Show that $\epsilon^m = \epsilon^{2n}$ if and only if

$$\epsilon^n - 1|\epsilon^m - \epsilon^n \wedge \epsilon^m - \epsilon^n|\epsilon^n - 1.$$

(c) Show: If $x \in \mathcal{O} \setminus \{0\}$ then there is an $n \neq 0$ such that $x|\epsilon^n - 1$ (hint: if x is not a unit then the ring $\mathcal{O}/(x)$ is finite, hence the natural image of ϵ has finite multiplicative order).

(d) Let $\phi(x)$ denote the formula

$$x \neq 0 \wedge x \neq 1 \wedge x \neq -1 \wedge x|\epsilon^n - 1 \wedge x - 1|\epsilon^n - 1 \wedge x + 1|\epsilon^n - 1.$$

Assume that if $n \neq 0$ and $\phi(x)$ is true then $2|x^2| < \epsilon^n$ ($|w|$ is the absolute value of w).

Show: For $x, y \in \mathcal{O} \setminus \{0, 1, -1\}$ the following is true: $y = x^2$ if and only if for some $n \in \mathbb{Z} \setminus \{0\}$ we have

$$\phi(x) \wedge \phi(y) \wedge \epsilon^n - x|\epsilon^{2n} - y.$$

(e) Consider known the following: The relation $x \neq 0$ is positive-existential in $L_| \cup \{\epsilon\}$.

Use (a)-(d) to produce a formula $\psi(x, y)$ of $L_| \cup \{\epsilon\}$ such that for any $x, y \in \mathcal{O}$ the following is true in $\mathcal{O}$:

$$\psi(x, y) \leftrightarrow y = x^2.$$

Thus squaring over $\mathcal{O}$ is positive-existential in $L_| \cup \{\epsilon\}$. Prove that this implies that multiplication over $\mathcal{O}$ is positive-existential in $L_| \cup \{\epsilon\}$. Conclude that the positive-existential theory of $L_| \cup \{\epsilon\}$ is undecidable.

1.3 Addition and exponentiation

The following are results from [58].

The first-order theory of $\mathbb{N}$ in the language $L = L_P \cup \{\exp_2\}$, where $\exp_2$ is the function which sends a natural number n to 2^n, is decidable.

Towards proving this, one has to study the behavior of 'exponential polynomials', e.g. of the form $2^{2^{3x-1}} - 2^{5x+5} + 1$. We will not go into the details.

Exercise 1.4 Show that there is an algorithm which, given any diophantine equation $f(x) = 0$ (x is a tuple of m variables and f a polynomial in x over $\mathbb{Z}$), decides whether that has or does not have solutions in the set $\{2^n : n \in \mathbb{N}\}^m$.

Remark By results of [33] the statement of the Exercise is true for any finitely generated multiplicative (i.e. closed under multiplication) set, e.g. the set $\{2^n 3^m : n, m \in \mathbb{N}\}$.

1.4 The analogue of Hilbert's tenth problem for algebraic groups

The following (among several) result can be found in [30].

Theorem 1.5 *There is an algorithm which, given any algebraic group G defined over $\mathbb{Z}$ and any positive integer k, decides whether $G(\mathbb{Z})$ contains a subgroup of index k.*

Two examples of such groups are:

1. $\mathbb{Z}$, $\mathbb{Z} \times \mathbb{Z}$ etc. with the group structure of component-wise addition.

2. The solution set of the equation $X^2 - dY^2 = 1$ where d is a square-free integer; the group law $\oplus$ is defined by

$$(x_1, y_1) \oplus (x_2, y_2) = (x_1x_2 + dy_1y_2, x_1y_2 + x_2y_1).$$

This justifies the claim that 'the more structure one has, the more likely decidability is'.

1.5 The results of Ax for 'almost all primes'

In [1] Ax gave an algorithm which, for any given diophantine equation, tests whether the equation has a solution modulo every prime number. In [2] he extended this to an algorithm that decides the truth of a sentence of L_r in all finite fields. The proofs make use of Weil's result on the congruence zeta function and of Čebotarev's density theorem.

For further results in this direction see [27] and [65].

1.6 Model-completeness

A set (subset of a power of $\mathbb{Z}$ or any effectively constructible countable set) is *recursive* if membership in it can be tested by an algorithm; it is *recursively enumerable* if its elements can be listed (eventually all) by an algorithm.

A theory of a language (finite or countable-and-recursive) L is a subset of the set of sentences of L. The theory of a structure (model) $\mathcal{A}$ in a language L is *model-complete* if every formula of L is equivalent to an existential formula of L over $\mathcal{A}$.

The following is a classical decidability argument in Model Theory:
Assume:

(a) $\mathcal{A}$ is a countable and recursive structure in the countable language L.

(b) The theory of $\mathcal{A}$ in L is *effectively model-complete*, that is, there is an algorithm which to any sentence of L associates an equivalent (over $\mathcal{A}$) existential sentence.
Conclude: The theory of $\mathcal{A}$ in L is decidable.

Proof Let σ be a sentence of L. Find an existential sentence, say $\exists x\ \phi(x)$, equivalent to σ, and an existential sentence, say $\exists y\ \psi(y)$, equivalent to $\neg\sigma$ where x is a tuple of m variables, y is a tuple of n variables and the formulas ϕ and ψ are quantifier-free.

Enumerate the tuples of m elements of $\mathcal{A}$ and the tuples of n elements of $\mathcal{A}$. By day plug in $\phi(x)$ tuples of m elements and check whether they make it true. By night do similarly with ψ.

One of σ or $\neg\sigma$ is true, hence either we will satisfy ϕ on a day (in which case σ is true) or we will satisfy ψ on a night (in which case $\neg\sigma$ is true). □

Exercise 1.6 (a) Prove that if in a theory T every universal formula is equivalent to an existential one then T is model-complete.

Consider the language L which extends the language of rings by a symbol $\geq$ for the ordinary ordering relation. A polynomial (or rational function) $f(x)$ in the m variables x, with coefficients in $\mathbb{R}$, is *positive-definite* if for all $a \in \mathbb{R}^m$ we have $f(a) \geq 0$.

(b) Show that every positive-definite polynomial in one variable x is the sum of two squares of polynomials of x with coefficients in $\mathbb{R}$. (This admits a generalization to m variables: A positive-definite rational function in m variables is equal to the sum of 2^m-many squares of rational functions; this is a positive answer to Hilbert's 17-th problem.)

(c) Use the positive answer to Hilbert's 17-th problem mentioned under (b) to prove that if $f(x_1, \ldots, x_m) \in \mathbb{Z}[x_1, \ldots, x_m]$ is a polynomial, then the universal formula

$$\forall x_1 \ldots \forall x_m \; f(x_1, \ldots, x_m) \geq 0$$

(where the quantifiers range over $\mathbb{R}$) is equivalent to an existential formula.

1.7 Complete theories

Suppose that T is a theory of the recursive language L. We say that T is *complete* if for any sentence σ of L either σ or $\neg\sigma$ is a consequence of T. Then one sees easily that

Theorem 1.7 *Assume that T is a recursively enumerable and complete theory of the recursive language L. Then there is an algorithm which tests any given sentence σ of L for being or not being a consequence of T.*

Exercise 1.8 Give an outline of a proof of the Theorem.

For example, the theory of algebraically closed fields of a fixed characteristic (the axioms for a field, together with axioms which fix the characteristic, together with axioms which state, for each n, "every polynomial of degree n has a zero"), which is known to be complete, has a recursive set of consequences.

Exercise 1.9 Describe in detail the sentences (up to equivalence) of the latter theory.

The latter Theorem transforms the question of whether the set of consequences of a given theory is decidable (i.e. recursive) to the question of whether the theory is complete (whenever that is true). This is a very common approach in Model Theory to decidability questions.

It is obvious that the latter Theorem may be not true if T is not complete. Examples of this kind exist in the bibliography.

For relevant bibliography see [9].

1.8 Power-series and germs of analytic functions: existential decidability and Artin approximation

The ring-theories of $\mathbb{Q}_p$ (p is a prime number) are decidable (results of Nerode, Ax and Kochen [3], Ershov [26], also cf. [42], [11], and [19]).

If F is a decidable field of characteristic 0 then the ring-theory of the field of formal fractional power series $F((T))$ in one variable T (and that of $F[[T]]$) is decidable ([36] and [70]).

But if T is two or more variables then we have undecidability ([16]).

For more relevant results see [5] and [22].

We wish to address the question of the decidability of the existential theory of a ring $\mathcal{H}_z(D)$ of functions of a tuple of variables z, analytic on a set D (i.e. analytic on some open super-set of D). We will address first the question for $D =$(a singleton), say $D = \{0\}$. Already in [36] it was shown that the ring-theory of $\mathcal{H}_z(\{0\})$ is decidable. But the techniques of that paper do not transfer to bigger D, so we will look into other approaches. One approach is via 'Approximation Properties'.

Definition 1.10 (i) Let R be a local ring and $\hat{R}$ be its completion. We say that R has the *Approximation Property* if every system of polynomial equations over R, which has a solution in $\hat{R}$, has a solution in R.

(ii) $\mathbb{C}\{z_1, \ldots, z_q\}$ is the ring of formal power series, in the variables $z_1, \ldots, z_q$ over $\mathbb{C}$, which converge in some neighborhood of the origin.

(iii) Let k be a field. Then $k\langle z_1, \ldots, z_q\rangle$ denotes the ring of formal power series in $z_1, \ldots, z_q$ over k which are algebraic over $k[z_1, \ldots, z_q]$.

Artin proved:

Theorem 1.11 *Let k be any field. The following two rings have the Approximation Property:*

(i) $k\langle z_1, \dots, z_q\rangle$ and

(ii) $\mathbb{C}\{z_1, \dots, z_q\}$.

Theorem 1.12 *Let k be an arbitrary field. Then, a system of polynomial equations, with coefficients in $k[[z_1, \dots, z_q]]$, has solutions in $k[[z_1, \dots, z_q]]$ if and only if, for any $n \in \mathbb{N}^+$, it has solutions modulo $(z_1, \dots, z_q)^n$.*

By Theorem 1.11, a system of equations, with coefficients in $\mathbb{C}[z]$, has solutions in $\mathcal{H}_z(\{0\})$ if and only if it has solutions in $\mathbb{C}\langle z\rangle$.
A much stronger version of Theorem 1.12, suitable for algorithmic computation of solutions of equations, is true: for each system of equations, there is a constant N, depending on the system (actually: algorithmically computable and depending only on the number of variables and the degrees of the involved polynomials), such that the system has solutions in $k[[z]]$ if and only if it has solutions modulo $(z)^N$. Results of this type occur in the literature under the name 'Strong Approximation theorems'.
Strong Approximation implies decidability of the positive existential theory of $k[[z]]$ (in order to see whether a system of equations has solutions, compute the constant N of the previous paragraph and check whether the system has solutions modulo $(z)^N$). But, since our main interest is to investigate the possibility of adapting our arguments to $\mathcal{H}_z(D)$ for D not a singleton, and because there is no known analogue of Strong Approximation for these rings, we will now present a decidability result, based only on Theorems 1.11 and 1.12. We will show how one can use these Theorems in order to detect whether a given system of polynomial equations, with coefficients in $\tilde{\mathbb{Q}}[z]$ ($z = (z_1, \dots, z_n)$ and $\tilde{\mathbb{Q}}$ denotes the algebraic closure of $\mathbb{Q}$), has solutions over $\mathcal{H}_z(\{0\}^n)$. Let such a system be given; by Theorem 1.11 it suffices to check whether the system has solutions over $\mathbb{C}\langle z\rangle$. First, observe that if the system has solutions over $\mathbb{C}\langle z\rangle$ then it has solutions whose constant coefficients are algebraic numbers, that is, the system has solutions over $\tilde{\mathbb{Q}}\langle z\rangle$. So, here is an algorithm that decides whether the system has solutions over $\mathcal{H}_z(\{0\}^n)$: We run in parallel the following two processes. The first process lists the tuples of $\tilde{\mathbb{Q}}\langle z\rangle$ and determines whether each one of them is a solution. The second process determines whether the system has solutions modulo z^m for $m = 1, 2, \dots$ (observe that the reduction of the system modulo any

z^m reduces to solving a new system over $\tilde{\mathbb{Q}}$, which can be done by an algorithm, since the existential theory of any algebraically closed field, such as $\tilde{\mathbb{Q}}$, is decidable) . If the system has solutions over $\tilde{\mathbb{Q}}\langle z\rangle$ then the first process will find them, eventually. If the system has no solutions, then the second process will eventually find an m for which the system has no solution modulo z^m. We note that a similar algorithm works for systems of algebraic differential equations (cf. [23]).
It is evident that, in the domains where one has analogues of Theorems 1.11 and 1.12, one may expect decidability results similar to the above. Unfortunately, analogues of Theorem 1.12 in rings $\mathcal{H}_z(D)$, for D not a singleton, are not known. Theorem 1.11 has been extended to the very general case of an *excellent regular local Henselian* ring (by Spivakovski, see [61]). Surprisingly perhaps, there is a similar result for compact domains D by van den Dries:

Theorem 1.13 ([63]) *Assume that D is a compact subset of $\mathbb{C}$ and denote by $\mathbb{C}_D\langle z\rangle$ the ring of functions on D, in the variable z, which are algebraic over $\mathbb{C}(z)$ and analytic on an open superset of D. Let $x = (x_1, \ldots, x_m)$, $f_i \in \mathbb{C}_D\langle z\rangle[x]$. Let $\epsilon > 0$.*
Assume that the system of equations $\bigwedge_i f_i[x_1, \ldots, x_m]$ has a solution $\alpha = (\alpha_1, \ldots, \alpha_m)$ in $\mathcal{H}_z(D)$. Then it also has a solution $\beta = (\beta_1, \ldots, \beta_m)$ with $\beta_i \in \mathbb{C}_D\langle z\rangle$ and such that $|\beta - \alpha|_\infty < \epsilon$ where $|\cdot|_\infty$ denotes the supremum norm on D.

It is easy to see that Theorem 1.11 reduces the question of decidability of the existential theory of $\mathcal{H}_z(D)$, for D compact, to the similar question for the existential theory of the ring of algebraic functions which are analytic on D. But the problem remains open:

Question 1.14 Is the existential theory of $\mathcal{H}_z(D)$ decidable for $D = \mathbb{C}$? for D =(the open unit disc)? for D =(the closed unit disc)?

In a later section we give more information on this problem.

1.9 Positive-existential versus existential

In some cases we do have a decision method for the positive-existential theory of a ring but we do not know the similar result for the existential theory. Such is the case, for example, for the power series ring $\mathbb{F}_p[[t]]$ where t is one variable.

In [24] it is shown that if there is Resolution of Singularities in positive characteristic p then the positive-existential theory of $\mathbb{F}_p[[t]]$ is decidable.

Exercise 1.15 Prove that for any prime $p \geq 3$ the set $\mathbb{F}_p[[t]]$ is positive-existentially definable in the field $\mathbb{F}_p((t))$ in the following way:

For any $x \in \mathbb{F}_p((t))$,

$$x \in \mathbb{F}_p[[t]] \text{ if and only if } \exists y[1 + tx^2 = y^2].$$

Conclude that the existential theory of $\mathbb{F}_p[[t]]$ is decidable if and only if the positive-existential theory of $\mathbb{F}_p((t))$ is decidable.

A rough presentation of some ideas in the proof is the following:

Assume that V is a variety over $\mathbb{F}_p((t))$, given by a finite set of polynomial equations. We want to decide whether V has points rational over $\mathbb{F}_p((t))$.

Step 1: Resolve the singularities of V.

A Resolution of Singularities (RoS) of the variety V over the field $\mathbb{F}_p((t))$ is a non-singular variety W, together with a surjective, birational and proper morphism $f : W \to V$. A basic observation is:

Observation: If every variety over a countable recursive field K has RoS then that is effective (that is, one can find a RoS).

This is done roughly as follows: List all possible pairs (W, f) and for each one of them, in the order of enumeration, check whether f maps onto V (this can be done effectively but we will not go into the details here). Since we have assumed that some RoS exists, at some point we will find one.

Existence of RoS is known (for any variety) over any field of zero characteristic (due to Hironaka), but unknown in positive characteristic. In what follows we will assume existence of RoS in characteristic p.

So we are reduced to the case in which the variety V is nonsingular, which we assume from now on.

Step 2: The variety V is described by a finite set of polynomial equations. By embedding V into the union of a finite set of affine spaces the existence of a point of V, rational over $\mathbb{F}_p((t))$, is translated into a finite set of questions each of which asks whether a system of equations, defining a nonsingular affine variety V', together with a set of inequations which claims that the considered point is not on a variety Y, has solutions over $\mathbb{F}_p[[t]]$. For example, if V were a plane curve, given by the equation $f(x_1, x_2) = 0$ (we work in affine space, the projective case is similar), then the existence of points of V, rational over $\mathbb{F}_p((t))$, is equivalent to the existence of points of the varieties

$[f(x_1^\delta, x_2^\epsilon) = 0 \wedge x_1 \diamond 0 \wedge x_2 \diamond 0]$, for $\delta, \epsilon \in \{-1, 1\}$ and where $\diamond$ is either $\neq$ or no relation (after clearing denominators in each system).

The next step depends, first on the non-singularity of V', and second on the Hensel-Rychlik Lemma, which is a sort of formal analogue of the Implicit Function Theorem for real functions.

Step 3: We consider the varieties V' and Y defined in the previous step. Using the Hensel-Rychlik Lemma, one proves that if V' has points rational over $\mathbb{F}_p[[t]]$, then it has points which are not points of Y, except if that is formally impossible, that is, the variety V' (which is here assumed irreducible) is contained in Y over an algebraic closure of $\mathbb{F}_p((t))$. To see the point, consider the similar situation for varieties V' and Y over the real or complex numbers; say V' is given by $f(x, y) = 0$ and Y is given by $y \neq 0$. If V' has a point (x_0, y_0), then that point is a non-singular point (since V' is non-singular), hence either the partial derivative of f, f_y, with respect to y, or f_x, is non-zero at (x_0, y_0); say it is f_y. Then, by the Implicit Function Theorem, one can vary y around y_0 and still get an x so that $f(x, y) = 0$; hence one obtains a solution for which $y \neq 0$. A similar argument holds if f_x is non-zero at (x_0, y_0). The exceptional case of 'formal impossibility' is illustrated by the example in which the varieties V and Y coincide; this can be decided algorithmically by examining whether the variety $V \setminus Y$ has any points over an algebraic closure of the field $\mathbb{F}_p((t))$ (observe that this can be decided by an effective elimination of quantifiers for algebraically closed fields of characteristic p, then compare Exercise 1.1).

Hence the question of existence of points of $V' \setminus Y$, rational over $\mathbb{F}_p[[t]]$, is reduced to the existence of points of V' (which is defined by equations only) over $\mathbb{F}_p[[t]]$.

Step 4: A variance of Greenberg's Theorem, due to Denef and Lipshitz in [23], gives an effective way to examine a system of equations for possessing a solution over $\mathbb{F}_p[[t]]$, that is to say, the positive-existential theory of $\mathbb{F}_p[[t]]$ is decidable. Hence one can check whether V' has points over $\mathbb{F}_p[[t]]$.

2 Part B: A qualitative analogue of HTP and the Conjectures of S. Lang

We want to address the following

Question 2.1 *Is there an algorithm which, given any variety V over*

$\mathbb{Q}$, decides whether V has infinitely many points over some number field K?

For example, due to the proof by Faltings of Mordell's Conjecture, the question has a positive answer for curves:
Any curve of geometric genus ≥ 2 has only a finite number of points over any given number field K.

And it is known that curves of genus 0 or 1 have infinitely many points over some number field K. Hence the algorithm is: Compute the genus of the curve V (the genus is a geometric invariant and known to be computable). If it is ≤ 1 reply YES, otherwise NO.

In [37] (and more extensively in [68] and from a different point of view in [8]) S. Lang announced a number of conjectures, which, if true, would imply that varieties that have infinitely many points over some number field are characterized by certain geometric properties. As an example,

Conjecture 2.2 (Lang) *Any hyperbolic variety has only a finite number of points over any number field.*

A variety V is *hyperbolic* if every complex analytic map $f : \mathbb{C} \to V$ is constant (this is one of several equivalent definitions). For example, the (irreducible) curves (varieties of dimension 1) that are hyperbolic are precisely those with geometric genus ≥ 2.

What are the implications of this for Question 2.1? If true, Conjecture 2.2 reduces Question 2.1 to deciding whether a given variety has certain geometric properties. Some of those properties (e.g. genus) are known to be decidable; but some other properties are not known (to be decidable or undecidable). As an example of the second kind (to the best of the authors' knowledge), we ask

Question 2.3 *Is there an algorithm which, given a variety V over $\mathbb{Q}$, decides whether that is hyperbolic?*

In the following sub-section we will address this question.

Even if the answer to it turns out to be positive, there still remain certain geometric properties which have to be decidable if the answer to Question 2.1 is positive.

As we will see in the next sub-section, there is an undecidability result for some type of 'geometric problem'; but it is probably too early to conjecture that Question 2.1 has a negative answer.

Now we give a more precise account of the conjectures of Lang and Vojta. The main notions involved are the following properties of an algebraic variety V, defined over $\mathbb{Q}$ (or a number field):

1. V has only finitely many points over any number field.
2. V is hyperbolic
3. V is Kobayashi hyperbolic. This means that V is hyperbolic and is, in addition, equipped with a metric with some special properties ([37]).
4. V and all its subvarieties are *of general type* (a notion defined in terms of algebraic geometry).

The union of the conjectures connecting these properties is that all four properties are equivalent. The only part which has been proved is that property 2 (hyperbolic varieties) is equivalent to property 3 (Kobayashi hyperbolic varieties); this is due to Brody [7]. All the other equivalences are conjectures.

(We are indebted to Professor Campana for explaining to us some of the details for these notions).

2.1 A language for geometric problems

When we study the decidability question for a ring of functions of one (or more) variable z, we usually augment the language of rings by a name (constant symbol) for z. This has as a result that the varieties that we study have coefficients in $\mathbb{Z}[z]$ (or a bigger ring) and are not invariant under even the most elementary geometric transformations, for example $z \mapsto z + \beta$ with β in the base field (field of constant functions). Since we want a language that does not have this defect, we define the language

$$L_T = \{+, \cdot\,; =, T\,; 0, 1\}$$

which augments the language of rings by the one-place predicate T, which will be interpreted by

$$T(x) \leftrightarrow \text{ (the function } x \text{ is not a constant).}$$

Notice that, given a variety V through a finite set of defining polynomials, the sentence '*The variety V is hyperbolic*' can be expressed in L_T over $\mathcal{H}_z(\mathbb{C})$.

We ask:

Question 2.4 Is the theory (resp. existential theory, positive-existential theory) of L_T over $\mathcal{H}_z(\mathbb{C})$ decidable? over $\mathbb{C}(z)$? over $\mathbb{C}[z]$?

The only known relevant results are given by the following two theorems:

Theorem 2.5 ([50]) *The positive-existential theory in L_T of a polynomial ring $F[z]$ over a field F is undecidable.*

Theorem 2.6 ([56]) *The existential theory in L_T of the ring $\mathcal{H}_z(U)$ is decidable, where U is either the open or the closed unit disc.*

The analogous question for fields of rational functions and for $\mathcal{H}_z(\mathbb{C})$ are open.

An outline of the proof of Rubel's Theorem 2.6 is:

Let V be an affine variety defined over $\mathbb{Q}$ (or $\bar{\mathbb{Q}}$), defined by a finite set of equations $f_i(X) = 0$ ($X = (X_1, \ldots, X_m)$ is a tuple of variables and $f_i \in \mathbb{Q}[X]$).

We want to determine whether V has points over $\mathcal{H}_z(U)$. We consider the system of differential equations and inequations

$$S \; : \; (\bigwedge_i f_i(X) = 0) \wedge (\bigwedge_j \frac{dX_j}{dz} \neq 0).$$

One examines S for solutions in a differentially closed field which extends $\mathcal{H}_z(U)$. This is effective by results of Seidenberg. If the answer is negative then we are done: the answer over $\mathcal{H}_z(U)$ is also negative. If the answer is positive, then, by a theorem of Ritt, S has a solution $\bar{X}(z)$ such that each $\bar{X}_i$ is analytic in a neighborhood of $z = 0$. Substitute z by cz, for a suitable constant c, to obtain the solution $\bar{X}(cz)$ (here we need the fact that the f_i are polynomials over $\mathbb{Q}$, hence their coefficients do not involve z) which is analytic on U.

2.2 A geometric problem

Let K be a number field. We consider the following problem:

Question 2.7 *Is there an algorithm to decide whether an arbitrary variety over K contains a rational curve?*

Note that for varieties of dimension one, the answer to this question is positive. Indeed if V is a variety of dimension one, then V will contain a

rational curve if and only if one of its irreducible components is a curve of genus 0 and this curve contains a K-rational point. Since, given a variety over K we can explicitly determine all its irreducible components, and given the fact that the genus of a curve is computable, one easily gets the algorithm asked for in the question.
For higher dimensional varieties the problem is open. However the following conjecture gives some information for surfaces:

Conjecture (S. Lang) *Let V be a variety of general type defined over a number field K. Then there exists a proper closed subvariety W, defined over K such that for any number field L containing K, $V(L) \setminus W(L)$ is finite.*

Let V be a surface of general type, defined over K. If V contains a rational curve C which has a parametrization defined over K, then $V(K)$ is infinite.
Conversely, suppose that V contains infinitely many points, then by Lang's conjecture there exists a subvariety, i.e. a one-dimensional variety, such that at least one of its irreducible components is a curve which contains infinitely many points (i.e. this curve is either rational or an elliptic curve of positive rank).
Summarizing we have: determining whether a surface contains a curve of genus at most 1 is as difficult as determining whether the surface has infinitely many points or not.
In view of the results in the previous section, one is lead to believe that determining whether a variety defined over K contains a K-rational curve is as difficult as HTP(K).

Remark 2.8 It is tempting to use induction on the dimension to use the similar argument for arbitrary varieties. Unfortunately this does not work: the subvariety W of V which Lang's conjecture asserts to exist if V is of general type, needs itself not be of general type.

2.3 Varieties with an infinite number of points over a ring

Here we address the decision problem of whether a given variety has an infinite number of points over a fixed ring.

Consider a commutative domain R whose fraction field is not algebraically closed. Let R_0 be a infinite recursive subring of R with fraction field k_0. Suppose further that the algebraic closure of k_0 is not contained in R. Denote the cardinality of R by $|R|$, and for any polynomial f in several variables, with coefficients in k_0, write $N_R(f)$ for the cardinality of the solution set of f over R.

Let $\mathcal{S}$ be an arbitrary proper subset of $\mathbb{N} \cup \{|R|\}$.

Theorem 2.9 *Consider the following two decision problems:*

(a) Is there an algorithm to decide for an arbitrary polynomial with coefficients in k_0 whether $N_R(f) \in \mathcal{S}$?

(b) Is there an algorithm to decide for an arbitrary polynomial with coefficients in k_0 whether $N_R(f) = 0$?

A negative answer to (b) implies a negative answer to (a).

This implies that deciding whether a polynomial has infinitely or finitely many solutions in R is as difficult as deciding whether its has a solution or not. This result was proved by M. Davis for $R = \mathbb{Z}$ and was generalized by the second author.

3 Part C: Hilbert's tenth problem for analytic and meromorphic functions

We will look more closely into Question 1.14. Let $L_{z,C}$ denote the language which extends the language of rings by a constant symbol for the variable z and by the predicate C which is interpreted by '$C(x)$ if and only if the function x is constant. The main existing results are:

Theorem 3.1 (R. Robinson [55]) *Assume that D contains a real open interval. Then the first order theory of $\mathcal{H}_D(z)$ in $L_{z,C}$ is undecidable.*

Proof First we state:

Lemma 3.2 (Huuskonen [32]) *Assume that $D \subset \mathbb{C}$ has nonempty interior. Then the set of constants (i.e. of constant functions) in $\mathcal{H}_z(D)$ is first order definable in the language $\{+, \cdot\, ; 0, 1, z\}$.*

For simplicity work in the case $q = 1$ so that $z = z_1$ and assume that the line segment $[0, 1]$ is contained in D; the general case is left to the reader. The following formula is equivalent to $\alpha \in \mathbb{N} \setminus \{0\}$:

$$C(\alpha) \wedge \exists\, x\,[x(0) = 1 \wedge x(1) = 0]$$
$$\wedge\ \forall\, y\,[(y \in \mathbb{C} \wedge y \neq 0 \wedge y + 1 \neq 0)$$
$$\rightarrow \left(x\left(\tfrac{1}{y}\right) = 0 \rightarrow \left(x\left(\tfrac{1}{y+1}\right) = 0\right) \vee y = \alpha\right)\Big].$$

Of course the expressions of the form $\frac{1}{z}$ do not belong to L_z but they can be replaced by new variables w preceded by $\exists w:\ z \cdot w = 1$.
The $\leftarrow$ direction holds because if $\alpha \notin \mathbb{N}$, then the formula implies that each point $\frac{1}{n}$, with $n \in \mathbb{N}\setminus\{0\}$, is a zero of the analytic function x, while $x(0) = 1$, which is impossible by the continuity of x; for the $\rightarrow$ direction, observe that, for $n \in \mathbb{N} \setminus \{0\}$, the formula is realized by taking

$$x = (-1)^{n-1}(n-1)!(z-1)(z-\frac{1}{2})\cdots(z-\frac{1}{n-1}).$$

□

It is obvious that the proof of this Theorem shows the similar result for any ring of functions which are continuous on $[0, 1]$, containing the identity function.

Theorem 3.3 ([41]) *The positive-existential theory of the ring $\mathcal{H}_{p,z}(\mathbb{C}_p)$ of functions of the variable z which are analytic on the p-adic complex plane $\mathbb{C}_p$, in the language which extends the language of rings by a constant symbol for z, is undecidable.*

Theorem 3.4 (Vidaux, [67]) *The positive-existential theory of the field $\mathcal{M}_{p,z}(\mathbb{C}_p)$ of functions of the variable z which are meromorphic on the p-adic complex plane $\mathbb{C}_p$, in the language which extends the language of rings by a constant symbol for z and by a predicate symbol for the property 'the function x has $z = 0$ as a zero', is undecidable.*

The analogous questions over the field of complex numbers $\mathbb{C}$ are open.

4 Part D: Comments on the analogue of Hilbert's tenth problem for $\mathbb{Q}$

As mentioned earlier, a major open problem in the area of decidability of existential theories of rings is the analogue of Hilbert's tenth problem for the field $\mathbb{Q}$ of rational numbers.

Question 4.1 *Does the analogue of Hilbert's tenth problem for $\mathbb{Q}$ have a negative answer?*

A naive approach might seek a possible positive answer to the following:

Question 4.2 *Is the set $\mathbb{Z}$ existentially definable in the field $\mathbb{Q}$?*

which, if true, would imply obviously a YES answer to Question 4.1. But the following observation seems to make such an expectation unlikely: All known examples of algebraic varieties over $\mathbb{Q}$ have the property that the real topological closure of the Zariski closure of their rational (over $\mathbb{Q}$) points has finitely many connected components.

In consequence Mazur asked whether this is true for all algebraic varieties ([44]). He also stated a more general similar statement (an analogue where the real topology is substituted by the p-adic topologies). These questions remain open. An implication of a possible positive answer to Mazur's Question would be that Question 4.2 has a negative answer: Finitely many components project onto finitely many components, hence existential sets of $\mathbb{Q}$ (being projections of varieties) would have only finitely many components, hence $\mathbb{Z}$ cannot be one of them. Actually the implications are much deeper (cf. [12]). Some of us doubt the truth of Mazur's Question (mainly because the analogue of the p-adic version fails in global fields of positive characteristic). But still, most (if not all) of us, expect the answer to Question 4.2 to be negative.

In [49] a programme is presented for proving Question 4.1. It is based on interpreting the rational integers as the points (over $\mathbb{Q}$) of an elliptic curve of rank 1 over $\mathbb{Q}$. Addition among points is given by the addition law on the curve. It therefore suffices to express existentially (in terms of the coordinates of the points) 'multiplication' among points. Since this seems inaccessible for the moment, one may, as a first step, try to define divisibility among points. Modulo conjectures (by Cornelissen and Everest) this may be existential. But even this does not suffice due to the fact that, as we saw, the existential theory of addition and divisibility over $\mathbb{Z}$ is decidable. Hence more (existentially definable) structure is needed. A proposal of [49] to this effect seems unlikely (by arguments of Cornelissen). But Cornelissen proposed a possible remedy: Look, not at the points of an elliptic curve, but to the points (over $\mathbb{Q}$) of an abelian variety with a group of points which, given a natural additional structure of multiplication, is isomorphic to a real quadratic extension of $\mathbb{Z}$ (say $\mathbb{Z}[\sqrt{5}]$). Such varieties do exist! If one can define existentially divisibility among points of such a variety, then a YES answer to Question 4.1 will result from the undecidability of the existential theory of addition and divisibility over $\mathbb{Z}[\sqrt{5}]$! See [13] for additional information on this type of approach.

Let us mention here that most, if not all, of the above approaches to resolve Question 4.1 would also disprove Mazur's Question.

Relevant material may be found also in [45] and the surveys [51] and [52].

References

[1] J. Ax, Solving diophantine equations modulo every prime, *Annals of Mathematics* 85-2 (1967), 161-183.

[2] J. Ax, The elementary theory of finite fields, *Annals of Mathematics* 88 (1968), 239-271.

[3] J. Ax and S. Kochen, Diophantine problems over local fields: III Decidable fields, *Annals of Mathematics* 83 (1966), 437-456.

[4] J. Becker, J. Denef and L. Lipshitz, Further remarks on the elementary theory of power series rings, in *Model theory of algebra and arithmetic (Proc. Conf., Karpacz, 1979)*, pp. 1–9, Lecture Notes in Math., 834, Springer, Berlin-New York, 1980.

[5] J. Becker, J. Denef, L. Lipshitz and L. van den Dries, Ultraproducts and approximation in local rings I, *Inventiones Mathematics* 51 (1979), 189-203.

[6] A. Bel'tyukov, Decidability of the universal theory of the natural numbers with addition and divisibility, *Seminars of the Steklov Math. Inst. (Leningrad)*, 60 (1976), 15-28.

[7] R. Brody, Compact manifolds in hyperbolicity, *Trans. Amer. Math. Soc.* 235 (1978), 213–219.

[8] F. Campana, Special varieties and classification theory: an overview. Monodromy and differential equations, (Moscow, 2001). *Acta Appl. Math.* 75 (2003), no. 1-3, 29–49.

[9] G. Cherlin, *Model theoretic Algebra*, Lecture Notes Math. 521 (1976), Springer.

[10] G. Cherlin and F. Point, On extensions of Presburger arithmetic, in *Proc. Fourth Easter conf. on model theory, Humboldt Univ. (1986)*, 17-34.

[11] P. Cohen, Decision procedures for real and p-adic fields, *Comm. Pure Appl. Math.* 22 (1969),131-151.

[12] G. Cornelissen and K. Zahidi, Topology of Diophantine Sets: Remark's on Mazur's Conjectures, in *Hilbert's tenth problem: relations with arithmetic and algebraic geometry (Ghent, 1999)*, Contemporary Mathematics 270 (2000), 253–260.

[13] G. Cornelissen and K. Zahidi, Complexity of undecidable formulae in the rationals and inertial Zsigmondy theorems for elliptic curves, ArXiv, `math.NT/0412473`, to appear in *Journal für die Reine und Angewandte Mathematik*

[14] M. Davis, Hilbert's tenth problem is unsolvable, *American Mathematical Monthly* 80, 233-269 (1973).

[15] M. Davis, Y. Matijasevic and J. Robinson, Hilbert's tenth problem. Dio-

phantine equations: positive aspects of a negative solution, *Proc. Sympos. Pure Math.* 28 (1976), Amer. Math. Soc. 323-378.

[16] F. Delon, Indécidabilité de la théorie des anneaux de séries formelles à plusieurs variables, *Fund. Math. CXII* (1981), 215-229.

[17] J. Denef, The diophantine problem for polynomial rings and fields of rational functions, *Transactions of the American Mathematical Society*, 242(1978),391-399.

[18] J. Denef, The diophantine problem for polynomial rings of positive characteristic, in *Logic Colloquium 78*, North Holland (1984), 131-145.

[19] J. Denef, p-adic semi-algebraic sets and cell decomposition, *J. Reine Angew. Math.* 369 (1986), 154-166.

[20] J. Denef and M. Gromov (communication by G. Cherlin), The ring of analytic functions in the disk has undecidable theory, 1985 (letter)

[21] J. Denef and L. van den Dries, *p*-adic and real subanalytic sets, *Ann. Math.*, 128 (1988), 79-138.

[22] J. Denef and L. Lipshitz, Ultraproducts and Approximation in local rings II, *Math. Ann.* 253 (1984)

[23] J. Denef and L. Lipshitz, Power series solutions of algebraic differential equations, *Math. Ann.* 267 (1980), 1-28.

[24] J. Denef and H. Schoutens, On the decidability of the existential theory of $\mathbb{F}_{p[[t]]}$, *Fields Inst. Comm.*, 33 (2003), 43-59

[25] M. Eichler, *Introduction to the theory of algebraic numbers and functions*, Academic Press, 1966.

[26] Yu. Ershov, On elementary theories of local fields, *Algebra i Logika* 4 (1965), 5-30.

[27] M. Fried and G. Sacerdote, Solving Diophantine problems over all residue class fields of a number field and all finite fields *Ann. of Math.* 104 (1976), 203-233.

[28] M. Greenberg, Strictly local solutions of diophantine equations, *Pacific J. Math.*, 51 (1974), 143-153.

[29] F. Grunewald and D. Segal, The solubility of certain decision problems in arithmetic and algebra, *Bull. Amer. Math. Soc.* 1-6 (1979), 915-918.

[30] F. Grunewald and D. Segal, Some general algorithms I and II, *Ann. Math.*, 112 (1980), 531-617.

[31] C. Ward Henson and L. Rubel, Some applications of Nevanlinna Theory to mathematical logic: identities of exponential functions, *Trans. Amer. Math. Soc.* 282 (1984), 1-32 (also: corrections).

[32] T. Huuskonen, Constants are definable in fields of analytic functions, *Proc. Amer. Math. Soc.* 122 (1994), 697-702.

[33] N. Katz and S. Lang, Finiteness theorems in geometric classfield theory. (With an appendix by Kenneth A. Ribet), *Enseign. Math. (2)* 27-3,4 (1981), 285-319.

[34] K. Kim and F. Roush, An approach to rational diophantine undecidability, *Proc. Asian Math. Conf.*, World Scient. Press, Singapore (1992), 242-257.

[35] K. Kim and F. Roush, Diophantine undecidability of $\mathbb{C}(t_1, t_2)$, *J. Algebra* 150 (1992), 35-44.

[36] S. Kochen, The model theory of local fields, in *Proc. Internat. Summer Inst. and Logic Colloq., Kiel, 1974*, 384–425, Springer Lecture Notes in Math. 499 (1975).

[37] S. Lang, Hyperbolic diophantine analysis, *Bulletin of the American Math-*

ematical Society 14, 159-205 (1986).

[38] L. Lipshitz, The diophantine problem for addition and divisibility, *Transactions of the American Mathematical Society* 235, (1978), 271-283.

[39] L. Lipshitz, Undecidable existential problems for addition and divisibility in algebraic number rings, II, *Proceedings of the American Mathematical Society,* 64 (1977), 122-128.

[40] L. Lipshitz, Undecidable existential problems for addition and divisibility in algebraic number rings, *Transactions of the American Mathematical Society,* 241 (1978), 121-128.

[41] L. Lipshitz and T. Pheidas, An analogue of Hilbert's Tenth Problem for p-adic entire functions, *The Journal of Symbolic Logic*, 60-4 (1995), 1301-1309

[42] A. Macintyre, On definable subsets of p-adic fields, *J. Symb. Logic* 41 (1976), 605-610.

[43] Y. Matijasevich, Enumerable sets are diophantine, *Doklady Akademii Nauka SSSR,* 191(1970), 272-282.

[44] B. Mazur, The topology of rational points, *Journal of Experimental Mathematics*, 1-1 (1992), 35-45.

[45] B. Mazur, Questions of decidability and undecidability in number theory, *The Journal of Symbolic Logic*, 59-2 (1994), 353-371.

[46] C. Michaux and R. Villemaire, Presburger arithmetic and recognizability of sets of natural numbers by automata: new proofs of Cobham's and Semenov's theorems, *Ann. Pure Appl. Logic* 77 (1996), 251–277.

[47] T. Pheidas, Extensions of Hilbert's Tenth Problem, *The Journal of Symbolic Logic,* 59-2 (1994), 372-397.

[48] T. Pheidas, Endomorphisms of elliptic curves and undecidability in function fields of positive characteristic, *Journal of Algebra* 273 (2004), no. 1, 395-411.

[49] T. Pheidas, An effort to prove that the existential theory of Q is undecidable, *Contemporary Mathematics* 270 (2000), 237-252.

[50] T. Pheidas and K. Zahidi, Undecidable existential theories of polynomial rings and function fields, *Communications in Algebra*, 27-10 (1999), 4993-5010.

[51] T. Pheidas and K. Zahidi, Undecidability of existential theories of rings and fields: A survey, *Contemporary Mathematics,* 270 (2000), 49-106.

[52] B. Poonen, Hilbert's Tenth Problem over rings of number-theoretic interest, obtainable from `http://math.berkeley.edu/~poonen/`

[53] J. Robinson, Definability and decision problems in arithmetic, Journ. Symb. Logic 14 (1949), 98-114.

[54] J. Robinson, Existential definability in arithmetic, *Trans. Amer. Math. Soc.* 72 (1952), 437-449.

[55] R. Robinson, Undecidable rings, *Trans. Amer. Math. Soc.* 70 (1951), 137.

[56] L. Rubel, An essay on diophantine equations for analytic functions, *Expositiones Mathematicae*, 14(1995),81-92.

[57] R. Rumely, Arithmetic over the ring of all algebraic integers, *Journ. Reine und Angew. Math.* 368 (1986), 127-133.

[58] A. Semenov, Logical theories of one-place functions on the set of natural numbers, *Math. USSR Izvestija*, 22 (1984), 587-618.

[59] A. Shlapentokh, Hilbert's tenth problem over number fields, a survey, *Contemporary Mathematics,* 270 (2000), 107-137.

[60] A. Seidenberg, Constructions in Algebra, *Trans. Amer. Math. Soc.* 197

(1974), 273–313.

[61] M. Spivakovski, A new proof of D. Popescu's theorem on smoothing of ring homomorphisms, *J. Amer. Math. Soc.* 12, 381-444

[62] A. Tarski, *A decision method for elementary algebra and geometry*, RAND Corporation, Santa Monica, Calif. (1948).

[63] L. van den Dries, A specialization theorem for analytic functions on compact sets, *Proceedings Koninklijke Nederlandse Academie van Weteschappen (A)*, 85-4 (1988),391-396.

[64] L. van den Dries, Elimination theory for the ring of algebraic integers, *Journal für die reine und angewandte Mathematik (Crelles Journal)*, 388 (1988), 189-205.

[65] L. van den Dries, A remark on Ax's theorem on solvability modulo primes, *Math. Z.* 208 (1991), 65-70.

[66] L. van den Dries, Analytic Ax-Kochen-Ersov theorems, in *Proceedings of the International Conference on Algebra, Part 3 (Novosibirsk, 1989)*, 379–398, Contemp. Math. 131, Amer. Math. Soc., Providence, RI, 1992.

[67] X. Vidaux, An analogue of Hilbert's 10th problem for fields of meromorphic functions over non-Archimedean valued fields, *Journal of Number Theory*, 101 (2003), 48-73.

[68] P. Vojta, *Diophantine approximations and value distribution theory*, Lecture Notes in Mathematics, Springer-Verlag, 1239 (1987)

[69] P. Vojta, Diagonal quadratic forms and Hilbert's Tenth Problem, *Contemporary Mathematics* 270, 261-274 (2000).

[70] V. Weispfenning, Quantifier elimination and decision procedures for valued fields in *Models and Sets*, Lect. Notes Math. 1103, Springer-Verlag (1984), 419-472.

Hilbert's Tenth Problem for function fields of characteristic zero

Kirsten Eisenträger
The Pennsylvania State University

Summary

In this article we outline the methods that are used to prove undecidability of Hilbert's Tenth Problem for function fields of characteristic zero. Following Denef we show how rank one elliptic curves can be used to prove undecidability for rational function fields over formally real fields. We also sketch the undecidability proofs for function fields of varieties over the complex numbers of dimension at least 2.

1 Introduction

Hilbert's Tenth Problem in its original form was to find an algorithm to decide, given a polynomial equation $f(x_1, \ldots, x_n) = 0$ with coefficients in the ring $\mathbb{Z}$ of integers, whether it has a solution with $x_1, \ldots, x_n \in \mathbb{Z}$. Matijasevič ([Mat70]), based on work by Davis, Putnam and Robinson ([DPR61]), proved that no such algorithm exists, *i.e.* Hilbert's Tenth Problem is undecidable. Since then, analogues of this problem have been studied by asking the same question for polynomial equations with coefficients and solutions in other commutative rings R. We will refer to this as *Hilbert's Tenth Problem over R*. Perhaps the most important unsolved question in this area is the case $R = \mathbb{Q}$. There has been recent progress by Poonen ([Poo03]) who proved undecidability for large subrings of $\mathbb{Q}$. The function field analogue, namely Hilbert's Tenth Problem for the function field k of a curve over a finite field, is undecidable. This was proved by Pheidas ([Phe91]) for $k = \mathbb{F}_q(t)$with q odd, and by Videla ([Vid94]) for $\mathbb{F}_q(t)$ with q even. Shlapentokh ([Shl00]) generalized Pheidas' result to finite extensions of $\mathbb{F}_q(t)$ with q odd and to certain function fields over possibly infinite constant fields of odd characteristic, and the remaining cases in characteristic 2 are treated in [Eis03]. Hilbert's Tenth Problem is also known to be undecidable for several rational function fields of characteristic zero: In 1978 Denef proved the undecidability of Hilbert's Tenth Problem for rational function fields $K(t)$ over formally real fields K ([Den78]). Kim and Roush ([KR92]) showed that the problem is undecidable for the purely transcendental function fields $\mathbb{C}(t_1, t_2)$ and $\overline{\mathbb{F}}_p(t_1, t_2)$. In [Eis04] this was generalized to finite extensions of $\mathbb{C}(t_1, \ldots, t_n)$ with $n \geq 2$. The problem is also known to be undecidable for function fields over p-adic fields ([KR95], [Eis07], [MB05]). In Hilbert's Tenth Problem the coefficients of the equations have to be input into a Turing machine, so when we consider the problem for uncountable rings we restrict the coefficients to a subring R' of R which is finitely generated as a $\mathbb{Z}$-algebra. We say that *Hilbert's Tenth Problem for R with coefficients in R'* is undecidable if there is no algorithm that decides whether or not multivariate polynomial equations with coefficients in R' have a solution in R. In this paper we will discuss the undecidability proofs for $\mathbb{R}(t)$, and $\mathbb{C}(t_1, t_2)$. In these cases, we consider polynomials with coefficients in $\mathbb{Z}[t]$ and $\mathbb{Z}[t_1, t_2]$, respectively.

The biggest open problems for function fields are Hilbert's Tenth Problem for $\mathbb{C}(t)$ and for $\overline{\mathbb{F}}_p(t)$.

2 Approach for function fields of characteristic zero

2.1 Preliminaries

Before we can describe the approach that is used in characteristic zero we need the following definition. All the rings we consider are commutative with 1.

Definition 2.1 Let R be a ring. A subset $Q \subseteq R^k$ is *diophantine over R* if there exists a polynomial $f(x_1, \ldots, x_k, y_1, \ldots, y_m) \in R[x_1, \ldots, x_k, y_1, \ldots, y_m]$ such that

$$Q = \{\vec{x} \in R^k : \exists\, y_1, \ldots, y_m \in R : f(\vec{x}, y_1, \ldots, y_m) = 0\}.$$

Let R' be a subring of R and suppose that f can be chosen such that its coefficients are in R'. Then we say that Q is *diophantine over R with coefficients in R'*.

Example 1 The set of natural numbers $\mathbb{N}$ is diophantine over $\mathbb{Z}$. This follows from the fact that every natural number can be written as a sum of four squares, so

$$\mathbb{N} = \{a \in \mathbb{Z} : \exists y_1, \ldots, y_4 \in \mathbb{Z} : (y_1^2 + y_2^2 + y_3^2 + y_4^2 - a = 0)\}.$$

Example 2 The set of primes is diophantine over $\mathbb{Z}$. This follows from the proof of Hilbert's Tenth Problem for $\mathbb{Z}$, where it was shown that every recursively enumerable subset of $\mathbb{Z}$ is diophantine over $\mathbb{Z}$ ([DPR61, Mat70]). Clearly the prime numbers form a recursively enumerable subset of $\mathbb{Z}$.

2.2 Combining diophantine equations

If K is a formally real field and $P_1 = 0$, $P_2 = 0$ are two diophantine equations over $K(t)$ with coefficients in $\mathbb{Z}[t]$, then $P_1 = 0 \wedge P_2 = 0$ and $P_1 = 0 \vee P_2 = 0$ are also diophantine with coefficients in $\mathbb{Z}[t]$: we have

$$P_1 = 0 \wedge P_2 = 0 \leftrightarrow P_1^2 + tP_2^2 = 0$$

and

$$P_1 = 0 \vee P_2 = 0 \leftrightarrow P_1 P_2 = 0.$$

The same holds for diophantine equations over $\mathbb{C}(t_1, t_2)$ with coefficients in $\mathbb{Z}[t_1, t_2]$:

$$P_1 = 0 \wedge P_2 = 0 \leftrightarrow P_1^2 + t_1 P_2^2 = 0 \text{ and } P_1 = 0 \vee P_2 = 0 \leftrightarrow P_1 P_2 = 0.$$

An argument similar to the one above can be made for many other rings whose quotient field is not algebraically closed, provided that the ring of coefficients is large enough.

The above argument shows that proving undecidability of Hilbert's Tenth Problem for $K(t)$ with coefficients in $\mathbb{Z}[t]$ is the same as proving that the positive existential theory of $K(t)$ in the language $\langle +, \cdot\,; 0, 1, t\rangle$ is undecidable. Similarly, Hilbert's Tenth Problem for $\mathbb{C}(t_1, t_2)$ with coefficients in $\mathbb{Z}[t_1, t_2]$ is undecidable if and only if the positive existential theory of $\mathbb{C}(t_1, t_2)$ in the language $\langle +, \cdot\,; 0, 1, t_1, t_2\rangle$ is undecidable.

2.3 Approach in characteristic zero

We can use a reduction argument to prove undecidability for a ring R of characteristic zero if we can give a diophantine definition of $\mathbb{Z}$ inside R. We have the following proposition:

Proposition 2.2 *Let R be an integral domain of characteristic zero. Let R' be a subring of R, which is finitely generated as a $\mathbb{Z}$-algebra and such that the fraction field of R does not contain an algebraic closure of R'. Assume that $\mathbb{Z}$ is a diophantine subset of R with coefficients in R'. Then Hilbert's Tenth Problem for R with coefficients in R' is undecidable.*

Proof Given a polynomial equation $f(x_1, \ldots, x_n) = 0$ over $\mathbb{Z}$ we can construct a system of polynomial equations over R with coefficients in R' by taking the original equation together with, for each $i = 1, \ldots, n$ an equation $g_i(x_i, \ldots) = 0$ involving x_i and a new set of variables, such that in any solution over R of the system, $g_i = 0$ forces x_i to be in $\mathbb{Z}$. In other words, the new system of equations has a solution over R if and only $f(x_1, \ldots, x_n)$ has a solution in $\mathbb{Z}$. Also, since the quotient field of R does not contain the algebraic closure of R', the system over R with coefficients in R' is equivalent to a single polynomial equation with coefficients in R' ([PZ00, p. 51]). □

Sometimes we cannot give a diophantine definition of the integers inside a ring R, but we might be able to construct a model of the integers inside R.

Definition 2.3 A *diophantine model* of $\langle \mathbb{Z}, 0, 1; +, \cdot \rangle$ over R is a diophantine subset $S \subseteq R^m$ equipped with a bijection $\phi : \mathbb{Z} \to S$ such that under ϕ, the graphs of addition and multiplication correspond to diophantine subsets of S^3.

Let R' be a subring of R. A *diophantine model* of $\langle \mathbb{Z}, 0, 1; +, \cdot \rangle$ over R *with coefficients in* R' is a diophantine model of $\langle \mathbb{Z}, 0, 1; +, \cdot \rangle$, where in addition S and the graphs of addition and multiplication are diophantine over R with coefficients in R'.

A similar argument as for Proposition 2.2 can be used to prove the following

Proposition 2.4 *Let R, R' be as in Proposition 2.2. If we have a diophantine model of $\langle \mathbb{Z}, 0, 1; +, \cdot \rangle$ over R with coefficients in R', then Hilbert's Tenth Problem for R with coefficients in R' is undecidable.*

3 Fields of rational functions over formally real fields

In this section we will give an outline of Denef's theorem:

Theorem 3.1 ([Den78]) *Let K be a formally real field, i.e. -1 is not the sum of squares. Hilbert's Tenth Problem for $K(t)$ with coefficients in $\mathbb{Z}[t]$ is undecidable.*

We follow Denef's proof and use elliptic curves to construct a model of $\langle \mathbb{Z}, 0, 1; +, \cdot \rangle$ in $K(t)$. Denef actually gives a diophantine definition of $\mathbb{Z}$ inside $K(t)$, but we will construct a model of the integers, because this approach is slightly shorter and it is used in subsequent papers ([KR92], [KR95], [Eis07]).

To construct the diophantine model of the integers we first have to obtain a set S which is diophantine over $K(t)$ and which has a natural bijection to $\mathbb{Z}$.

3.1 Obtaining S

Let E be an elliptic curve defined over $\mathbb{Q}$ without complex multiplication and with Weierstrass equation

$$y^2 = x^3 + ax + b. \tag{1}$$

The points (x, y) satisfying equation (1) together with the point at infinity form an abelian group, and the group law is given by equations. We will now look at a twist of E, the elliptic curve $\mathcal{E}$, which is defined to be the smooth projective model of

$$(t^3 + at + b)Y^2 = X^3 + aX + b.$$

This is an elliptic curve defined over the rational function field $\mathbb{Q}(t)$. An obvious point on $\mathcal{E}$ which is defined over $\mathbb{Q}(t)$ is the point $P_1 := (t, 1)$. Denef proved:

Theorem 3.2 [Den78, p. 396] *The point P_1 has infinite order and generates the group $\mathcal{E}(K(t))$ modulo points of order 2.*

The elliptic curve $\mathcal{E}$ is a projective variety, but any projective algebraic set can be partitioned into finitely many affine algebraic sets, which can then be embedded into a single affine algebraic set. This implies that the set

$$\mathcal{E}(K(t)) = \{(X,Y) : X, Y \in K(t) \wedge (t^3+at+b)Y^2 = X^3+aX+b\} \cup \{\mathbf{O}\}$$

is diophantine over $K(t)$, since we can take care of the point at infinity $\mathbf{O}$ of $\mathcal{E}$.

We can express by polynomial equations that a point on the elliptic curve is of the form $2 \cdot P$, where P is another point on the curve. Hence the set

$$\begin{aligned} S' =& \{(X_{2n}, Y_{2n}) : n \in \mathbb{Z}\} \\ =& \{(x, y) \in (K(t))^2 : \\ & \exists\, u, v \in K(t) : (u, v) \in \mathcal{E}(K(t)) \wedge (x, y) = 2(u, v)\} \end{aligned}$$

is diophantine over $K(t)$ with coefficients in $\mathbb{Z}[t]$. Then the set

$$\begin{aligned} S'' &= \{(X_n, Y_n) : n \in \mathbb{Z}\} \\ &= \{(x, y) \in (K(t))^2 : \exists\, n \in \mathbb{Z} : \\ &((x, y) = (X_{2n}, Y_{2n}) \vee (x, y) = (X_{2n}, Y_{2n}) + P_1)\} \end{aligned}$$

is diophantine over $K(t)$ with coefficients in $\mathbb{Z}[t]$ as well.

Let $P_n := n \cdot (t, 1) = (X_n, Y_n)$ for $n \in \mathbb{Z} - \{0\}$, and let $P_0 := \mathbf{O}$. The set $S' \cup S''$ is equal to the set $\{P_n : n \in \mathbb{Z}\}$. Let $Z_n := \frac{X_n}{tY_n}$ for $n \in \mathbb{Z} - \{0\}$, and let $Z_0 := 0$. We define S to be the set

$$S = \{Z_n : n \in \mathbb{Z}\}.$$

Then S is diophantine over $K(t)$. Since $Z_n \in \mathbb{Q}(t)$, we can consider Z_n as a function on the projective line $\mathbf{P}^1_{\mathbb{Q}} = \mathbb{Q} \cup \{\infty\}$. Denef ([Den78, p. 396]) proved the following proposition:

Proposition 3.3 *Considered as a function on $\mathbf{P}^1_{\mathbb{Q}}$, Z_n takes the value n at infinity.*

For $n \neq m$, we have $Z_n \neq Z_m$, and so by associating the point Z_n to an integer n we obtain an obvious bijection between $\mathbb{Z}$ and S. This is the set that we will use for the diophantine model of $\langle \mathbb{Z}, 0, 1; +, \cdot \rangle$.

3.2 Existentially defining multiplication and addition

The bijection $\phi : \mathbb{Z} \to S$ given by $\phi(n) = Z_n$ induces multiplication and addition laws on the set S, and it remains to show that the graphs of addition and multiplication on S are diophantine over $K(t)$. This means that we have to show that the sets

$$S_{add} := \{(Z_n, Z_m, Z_\ell) \in S^3 : n + m = \ell\}$$

and

$$S_{mult} := \{(Z_n, Z_m, Z_\ell) \in S^3 : n \cdot m = \ell\}$$

are diophantine over $K(t)$. Since addition of points on the elliptic curve is given by equations involving rational functions of the coordinates of the points, it follows easily that the set S_{add} is diophantine over $K(t)$:

$$Z_n + Z_m = Z_\ell \leftrightarrow \exists (X,Y),(X',Y'),(X'',Y'') \in \mathcal{E}(K(t)):$$
$$\left(Z_n = \frac{X}{tY}, Z_m = \frac{X'}{tY'}, Z_\ell = \frac{X''}{tY''} \wedge (X,Y) + (X',Y') = (X'',Y'') \right).$$

The difficult part is showing that S_{mult} is diophantine.

3.3 Defining multiplication

We define the discrete valuation $\text{ord}_{t^{-1}} : K(t) \to \mathbb{Z} \cup \{\infty\}$ by $\text{ord}_{t^{-1}} u = -\deg f + \deg g$, for $u \in K(t)^*$, $u = f/g$ with $f, g \in K[t]$. We let $\text{ord}_{t^{-1}}(0) = \infty$. Proposition 3.3 implies that for $n \neq 0$, $\text{ord}_{t^{-1}}(Z_n) = 0$ and $\text{ord}_{t^{-1}}(Z_n - n) > 0$.

The discrete valuation $\text{ord}_{t^{-1}}$ has the following properties: For $u \in K(t)$ we have

(1) $\text{ord}_{t^{-1}}(u) = 0$ if and only if u takes a nonzero value $a \in K$ at infinity.
(2) $\text{ord}_{t^{-1}}(u) > 0$ if and only if u takes the value zero at infinity.
(3) $\text{ord}_{t^{-1}}(u) < 0$ if and only if u takes the value infinity at infinity.

We will use the discrete valuation $\text{ord}_{t^{-1}}$ to existentially define multiplication of elements of S. This is done in the following theorem.

Theorem 3.4 *Assume that the set* $T' := \{u \in K(t) : \text{ord}_{t^{-1}}(u) > 0\}$ *is diophantine over* $K(t)$. *Then the set* S_{mult} *is diophantine over* $K(t)$, *i.e. we can existentially define multiplication of elements of* S.

Proof The theorem follows immediately from the following claim.
Claim Given $n, m, \ell \in \mathbb{Z}$ we have $n \cdot m = \ell$ if and only if $Z_n \cdot Z_m - Z_\ell \in T'$, *i.e.*

$$\text{ord}_{t^{-1}}(Z_n \cdot Z_m - Z_\ell) > 0.$$

Proof of Claim If $n \cdot m = \ell$, then by Proposition 3.3, $Z_n \cdot Z_m - Z_\ell$ takes the value $n \cdot m - \ell = 0$ at infinity, and hence $\text{ord}_{t^{-1}}(Z_n \cdot Z_m - Z_\ell) > 0$.

If $n \cdot m \neq \ell$, then $Z_n \cdot Z_m - Z_\ell$ takes the value $n \cdot m - \ell \neq 0$ at infinity. Hence $\text{ord}_{t^{-1}}(Z_n \cdot Z_m - Z_\ell) = 0$. This proves the claim.

Since we assumed that the set T' of all elements with positive valuation at t^{-1} was diophantine over $K(t)$ this proves the theorem. □

Remark 1 We can modify the set T' and still make the proof of Theorem 3.4 work. What we needed in the proof was a diophantine set T with the following properties:

(1) If $Z \in \mathbb{Q}(t)$ and $\text{ord}_{t^{-1}}(Z) > 0$, then $Z \in T$.
(2) If $Z \in K(t)$ and $Z \in T$, then $\text{ord}_{t^{-1}}(Z) > 0$.

This is enough because the functions Z_n are elements of $\mathbb{Q}(t)$.

3.4 How to obtain a diophantine definition for T

We will now define the set T that has the properties in Remark 1. By Theorem 3.4 this is enough to finish the proof of Theorem 3.1.

Consider the relation $\text{Com}(y)$ defined by

$$\text{Com}(y) \leftrightarrow y \in K(t) \wedge \exists x \in K(t) : y^2 = x^3 - 4.$$

Since $y^2 = x^3 - 3$ is a curve of genus 1, it does not admit a rational parameterization, and so if an element y satisfies $\text{Com}(y)$, then y lies in K. Also, Denef ([Den78]) showed that for every rational number z, there exists a rational number $y > z$ satisfying $\text{Com}(y)$. We are now ready to define the set T.

Theorem 3.5 *Define the set T by*

$$Z \in T \leftrightarrow \exists X_1, \ldots, X_5, y \in K(t) : \\ (\text{Com}(y) \wedge \\ (y-t)Z^2 + 1 = X_1^2 + X_2^2 + \cdots + X_5^2) . \tag{2}$$

Then T has the properties as in Remark 1.

Proof We follow the proof in [Den78]: We will first show that every element $Z \in T$ has positive order at t^{-1}.

Suppose there exist $X_1, \ldots, X_5, y$ in $K(t)$ as in Equation (2), and assume by contradiction that $\text{ord}_{t^{-1}}(Z) \leq 0$. Then $\deg Z \geq 0$, where $\deg Z$ denotes the degree of the rational function Z. Since y satisfies $\text{Com}(y)$, we have $y \in K$, which implies that $\deg(y - t) = 1$, and so $\deg((y - t)Z^2 + 1)$ is positive and odd. But the degree of the rational function $X_1^2 + \ldots X_5^2$ is even, since in a formally real field, a sum of squares is zero if and only if each term is zero, and hence there is no cancellation of the coefficients of largest degree. Hence the left-hand-side of (2) has odd degree, while the right-hand-side has even degree, contradiction.

To show that the set T satisfies the second property, let $Z \in \mathbb{Q}(t)$, and assume $\operatorname{ord}_{t^{-1}}(Z) > 0$. We want to show that $Z \in T$. Since $\operatorname{ord}_{t^{-1}}(Z) > 0$, we have $\operatorname{ord}_{t^{-1}}(tZ^2) > 0$, and so $tZ^2(r) \to 0$ as $|r| \to \infty$ ($r \in \mathbb{R}$). Hence we can find a natural number n, such that for real numbers r with $|r| > n$, we have $|tZ^2(r)| \leq 1/2$. Pick a rational number y with $y > n > 0$ and satisfying $\operatorname{Com}(y)$. Such a y exists by the discussion before Theorem 3.5. Then

$$((y-t)Z^2+1)(r) = yZ^2(r) - tZ^2(r) + 1 \geq yZ^2(r) - 1/2 + 1 > 0$$

for all $r \in \mathbb{R}$. By Pourchet's theorem, every positive definite rational function over $\mathbb{Q}$ can be written as a sum of five squares in $\mathbb{Q}(t)$. Hence there exist $X_1, \ldots, X_5 \in K(t)$ as desired. □

4 Function fields over the complex numbers in two variables

Unfortunately, the diophantine definition of the set T which defined the elements of positive order at t^{-1} and which was crucial for the proof of Theorem 3.1 only works for formally real fields.

For $\mathbb{C}(t_1, t_2)$ and finite extensions we will do something else that avoids defining order.

4.1 Hilbert's Tenth for the rational function field $\mathbb{C}(t_1, t_2)$

In this section we will outline the proof of the following

Theorem 4.1 ([KR92]) *Hilbert's Tenth Problem for* $\mathbb{C}(t_1, t_2)$ *with coefficients in* $\mathbb{Z}[t_1, t_2]$ *is undecidable.*

To prove undecidability of Hilbert's Tenth Problem for $K := \mathbb{C}(t_1, t_2)$ we will construct a diophantine model of the structure

$$\mathcal{S} := \langle \mathbb{Z} \times \mathbb{Z}, +, |, \mathcal{Z}, \mathcal{W} \rangle$$

in K (with coefficients in $\mathbb{Z}[t_1, t_2]$). Here $+$ denotes the usual component-wise addition of pairs of integers, $|$ represents a relation which satisfies

$$(n, 1) \mid (m, s) \Leftrightarrow m = ns,$$

and $\mathcal{Z}$ is a unary predicate which is interpreted as

$$\mathcal{Z}(n, m) \Leftrightarrow m = 0.$$

The predicate $\mathcal{W}$ is interpreted as

$$\mathcal{W}((m,n),(r,s)) \Leftrightarrow m = s \wedge n = r.$$

A *diophantine model* of $\mathcal{S}$ over K is a diophantine subset $S \subseteq K^n$ equipped with a bijection $\phi : \mathbb{Z} \times \mathbb{Z} \to S$ such that under ϕ, the graphs of addition, $|$, $\mathcal{Z}$, and $\mathcal{W}$ in $\mathbb{Z} \times \mathbb{Z}$ correspond to diophantine subsets of S^3, S^2, S, and S^2, respectively.

A *diophantine model* of $\mathcal{S}$ over K with coefficients in $\mathbb{Z}[t_1, t_2]$ is a model, where in addition S and the graphs of addition, $|$, $\mathcal{Z}$, and $\mathcal{W}$ are diophantine over K with coefficients in $\mathbb{Z}[t_1, t_2]$.

We will now show that constructing such a model is sufficient to prove undecidability of Hilbert's Tenth Problem for K. First we can show the following

Proposition 4.2 ([Eis04]) *The relation $\mathcal{W}$ can be defined entirely in terms of the other relations.*

Proof It is enough to verify that

$$\mathcal{W}((a,b),(x,y)) \Leftrightarrow (1,1) \mid ((x,y)+(a,b)) \wedge (-1,1) \mid ((x,y)-(a,b)).$$

□

As Pheidas and Zahidi ([PZ00]) point out we can existentially define the integers with addition and multiplication inside

$$\mathcal{S} = \langle \mathbb{Z} \times \mathbb{Z}, +, |, \mathcal{Z}, \mathcal{W} \rangle,$$

so $\mathcal{S}$ has an undecidable positive existential theory:

Proposition 4.3 *The structure $\mathcal{S}$ has an undecidable positive existential theory.*

Proof We interpret the integer n as the pair $(n, 0)$. The set $\{(n,0) : n \in \mathbb{Z}\}$ is existentially definable in $\mathcal{S}$ through the relation $\mathcal{Z}$. Addition of integers n, m corresponds to the addition of the pairs $(n, 0)$ and $(m, 0)$. To define multiplication of the integers m and r, note that $n = mr$ if and only if $(m, 1) \mid (n, r)$, hence $n = mr$ if and only if

$$\exists a, b \ : ((m,0)+(0,1)) \mid ((n,0)+(a,b)) \wedge \mathcal{W}((a,b),(r,0)).$$

Since the positive existential theory of the integers with addition and multiplication is undecidable, $\mathcal{S}$ has an undecidable positive existential theory as well. □

The above proposition shows that in order to prove Theorem 4.1 it is enough to construct a diophantine model of $\mathcal{S}$ over K with coefficients in $\mathbb{Z}[t_1, t_2]$. In the next section we will construct this model.

4.2 Generating elliptic curves of rank one

As before, let $K := \mathbb{C}(t_1, t_2)$. Our first task is to find a diophantine set A over K which is isomorphic to $\mathbb{Z} \times \mathbb{Z}$ as a set. Following Kim and Roush ([KR92]) we will obtain such a set by using the K-rational points on two elliptic curves which have rank one over K. The same argument as in Theorem 3.2 shows that the following proposition holds:

Proposition 4.4 *Let E be an elliptic curve over $\mathbb{Q}$ without complex multiplication and with Weierstrass equation $y^2 = x^3 + ax + b$, where $a, b \in \mathbb{Q}$ and $b \neq 0$. Consider the twists $\mathcal{E}_1, \mathcal{E}_2$ of E defined by*

$$\mathcal{E}_1 : (t_1^3 + at_1 + b)Y^2 = X^3 + aX + b$$

and

$$\mathcal{E}_2 : (t_2^3 + at_2 + b)Y^2 = X^3 + aX + b.$$

The point $(t_i, 1) \in \mathcal{E}_i(K)$ has infinite order for $i = 1, 2$, and $(t_i, 1)$ generates $\mathcal{E}_i(K)$ modulo points of order 2.

To be able to define a suitable set S which is isomorphic to $\mathbb{Z} \times \mathbb{Z}$ we need to work in an algebraic extension F of K. Let $F := \mathbb{C}(t_1, t_2)(h_1, h_2)$, where h_i is defined by $h_i^2 = t_i^3 + at_i + b$, for $i = 1, 2$.

To prove that the positive existential theory of K in the language $\langle +, \cdot\, ; 0, 1, t_1, t_2 \rangle$ is undecidable, it is enough to prove that the positive existential theory of F in the language $\langle +, \cdot\, ; 0, 1, t_1, t_2, h_1, h_2, \mathcal{P} \rangle$ is undecidable, where $\mathcal{P}$ is a predicate for the elements of the subfield K ([PZ00, Lemma 1.9]). So from now on we will work with equations over F.

Over F both $\mathcal{E}_1$ and $\mathcal{E}_2$ are isomorphic to E. There is an isomorphism between $\mathcal{E}_1$ and E that sends $(x, y) \in \mathcal{E}_1$ to the point $(x, h_1 y)$ on E. Under this isomorphism the point $(t_1, 1)$ on $\mathcal{E}_1$ corresponds to the point $P_1 := (t_1, h_1)$ on E. Similarly there is an isomorphism between $\mathcal{E}_2$ and E that sends the point $(t_2, 1)$ on $\mathcal{E}_2$ to the point $P_2 := (t_2, h_2)$ on E.

So the element $(n, m) \in \mathbb{Z} \times \mathbb{Z}$ corresponds to the point $nP_1 + mP_2 \in E(F)$. As in Section 3.1, we can take care of the point at ∞ on the curve E.

The set of points $\mathbb{Z}P_1 \times \mathbb{Z}P_2 \subseteq E(F)$ is existentially definable in our language, because we have a predicate for the elements of K : Since $\mathcal{E}_1$ has 2-torsion, we first give a diophantine definition of $2 \cdot \mathbb{Z}P_1$ as in Section 3.1:

$$P \in 2 \cdot \mathbb{Z}P_1 \Leftrightarrow \exists\, x, y \in K\ (t_1^3 + at_1 + b)\, y^2 = x^3 + ax + b \wedge P = 2 \cdot (x, h_1 y)$$

Then $\mathbb{Z}P_1$ can be defined as

$$P \in \mathbb{Z}P_1 \Leftrightarrow (P \in 2 \cdot \mathbb{Z}P_1) \text{ or } (\exists\, Q \in 2 \cdot \mathbb{Z}P_1 \text{ and } P = Q + P_1)$$

Similarly we have a diophantine definition for $\mathbb{Z}P_2$. Hence the cartesian product $\mathbb{Z}P_1 \times \mathbb{Z}P_2 \subseteq E(F)$ is existentially definable, since addition on E is existentially definable.

4.3 Existential definition of $+$ and $\mathcal{Z}$

The unary relation $\mathcal{Z}$ is existentially definable, since this is the same as showing that the set $\mathbb{Z}P_1$ is diophantine, which was done above. Addition of pairs of integers corresponds to addition on the cartesian product of the elliptic curves $\mathcal{E}_i$ (as groups), hence it is existentially definable. Since $\mathcal{W}$ can be defined in terms of the other relations, it remains to define the divisibility relation $|$.

4.4 Existential definition of $(m, 1) \mid (n, r)$

In the following $x(P)$ will denote the x-coordinate of a point P on E, and $y(P)$ will denote the y-coordinate of P. The following theorem gives the existential definition of $|$:

Theorem 4.5

$$\forall m \in \mathbb{Z}, n, r \in \mathbb{Z} - \{0\} :$$

$$(m, 1) \mid (n, r) \Leftrightarrow$$

$$(\exists\, z, w \in F^*\ x(nP_1 + rP_2)\, z^2 + x(mP_1 + P_2)\, w^2 = 1)$$

Clearly this definition is existential in $(m, 1)$ and (n, r). It is enough to give an existential definition of $|$ for $n, r \in \mathbb{Z} - \{0\}$, because we can handle the cases when n or r are zero separately.

Proof For the first implication, assume that $(m, 1) \mid (n, r)$, *i.e.* $n = mr$. Then both $x(nP_1 + rP_2) = x(r(mP_1 + P_2))$ and $x(mP_1 + P_2)$ are elements of $\mathbb{C}(x(mP_1+P_2), y(mP_1+P_2))$, which has transcendence degree one over $\mathbb{C}$. This means that we can apply the Tsen-Lang Theorem (Theorem 6.2 from the appendix) to the quadratic form

$$x(nP_1 + rP_2)z^2 + x(mP_1 + P_2)w^2 - v^2$$

to conclude that there exists a nontrivial zero (z, w, v) over $\mathbb{C}(x(mP_1 + P_2), y(mP_1 + P_2))$. From the theory of quadratic forms it follows that there exists a nontrivial zero (z, w, v) with $z \cdot w \cdot v \neq 0$.

For the other direction, suppose that $n \neq mr$ and assume by contradiction that there exist $z, w \in F^*$ with

$$x(nP_1 + rP_2)\, z^2 + x(mP_1 + P_2)\, w^2 = 1. \tag{3}$$

Claim There exists a discrete valuation $w_m : F^* \twoheadrightarrow \mathbb{Z}$ such that $w_m(x(mP_1 + P_2)) = 1$ and such that $w_m(x(nP_1 + rP_2)) = 0$.

Proof of Claim Let $P_2' = mP_1 + P_2 = (t_2', h_2')$. Remember that $F = \mathbb{C}(t_1, t_2, h_1, h_2)$. Then

$$F = \mathbb{C}(t_1, t_2, h_1, h_2) = \mathbb{C}(t_1, h_1, t_2', h_2') = \mathbb{C}(x(P_1), y(P_1), x(P_2'), y(P_2')),$$

since $t_2 = x(P_2' - mP_1)$ and $h_2 = y(P_2' - mP_1)$.

Now let $w_m : F^* \twoheadrightarrow \mathbb{Z}$ be a discrete valuation which extends the discrete valuation w of $\mathbb{C}(t_1, h_1)(t_2')$ associated to t_2'. The valuation w is the discrete valuation that satisfies $w(\gamma) = 0$ for all $\gamma \in \mathbb{C}(t_1, h_1)$ and $w(t_2') = 1$.

Let $s := n - mr$. By assumption $s \neq 0$. We have $nP_1 + rP_2 = sP_1 + rP_2'$. The residue field of w_m is $\mathbb{C}(t_1, h_1)$. Let $x_{s,r}$ denote the image of $x(sP_1 + rP_2')$ in the residue field of w_m. Then $x_{s,r} = x\left(s(t_1, h_1) + r(0, \pm\sqrt{b})\right)$. We can show that this x-coordinate cannot be zero, which will imply that $w_m(x(nP_1 + rP_2)) = w_m(x(sP_1 + rP_2')) = 0$: The point $(t_1, h_1) \in E(\mathbb{C}(t_1, h_1))$ has infinite order, t_1 is transcendental over $\mathbb{C}$, and all points of E whose x-coordinate is zero are defined over $\mathbb{C}$. Since $s \neq 0$, this implies that $x\left(s(t_1, h_1) + r(0, \pm\sqrt{b})\right) \neq 0$. This proves the claim.

Since z, w satisfy Equation (3) it easily follows that $x_{s,r}$ is a square in the residue field. This will give us a contradiction:

We have $x_{s,r} = x\left(s(t_1, h_1) + r(0, \pm\sqrt{b})\right)$. Since the residue field of w_m is $\mathbb{C}(t_1, h_1)$, which is the function field of E, we can consider $x_{s,r}$ as a function $E \to \mathbf{P}^1_{\mathbb{C}}$. Then $x_{s,r}$ corresponds to the function on E which can be obtained as the composition $P \mapsto sP + r(0, \sqrt{b}) \mapsto x(sP + r(0, \sqrt{b}))$. The x-coordinate map is of degree 2 and has two distinct zeros, namely $(0, \sqrt{b})$ and $(0, -\sqrt{b})$. The map $E \to E$ which maps P

to $(sP + r(0, \sqrt{b}))$ is unramified since it is the multiplication-by-s map followed by a translation. Hence the composition of these two maps has $2s^2$ simple zeros. In particular, it is not a square in $\mathbb{C}(t_1, h_1)$. This completes the proof of the theorem. □

5 Generalization to finite extensions of $\mathbb{C}(t_1, t_2)$

In [Eis04] we proved the following theorem.

Theorem 5.1 *Let L be the function field of a surface over the complex numbers. There exist $z_1, z_2 \in L$ that generate an extension of transcendence degree 2 of $\mathbb{C}$ and such that Hilbert's Tenth Problem for L with coefficients in $\mathbb{Z}[z_1, z_2]$ is undecidable.*

Remark 2 This theorem also holds for transcendence degree ≥ 2, *i.e.* for finite extensions of $\mathbb{C}(t_1, \ldots, t_n)$, with $n \geq 2$, and the proof of the more general theorem can also be found in [Eis04]. To make our exposition as short as possible and to avoid extra notation, we will only discuss the transcendence degree 2 case here.

Our proof will proceed as the proof for $\mathbb{C}(t_1, t_2)$, *i.e.* we will construct a diophantine model of $\langle \mathbb{Z} \times \mathbb{Z}, +, |, \mathcal{Z}, \mathcal{W} \rangle$ in L. We will now give an outline of the steps that are needed in the proof.

5.1 Finding suitable elliptic curves of rank one

In the above undecidability proof for $\mathbb{C}(t_1, t_2)$ ([KR92]) we obtained a set that was in bijection with $\mathbb{Z} \times \mathbb{Z}$ by using the $\mathbb{C}(t_1, t_2)$-rational points on two elliptic curves which have rank one over $\mathbb{C}(t_1, t_2)$. However, the two elliptic curves that we used in Section 4.2 could have a rank higher than one over L, so we need to construct two new elliptic curves which will have rank one over L.

To do this we use a theorem by Moret-Bailly ([MB05, Theorem 1.8]):

Theorem 5.2 *Let k be a field of characteristic zero. Let C be a smooth projective geometrically connected curve over k with function field F. Let Q be a finite nonempty set of closed points of C. Let E be an elliptic curve over k with Weierstrass equation $y^2 = x^3 + ax + b$ where $a, b \in k$ and $b \neq 0$. Let $f \in F$ be admissible for E, Q. For $\lambda \in k^*$ consider the twist $\mathcal{E}_{\lambda f}$ of E which is defined to be the smooth projective model of*

$$((\lambda f)^3 + a(\lambda f) + b)\, y^2 = x^3 + ax + b.$$

Then the natural homomorphism $\mathcal{E}(k(T)) \hookrightarrow \mathcal{E}_{\lambda f}(F)$ induced by the inclusion $k(T) \hookrightarrow F$ that sends T to λf is an isomorphism for infinitely many $\lambda \in \mathbb{Z}$.

We will not define here what it means to be admissible, but we will only state that given C, E, Q as above, admissible functions exists, and if f is admissible for C, E, Q, then for all but finitely many $\lambda \in k^*$, λf is still admissible.

Now we can state our theorem that allows us to obtain two suitable elliptic curves of rank one over L:

Theorem 5.3 *Let L be a finite extension of $\mathbb{C}(t_1, t_2)$. Let $E/\mathbb{C}$ be an elliptic curve with Weierstrass equation*

$$y^2 = x^3 + ax + b,$$

where $a, b \in \mathbb{C}$, $b \neq 0$. Assume that E does not have complex multiplication. There exist $z_1, z_2 \in L$ such that $\mathbb{C}(z_1, z_2)$ has transcendence degree 2 over $\mathbb{C}$ and such that the two elliptic curves $\mathcal{E}_1, \mathcal{E}_2$ given by the affine equations $\mathcal{E}_1 : (z_1^3 + az_1 + b)\, y^2 = x^3 + ax + b$ and $\mathcal{E}_2 : (z_2^3 + az_2 + b)\, y^2 = x^3 + ax + b$ have rank one over L with generators $(z_1, 1)$ and $(z_2, 1)$, respectively (modulo 2-torsion).

Proof This is proved in [Eis04]. In the proof we apply Theorem 5.2 with k chosen to be the algebraic closure of $\mathbb{C}(t_2)$ inside L. □

5.2 Diophantine definition of the relation |

To existentially define the relation |, we need an elliptic curve E as in Theorem 5.3 with the additional property that the point $(0, \sqrt{b})$ has infinite order. So from now on we fix E to be the smooth projective model of $y^2 = x^3 + x + 1$. This curve does not have complex multiplication, and the point $(0, 1)$ has infinite order (see the curve 496A1 in [Cre97]). We fix z_1, z_2 as in Theorem 5.3.

As before, let $F := \mathbb{C}(z_1, z_2)(h_1, h_2)$, where h_i is defined by $h_i^2 = z_i^3 + az_i + b$, for $i = 1, 2$. Let $M := L(h_1, h_2)$. Over M, the elliptic curves $\mathcal{E}_1$ and $\mathcal{E}_2$ are isomorphic to E. Let $P_1 := (z_1, h_1)$, $P_2 := (z_2, h_2)$ be the two points on E as before. ¿From now on we will work with equations over M. To give a diophantine definition we would like to prove an analogue of Theorem 4.5. To make this theorem work we have to introduce extra equations.

Let $\alpha := [M : F]$. We have the following theorem:

Theorem 5.4 [Eis04] *There exists a finite set $U \subseteq \mathbb{Z}$ such that for all $m \in \mathbb{Z} - U$ we have: for all $n, r \in \mathbb{Z} - \{0\}$*

$$\begin{aligned}
&(m,1) \mid (n,r) \Leftrightarrow \\
&(\exists y_0, z_0 \in M^* \;\; x(nP_1 + rP_2)\, y_0^2 + x(mP_1 + P_2)\, z_0^2 = 1 \\
&\wedge \exists y_1, z_1 \in M^* \;\; x(2nP_1 + 2rP_2)\, y_1^2 + x(mP_1 + P_2)\, z_1^2 = 1 \\
&\qquad \cdots \\
&\wedge \exists y_\alpha, z_\alpha \in M^* \;\; x(2^\alpha nP_1 + 2^\alpha rP_2)\, y_\alpha^2 + x(mP_1 + P_2)\, z_\alpha^2 = 1\,)\,.
\end{aligned}$$

Outline of Proof. By the same argument as in the proof of Theorem 4.5, if $n = mr$, then the $\alpha + 1$ equations can all be satisfied.

For the other direction, the exceptional set U is necessary here because as in the proof for $\mathbb{C}(t_1, t_2)$, for each m, we construct a discrete valuation $w_m : M^* \twoheadrightarrow \mathbb{Z}$. This valuation w_m extends a certain other discrete valuation $v_m : F^* \twoheadrightarrow \mathbb{Z}$. We have to exclude all integers m, for which $w_m | v_m$ is ramified, and we define U to be this set of integers. Then U is finite by Theorem 6.1 from the appendix.

Assume that $n \neq mr$, and let $s := n - mr$. Assume by contradiction that we can satisfy all $\alpha + 1$ equations. We can show that for all $m \in \mathbb{Z} - U$ there exists a discrete valuation $w_m : M^* \twoheadrightarrow \mathbb{Z}$ such that $w_m(x(mP_1 + P_2)) = 1$ and such that $w_m(x(knP_1 + krP_2)) = 0$ for $k = 1, 2, 4, \ldots, 2^\alpha$. Let $P_2' = mP_1 + P_2$, and denote by $x_{s,r}$ the image of $x(sP_1 + rP_2') = x(nP_1 + rP_2)$ in the residue field ℓ of w_m. The proof of Theorem 5.4 proceeds by first showing that the elements $x_{s,r}\,,\ \ldots,\ x_{2^\alpha s\,,\,2^\alpha r}$ are not squares in $\mathbb{C}(z_1, h_1)$, and then by proving that the images of $x_{s,r}, \ldots, x_{2^\alpha s, 2^\alpha r}$ in

$$V := [(\ell^*)^2 \cap \mathbb{C}(z_1, h_1)^*]/(\mathbb{C}(z_1, h_1)^*)^2 \text{ are distinct.}$$

But using Kummer theory one can show that since $[\ell : \mathbb{C}(z_1, h_1)] \leq \alpha$, the size of V is bounded by α as well. This gives us the desired contradiction. □

Once we have Theorem 5.4, it is easy to define the relation $|$ for all $m \in \mathbb{Z}$ as follows:

Let m_0 be a fixed element in $\mathbb{Z} - U$, and let d be a positive integer such that $U \subseteq (m_0 - d, m_0 + d)$. Since $n = mr \Leftrightarrow dn + m_0 r = dmr + m_0 r = (dm + m_0)r$, we have

$$(m,1) \mid (n,r) \Leftrightarrow (dm + m_0, 1) \mid (dn + m_0 r, r),$$

and we can just work with that formula instead. So

$$(m,1) \mid (n,r) \Leftrightarrow \exists a,b (dm+m_0,1) \mid ((dn,r)+m_0(a,b)) \wedge \mathcal{W}((a,b),(0,r)).$$

It is an easy exercise to show that $\mathcal{W}$ is existentially definable using Theorem 5.4. Since m_0 is a fixed integer, this together with Theorem 5.4 implies that the last expression is existentially definable in $(m,1)$ and (n,r).

6 Appendix

In this section we state two theorems that we needed in Sections 4 and 5.

Theorem 6.1 *Let L and K be function fields of one variable with constant fields C_L and C_K, respectively, such that L is an extension of K. If L is separably algebraic over K, then there are at most a finite number of places of L which are ramified over K.*

Proof This theorem is proved on p. 111 of [Deu73] when $C_L \cap K = C_K$, and the general theorem also follows. □

Theorem 6.2 *Tsen-Lang Theorem. Let K be a function field of transcendence degree j over an algebraically closed field k. Let $f_1, \cdots, f_r$ be forms in n variables over K, of degrees $d_1, \cdots, d_r$. If*

$$n > \sum_{i=1}^{r} d_i^j$$

then the system $f_1 = \cdots = f_r = 0$ has a non-trivial zero in K^n.

Proof This is proved in Proposition 1.2 and Theorem 1.4 in Chapter 5 of [Pfi95]. □

References

[Cre97] J. E. Cremona, *Algorithms for modular elliptic curves*, Cambridge University Press, Cambridge, second edition, 1997.

[Den78] J. Denef, The Diophantine problem for polynomial rings and fields of rational functions, *Trans. Amer. Math. Soc.*, 242 (1978), 391–399.

[Deu73] M. Deuring, *Lectures on the theory of algebraic functions of one variable*, Springer-Verlag, Berlin, 1973, Lecture Notes in Mathematics, Vol. 314.

[DPR61] M. Davis, H. Putnam, and J. Robinson, The decision problem for exponential diophantine equations, *Ann. of Math. (2)*, 74 (1961), 425–436.

[Eis03] K. Eisenträger, Hilbert's Tenth Problem for algebraic function fields of characteristic 2, *Pacific J. Math.*, 210 (2003) no. 2, 261–281.

[Eis04] K. Eisenträger, Hilbert's Tenth Problem for function fields of varieties over **C**, *Int. Math. Res. Not.*, 59 (2004), 3191–3205.

[Eis07] K. Eisenträger, Hilbert's Tenth Problem for function fields of varieties over number fields and p-adic fields, *J. Algebra* 310 (2007), no. 2, 775–792.

[KR92] K. H. Kim and F. W. Roush, Diophantine undecidability of $\mathbf{C}(t_1, t_2)$, *J. Algebra*, 150 (1992) no. 1, 35–44.

[KR95] K. H. Kim and F. W. Roush, Diophantine unsolvability over p-adic function fields, *J. Algebra*, 176 (1995) no. 1, 83–110.

[Mat70] Yu. V. Matijasevič, The Diophantineness of enumerable sets, *Dokl. Akad. Nauk SSSR*, 191 (1970), 279–282.

[MB05] L. Moret-Bailly, Elliptic curves and Hilbert's Tenth Problem for algebraic function fields over real and p-adic fields, *J. Reine Angew. Math.*, 587 (2005), 77–143.

[Pfi95] A. Pfister, *Quadratic forms with applications to algebraic geometry and topology*, volume 217 of London Mathematical Society Lecture Note Series, Cambridge University Press, Cambridge, 1995.

[Phe91] Th. Pheidas, Hilbert's tenth problem for fields of rational functions over finite fields, *Invent. Math.*, 103 (1991) no. 1, 1–8.

[Poo03] B. Poonen, Hilbert's tenth problem and Mazur's conjecture for large subrings of $\mathbb{Q}$, *J. Amer. Math. Soc.*, 16 (2003) no. 4, 981–990 (electronic).

[PZ00] Th. Pheidas and K. Zahidi, Undecidability of existential theories of rings and fields: a survey, In *Hilbert's tenth problem: relations with arithmetic and algebraic geometry (Ghent, 1999)*, pages 49–105. Amer. Math. Soc., Providence, RI, 2000.

[Shl00] A. Shlapentokh, Hilbert's tenth problem for algebraic function fields over infinite fields of constants of positive characteristic, *Pacific J. Math.*, 193 (2000) no. 2, 463–500.

[Vid94] C. R. Videla, Hilbert's tenth problem for rational function fields in characteristic 2, *Proc. Amer. Math. Soc.*, 120 (1994) no. 1, 249–253.

First-order characterization of function field invariants over large fields

Bjorn Poonen[†]
University of California, Berkeley

Florian Pop[‡]
University of Pennsylvania

1 Introduction

Definition 1.1 A field k is *large* if every smooth curve with a k-point has infinitely many k-points [Pop96, p. 2].

This condition is equivalent to the condition that k be existentially closed in the Laurent series field $k((t))$ [Pop96, Proposition 1.1]. It is in some sense opposite to the "Mordellic" properties satisfied by number fields, over which curves of genus greater than 1 have finitely many rational points [Fal83].

If p is any prime number, then any *p-field* (field for which all finite extensions are of p-power degree) is large [CT00, p. 360]. In particular, separably closed fields and real closed fields are large. Other examples of large fields include henselian fields and PAC fields. (PAC stands for pseudo-algebraically closed: a *PAC* field is one over which every geometrically integral variety has a rational point. See [FJ05, Chapter 11] for further properties of these fields.) For further examples of large fields, see [Pop96]. An algebraic extension of an large field is large [Pop96, Proposition 1.2].

Definition 1.2 Let k be a field. A *function field* over k is a finitely generated extension K of k with $\operatorname{trdeg}(K|k) > 0$.

Definition 1.3 The *constant field* of a field K finitely generated over k is the relative algebraic closure of k in K.

† B.P. was supported by NSF grant DMS-0301280 a Packard Fellowship, and the Miller Institute for Basic Research in Science. He thanks the Isaac Newton Institute for hosting a visit in the summer of 2005.
‡ F.P. was supported by NSF grant DMS-0401056. He would also like to thank the Isaac Newton Institute for hosting a 2005 summer visit.

Theorem 1.4 *There exists a formula $\phi(t)$ that when interpreted in a field K finitely generated over an large field k defines the constant field.*

Theorem 1.5 *For each of the following classes of fields, there is a sentence that is true for fields in that class and false for fields in the other five classes:*

(1) *finite and large fields*
(2) *number fields*
(3) *function fields over finite fields*
(4) *function fields over large fields of characteristic* > 0
(5) *function fields over large fields of characteristic* 0
(6) *function fields over number fields*

Remark 1.6 It is impossible to distinguish all finite fields from all large fields with a single sentence, since a nontrivial ultraproduct of finite fields is large.

Finally, we have a few theorems characterizing algebraic dependence. Some of these require that the ground field k be "2-cohomologically well behaved" in the sense of Definition 5.1 in Section 5. The following theorems will be proved in Section 5.

Theorem 1.7 *There exists a formula $\phi_n(t_1, \ldots, t_n)$ such that for every K finitely generated over a real closed or separably closed field k, and every $t_1, \ldots, t_n \in K$, the formula holds if and only if $t_1, \ldots, t_n$ are algebraically dependent over k.*

Theorem 1.8 *Let k be a 2-cohomologically well behaved field. Let $K|k$ be a finitely generated extension. Then there exists a first order formula (depending on K and k) with r free variables, in the language of fields augmented by a predicate for a subfield, that when interpreted for elements $t_1, \ldots, t_r \in K$ with the subfield being k holds if and only if the elements are algebraically independent over k.*

Corollary 1.9 *Let k be a finite field, a number field, or a 2-cohomologically well behaved large field. Then there exists a first order formula (depending on K and k) with r free variables, in the language of fields, that when interpreted for elements $t_1, \ldots, t_r \in K$ holds if and only if the elements are algebraically independent over k.*

Proof Theorem 1.4 of [Poo07] handles the case where k is finite or a number field. If k is large, combine Theorems 1.4 and 1.8. □

Remark 1.10 We do not know if Theorem 1.8 and Corollary 1.9 can be made uniform in k and K, i.e., whether the formula can be chosen independent of k and K.

2 Defining the constants

In this section we prove Theorem 1.4.

Lemma 2.1 *Let k be an infinite field of characteristic p. Let S_0 be a finite subset of k, and let $S = \{s^{p^n} : s \in S_0, n \in \mathbb{N}\}$. Then $k - S$ is infinite.*

Proof If k is algebraic over $\mathbb{F}_p$, then S is finite, so $k - S$ is infinite. Otherwise, choose $t \in k$ transcendental over $\mathbb{F}_p$; then for a given $s \in k$, the set $\{s^{p^n} : n \in \mathbb{N}\}$ contains at most one element of $\{t^\ell : \ell \text{ is prime}\}$, so $k - S$ is infinite. □

An algebraic family of curves $C \to B$ over an irreducible k-curve B is called *isotrivial* if over some finite extension of the function field of B, the generic fiber becomes birational to the base extension of a curve over a finite extension of k. This is equivalent to the condition that the rational map from B to the moduli space of curves be constant. So if a family is non-isotrivial, each isomorphism class of curves occurs at most finitely often among the fibers of the family. We will consider the case $B = \mathbb{A}^1$, and write $\{C_a\}$ to denote a family: here C_a denotes the fiber above $a \in B(k)$.

Lemma 2.2 *Let k be an infinite field. Let V be a k-variety. Let $\{C_a\}$ be a non-isotrivial family of curves of genus ≥ 2 over k with parameter a. Then there exist infinitely many $a \in k$ such that all rational maps from V to C_a are constant.*

Proof Let p be the characteristic of k. A theorem of Severi [Sam66, Théorème 2] states that there are only finitely many fields L between k and the function field K of V such that L is the function field of a curve of genus ≥ 2 over k and K is separable over L. Thus the set S of $a \in k$ such that C_a admits a non-constant rational map from V is a finite set

S_0 together with (if $p > 0$) the p^n-th powers of the elements of S_0 for all $n \in \mathbb{N}$. By Lemma 2.1, $k - S$ is infinite. $\square$

Proof of Theorem 1.4. Without loss of generality we may assume that k is relatively algebraically closed in K. The discriminant of $x^5 + ax + 1$ (with respect to x) is $256a^5+3125$; if $\operatorname{char} k \notin \{2, 5\}$, this is a nonconstant squarefree polynomial in a, so the family of affine curves $C_a \colon y^2 = x^5 + ax+1$ has both smooth and nodal curves, and is therefore non-isotrivial. If $\operatorname{char} k = 5$, the family $C_a \colon y^2 = x^7 + ax + 1$ is non-isotrivial for the same reason; and if $\operatorname{char} k = 2$, the family $C_a \colon y^2 + y = x^5 + ax$ is non-isotrivial, since a direct calculation (using the fact that the unique Weierstrass point must be preserved) shows that no two members of this family are isomorphic over an algebraic closure of k. The projection $x \colon C_a \to \mathbb{A}^1$ is étale above $0 \in \mathbb{A}^1(k)$.

For $a \in K$, define

$$S_a := \left\{ \frac{x_1}{x_2} : (x_1, y_1), (x_2, y_2) \in C_a(K) \text{ with } x_2 \neq 0 \right\}.$$

(A very similar definition was used in the proof of [Koe02, Theorem 2].) We have

(1) If $a \in k$, then $k \subseteq S_a$. *Proof:* Let $f(x, y) = 0$ be the equation of C_a in $\mathbb{A}^2$. Let $c \in k$. The map $(x_1, x_2) \colon C_a \times C_a \to \mathbb{A}^2$ is étale above $(0, 0)$, so the point $(x_1, y_1, x_2, y_2) = (0, 1, 0, 1)$ on the inverse image Y of the line $x_1 = cx_2$ in $C_a \times C_a$ is smooth. Since k is large, Y has infinitely many other k-points, so $c \in S_a$.

(2) There exists $a_0 \in k$ such that $S_{a_0} = k$. *Proof:* Let V be an integral k-variety with function field K. Lemma 2.2 gives $a_0 \in k$ such that there is no nonconstant rational map $V \dashrightarrow C_{a_0}$ over k. Equivalently, $C_a(K) = C_a(k)$. So $S_{a_0} \subseteq k$, and we already know the opposite inclusion.

(3) If $a \in K - k$, then S_a is finite. *Proof:* By the function field analogue of the Mordell conjecture [Sam66, Théorème 4], $C_a(K)$ is finite, so $S_a(K)$ is finite.

Let A be the set of $a \in K$ such that S_a is a field containing a. Let $L := \bigcap_{a \in A} S_a$. Then L is uniformly definable by a formula. By (3), $A \subseteq k$ (a finite field cannot contain an element transcendental over k). Now by (1) and (2), $L = k$. $\square$

Remark 2.3 Suppose K is finitely generated over a field k, and k is relatively algebraically closed in K. By the Weil conjectures applied to

Y, there exists an explicit positive integer m such that (1) is true also in the case where k is a finite field of size $> m$. Let S'_a be the union of S_a with the set of zeros of $x^q - x$ in K for all $q \in \{2, 3, \ldots, m\}$. Let (1)', (2)', (3)' be the statements analogous to (1), (2), (3) but with S'_a in place of S_a. Then (1)', (2)', (3)' remain true for large k, but now (1)' and (3)' hold also for finite k.

3 Some facts about quadratic forms

Proposition 3.1 *Let $q(x_1, \ldots, x_n)$ be a quadratic form over a field K, and let L be a finite extension of K of odd degree. If q has a nontrivial zero over L, then q has a nontrivial zero over K.*

Proof This is well known: see [Lan02, Chapter V, Exercise 28]. □

Corollary 3.2 *Let K be a field of characteristic not 2. Let q be a quadratic form over K. Let L be a purely inseparable extension of K. If q has a nontrivial zero over L, then q has a nontrivial zero over K.*

Proof If q has a nontrivial zero over L, the coordinates of this zero generate a finite purely inseparable extension of K, so we may assume $[L : K] < \infty$. Now the result follows from Proposition 3.1. □

For nonzero a, let $\langle\langle a \rangle\rangle$ denote the quadratic form $x^2 + ay^2$ and let $\langle\langle a_1, \ldots, a_n \rangle\rangle = \langle\langle a_1 \rangle\rangle \otimes \cdots \otimes \langle\langle a_n \rangle\rangle$ be the n-fold Pfister form.

Lemma 3.3 *Let k be a field, and let V be an integral k-variety with function field K. Suppose that v is a regular point on V, and that $t_1, \ldots, t_m$ are part of a system of local parameters at v. Let q be a diagonal quadratic form over k having no nontrivial zero over the residue field of v. Then $q \otimes \langle\langle t_1, \ldots, t_m \rangle\rangle_d$ has no nontrivial zero over K.*

Proof This result is essentially contained in [Pop02]. The proof is given again in Lemma A.5 in [Poo07]. □

Lemma 3.4 *Let ℓ be a field of characteristic not 2. Let L be a finitely generated extension of ℓ. Suppose that every 3-fold Pfister form $\langle\langle a, b, c \rangle\rangle$ over L has a nontrivial zero. Then*

(1) $\operatorname{trdeg}(L|\ell) \leq 2$.

(2) *If moreover L admits a valuation that is trivial on $\ell^\times$ such that ℓ maps isomorphically to the residue field, and not every element of ℓ is a square in ℓ, then* $\operatorname{trdeg}(L|\ell) \leq 1$.

Proof

(1) Let $t_1, \ldots, t_d$ be a transcendence basis for $L|\ell$. Let K be the maximal separable extension of $\ell(t_1, \ldots, t_d)$ contained in L. Let V be an integral variety over ℓ with function field K. Replacing V by an open subset if necessary, we may assume that $(t_1, \ldots, t_d)\colon V \to \mathbb{A}^n_\ell$ is étale. If ℓ is infinite, choose $(a_1, \ldots, a_d) \in \mathbb{A}^n(\ell)$ in the image of V; then by Lemma 3.3, $\langle\langle t_1 - a_1, \ldots, t_d - a_d\rangle\rangle$ has no nontrivial zero over K, and hence by Corollary 3.2, no nontrivial zero over L. If ℓ is finite, choose $(a_1, \ldots, a_d) \in \mathbb{A}^n(\ell')$ in the image of V for some $\ell'|\ell$ of odd degree, and repeat the previous argument with the minimal polynomial $P_{a_i}(t_i)$ of a_i over ℓ in place of $t_i - a_i$. In either case, this Pfister form contradicts the hypothesis if $d \geq 3$. Thus $d \leq 2$.

(2) Suppose not. Then by (1), $\operatorname{trdeg}(L|\ell) = 2$. By the resolution of singularities for surfaces (see e.g. [Abh69]), we may choose a regular projective surface V over ℓ with function field L. The center of the given valuation on V is an ℓ-rational point $v \in V(\ell)$; hence v is actually a smooth point of V. Choose local parameters u_1, u_2 at v. Let $\alpha \in \ell$ be a non-square. By Lemma 3.3, $\langle\langle -\alpha, u_1, u_2\rangle\rangle$ has no nontrivial zero over L. This contradicts the hypothesis. □

Lemma 3.5 *Let X be a variety over an infinite field k. There exists an integer m such that the points on X of degree $\leq m$ over k are Zariski dense in X.*

Proof The desired property depends only on the birational class of X over $\overline{k}$. Therefore, enlarging k, we may reduce to the case where X is a geometrically integral closed hypersurface in $\mathbb{P}^n$. Choose $P \in (\mathbb{P}^n - X)(k)$. Projection from Q determines a generically finite rational map from X to $\mathbb{P}^{n-1}$, and the fibers above k-points in a Zariski dense open subset of $\mathbb{P}^{n-1}$ contain points of bounded degree. These points are Zariski dense in X. □

4 Distinguishing classes of fields

Proposition 4.1 *There is a sentence ϕ that is true for finite fields and large fields, false for function fields over any field, and false for number fields.*

Proof Let K be a field. Define S'_a as in Remark 2.3. Let ϕ be the sentence saying that $S'_a = K$ for all $a \in K$. This is true if K is finite or large.

If K is a function field, then (3)' (whose proof is valid over any k) shows that for some a, the set S'_a is finite. If K is a number field, then S'_a is finite for all but finitely many a, by the Mordell conjecture [Fal83] applied to C_a. In both these cases, there exists $a \in K$ with $S'_a \neq K$. □

We can generalize Theorem 1.4 to include finitely generated extensions of finite fields:

Proposition 4.2 *There exists a formula that for K finitely generated over a finite or large field k defines the constant field.*

Proof We may assume that k is relatively algebraically closed in K. We use the notation of the proof of Theorem 1.4 and Remark 2.3. Let A' be the set of $a \in K$ such that S'_a is a field containing a. Let $k_1 := \bigcap_{a \in A'} S'_a$. Theorem 1.3 of [Poo07] gives a formula that defines the constant subfield if K is finitely generated over a finite field; over any field K, let k_2 be the subset it defines. Define

$$\tilde{k} := \begin{cases} k_1, & \text{if } S'_a \supseteq k_1 \text{ for every } a \in k_1, \\ k_2, & \text{otherwise.} \end{cases}$$

The subset $\tilde{k}$ is definable by a uniform formula; we claim that $\tilde{k} = k$.

If k is large, then by the proof of Theorem 1.4, $k_1 = k$, and $\tilde{k} = k_1 = k$.

Now suppose k is finite, so $k_2 = k$. The set k_1 is a field (since it is an intersection of fields), and it contains k by Remark 2.3. If $k_1 = k$, then $\tilde{k} = k$. If $k_1 \supsetneq k$, and $a \in k_1 - k$, then by (3) (above Remark 2.3), S'_a is finite, so it cannot contain k_1; thus $\tilde{k} = k_2 = k$. □

Proposition 4.3 *There exists a sentence that is true for function fields over finite or large fields and false for number fields and function fields over number fields.*

Proof Use the sentence that says that the formula in Proposition 4.2 defines a field satisfying the sentence of Proposition 4.1. □

Proposition 4.4 *There is a sentence in the language of rings extended by a unary predicate that when interpreted in a function field K over a field k (not necessarily relatively algebraically closed) with the unary predicate defining k is true if and only if k is finite.*

Proof By [Poo07, Remark 5.1], there is a formula $\phi(x, y)$ in the language of rings such that when it is interpreted in a function field K with finite constant field ℓ,

$$\{y \in K : \phi(x, y)\} = \ell[x]$$

for each $x \in K$. By [Rum80], there is a formula ψ defining a family of subsets that when interpreted in $\ell(x)$ for ℓ finite gives exactly the family of nontrivial valuation rings in $\ell(x)$ (possibly with repeats).

Now let K be a function field over an arbitrary field k. We claim that k is finite if and only if for some $x \in K$ the following hold:

(1) The set R defined by $\phi(x, \cdot)$ is a ring containing k and x.
(2) The family $\mathcal{F}$ defined by ψ interpreted over the fraction field L of R defines a set of nontrivial valuation rings in L, each containing k.
(3) The intersection of the valuation rings in $\mathcal{F}$ is a field ℓ.
(4) The element x is not in ℓ.
(5) The field ℓ maps isomorphically to the residue field of some valuation ring in $\mathcal{F}$.
(6) If $2 = 0$, then $[L : L^2] = 2$.
(7) If $2 \neq 0$, then every 3-fold Pfister form $\langle\langle a, b, c\rangle\rangle$ over L has a nontrivial zero, and some element of ℓ is not a square in ℓ.
(8) The intersection of the rings in $\mathcal{F}$ containing R equals R.
(9) Every ideal $aR + bR$ of R generated by two elements is principal.
(10) The elements $x - a$ for $a \in \ell$ are irreducible, and generate pairwise distinct ideals of R.
(11) There exists a nonzero $f \in R$ divisible in R by $x - a$ for all $a \in \ell$.

(These conditions can be expressed by a first order sentence in the language of rings with a predicate for k.)

If k is finite, and $x \in K$ is not in the constant field ℓ of K, then $R = \ell[x]$ for a finite field ℓ, and conditions (1)–(11) hold.

Conversely, suppose that conditions (1)–(11) hold for some $x \in K$. If char$K = 2$, then (6) implies $\mathrm{trdeg}(L|\ell) \leq 1$. If char$K \neq 2$, then (5) and (7) imply that $\mathrm{trdeg}(L|\ell) \leq 1$, by Lemma 3.4. Thus in every case, $\mathrm{trdeg}(L|\ell) \leq 1$. By (3), ℓ is an intersection of valuation rings, so it is relatively algebraically closed in L. By (4), $x \in L - \ell$, so $\mathrm{trdeg}(L|\ell) = 1$. Since L is a function field over k and $k \subseteq \ell$, L is a function field of transcendence degree 1 over ℓ. By (8), R is integrally closed in L; in particular it contains the integral closure R_0 of $\ell[x]$ in L. Thus R_0 is a Dedekind domain with fraction field L. Any ring between a Dedekind domain and its fraction field is itself a Dedekind domain, so R is a Dedekind domain. By (9), R is a principal ideal domain, and hence a unique factorization domain. Now (10) and (11) imply that ℓ is finite. So k is finite. □

Proposition 4.5 *There is a sentence that is true for function fields over finite fields and false for function fields over large fields.*

Proof Combine Propositions 4.2, and 4.4. □

Proposition 4.6 *There exists a sentence that for a function field K over a finite or large field is true if and only if* char$K = 0$.

Before beginning the proof of Proposition 4.6, we need a few definitions and a lemma. If M is an Abelian group and $n \geq 1$, let $M[n]$ be the kernel of the multiplication-by-n map $M \to M$. Also define $M_{\mathrm{tors}} := \bigcup_{n \geq 1} M[n]$. If E: $y^2 = f(x)$ is an elliptic curve over a field K of characteristic $\neq 2$, and $t \in K$, then the twisted elliptic curve E_t is defined by $f(t)y^2 = f(x)$ over K. We will use the following, which is essentially a special case of a result of Moret-Bailly.

Lemma 4.7 *Let k be a field of characteristic 0. Let K be a function field over k. Let E: $y^2 = f(x)$ be an elliptic curve over k, where f is a cubic polynomial. Then there are infinitely many $t \in K$ with $f(t) \in K^\times - k^\times K^{\times 2}$ such that $E_t(K)$ is a finitely generated Abelian group with* $\mathrm{rk} E_t(K) = \mathrm{rk}\,\mathrm{End}_K(E)$.

Proof We may enlarge k to assume that K is the function field of a geometrically irreducible curve over k. Replacing $f(x)$ by $f(x+c)$ for suitable $c \in k$, we may assume that $f(0) \neq 0$.

We use the notions "admissible", "Good", and "Good" defined in [MB05, §1.5]. Let Γ be the smooth projective model of the curve $y^2 =$

$x^4 f(1/x)$; cf. [MB05, 1.4.5(ii)]. By [MB05, 2.3.1], there exists $g \in K - k$ that is admissible for Γ. By [MB05, 1.8(ii) and 1.4.7], $\mathrm{Good}(k) \cap \mathbb{Z}$ is infinite.

We claim that for any $\lambda \in \mathrm{Good}(k) \cap \mathbb{Z}$, the value $t := \frac{1}{\lambda g}$ satisfies the required conditions. For such λ and t, we have $\lambda \in \mathrm{Good}(k)$ by [MB05, 1.5.4(i)]; thus $E'\colon (\lambda g)^4 f(\frac{1}{\lambda g}) y^2 = f(x)$ is an elliptic curve over K such that $E'(K)$ is finitely generated and $\mathrm{rk} E'(K) = \mathrm{rk}\mathrm{End}_K(E)$. By definition, E' is isomorphic to E_t.

Let $K\overline{k}$ be a compositum of K with an algebraic closure of $\overline{k}$ over k. If $f(t)$ were in $k^\times K^{\times 2}$, then E' would be isomorphic over $K\overline{k}$ to E, so $E'(K\overline{k}) \simeq E(K\overline{k}) \supseteq E(\overline{k})$ would not be finitely generated, contradicting the definition of $\mathrm{Good}(k)$. □

Proof of Proposition 4.6. Use $\neg\phi$, where ϕ is a sentence equivalent to the following: $2 = 0$ or there exists an extension L of K with $[L : K] \leq 2$ such that for ℓ the subset defined by the formula of Proposition 4.2 applied to L, there exist distinct $e_1, e_2, e_3 \in \ell$ such that if we write $f(x) := (x-e_1)(x-e_2)(x-e_3)$, then for all $t \in L$ with $f(t) \in L^\times - \ell^\times L^{\times 2}$, the twist E_t of $E\colon y^2 = f(x)$ satisfies $\#E_t(L)/2E_t(L) \geq 64$. For the K we are interested in, L is a function field over a finite or large field, so ℓ is the constant field of L.

If $\mathrm{char} K = 2$, then ϕ is true. Now suppose K is a function field over an large field of characteristic $p > 2$. Let L be a compositum of K with $\mathbb{F}_{p^2}$. Let E be an elliptic curve over $\mathbb{F}_p$ with $\#E(\mathbb{F}_p) = p + 1$. Then the p^2-Frobenius endomorphism of E is multiplication by $-p$, so $\mathrm{rk}\mathrm{End}_{\mathbb{F}_{p^2}}(E) = 4$, and $E[2] \subseteq E(\mathbb{F}_{p^2})$. The curve $E_{\mathbb{F}_{p^2}}$ has an equation $y^2 = f(x)$ where $f(x) := (x-e_1)(x-e_2)(x-e_3)$ with distinct $e_1, e_2, e_3 \in \mathbb{F}_{p^2} \subseteq \ell$. Suppose $t \in L$ satisfies $f(t) \in L^\times - \ell^\times L^{\times 2}$. The restriction on t implies that E_t is not isomorphic over L to an elliptic curve over ℓ, so $E_t(L)$ is finitely generated. Quadratic twists of an elliptic curve have the same endomorphism ring, so the ring $\mathcal{O} := \mathrm{End}_L(E_t)$ is a maximal order in a non-split quaternion algebra $\mathbb{H}$ over $\mathbb{Q}$. Since $E_t(L) \otimes \mathbb{Q}$ is an $\mathbb{H}$-vector space, $4 \mid \mathrm{rk}_{\mathbb{Z}} E_t(L)$. The point $(t, 1) \in E_t(L)$ has infinite order, since under the $L(\sqrt{f(t)})$-isomorphism $E_t \to E$ mapping (x, y) to $(x, y\sqrt{f(t)})$ it corresponds to a point of E whose x-coordinate is transcendental over ℓ. Thus $\mathrm{rk}_{\mathbb{Z}} E_t(L) > 0$, so $\mathrm{rk}_{\mathbb{Z}} E_t(L) \geq 4$. Also, $E_t[2] \subseteq E_t(L)$, so $\#E_t(L)/2E_t(L) \geq 2^2 \cdot 2^4 = 64$.

Now suppose that K is a function field over an large field of characteristic 0. Suppose L is an extension with $[L : K] \leq 2$, and $e_1, e_2, e_3 \in \ell$ are distinct. By Lemma 4.7 applied to L over ℓ, there exists $t \in L$

with $f(t) \notin \ell^\times L^{\times 2}$ such that $E_t(L)$ is finitely generated with $\mathrm{rk}E_t(L) = \mathrm{rkEnd}_L(E)$. Since $\mathrm{rkEnd}_L(E) \in \{1, 2\}$, and since $E_t(L)_{\mathrm{tors}}$ is generated by at most 2 elements, we get $\#E_t(L)/2E_t(L) \leq 2^2 \cdot 2^2 = 16$. □

Proof of Theorem 1.5. Taking $d = 0$ in the first claim of Theorem 1.5(3) of [Pop02] gives a sentence that is true for number fields and false for function fields over number fields. Combining this with Propositions 4.1, 4.3, 4.5, and 4.6 gives the result. □

5 Detecting algebraic dependence

We begin by recalling the following general facts: Let E be an arbitrary field of characteristic $\neq 2$. In particular, $\mu_2 = \{\pm 1\}$ is contained in E. We denote by G_E the absolute Galois group of E, and view μ_2 as a G_E-module.

1) Let $\mathrm{cd}_2^0(E) \in \mathbb{N} \cup \{\infty\}$ be the supremum over all the natural numbers n such that $\mathrm{H}^n(E, \mu_2) \neq 0$. Since the 2-cohomological dimension $\mathrm{cd}_2(E)$ is defined similarly, but the supremum is taken over all possible 2-torsion G_k-modules, one has

$$\mathrm{cd}_2^0(E) \leq \mathrm{cd}_2(E).$$

Also define $\mathrm{vcd}_2(E) := \mathrm{cd}_2(E(\sqrt{-1}))$.

2) Recall the Milnor Conjecture (proved by Voevodsky et al.) It asserts that the nth cohomological invariant

$$e_n \colon I_n(E)/I_{n+1}(E) \to \mathrm{H}^n(E, \mu_2),$$

which maps each n-fold Pfister form $\langle\langle a_1, \ldots, a_n \rangle\rangle$ to the cup product $(-a_1) \cup \cdots \cup (-a_n)$, is a well defined isomorphism. Using the Milnor Conjecture one can describe $\mathrm{cd}_2^0(E)$ via the behavior of Pfister forms as follows: $n > \mathrm{cd}_2^0(E)$ if and only if every n-fold Pfister form over E represents 0 over E.

3) There exists a field E with $\mathrm{cd}_2^0(E) < \mathrm{cd}_2(E)$. For instance, let E be a maximal pro-2 Galois extension of a global or local field of characteristic $\neq 2$. Then every element of E is a square, so $\mathrm{cd}_2^0(E) = 0$. On the other hand, since the Sylow 2-groups of G_E are non-trivial, one has $\mathrm{cd}_2(E) > 0$ by [Ser02, §I.3.3, Corollary 2].

Definition 5.1 A field E is said to be 2-*cohomologically well behaved* if $\mathrm{char}E \neq 2$ and for every finite extension $E'|E$ containing $\sqrt{-1}$ one has $\mathrm{cd}_2^0(E') = \mathrm{cd}_2(E') < \infty$.

Remark 5.2 If E is 2-cohomologically well behaved, and $E'|E$ is a finite extension containing $\sqrt{-1}$, then

$$\mathrm{cd}_2^0(E') = \mathrm{vcd}_2(E') = \mathrm{vcd}_2(E),$$

since $\mathrm{cd}_2(E') = \mathrm{cd}_2(E(\sqrt{-1}))$ by [Ser02, §II.4.2, Proposition 10].

Example/Fact 5.3 The following fields, when of characteristic $\neq 2$, are 2-cohomologically well behaved:

- separably closed fields (trivial),
- finite fields (follows from [Ser02, II.§3]),
- local fields (follows from [Ser02, II.§4.3]),
- number fields (follows from [Ser02, II.§4.4]), and
- finitely generated fields (follows from the above and Proposition 5.4 below).

Proposition 5.4 *If E is 2-cohomologically well behaved, and E' is a function field over E, then E' is 2-cohomologically well behaved and* $\mathrm{vcd}_2(E') = \mathrm{vcd}_2(E) + \mathrm{trdeg}(E'|E)$.

Proof We may assume $\sqrt{-1} \in E$. The case $\mathrm{trdeg}(E'|E) = 0$ follows from Remark 5.2. By induction on $\mathrm{trdeg}(E'|E)$, it will suffice to prove that $\mathrm{cd}_2^0(E') = \mathrm{vcd}_2(E) + 1$ for every extension $E'|E$ with $\mathrm{trdeg}(E'|E) = 1$. We may assume that E' is separably generated over E. Let X be a curve over E with function field E', let P be a smooth point on X, let κ be the residue field of P, and let $t \in E'$ be a uniformizer at P. Let $n = \mathrm{cd}_2^0(\kappa) = \mathrm{vcd}_2(E)$. By definition, there exists an n-fold Pfister form $\langle\langle \bar{a}_1, \dots, \bar{a}_n \rangle\rangle$ that does not represent 0 over κ. Lift each $\bar{a}_i$ to an a_i in the local ring at P. Then $\langle\langle a_1, \dots, a_n, t \rangle\rangle$ does not represent 0 over E'. Thus $\mathrm{cd}_2^0(E') \geq \mathrm{vcd}_2(E) + 1$. On the other hand, $\mathrm{cd}_2^0(E') \leq \mathrm{vcd}_2(E') = \mathrm{vcd}_2(E) + 1$ by [Ser02, §II.4.2, Proposition 11], so we have equality. □

Proposition 5.5 *Let k be a field which is 2-cohomologically well behaved, and let $e = \mathrm{vcd}_2(k)$. Let $K|k$ be a finitely generated extension. Then the following hold:*

(1) *For each $n \in \mathbb{Z}_{\geq 0}$, there exists a sentence ϕ_n in the language of fields (depending on e) such that ϕ_n is true in K if and only if* $\mathrm{trdeg}(K|k) = n$.

One can take ϕ_n to be the following sentence: Every $(e+n+1)$*-fold Pfister form over* $K[\sqrt{-1}]$ *represents* 0*, but there exist* $(e+n)$*-fold Pfister forms over* $K[\sqrt{-1}]$ *which do not represent* 0.

(2) *For elements $t_1, \dots, t_r \in K^\times$, the following are equivalent:*

(a) *$(t_1, \dots, t_r)$ are algebraically independent over k.*

(b) *There exist a finite separable extension $l|k$ (depending on $t_1, \dots, t_r$) containing $\sqrt{-1}$ and $a_1, \dots, a_e, b_1, \dots, b_r \in l^\times$ such that $\langle\langle a_1, \dots, a_e, t_1 - b_1, \dots t_r - b_r \rangle\rangle$ does not represent 0 over Kl.*

Proof

(1) By the discussion preceding Proposition 5.5 we have:

$$\mathrm{cd}_2^0(K[\sqrt{-1}]) = \mathrm{cd}_2^0(k[\sqrt{-1}]) + \mathrm{trdeg}(K|k) = e + \mathrm{trdeg}(K|k).$$

Now use the characterization of cd_2^0 in terms of Pfister forms.

(2), (b) $\Rightarrow$ (a): Suppose for the sake of obtaining a contradiction that $(t_1, \dots, t_r)$ is algebraically dependent over k. Let $L = l(t_1, \dots, t_r) \subset Kl$. Since $\sqrt{-1} \in l$, we have:

$$\mathrm{cd}_2(L) = \mathrm{cd}_2(l) + \mathrm{trdeg}(L|l) = e + \mathrm{trdeg}(L|l) < e + d.$$

Thus by the discussion above, every $(e+d)$-fold Pfister form over L represents 0 over L. In particular, for all $(a_i)_i$ and $(b_j)_j$ as in (b), the resulting $(e+d)$-fold Pfister form $\langle\langle a_1, \dots, a_e, t_1 - b_1, \dots, t_r - b_r \rangle\rangle$ represents 0 over L. Since $L \subseteq Kl$, it follows that $\langle\langle a_1, \dots, a_e, t_1 - b_1, \dots t_r - b_r \rangle\rangle$ represents 0 over Kl, a contradiction!

(2), (a) $\Rightarrow$ (b): The proof is an adaptation from and similar to [Pop02], Section 1. By extending the list $\mathcal{T} := (t_1, \dots, t_r)$, we may assume that it is a transcendence basis for $K|k$. Let $K_0|k(\mathcal{T})$ be the relative separable closure of $k(\mathcal{T})$ in K. Thus $\mathcal{T}$ is a separable transcendence basis of $K_0|k$, and $K|K_0$ is a finite purely inseparable field extension. Further let R be the integral closure of $k[\mathcal{T}]$ in K_0, and let $X = \mathrm{Spec}R$. The k-embedding $k[\mathcal{T}] \hookrightarrow R$ defines a finite k-morphism $\phi\colon X \to \mathrm{Spec}k[\mathcal{T}] = \mathbb{A}_k^r$. Further, since $K_0|k(\mathcal{T})$ is separable, the k-morphism ϕ is generically étale. Therefore, ϕ is étale on a Zariski dense open subset $U \subset X'$. We choose a finite separable extension $l|k$ containing $\sqrt{-1}$ such that $U(l)$ is non-empty. Choose $x \in U(l)$, and let $b := (b_1, \dots, b_r) = \phi(x)$ be its image in $\mathbb{A}_k^r(l) = l^r$. Then $t_1 - b_1, \dots, t_r - b_r$ are local parameters at x. Since $\mathrm{cd}_2^0 l = e$, we may choose $a_1, \dots, a_e \in l^\times$ such that $\langle\langle a_1, \dots, a_e \rangle\rangle$ has no nontrivial zero over l. Then by Lemma 3.3, $\langle\langle a_1, \dots, a_e, t_1 - b_1, \dots, t_r - b_r \rangle\rangle$ has no nontrivial zero over $K_0 l$. By Corollary 3.2, $\langle\langle a_1, \dots, a_e, t_1 - b_1, \dots, t_r - b_r \rangle\rangle$ has no nontrivial zero over Kl. □

Proof of Theorem 1.8. Theorem 1.4 of [Poo07] handles the case where k is finite, so assume that k is infinite. By replacing k with a finite extension k' and simultaneously replacing K with Kk' (these extensions can be interpreted over (K, k)), we may assume that K is the function field of a geometrically integral variety X over k where $\sqrt{-1} \in k$, and by Lemma 3.5 we may assume that the points of degree $\leq m$ on X are Zariski dense. Now, by the same proof as in Proposition 5.5(2), $t_1, \dots, t_r$ are algebraically independent over k if and only if there exists an extension $l|k$ of degree $\leq m$ such that there exist $a_1, \dots, a_e, b_1, \dots, b_r \in l^\times$ such that $\langle\langle a_1, \dots, a_e, t_1 - b_1, \dots, t_r - b_r \rangle\rangle$ has no nontrivial zero over Kl. The preceding statement is expressible as a certain first order formula evaluated at $t_1, \dots, t_r$. $\square$

Unfortunately, in the case char $= 2$ we do not have at our disposal an easy way to relate $\operatorname{trdeg}(K|k)$ to (some) well understood invariants (say similar to the cohomological dimension). In the case k is separably closed, one can though employ the theory of $C_i^{(p)}$ fields. Recall that a field E is said to be a $C_i^{(p)}$ field, if every system of homogeneous forms

$$f_\rho(X_1, \dots, X_n) \quad (\rho = 1, \dots, r)$$

has a non-trivial common zero, provided the degrees d_ρ of the forms satisfy: $n > \sum_\rho d_\rho^i$ and $(p, d_\rho) = 1$ for all ρ.

The following are well known facts about $C_i^{(p)}$ fields, see e.g., [Pfi95]:

1) Suppose that E is a p-field, i.e., every finite extension $E'|E$ has degree a power of p. Then E is a $C_0^{(p)}$ field.

2) If E is a $C_i^{(p)}$ field, then every finite extension $E'|E$ is again a $C_i^{(p)}$ field.

3) If E is a $C_i^{(p)}$ field, then the rational function field $E(t)$ in one variable over E is an $C_{i+1}^{(p)}$ field.

In particular, if k is a $C_i^{(p)}$ field, and $K|k$ is a function field with $\operatorname{trdeg}(K|k) = d$, then K is a $C_{i+d}^{(p)}$ field.

Now let $K|k$ be a function field. For every integer $\ell \geq 2$ and every system $\underline{t} = (t_1, \dots, t_r)$ of elements of $K^\times$, let

$$q^{(\ell)}_{(t_1,\dots,t_r)} = \sum_{\underline{i}} \underline{t}^{\underline{i}} X_{\underline{i}}^\ell$$

be the "generalized Pfister form" of degree ℓ in ℓ^r variables as introduced in [Pop02], Section 1, p. 388. Here $\underline{i}$ is a multi-index $\underline{i} = (i_1, \dots, i_r)$, with $0 \leq i_j < \ell$.

Proposition 5.6 *Let k be a p-field. Let $K|k$ be a function field. Suppose $\ell \geq 2$ and $(\ell, \mathrm{char}(K)) = (\ell, p) = 1$. Then:*

(1) *For every $r > \mathrm{trdeg}(K|k)$, and every system $(t_1, \ldots, t_r)$ of elements of $K^\times$, the form $q^{(\ell)}_{(t_1,\ldots,t_r)}$ defined above represents 0 over K.*

(2) *For a given system $(t_1, \ldots, t_r)$ the following conditions are equivalent:*

   (a) *$(t_1, \ldots, t_r)$ is algebraically independent over k.*

   (b) *there exist $b_1, \ldots, b_r \in k$ such that $q^{(\ell)}_{(t_1-b_1,\ldots,t_r-b_r)}$ does not represent 0 over K.*

(3) *In particular, for each $n \in \mathbb{Z}_{\geq 0}$ there exists a sentence in the language of fields that holds for K if and only if $\mathrm{trdeg}(K|k) = n$.*

Thus given algebraically independent elements $x_1, \ldots, x_r \in K$ over k, the relative algebraic closure L of $k(x_1, \ldots, x_r)$ in K is described by a predicate in one variable x as follows:

$$L = \{\, x \in K \mid (x_1, \ldots, x_r, x) \textit{ is not algebraically independent over } k \,\}$$

Proof of Proposition 5.6.

(1): By the discussion above, K is a $C^{(p)}_d$ field for $d = \mathrm{trdeg}(K|k)$.

(2), (b) $\Rightarrow$ (a): Let $L = k(t_1, \ldots, t_r)$. If $t_1, \ldots, t_r$ are algebraically dependent, then $\mathrm{trdeg}(L|k) < r$, so by (1), any form $q^{(\ell)}_{(t_1-b_1,\ldots,t_r-b_r)}$ represents 0 over L, and hence represents 0 over K.

(2), (a) $\Rightarrow$ (b): The proof is very similar to the proof of the corresponding implication in Proposition 5.5. The relative separable closure K_0 of $k(t_1, \ldots, t_r)$ in K is the function field of an étale cover $U \to \mathbb{A}^r_k$ with the morphism being given by $(t_1, \ldots, t_r)$. Choose $(b_1, \ldots, b_n) \in \mathbb{A}^r(k)$ and a closed point $u \in U$ above it. If l is the residue field of u, then K_0 embeds into the iterated Laurent power series field $\Lambda := l((t_r - b_r)) \ldots ((t_1 - b_1))$, and K embeds into a purely inseparable finite extension Λ' of Λ. The field Λ has a natural valuation v whose value group is $\mathbb{Z}^r$ ordered lexicographically, generated by $v(t_i - b_i)$ for $1 \leq i \leq r$. The values of the coefficients of $q^{(\ell)}_{(t_1-b_1,\ldots,t_r-b_r)}$ are distinct modulo ℓ (they even form a system of representatives for $\mathbb{Z}^r/\ell\mathbb{Z}^r$). If we extend v to a valuation v' on Λ', the value group G of v' contains $\mathbb{Z}^r$ with index prime to ℓ, so the v'-valuations of these coefficients have distinct images in $G/\ell G$. Now for any non-zero system of elements $\underline{x} = (x_i)_i$ from $K \subseteq \Lambda'$,

$q^{(\ell)}_{(t_1-b_1,\ldots,t_r-b_r)}(\underline{x})$ is a sum of elements having distinct v'-valuations (distinct even modulo ℓ). So $q^{(\ell)}_{(t_1-b_1,\ldots,t_r-b_r)}(\underline{x}) \neq 0$.

The remaining assertions of Proposition 5.6 are clear. □

Proof of Theorem 1.7.

Case 1: $\mathrm{char}(k) \neq 2$.

If k is either real closed or separably closed, then $l := k[\sqrt{-1}]$ is the unique finite separable field extension of k containing $\sqrt{-1}$. Thus the result follows from Proposition 5.5 (2).

Case 2: $\mathrm{char}(k) = 2$.

Then k is a 2-field, so it is a $C_0^{(2)}$ field. To conclude, one applies Proposition 5.6 with $p = 2$ and $\ell = 3$. □

Acknowledgments

We thank Laurent Moret-Bailly for some discussions of his paper [MB05].

References

[Abh69] S. S. Abhyankar, Resolution of singularities of algebraic surfaces, in *Algebraic Geometry (Internat. Colloq., Tata Inst. Fund. Res., Bombay, 1968)*, 1 - 11, Oxford Univ. Press, London, 1969.

[CT00] J.-L. Colliot-Thélène, Rational connectedness and Galois covers of the projective line, *Ann. of Math. (2)*, **151** (2000), no. 1, 359 - 373.

[Fal83] G. Faltings, Endlichkeitssätze für abelsche Varietäten über Zahlkörpern, *Invent. Math.*, g **73** (1983), no. 3, 349 - 366, (German); Engl. transl. Finiteness theorems for abelian varieties over number fields, in *Arithmetic geometry (Storrs, Conn., 1984)*, 9 – 27, (1986), Erratum in: *Invent. Math.* **75** (1984), 381.

[FJ05] M. D. Fried, M. Jarden, *Field arithmetic*, 2nd. ed., Ergebnisse der Mathematik und ihrer Grenzgebiete. 3. Folge. A Series of Modern Surveys in Mathematics , vol. 11, Springer-Verlag, Berlin, (2005).

[Koe02] J. Koenigsmann, Defining transcendentals in function fields, *J. Symbolic Logic*, **67** (2002), no. 3, 947 – 956.

[Lan02] S. Lang, *Algebra*, 3rd ed., Graduate Texts in Mathematics, vol. 211, Springer-Verlag, New York, 2002.

[MB05] L. Moret-Bailly, Elliptic curves and Hilbert's tenth problem for algebraic function fields over real and p-adic fields, *J. Reine Angew. Math.*, **587** (2005), 77 – 143.

[Pfi95] A. Pfister, *Quadratic forms with applications to algebraic geometry and topology*, London Mathematical Society Lecture Note Series, vol. 217, Cambridge University Press, Cambridge, 1995.

[Poo07] B. Poonen, Uniform first-order definitions in finitely generated fields, *Duke Math. J.* **138** (2007), no. 1, 1–21.

[Pop96] F. Pop, Embedding problems over large fields, *Ann. of Math. (2)*, **144** (1996), no. 1, 1 - 34.

[Pop02] F. Pop, Elementary equivalence versus isomorphism, *Invent. Math.*, **150** (2002), no.2, 385 - 408.

[Rum80] R. S. Rumely, Undecidability and definability for the theory of global fields, *Trans. Amer. Math. Soc.*, **262** (1980), no. 1, 195 - 217.

[Sam66] P. Samuel, Compléments à un article de Hans Grauert sur la conjecture de Mordell (French), *Inst. Hautes Études Sci. Publ. Math.*, **29** (1966), 55 - 62.

[Ser02] J.-P. Serre, *Galois cohomology*, Springer Monographs in Mathematics, Corrected reprint of the 1997 English edition, Springer-Verlag, Berlin, 2002.

Nonnegative solvability of linear equations in ordered Abelian groups

Philip Scowcroft†

Summary

In any ordered Abelian group, the projection of a finite intersection of generalized halfspaces is a finite intersection of generalized halfspaces. The generalized halfspaces making this result possible were introduced in [7], which showed that regular groups obey the result. Just as before, the result implies a generalization of Farkas' Lemma. The result amounts to a special quantifier-elimination theorem, which is uniform in parameters in a fashion described below.

1 Introduction

This paper generalizes, to arbitrary ordered Abelian groups, the closure under projection of the class of finite intersections of halfspaces. The result rests on a generalization, of the notion of halfspace, introduced in ([7], Section 4). Just as a halfspace is the solution set of a homogeneous weak linear inequality, so a generalized halfspace is the solution set of a so-called congruence inequality, which combines a weak linear inequality with a congruence in a special way described in Section 2. [7] uses model-theoretic arguments to show that in any regular group, the class of finite intersections of generalized halfspaces is closed under projection. The language $\mathcal{L} = \{+, -, \leq, 0\}$ of ordered Abelian groups is expanded by new predicate symbols for congruence inequalities, and [7] applies a model-theoretic test for when a formula is equivalent, modulo a given theory, to a conjunction of atomic formulas. The mathematical challenge in [7] is to extend a congruence-inequality-preserving homomorphism, from

† I am grateful to the Isaac Newton Institute for Mathematical Sciences for its support while this paper was written.

a substructure of a direct product of regular groups into a sufficiently saturated regular group, to the entire direct product.

The problem here may be attacked in the same way, though since the target group and the factors of the product are ordered Abelian groups which may not be regular, the present argument has several new features. One concerns reliance on a conservative extension, of the theory of ordered Abelian groups, in which one may discuss certain convex subgroups of an ordered Abelian group as well as their regular quotients (see Section 2). Though this new structure of convex subgroups will not be preserved by the homomorphisms under consideration, it does provide the target group with extra structure permitting arguments by induction. In the attempt to show that certain conjunctions

$$\bigwedge_{j\in J} g(A_j) + r_j u \leq^{D_j,H_j}_{g(P_j)+t_j u} g(B_j) + s_j u$$

of congruence inequalities[1] may be satisfied in the target group, one argues by induction on the number of distinct convex subgroups, represented in the new expanded language, corresponding to the elements $g(A_j)$ and $g(B_j)$. The basis of this induction would follow from the arguments of ([7], Section 4) were it not for the presence of the parameters $g(\overline{P}_j)$, which may not belong to the convex subgroups containing the $g(A_j)$'s and $g(B_j)$'s. So Section 3 extends the arguments of ([7], Section 4) to handle a target group in which a convex regular subgroup contains the $g(A_j)$'s and $g(B_j)$'s. .

This extension reveals another new aspect of the present proof. To realize a conjunction of congruence inequalities in the target group, one realizes a related conjunction in a subgroup, an image of the original conjunction in a quotient group, and then combines these two solutions to obtain a solution to the original conjunction in the original target group. Because the induction step similarly combines solutions in subgroups with solutions in quotients, the induction hypothesis must concern the solvability of systems of congruence inequalities, not merely in the original target group, but also in quotients by certain definable convex subgroups. After Section 4 treats conjunctions of congruence inequalities in a special normal form, Section 5 completes the argument and also analyzes an example revealing limits to improvements of the result.

Having shown that the existential quantification of a conjunction of

1 Again, see Section 2 for the definition of congruence inequalities $\leq^{D,H}$.

congruence inequalities is equivalent to a conjunction of congruence inequalities in any ordered Abelian group, one next notes that the coefficients in the final conjunction depend in a special way on the coefficients of the original existential formula. This result is inspired by work of van den Dries and Holly [2] on a two-sorted version of Presburger arithmetic (without order; see [10] for a related result on Presburger arithmetic). They expand the language of Abelian groups to a two-sorted language, featuring a new sort for scalars, in which the theory of torsion-free modules over Bézout domains has a universal axiomatization. They show that modulo this theory, every formula without quantifiers governing scalar variables is equivalent to a quantifier-free formula. Similarly, in an expansion of $\mathcal{L}$ to a two-sorted language featuring a sort for scalars, one may formulate axioms for certain torsion-free modules over ordered Bézout domains, and then show that if $\varphi(x_1, \ldots, x_l, \ldots, x_m, \sigma_1, \ldots, \sigma_k)$ is a conjunction of atomic formulas in the module variables x and scalar variables σ, then k-dimensional parameter space may be partitioned into finitely many quantifier-free definable pieces, over each of which $\exists x_{l+1} \ldots \exists x_m \varphi$ is equivalent to a conjunction of congruence inequalities. Because any ordered Abelian group forms an ordered $\mathbb{Z}$-module of the special kind captured by the axioms, this result reveals a special uniformity in the result proved in Sections 3-5.

It implies a version of Farkas' Lemma ([3], p. 5) for commutative ordered domains, and Section 7 indicates how the arguments of ([7], Section 4) apply in the present context.

The Conclusion, finally, mentions some open problems.

In Sections 2-3 and 7 the reader may need a copy of [7] to follow the argument. Any model-theoretic techniques exploited here are explained in [1] and in [6].

2 Preliminaries on ordered Abelian groups

Let $\mathcal{L} = \{+, -, \leq, 0\}$ be the language of ordered Abelian groups and T be the $\mathcal{L}$-theory of ordered Abelian groups. Much as in ([7], Section 4) one may expand $\mathcal{L}$ to languages $\mathcal{L}_{\text{con}}$ and $\mathcal{L}_{\text{in}}$ and T to definitional expansions T_{con} and T_{in} in these richer languages. $\mathcal{L}_{\text{con}}$ results from $\mathcal{L}$ through adjunction of binary relation symbols $\equiv_k$, for each integer $k \geq 2$, to represent congruences, and T_{con} is the expansion of T by definitions

$$\forall x, y\, (x \equiv_k y \leftrightarrow \exists z\, (x = y + kz))$$

for each $k \geq 2$. Any formula $t \equiv_k u$ is called a congruence of modulus k. $\mathcal{L}_{\text{in}}$ is the expansion of $\mathcal{L}_{\text{con}}$ by relation symbols $\leq^{D,H}$ corresponding to so-called basic pairs (D, H) of $\mathcal{L}_{\text{con}}$-formulas of the following kind. When $\overline{x} = (x_1, \ldots, x_m)$ and $\overline{y} = (y_1, \ldots, y_n)$ are disjoint lists of distinct variables, a basic pair (D, H) consists of a conjunction $D(\overline{x})$ of inequalities

$$x_i \leq x_j$$

—where the graph $\{(i,j) : x_i \leq x_j \text{ is a conjunct of } D\}$ has no cycles—and $H(\overline{x}; \overline{y})$ is a conjunction of congruences between $\mathcal{L}_{\text{con}}$-terms in the variables from $\overline{x}$ and $\overline{y}$. $\leq^{D,H}$ is an $(n+2)$-place relation symbol, and T_{in} is the $\mathcal{L}_{\text{in}}$-theory obtained from T_{con} by adding all axioms

$$\forall u, v, \overline{y}\, (u \leq^{D,H}_{\overline{y}} v \leftrightarrow \exists \overline{x}\, (u \leq \overline{x} \leq v \ \wedge\ D(\overline{x}) \ \wedge\ H(\overline{x}; \overline{y}))).$$

Any formula $r \leq^{D,H}_{\overline{t}} s$ is called a congruence inequality.

Let

$$\mathcal{L}^+ = \mathcal{L}_{\text{in}} \cup \{L, U\},$$

where L and U are new binary relation symbols. In what follows $L(s,t)$ ($U(s,t)$) will be written $s \in L_t$ ($s \in U_t$), and L_t (U_t) will be used to abbreviate $\{s : s \in L_t\}$ ($\{s : s \in U_t\}$). T^+ is the $\mathcal{L}^+$-theory which extends T_{in} by axioms guaranteeing that

$\forall x\, (L_x \text{ and } U_x \text{ are convex subgroups})$
$\forall x\, (L_x \subseteq U_x)$
$\forall x\, (x \in U_x \ \wedge\ (x \in L_x \rightarrow x = 0))$
$U_0 = \{0\} = L_0$
$\forall x\, (U_x / L_x \text{ is } n\text{-regular})$ (for each $n \geq 2$)
$\forall x, y\, [(L_x \subset U_y \rightarrow U_x \subseteq U_y) \ \wedge\ (L_x \subset L_y \rightarrow U_x \subseteq L_y)$
$\qquad \wedge\ (L_x \subset U_y \rightarrow L_x \subseteq L_y) \ \wedge\ (U_x \subset U_y \rightarrow U_x \subseteq L_y)]$

(an ordered Abelian group is n-regular just in case every interval with at least n elements contains an n-divisible element).

Proposition 2.1 *T^+ is a conservative extension of T_{in}.*

Proof One need show merely that every $\mathcal{M} \models T_{\text{in}}$ may be expanded to a model $\mathcal{M}^+$ of T^+. Let

$$U_0^{\mathcal{M}^+} = \{0^{\mathcal{M}}\} = L_0^{\mathcal{M}^+},$$

and for $m \neq 0$ in M let $U_m^{\mathcal{M}^+}$ ($L_m^{\mathcal{M}^+}$) be the smallest (largest) convex subgroup of M that does (not) contain m. Then each $U_m^{\mathcal{M}^+} / L_m^{\mathcal{M}^+}$ lacks nontrivial convex subgroups, and so is Archimedean and thus n-regular for all $n \geq 2$ ([9], pp. 58–59). □

The target of the homomorphism discussed in the Introduction will be a sufficiently saturated model of T^+, and the next few results show that such a model may be regarded as a lexicographic sum. First, a purely group-theoretic result.

Lemma 2.2 *Let $\mathcal{H}$ be an infinite Abelian group $\aleph_1$-saturated with respect to a language containing the language $\mathcal{L}_g = \{+, -, 0\}$ of Abelian groups. If $G \subseteq H$ is a pure subgroup definable in $\mathcal{H}$ and $\pi : H \to H/G$ is the corresponding quotient homomorphism, then there is a subgroup S of H such that $H = S \oplus G$ as groups and $\pi \restriction S$ is a group isomorphism between S and H/G.*

Proof If $\mathcal{G}$ is the substructure of $\mathcal{H} \restriction \mathcal{L}_g$ with domain G, then $\mathcal{G}$ is $\aleph_1$-saturated and also pp-normal in the sense of ([6], p. 530). Thus $\mathcal{G}$ is atomic compact ([6], Theorem 10.7.3). Since the inclusion of $\mathcal{G}$ in $\mathcal{H} \restriction \mathcal{L}_g$ is pure ([6], p. 528), there is a homomorphism f of $\mathcal{H} \restriction \mathcal{L}_g$ onto $\mathcal{G}$ which is the identity on G ([6], Theorem 10.7.1(e)). $\ker f$ may serve as S. □

Now for ordered Abelian groups:

Lemma 2.3 *Let $\mathcal{H}$ and G be as in Lemma 2.2. If $\mathcal{H}$ is an ordered group, the language of $\mathcal{H}$ contains the language $\mathcal{L}$ of ordered Abelian groups, and G is convex in H, then as an ordered group H is the lexicographic sum[2] $S \overrightarrow{\times} G$, and $\pi \restriction S$ is an isomorphism of the ordered group S onto the ordered group H/G.*

Proof Because π is a homomorphism of ordered groups, the group isomorphism $\pi \restriction S$ is order preserving, and so is an isomorphism of ordered groups.

Suppose $y \in S$, $z \in G$, and $x = y + z$. If $x > 0$ but $y \neq 0$, then $y > 0$: otherwise, since $y \notin G$, which is convex in H, $y + z < 0$. Conversely, if $y > 0$, then since z belongs to the convex subgroup G of H while y does not, $x = y + z > 0$; while if $y = 0$ but $z > 0$, then of course $x = y + z > 0$. Thus H and $S \overrightarrow{\times} G$ have the same positive elements, and so have the same ordering. □

2 $S \overrightarrow{\times} G$ is ordered so that S comes first: i.e. for $s, s' \in S$ and $g, g' \in G$, $s+g \leq s'+g'$ just in case $s < s'$ or both $s = s'$ and $g \leq g'$.

Finally, one may consider what happens to the structure represented by $\mathcal{L}^+ - \mathcal{L}$ when one starts with a model of T^+:

Lemma 2.4 *Let $\mathcal{H}$ and G be as in Lemma 2.3. If $\mathcal{H} \models T^+$, G is $L_a^{\mathcal{H}}$ or $U_a^{\mathcal{H}}$ for some $a \in H$, and $\mathcal{S}$ is the substructure of $\mathcal{H}$ with domain S, then one may expand the ordered group H/G to an $\mathcal{L}^+$-structure isomorphic via $(\pi \restriction S)^{-1}$ to $\mathcal{S}$, and both are models of T^+.*

Proof First one expands H/G to a model $\mathcal{H}/G$ of T^+ so that π is an $\mathcal{L}^+$-homomorphism, and then one shows that $\pi \restriction S$ is an $\mathcal{L}^+$-isomorphism of $\mathcal{S}$ onto $\mathcal{H}/G$.

Because atomic formulas built from relation symbols of $\mathcal{L}_{\text{in}} - \mathcal{L}$ are equivalent mod T_{in} to positive existential $\mathcal{L}$-formulas, the natural interpretations in the ordered group H/G of the symbols in $\mathcal{L}_{\text{in}} - \mathcal{L}$ make π an $\mathcal{L}_{\text{in}}$-homomorphism. Thus $\pi \restriction S$ is an $\mathcal{L}_{\text{in}}$-homomorphism; and since $(\pi \restriction S)^{-1}$ is a homomorphism of ordered groups, it is also an $\mathcal{L}_{\text{in}}$-homomorphism. So $\pi \restriction S$ is an $\mathcal{L}_{\text{in}}$-isomorphism.

If $x \in H$, then $L_x^{\mathcal{H}}$ and G are convex subgroups of H, one is contained in the other, and $L_x^{\mathcal{H}} + G$ is a convex subgroup of H that contains G. So $(L_x^{\mathcal{H}} + G)/G = \pi(L_x^{\mathcal{H}})$ is a convex subgroup of H/G.

If $y \in H$ and $\pi(x) = \pi(y)$, one may show as follows that $\pi(L_x^{\mathcal{H}}) = \pi(L_y^{\mathcal{H}})$. Since $\pi(x) = \pi(y)$, $x - y \in G$; and without loss of generality $x \neq y$. If $x = 0$, $y \in G - \{0\}$ and $y \notin L_y^{\mathcal{H}}$: so the convex subgroup $L_y^{\mathcal{H}}$ is contained in the convex subgroup G, and

$$\pi(L_x^{\mathcal{H}}) = \{0\} = \pi(G) = \pi(L_y^{\mathcal{H}}).$$

Suppose now that neither x nor y is 0. If either belongs to G, the previous argument yields the desired conclusion: so, assume that neither belongs to G. If $G = L_a^{\mathcal{H}}$, then by the axioms for T^+

$$L_x^{\mathcal{H}} \subset G = L_a^{\mathcal{H}} \Rightarrow x \in U_x^{\mathcal{H}} \subseteq L_a^{\mathcal{H}} = G:$$

so $L_x^{\mathcal{H}} \not\subset G$ and $G \subseteq L_x^{\mathcal{H}}$. If $G = U_a^{\mathcal{H}}$, then by the axioms for T^+

$$L_x^{\mathcal{H}} \subset G = U_a^{\mathcal{H}} \Rightarrow x \in U_x^{\mathcal{H}} \subseteq U_a^{\mathcal{H}} = G:$$

so $L_x^{\mathcal{H}} \not\subset G$ and $G \subseteq L_x^{\mathcal{H}}$. Thus $G \subseteq L_x^{\mathcal{H}}$, and a similar argument shows that $G \subseteq L_y^{\mathcal{H}}$. If $L_x^{\mathcal{H}} \subset L_y^{\mathcal{H}}$, then $x \in U_x^{\mathcal{H}} \subseteq L_y^{\mathcal{H}}$ by the axioms for T^+, and since $x - y \in G \subseteq L_y^{\mathcal{H}}$, $y \in L_y^{\mathcal{H}}$ and $y = 0$, contrary to hypothesis. Thus $L_x^{\mathcal{H}} \not\subset L_y^{\mathcal{H}}$, $L_y^{\mathcal{H}} \not\subset L_x^{\mathcal{H}}$ by a similar argument, and $L_x^{\mathcal{H}} = L_y^{\mathcal{H}}$, from which the desired conclusion follows.

Similar arguments show that when $x \in H$, $U_x^{\mathcal{H}} + G$ is a convex subgroup of H that contains G, $(U_x^{\mathcal{H}}+G)/G = \pi(U_x^{\mathcal{H}})$ is a convex subgroup of H/G, and $\pi(U_x^{\mathcal{H}}) = \pi(U_y^{\mathcal{H}})$ when $\pi(x) = \pi(y)$. One may therefore interpret L and U in H/G by letting

$$L_{\pi(x)}^{\mathcal{H}/G} = \pi(L_x^{\mathcal{H}}) \text{ and } U_{\pi(x)}^{\mathcal{H}/G} = \pi(U_x^{\mathcal{H}})$$

for $x \in H$, and this definition makes π an $\mathcal{L}^+$-homomorphism.

The next challenge is to show that $\mathcal{H}/G \models T^+$. Each $L_{\pi(x)}^{\mathcal{H}/G}$ and $U_{\pi(x)}^{\mathcal{H}/G}$ is a convex subgroup of H/G, and since $L_x^{\mathcal{H}} \subseteq U_x^{\mathcal{H}}$ and $x \in U_x^{\mathcal{H}}$,

$$L_{\pi(x)}^{\mathcal{H}/G} = \pi(L_x^{\mathcal{H}}) \subseteq \pi(U_x^{\mathcal{H}}) = U_{\pi(x)}^{\mathcal{H}/G}$$

and $\pi(x) \in \pi(U_x^{\mathcal{H}}) = U_{\pi(x)}^{\mathcal{H}/G}$. If $\pi(x) \in L_{\pi(x)}^{\mathcal{H}/G} = \pi(L_x^{\mathcal{H}})$, then there is $y \in L_x^{\mathcal{H}}$ with $x - y \in G$. If $x \notin G$, a convex subgroup, then $G \subset U_x^{\mathcal{H}}$, and since G is of the form $L_a^{\mathcal{H}}$ or $U_a^{\mathcal{H}}$, $G \subseteq L_x^{\mathcal{H}}$ by the axioms of T^+, $x = (x - y) + y \in L_x^{\mathcal{H}}$, and $x = 0 \in G$: contradiction. Thus $\pi(x) = 0$ if $\pi(x) \in L_{\pi(x)}^{\mathcal{H}/G}$. So $L_{\pi(0)}^{\mathcal{H}/G} = \{0\}$ and $U_{\pi(0)}^{\mathcal{H}/G} = \pi(U_0^{\mathcal{H}}) = \pi(\{0\}) = \{0\}$. When verifying the n-regularity of $U_{\pi(x)}^{\mathcal{H}/G}/L_{\pi(x)}^{\mathcal{H}/G}$ one may therefore assume that $\pi(x) \neq 0$: i.e., $x \notin G$. But in this case the convex subgroup G is properly contained in $U_x^{\mathcal{H}}$, and so contained in $L_x^{\mathcal{H}}$ because G is of the form $L_a^{\mathcal{H}}$ or $U_a^{\mathcal{H}}$, and

$$\frac{U_{\pi(x)}^{\mathcal{H}/G}}{L_{\pi(x)}^{\mathcal{H}/G}} = \frac{(U_x^{\mathcal{H}}+G)/G}{(L_x^{\mathcal{H}}+G)/G} = \frac{U_x^{\mathcal{H}}/G}{L_x^{\mathcal{H}}/G} \cong \frac{U_x^{\mathcal{H}}}{L_x^{\mathcal{H}}}$$

is n-regular. When checking that $U_{\pi(x)}^{\mathcal{H}/G} \subseteq U_{\pi(y)}^{\mathcal{H}/G}$ if $L_{\pi(x)}^{\mathcal{H}/G} \subset U_{\pi(y)}^{\mathcal{H}/G}$, one may again assume that $\pi(x) \neq 0$: but then $\pi(y) \neq 0$, and as above $G \subseteq L_x^{\mathcal{H}} \subseteq U_x^{\mathcal{H}}$ and $G \subseteq L_y^{\mathcal{H}} \subseteq U_y^{\mathcal{H}}$. Because π establishes a bijective inclusion-preserving correspondence between the subgroups of H containing G and the subgroups of H/G, $\pi(L_x^{\mathcal{H}}) \subset \pi(U_y^{\mathcal{H}})$ implies that $L_x^{\mathcal{H}} \subset U_y^{\mathcal{H}}$, which implies $U_x^{\mathcal{H}} \subseteq U_y^{\mathcal{H}}$ by the axioms of T^+; so $U_{\pi(x)}^{\mathcal{H}/G} = \pi(U_x^{\mathcal{H}}) \subseteq \pi(U_y^{\mathcal{H}}) = U_{\pi(y)}^{\mathcal{H}/G}$. Similar arguments complete the demonstration that $\mathcal{H}/G \models T^+$.

Because $\pi \restriction S$ is an $\mathcal{L}_{\text{in}}$-isomorphism and an $\mathcal{L}^+$-homomorphism, it will be an $\mathcal{L}^+$-isomorphism just in case for all $x, y \in S$

$$\pi(x) \in L_{\pi(y)}^{\mathcal{H}/G} \Rightarrow x \in L_y^{\mathcal{H}}$$

and

$$\pi(x) \in U_{\pi(y)}^{\mathcal{H}/G} \Rightarrow x \in U_y^{\mathcal{H}}.$$

If $\pi(x) \in L^{\mathcal{H}/G}_{\pi(y)}$, then there is $z \in G$ with $x - z \in L^{\mathcal{H}}_y$. If $y = 0$, then $x - z \in L^{\mathcal{H}}_0 = \{0\}$, $x = z \in G$, and $x \in S \cap G = \{0\}$: so $x \in L^{\mathcal{H}}_y$. If $y \neq 0$, then since $y \in S$ and $H = S \overrightarrow{\times} G$, $|y| > G$, $G \subset U^{\mathcal{H}}_y$, and so $G \subseteq L^{\mathcal{H}}_y$ because G is of the form $L^{\mathcal{H}}_a$ or $U^{\mathcal{H}}_a$; thus

$$x = z + (x - z) \in G + L^{\mathcal{H}}_y = L^{\mathcal{H}}_y.$$

A similar argument establishes the other conditional and completes the proof of the Lemma. □

Given Proposition 2.1, one may show that $\mathcal{L}_{\text{in}}$-formulas are equivalent modulo T_{in} by showing that they are equivalent modulo T^+. To avoid showing merely that the existential quantification of a conjunction of atomic $\mathcal{L}_{\text{in}}$-formulas is equivalent to a conjunction of atomic $\mathcal{L}^+$-formulas, one may generalize the model-theoretic test of [7] as follows.

Lemma 2.5 *Let $\mathcal{L}$ be a first-order language, T be a set of $\mathcal{L}$-sentences, $\varphi(x_1, \ldots, x_n)$ be an $\mathcal{L}$-formula, with free variables among those displayed, such that $T \models \exists \overline{x} \varphi(\overline{x})$, and $\mathcal{L}' \subseteq \mathcal{L}$ contain at least one constant symbol. The following conditions are equivalent:*

(i) *$\varphi(\overline{x})$ is equivalent, modulo T, to a conjunction of atomic $\mathcal{L}'$-formulas.*

(ii) *Suppose that for all i in the nonempty set I, there is an n-tuple $\overline{a}_i$ from $\mathcal{M}_i \models T$ with $\mathcal{M}_i \models \varphi[\overline{a}_i]$. If $\mathcal{N} \subseteq \prod_{i \in I} \mathcal{M}_i$, $\overline{a} = (\overline{a}_i)_{i \in I}$ belongs to N^n, $\mathcal{H} \models T$, and $g : \mathcal{N} \upharpoonright \mathcal{L}' \to \mathcal{H} \upharpoonright \mathcal{L}'$ is a homomorphism, then $\mathcal{H} \models \varphi[g(\overline{a})]$.*

Proof The argument follows that for Lemma 2.1 in [7]. The only changes are as follows: Γ is the set of atomic $\mathcal{L}'$-formulas, with free variables among $x_1, \ldots, x_n$, that are implied by $\varphi(\overline{x})$, modulo T; Γ' is the set of atomic $\mathcal{L}'$-formulas, with free variables among $x_1, \ldots, x_n$, that do not belong to Γ; the third paragraph discusses atomic $\mathcal{L}'$-formulas $\delta(\overline{x})$; and in the fourth paragraph, the constant symbol d comes from $\mathcal{L}'$. □

Eventually Lemma 2.5 will be applied to $\mathcal{L} = \mathcal{L}^+$, $\mathcal{L}' = \mathcal{L}_{\text{in}}$, and $\varphi(\overline{x}) = \exists \overline{y} \psi(\overline{x}, \overline{y})$, where ψ is a conjunction of atomic $\mathcal{L}_{\text{in}}$-formulas.

3 (g, a)-systems

The next two sections will discuss how the solvability of certain finite systems of congruence inequalities may be transferred between $\mathcal{L}^+$-

structures connected by an $\mathcal{L}_{\text{in}}$-homomorphism. For all i in the nonempty index set I, let $\mathcal{M}_i \models T^+$, $\mathcal{N} \subseteq \prod_{i\in I} \mathcal{M}_i = \mathcal{M}$, $\mathcal{H} \models T^+$ be $\aleph_1$-saturated, and $g : \mathcal{N} \restriction \mathcal{L}_{\text{in}} \to \mathcal{H} \restriction \mathcal{L}_{\text{in}}$ be a homomorphism. For each j in the finite set J, let $\psi_j(u)$ be the congruence inequality

$$A_j + r_j u \leq^{D_j, H_j}_{P_j + t_j u} B_j + s_j u,$$

where the A's, B's, and P's come from $\mathcal{N}$ and the r's, s's, and t's come from $\mathbb{Z}^3$. The g-image of $\Psi = \{\psi_j(u) : j \in J\}$ is $g(\Psi) = \{g(\psi_j(u)) : j \in J\}$, where $g(\psi_j(u))$ is

$$g(A_j) + r_j u \leq^{D_j, H_j}_{g(P_j) + t_j u} g(B_j) + s_j u.$$

Definition 3.1 If $a \in H$, $\{\psi_j(u) : j \in J\}$ is a (g, a)-system just in case $\{\psi_j(u) : j \in J\}$ is satisfiable in $\mathcal{M}$ and every $g(A_j)$ and $g(B_j)$ belongs to $U^{\mathcal{H}}_a$.

The rest of this section is devoted to the proof of

Proposition 3.2 *If $a \in H$ and $L^{\mathcal{H}}_a = \{0\}$, then the g-image of any (g, a)-system is satisfiable in $\mathcal{H}$.*

Proof Because $U^{\mathcal{H}}_u \cong U^{\mathcal{H}}_u / \{0\} = U^{\mathcal{H}}_a / L^{\mathcal{H}}_a$ is regular, the desired conclusion generalizes Lemmas 3.2, 3.4, and 4.3-4.6 of [7]. But one cannot simply quote the earlier results here because they would assume that the parameters $g(\overline{P}_j)$ belong to $U^{\mathcal{H}}_a$. Since much of the argument for Lemmas 4.3-4.6 of [7] applies in the present context, the rest of the proof will point out just the changes needed in the earlier arguments. Though the most important changes handle parameters $g(\overline{P}_j)$ not necessarily in $U^{\mathcal{H}}_a$, minor changes also are needed to handle simultaneously the following two cases: $U^{\mathcal{H}}_a$ is a $\mathbb{Z}$-group (treated explicitly in [7], Lemmas 4.3-4.6); $U^{\mathcal{H}}_a$ is dense regular.

Assume now that the system Ψ given above is a (g, a)-system, and let $U \in M$ satisfy Ψ in $\mathcal{M}$.

Lemma 3.3 *If $r_j = s_j = 0$ for all $j \in J$, then some $\mathbb{U} \in H$ satisfies $g(\Psi)$.*

Proof Follow the proof of Lemma 4.3 in [7], with changes as noted below.

At the first mention of the Chinese Remainder Theorem, note that one

3 Useful notational conventions about congruence inequalities are described in Note 5 of ([7], p. 3550).

may exploit it, in any torsion-free Abelian group, to eliminate existential quantifiers applied to conjunctions of congruences. The applications are not limited, as in Section 4 of [7], to (products of) $\mathbb{Z}$-groups.

In the definition of J_∞, replace "is infinite" by "> 0" when $U_a^{\mathcal{H}}$ is dense regular.

Note that $M = g(\Delta)$ and the $g(\Delta_i)$'s are 0 when $U_a^{\mathcal{H}}$ is dense regular. In this case the witnesses $\overline{\mathbb{X}}_i$, $\overline{\mathbb{X}}_i'$ introduced slightly later are zero.

When witnesses $\overline{\mathbb{X}}_i$ are found in $\mathcal{H}$ to the truth of K', one must argue more carefully to show that they may be chosen in certain intervals in a certain order: for the parameters $g(\overline{P}_i)$ (with $i \in J$) in K' may not belong to the regular group $U_a^{\mathcal{H}}$. Since $\mathcal{H}$ is $\aleph_1$-saturated, one may apply Lemma 2.4 to $\mathcal{H}$ and $G = U_a^{\mathcal{H}}$ to obtain $\mathcal{S} \subseteq \mathcal{H}$. If $\pi_1 : H \to S$ and $\pi_2 : H \to G$ are the group-theoretic homomorphisms of H onto the lexicographic summands S and G, then π_1 is an $\mathcal{L}^+$-homomorphism, $\pi_1 \circ g : \mathcal{N} \restriction \mathcal{L}_{\text{in}} \to \mathcal{S} \restriction \mathcal{L}_{\text{in}}$, and Ψ is a $(\pi_1 \circ g, 0)$-system. Repeating the argument made so far, but with $\pi_1 \circ g : \mathcal{N} \restriction \mathcal{L}_{\text{in}} \to \mathcal{S} \restriction \mathcal{L}_{\text{in}}$ rather than $g : \mathcal{N} \restriction \mathcal{L}_{\text{in}} \to \mathcal{H} \restriction \mathcal{L}_{\text{in}}$, one finds that the new set $J_\infty^{\mathcal{S}}$ is empty and that in $\mathcal{S} \subseteq \mathcal{H}$,

$$K'(\overline{0}_i; \pi_1 \circ g(\overline{P}_i), 0)_{i\in J},$$

where $\overline{0}_i$ is a zero vector with the same number of entries as $\overline{\mathbb{X}}_i$. Because the system of congruences

$$K'(\overline{\mathbb{X}}_i; g(\overline{P}_i), g(A_i))_{i\in J}$$

is true in $\mathcal{H}$ and $\pi_2 : H \to G$ is a group homomorphism,

$$K'(\pi_2(\overline{\mathbb{X}}_i); \pi_2 \circ g(\overline{P}_i), \pi_2 \circ g(A_i))_{i\in J};$$

and when $i \in J_{<\infty}$

$$\pi_2(\overline{\mathbb{X}}_i) = \overline{\mathbb{X}}_i$$

because $g(A_i) \leq \overline{\mathbb{X}}_i \leq g(B_i)$ and $g(A_i), g(B_i) \in U_a^{\mathcal{H}}$. Thus

$$K'(\overline{0}_i + \pi_2(\overline{\mathbb{X}}_i); \pi_1 \circ g(\overline{P}_i) + \pi_2 \circ g(\overline{P}_i), 0 + \pi_2 \circ g(A_i))_{i\in J}.$$

The definition of π_1 and π_2 makes

$$\pi_1 \circ g(\overline{P}_i) + \pi_2 \circ g(\overline{P}_i) = g(\overline{P}_i),$$

and $\pi_2 \circ g(A_i) = g(A_i)$ because $g(A_i) \in U_a^{\mathcal{H}} = G$. Thus

$$K'(\pi_2(\overline{\mathbb{X}}_i); g(\overline{P}_i), g(A_i))_{i\in J},$$

where each $\pi_2(\overline{\mathbb{X}}_i)$ comes from G, and $\pi_2(\overline{\mathbb{X}}_i) = \overline{\mathbb{X}}_i$ when $i \in J_{<\infty}$.

For $i \in J_\infty$, one may therefore find elements in the infinite interval $(g(A_i), g(B_i))$ of the regular group $U_a^{\mathcal{H}} = G$ congruent to the entries of the vector $\pi_2(\overline{\mathbb{X}}_i)$ from G, and in the order dictated by D_i. So as in [7], the argument is complete. □

By repeating the argument for Lemma 4.4 of [7], one concludes that

Lemma 3.4 *If every $r_i = s_i$, then some $\mathbb{U} \in H$ satisfies $g(\Psi)$ in $\mathcal{H}$.*

The earlier argument needs some supplementation when one proves

Lemma 3.5 *If no two $r_i - s_i$'s have strictly different signs, then some $\mathbb{U} \in H$ satisfies $g(\Psi)$ in $\mathcal{H}$.*

Proof Proceed as in the proof of Lemma 4.5 from [7] through the demonstration that

$$\bigwedge_{l \in J_+} H_l(\overline{\mathbb{X}}_l + s_l\mathbb{U}; \overline{g(P_l) + t_l\mathbb{U}}).$$

Let P be the product of the moduli in these congruences. When $i \in J_+$, one may add P-divisible elements to the entries of $\overline{\mathbb{X}}_i$ to guarantee that they are ordered as D_i demands and that all are at most $g(D_i)$. Because $r_i - s_i > 0$ when $i \in J_+$, one may then subtract from $\mathbb{U}$ a nonnegative P-divisible element to obtain $\mathbb{V} \in H$ with

$$g(A_i) + (r_i - s_i)\mathbb{V} \leq \overline{\mathbb{X}}_i \text{ for all } i \in J_+;$$

if at the same time one replaces each $\overline{\mathbb{X}}_j$ with $j \in J_=$ by $\mathbb{X}_j + r_j(\mathbb{V} - \mathbb{U})$, this new vector still serves as a witness for $g(\psi_j(u))$. So the desired conclusion holds. □

More changes are needed in the proof of

Lemma 3.6 *If at least two $r_i - s_i$'s have strictly different signs, then some $\mathbb{U} \in H$ satisfies $g(\Psi)$ in $\mathcal{H}$.*

Proof Proceed as in the proof of Lemma 4.6 in [7] until the examination of the case in which $\mathfrak{M} - \mathfrak{m}$ is infinite (and note that when $U_a^{\mathcal{H}}$ is dense regular, "$\mathfrak{M} - \mathfrak{m}$ is infinite" means that $\mathfrak{M} - \mathfrak{m} > 0$). Use the Chinese Remainder Theorem to find a conjunction $K(\overline{x}_i; \overline{p}_j, a_j, b_j, u)_{i \in J_=, j \in J}$ of congruences equivalent to

$$\exists(\overline{x}_l)_{l\in J_-\cup J_+}\Big(\bigwedge_{j\in J_-} H_j(\overline{x}_j+(b_j+r_ju);\overline{p_j+t_ju}) \\ \wedge\bigwedge_{j\in J_+} H_j(\overline{x}_j+(a_j+s_ju);\overline{p_j+t_ju}) \\ \wedge\bigwedge_{j\in J_=} H_j(\overline{x}_j+r_jNu;\overline{p_j+t_jNu})\Big)$$

in any torsion-free Abelian group. When $i\in J_=$ there is $\widetilde{X}_i$ from $\mathcal{M}$ with

$$A_i+r_iNU\le\widetilde{X}_i\le B_i+r_iNU\ \wedge\ D_i(\widetilde{X}_i)\ \wedge\ H_i(\widetilde{X}_i;\overline{P_i+t_iNU});$$

letting $\widehat{X}_i=\widetilde{X}_i-r_iNU$, one finds that

$$\bigwedge_{i\in J_=} A_i\le\widehat{X}_i\le B_i\ \wedge\ D_i(\widehat{X}_i)\ \wedge\ K(\widehat{X}_j;\overline{P}_l,A_l,B_l,U)_{j\in J_=,l\in J}.$$

So by the proof of Lemma 3.3 there are $\mathbb{U}$ and $\overline{\mathbb{X}}_i$ from H, for $i\in J_=$, with

$$\bigwedge_{i\in J_=} g(A_i)\le\overline{\mathbb{X}}_i\le g(B_i)\ \wedge\ D_i(\overline{\mathbb{X}}_i)\ \wedge \\ K(\overline{\mathbb{X}}_j;g(\overline{P}_l),g(A_l),g(B_l),\mathbb{U})_{j\in J_=,l\in J}.$$

Thus the $\overline{\mathbb{X}}_i$ come from $U_a^{\mathcal{H}}$. By the choice of K there are $\overline{\mathbb{X}}_i$ from H, for $i\in J_-\cup J_+$, with

$$\bigwedge_{j\in J_-} H_j(\overline{\mathbb{X}}_j+(g(B_j)+r_j\mathbb{U});\overline{g(P_j)+t_j\mathbb{U}}) \\ \wedge\bigwedge_{j\in J_+} H_j(\overline{\mathbb{X}}_j+(g(A_j)+s_j\mathbb{U});\overline{g(P_j)+t_j\mathbb{U}}) \\ \wedge\bigwedge_{j\in J_=} H_j(\overline{\mathbb{X}}_j+r_jN\mathbb{U};\overline{g(P_j)+t_jN\mathbb{U}}).$$

Apply Lemma 2.4 to $\mathcal{H}$ and $G=U_a^{\mathcal{H}}$ to obtain $\mathcal{S}\subseteq\mathcal{H}$ and the projections $\pi_1:H\to S$ and $\pi_2:H\to G$ as in the proof of Lemma 3.3. Because $\pi_1\circ g:\mathcal{N}\restriction\mathcal{L}_{\text{in}}\to\mathcal{S}\restriction\mathcal{L}_{\text{in}}$, Ψ is a $(\pi_1\circ g,0)$-system, to which one may apply the argument of Lemma 4.6 from [7]. In the present context $J^{\mathcal{S}}_{\pm}=J_\pm$, $J^{\mathcal{S}}_{=}=J_=$, and $N^{\mathcal{S}}=N$, but $\mathfrak{M}^{\mathcal{S}}=\mathfrak{m}^{\mathcal{S}}$ and $(J^\infty)^{\mathcal{S}}=\emptyset$: so one skips the "$\mathfrak{M}^{\mathcal{S}} - \mathfrak{m}^{\mathcal{S}}$ is infinite" case, and $\mathfrak{L}^{\mathcal{S}} = \mathfrak{U}^{\mathcal{S}} = \mathfrak{P}^{\mathcal{S}} = J_- \times J_+ = \mathfrak{P}$. Fixing $(i_0, j_0) \in \mathfrak{P}$, one sees that

$$\begin{aligned}
\bigwedge_{(i,j)\in\mathfrak{P}} & 0 \le \overline{X}_i - Q_{(i,j)} \le V - Q_{(i,j)} \le \overline{X}_j - Q_{(i,j)} \le R_{(i,j)} - Q_{(i,j)} \\
& \wedge\ D_i(\overline{X}_i - Q_{(i,j)}) \ \wedge\ D_j(\overline{X}_j - Q_{(i,j)}) \\
& \wedge\ H_i^*((\overline{X}_i - Q_{(i,j)}) + Q_{(i,j)}, (V - Q_{(i,j)}) + Q_{(i,j)}; \overline{P}_i, W_i) \\
& \wedge\ H_j^*((\overline{X}_i - Q_{(i,j)}) + Q_{(i,j)}, (V - Q_{(i,j)}) + Q_{(i,j)}; \overline{P}_j, W_j) \\
& \wedge\ (V - Q_{(i,j)}) + Q_{(i,j)} \equiv_N 0 \\
\wedge \bigwedge_{k\in J_=} & 0 \le \overline{X}_k \le B_k - A_k \ \wedge\ D_k(\overline{X}_k) \\
& \wedge\ H_k^*(\overline{X}_k, (V - Q_{(i_0,j_0)}) + Q_{(i_0,j_0)}; \overline{P}_k, W_k).
\end{aligned}$$

Letting

$$\begin{aligned}
\overline{X}_1^{(i,j)} &= \overline{X}_i - Q_{(i,j)} \\
\overline{X}_2^{(i,j)} &= \overline{X}_j - Q_{(i,j)} \\
V^{(i,j)} &= V - Q_{(i,j)}
\end{aligned}$$

for $(i,j) \in \mathfrak{P}$, one may write

$$\begin{aligned}
\bigwedge_{(i,j)\in\mathfrak{P}} & 0 \le \overline{X}_1^{(i,j)} \le V^{(i,j)} \le \overline{X}_2^{(i,j)} \le R_{(i,j)} - Q_{(i,j)} \\
& \wedge\ D_i(\overline{X}_1^{(i,j)}) \ \wedge\ D_j(\overline{X}_2^{(i,j)}) \\
& \wedge\ H_i^*(\overline{X}_1^{(i,j)} + Q_{(i,j)}, V^{(i,j)} + Q_{(i,j)}; \overline{P}_i, W_i) \\
& \wedge\ H_j^*(\overline{X}_2^{(i,j)} + Q_{(i,j)}, V^{(i,j)} + Q_{(i,j)}; \overline{P}_j, W_j) \\
& \wedge\ V^{(i,j)} + Q_{(i,j)} \equiv_N 0 \\
\wedge \bigwedge_{k\in J_=} & 0 \le \overline{X}_k \le B_k - A_k \ \wedge\ D_k(\overline{X}_k) \\
& \wedge\ H_k^*(\overline{X}_k, V^{(i_0,j_0)} + Q_{(i_0,j_0)}; \overline{P}_k, W_k).
\end{aligned}$$

For $(i,j) \in \mathfrak{P}$ let $\overline{x}_1^{(i,j)}$, $\overline{x}_2^{(i,j)}$ be lists of distinct new variables—with the same number of entries as $\overline{x}_i$, $\overline{x}_j$, respectively—and $v^{(i,j)}$, $q_{(i,j)}$ be new variables. Let $D(\overline{x}_1^{(i,j)}, \overline{x}_2^{(i,j)}, v^{(i,j)}, \overline{x}_k)_{(i,j)\in\mathfrak{P}, k\in J_=}$ be the formula

$$\bigwedge_{(i,j)\in\mathfrak{P}} D_i(\overline{x}_1^{(i,j)}) \wedge \overline{x}_1^{(i,j)} \le v^{(i,j)} \le \overline{x}_2^{(i,j)} \wedge D_j(\overline{x}_2^{(i,j)}) \wedge \bigwedge_{k\in J_=} D_k(\overline{x}_k),$$

$H(\overline{x}_1^{(i,j)}, \overline{x}_2^{(i,j)}, \overline{x}_k, v^{(i,j)}; \overline{p}_i, \overline{p}_j, w_i, w_j, q_{(i,j)})_{(i,j)\in\mathfrak{P}, k\in J_=}$ be the formula

$$\begin{aligned}\bigwedge_{(i,j)\in\mathfrak{P}} & H_i^*(\overline{x}_1^{(i,j)} + q_{(i,j)}, v^{(i,j)} + q_{(i,j)}; \overline{p}_i, w_i) \\ & \wedge\ H_j^*(\overline{x}_2^{(i,j)} + q_{(i,j)}, v^{(i,j)} + q_{(i,j)}; \overline{p}_j, w_j) \\ & \wedge\ v^{(i,j)} + q_{(i,j)} \equiv_N 0 \\ \wedge \bigwedge_{k\in J_=} & H_k^*(\overline{x}_k, v^{(i_0,j_0)} + q_{(i_0,j_0)}; \overline{p}_k, w_k),\end{aligned}$$

and

$$\theta = \sum_{(i,j)\in\mathfrak{P}} (R_{(i,j)} - Q_{(i,j)}) + \sum_{k\in J_=} B_k - A_k.$$

Then

$$0 \leq^{D,H}_{(\overline{P}_i, \overline{P}_j, \overline{P}_k, W_i, W_j, W_k, Q_{(i,j)})_{(i,j)\in\mathfrak{P}, k\in J_=}} \theta.$$

Applying the homomorphism $\pi_1 \circ g : \mathcal{N} \restriction \mathcal{L}_{\text{in}} \to \mathcal{S} \restriction \mathcal{L}_{\text{in}} \subseteq \mathcal{H} \restriction \mathcal{L}_{\text{in}}$ and noting that $\pi_1 \circ g$ sends the A's, B's, Q's, and R's to zero, one finds that

$$\bigwedge_{(i,j)\in\mathfrak{P}} H_i^*(\overline{0}, 0; \pi_1 \circ g(\overline{P}_i), 0) \ \wedge\ H_j^*(\overline{0}, 0; \pi_1 \circ g(\overline{P}_j), 0) \\ \wedge \bigwedge_{k\in J_=} H_k^*(\overline{0}, 0; \pi_1 \circ g(\overline{P}_k), 0)$$

and so

$$\bigwedge_{(i,j)\in\mathfrak{P}} H_i(\overline{0}; \pi_1 \circ g(\overline{P}_i)) \ \wedge\ H_j(\overline{0}; \pi_1 \circ g(\overline{P}_j)) \ \wedge \bigwedge_{k\in J_=} H_k(\overline{0}; \pi_1 \circ g(\overline{P}_k)).$$

Applying π_2 to the congruences obtained from Ψ viewed as a (g, a)-system, one finds that

$$\begin{aligned}\bigwedge_{(i,j)\in\mathfrak{P}} & H_i(\pi_2(\overline{\mathbb{X}}_i) + (g(B_i) + r_i\pi_2(\mathbb{U})); \overline{\pi_2 \circ g(P_i) + t_i\pi_2(\mathbb{U})}) \\ & \wedge\ H_j(\pi_2(\overline{\mathbb{X}}_j) + (g(A_j) + s_j\pi_2(\mathbb{U})); \overline{\pi_2 \circ g(P_j) + t_j\pi_2(\mathbb{U})}) \\ \wedge \bigwedge_{k\in J_=} & H_k(\overline{\mathbb{X}}_k + r_k N\pi_2(\mathbb{U}); \overline{\pi_2 \circ g(P_k) + t_k N\pi_2(\mathbb{U})}),\end{aligned}$$

where $\pi_2 \circ g(B_i) = g(B_i)$, $\pi_2 \circ g(A_j) = g(A_j)$, $\pi_2(\overline{\mathbb{X}}_k) = \overline{\mathbb{X}}_k$, and $\pi_2 \circ g(A_k) = g(A_k)$ because $g(B_i)$, $g(A_j)$, $\overline{\mathbb{X}}_k$, and $g(A_k)$ all come from $U_a^{\mathcal{H}} = G$. Adding corresponding congruences from the last two displays, and remembering that $\pi_1 + \pi_2$ is the identity map, one sees that

$$\bigwedge_{(i,j)\in\mathfrak{P}} H_i(\pi_2(\overline{\mathbb{X}}_i) + (g(B_i) + r_i\pi_2(\mathbb{U})); \overline{g(P_i) + t_i\pi_2(\mathbb{U})})$$

$$\wedge\ H_j(\pi_2(\overline{\mathbb{X}}_j) + (g(A_j) + s_j\pi_2(\mathbb{U})); \overline{g(P_j) + t_j\pi_2(\mathbb{U})})$$

$$\wedge \bigwedge_{k\in J_=} H_k(\overline{\mathbb{X}}_k + r_kN\pi_2(\mathbb{U}); \overline{g(P_k) + t_kN\pi_2(\mathbb{U})}),$$

where when $k \in J_=$

$$g(A_k) \leq \overline{\mathbb{X}}_k \leq g(B_k)\ \wedge\ D_k(\overline{\mathbb{X}}_k)$$

and so

$$\begin{aligned} g(A_k) + r_kN\pi_2(\mathbb{U}) &\leq \overline{\mathbb{X}}_k + r_kN\pi_2(\mathbb{U}) \\ &\leq g(B_k) + r_kN\pi_2(\mathbb{U})\ \wedge\ D_k(\overline{\mathbb{X}}_k + r_kN\pi_2(\mathbb{U})). \end{aligned}$$

When $(i,j) \in \mathfrak{P}$, $\pi_2(\overline{\mathbb{X}}_i)$, $\pi_2(\overline{\mathbb{X}}_j)$, and $\pi_2(\mathbb{U})$ come from $G = U_a^{\mathcal{H}}$: so with these new witnesses to the truth of the H_j's in $\mathcal{H}$, one may finish the argument, when $\mathfrak{M} - \mathfrak{m}$ is infinite, as in the proof of Lemma 4.6 from [7].

When $\mathfrak{M} - \mathfrak{m}$ is finite—or $\mathfrak{M} - \mathfrak{m}$, when $U_a^{\mathcal{H}}$ is dense regular—proceed as in the proof of Lemma 4.6 from [7], where in the definitions of $\mathfrak{L}$, $\mathfrak{U}$, and J^∞, "finite" means "zero" when $U_a^{\mathcal{H}}$ is dense regular. Note, in this case, that the large sum providing a lower bound for B is 0. So again when $U_a^{\mathcal{H}}$ is dense regular, every

$$g(\Delta_i) = g(\Delta_{(l,m)}) = 0 = g(\Delta),$$

and all the elements

$$\overline{\mathbb{X}}^{(l,m)}_{\pi_-(l)}, \overline{\mathbb{X}}^{(l,m),+}_{\pi_-(l)}, \overline{\mathbb{X}}^{(l,m)}_{\pi_+(m)}, \overline{\mathbb{X}}^{(l,m),+}_{\pi_+(m)}, \overline{\mathbb{X}}_i, \overline{\mathbb{X}}_i^+, \mathbb{V}_l, \mathbb{V}^+_{(l,m)}$$

are zero. So when $U_a^{\mathcal{H}}$ is dense regular, $g(Q_l) = g(Q_{l'})$ for all $l, l' \in \mathfrak{L}$, and one may still conclude that

$$\bigwedge_{l,l'\in\mathfrak{L}} \mathbb{V}_l + g(Q_l) = \mathbb{V}_{l'} + g(Q_{l'}) \equiv_N 0$$

$$\wedge \bigwedge_{\substack{(l,m),(l',m')\in\mathfrak{L}\times\mathfrak{U} \\ \pi_-(l)=\pi_-(l')}} \overline{\mathbb{X}}^{(l,m)}_{\pi_-(l)} + g(Q_l) = \overline{\mathbb{X}}^{(l',m')}_{\pi_-(l')} + g(Q_{l'}) \equiv_S \overline{\mathbb{X}}_{\pi_-(l)}$$

$$\wedge \bigwedge_{\substack{(l,m),(l',m')\in\mathfrak{L}\times\mathfrak{U} \\ \pi_+(m)=\pi_+(m')}} \overline{\mathbb{X}}^{(l,m)}_{\pi_+(m)} + g(Q_l) = \overline{\mathbb{X}}^{(l',m')}_{\pi_+(m')} + g(Q_{l'}) \equiv_S \overline{\mathbb{X}}_{\pi_+(m)}.$$

When $i \in J^\infty$ one cannot argue, simply from the fact that $g(B_i) - g(A_i)$ is infinite, that one may choose the witnesses $\overline{\mathbb{X}}_i$ to H_i^* to lie in the interval $(g(A_i), g(B_i))$ in a certain order. However, the argument up to this point allows one to state that

$$\begin{aligned}
&\bigwedge_{(l,m)\in\mathfrak{L}\times\mathfrak{U}} g(Q_l) \le \overline{\mathbb{X}}_{\pi_-(l)} \le N\mathbb{U} \le \overline{\mathbb{X}}_{\pi_+(m)} \le g(R_m)\\
&\qquad\qquad \wedge\ D_{\pi_-(l)}(\overline{\mathbb{X}}_{\pi_-(l)}) \ \wedge\ D_{\pi_+(m)}(\overline{\mathbb{X}}_{\pi_+(m)})\\
&\qquad\qquad \wedge\ H^*_{\pi_-(l)}(\overline{\mathbb{X}}_{\pi_-(l)}, N\mathbb{U}; g(\overline{P}_{\pi_-(l)}), g(W_{\pi_-(l)}))\\
&\qquad\qquad \wedge\ H^*_{\pi_+(m)}(\overline{\mathbb{X}}_{\pi_+(m)}, N\mathbb{U}; g(\overline{P}_{\pi_+(m)}), g(W_{\pi_+(m)}))\\
&\wedge \bigwedge_{i\in(J_- - \pi_-(\mathfrak{L}))\cup(J_+ - \pi_+(\mathfrak{U}))} H_i^*(\overline{\mathbb{X}}_i, N\mathbb{U}; g(\overline{P}_i), g(W_i))\\
&\wedge \bigwedge_{i\in J^{<\infty}} g(A_i) \le \overline{\mathbb{X}}_i + g(A_i) \le g(B_i)\ \wedge\ D_i(\overline{\mathbb{X}}_i)\\
&\qquad\qquad \wedge\ H_i^*(\overline{\mathbb{X}}_i, N\mathbb{U}; g(\overline{P}_i), g(W_i))\\
&\wedge \bigwedge_{i\in J^\infty} H_i^*(\overline{\mathbb{X}}_i, N\mathbb{U}; g(\overline{P}_i), g(W_i)).
\end{aligned}$$

Having proceeded as far as possible with Ψ viewed as a (g,a)-system, one may now introduce $G = U_a^{\mathcal{H}}$, $\mathcal{S} \subseteq \mathcal{H}$, $\pi_1 : H \to S$, and $\pi_2 : H \to G$ as above, and apply a similar analysis to the $(\pi_1 \circ g, 0)$-system Ψ to conclude that

$$\begin{aligned}
&\bigwedge_{(i,j)\in\mathfrak{P}} H_i^*(\overline{0}, 0; \pi_1 \circ g(\overline{P}_i), 0) \ \wedge\ H_j^*(\overline{0}, 0; \pi_1 \circ g(\overline{P}_j), 0)\\
&\qquad\qquad\qquad\qquad\qquad \wedge \bigwedge_{k\in J_=} H_k^*(\overline{0}, 0; \pi_1 \circ g(\overline{P}_k), 0).
\end{aligned}$$

When $(l,m) \in \mathfrak{L}\times\mathfrak{U}$ and $i \in J^{<\infty}$, $\overline{\mathbb{X}}_{\pi_-(l)}$, $\overline{\mathbb{X}}_{\pi_+(m)}$, and $\overline{\mathbb{X}}_i$ come from $U_a^{\mathcal{H}} = G$, as do $\mathbb{U}$ and every $g(W_j)$. So if one applies π_2 to the congruences involving these elements, adds corresponding congruences from the last display, and remembers that $\pi_1 + \pi_2$ is the identity, one concludes that

$$\bigwedge_{(l,m)\in\mathfrak{L}\times\mathfrak{U}} g(Q_l) \le \overline{\mathbb{X}}_{\pi_-(l)} \le N\mathbb{U} \le \overline{\mathbb{X}}_{\pi_+(m)} \le g(R_m)$$

$$\wedge\ D_{\pi_-(l)}(\overline{\mathbb{X}}_{\pi_-(l)})\ \wedge D_{\pi_+(m)}(\overline{\mathbb{X}}_{\pi_+(m)})$$

$$\wedge\ H^*_{\pi_-(l)}(\overline{\mathbb{X}}_{\pi_-(l)}, N\mathbb{U}; g(\overline{P}_{\pi_-(l)}), g(W_{\pi_-(l)}))$$

$$\wedge\ H^*_{\pi_+(m)}(\overline{\mathbb{X}}_{\pi_+(m)}, N\mathbb{U}; g(\overline{P}_{\pi_+(m)}), g(W_{\pi_+(m)}))$$

$$\wedge \bigwedge_{i\in(J_- - \pi_-(\mathfrak{L}))\cup(J_+ - \pi_+(\mathfrak{U}))} H^*_i(\pi_2(\overline{\mathbb{X}}_i), N\mathbb{U}; g(\overline{P}_i), g(W_i))$$

$$\wedge \bigwedge_{i\in J^{<\infty}} g(A_i) \le \overline{\mathbb{X}}_i + g(A_i) \le g(B_i)\ \wedge D_i(\overline{\mathbb{X}}_i)$$

$$\wedge\ H^*_i(\overline{\mathbb{X}}_i, N\mathbb{U}; g(\overline{P}_i), g(W_i))$$

$$\wedge \bigwedge_{i\in J^\infty} H^*_i(\pi_2(\overline{\mathbb{X}}_i), N\mathbb{U}; g(\overline{P}_i), g(W_i)).$$

All the $\pi_2(\overline{\mathbb{X}}_i)$'s for $i \in (J_- - \pi_-(\mathfrak{L})) \cup (J_+ - \pi_+(\mathfrak{U})) \cup J^\infty$ come from $G = U^{\mathcal{H}}_a$. So the objection, noted above, to the treatment in [7] of $i \in J^\infty$ no longer applies, and one completes the argument as in [7]. □

4 g-normal systems of finite rank

When $g : \mathcal{N} \restriction \mathcal{L}_{\text{in}} \to \mathcal{H} \restriction \mathcal{L}_{\text{in}}$ is as in the last section, this section introduces a normal form, for finite systems $\Psi = \{\psi_j(u) : j \in J\}$ of congruence inequalities over $\mathcal{N}$, allowing one to prove by induction that $g(\Psi)$ is solvable in $\mathcal{H}$ when Ψ is in normal form.

Definition 4.1 Let $\Psi = \{\psi_j(u) : j \in J\}$ be a finite system of congruence inequalities over $\mathcal{N}$ and $g : \mathcal{N} \restriction \mathcal{L}_{\text{in}} \to \mathcal{H} \restriction \mathcal{L}_{\text{in}}$ be as in Section 3. Ψ is in g-normal form of g-rank $k \in \mathbb{N}$ just in case the following conditions hold:

(1) There are $\overline{P}$ from $\mathcal{N}$ so that for each $j \in J$,

(I)$_j$ $\psi_j(u)$ is $0 \le^{D_j,H_j}_{\overline{P},u} B_j$

or (II)$_j$ $\psi_j(u)$ is $A_j \le^{D_j,H_j}_{\overline{P},u} u$

or (III)$_j$ $\psi_j(u)$ is $u \le^{D_j,H_j}_{\overline{P},u} E_j$.

(2) Ψ is satisfiable in $\mathcal{M}$.

(3) If formulas both of types (II) and (III) occur in Ψ, then some $A_j = 0$ and every $g(A_j) \le 0$.

(4) For $j \in J$ let

$$q_j = \begin{cases} g(B_j) & \text{if (I)}_j \\ g(A_j) & \text{if (II)}_j \\ g(E_j) & \text{if (III)}_j. \end{cases}$$

J is the disjoint union of nonempty sets $J_0, \ldots, J_k$, where

(i) $J_0 = \{j \in J : U^{\mathcal{H}}_{q_j} = \{0\}$ or $U^{\mathcal{H}}_{q_j}$ is minimal[4] in $\{U^{\mathcal{H}}_{q_l} : U^{\mathcal{H}}_{q_l} \neq \{0\}\}\}$.

(ii) If $j \in J_i$ and $m \in J_{i+1}$, then $U^{\mathcal{H}}_{q_j} \subseteq L^{\mathcal{H}}_{q_m}$.

(iii) If $j, m \in J_{i+1}$, then $U^{\mathcal{H}}_{q_j} = U^{\mathcal{H}}_{q_m}$.

Ψ is said to be constrained if formulas of types (II) and (III) occur in Ψ; otherwise Ψ is said to be unconstrained.

The following result shows that when trying to solve in $\mathcal{H}$ the images of finite systems of congruence inequalities satisfiable in $\mathcal{M}$, one may restrict attention to systems in normal form:

Lemma 4.2 *If the finite system $\Psi = \{\psi_j(u) : j \in J\}$ of congruence inequalities over $\mathcal{N}$ is satisfiable in $\mathcal{M}$, then there is a g-normal system $\Psi' = \{\psi'_j(u) : j \in J\}$ (of some finite g-rank) such that $g(\Psi)$ is satisfiable in $\mathcal{H}$ just in case $g(\Psi')$ is satisfiable in $\mathcal{H}$.*

Proof Suppose, as above, that each $\psi_j(u)$ is

$$A_j + r_j u \leq^{D_j, H_j}_{P_j + t_j u} B_j + s_j u,$$

where the A's, B's, and P's come from $\mathcal{N}$ and the r's, s's, and t's are integers. By translating the arguments $A_j + r_j u$ and $B_j + s_j u$ and making corresponding changes to the system H_j of congruences, one may assume that $r_j, s_j \geq 0$, that at most one is nonzero, that at most one of A_j, B_j is not 0, that $r_j = 0$ if $A_j \neq 0$, that $s_j = 0$ if $B_j \neq 0$, and that $A_j = 0$ if $r_j = s_j = 0$. By multiplying the arguments of different $\psi_j(u)$'s by positive integers and making corresponding changes to the systems H_j of congruences, one may assume that there is a fixed positive integer t such that for any j, $r_j = t$ if $r_j \neq 0$ and also $s_j = t$ if $s_j \neq 0$. Multiplying the moduli and variables in the systems H_j of congruences by t, one may assume that u occurs in an H_j only as a multiple of tu. Replacing tu everywhere by u and conjoining '$u \equiv_t 0$' to every H_j, one may assume that $t = 1$. The modified Ψ still is satisfiable in $\mathcal{M}$, and each $\psi_j(u)$

4 With respect to inclusion.

obeys one of $(\mathrm{I})_j - (\mathrm{III})_j$. If the new Ψ is constrained, then because it is satisfiable in $\mathcal{M}$

$$\max_j g(A_j) \leq \min_l g(E_l).$$

If one picks $j_0 \in J$ with

$$g(A_{j_0}) = \max_j g(A_j),$$

translates the arguments of the congruence inequalities obeying $(\mathrm{II})_j$ or $(\mathrm{III})_j$ through $-A_{j_0}$, replaces arguments $u - A_{j_0}$ by u and occurrences of u in any H_i by $u + A_{j_0}$, and makes corresponding changes to $\overline{x}$ in $H_j(\overline{x}; \overline{P}_j, u)$ when $(\mathrm{II})_j$ or $(\mathrm{III})_j$, one obtains a new system Ψ obeying (2) and (3). To obtain $k \in \mathbb{N}$ and $J_0, \ldots, J_k$ satisfying (4), define J_0 as above and

$$\begin{aligned} J'_i &= J - \cup_{l \leq i} J_l \\ \mathcal{C}_i &= \{U^{\mathcal{H}}_{q_j} : j \in J'_i\} \\ J_{i+1} &= \{j \in J'_i : U^{\mathcal{H}}_{q_j} \text{ is minimal in } \mathcal{C}_i\} \end{aligned}$$

for $i \geq 0$. Because the $U^{\mathcal{H}}_{q_j}$'s are linearly ordered by inclusion and J is finite, there is a least $k \in \mathbb{N}$ with $J'_k = \emptyset$, and J will be the disjoint union of nonempty sets $J_0, \ldots, J_k$ satisfying (i) and (iii). Every element of $\{U^{\mathcal{H}}_{q_j} : j \in J_0\}$ is a proper subgroup of every element of $\{U^{\mathcal{H}}_{q_j} : j \in J'_0\}$. If every element of $\{U^{\mathcal{H}}_{q_j} : j \in \cup_{l \leq i} J_l\}$ is a proper subgroup of every element of $\{U^{\mathcal{H}}_{q_j} : j \in J'_i\}$, then since the $j \in J_{i+1}$ are the $j \in J'_i$ with $U^{\mathcal{H}}_{q_j}$ minimal in $\mathcal{C}_i$, every element of $\{U^{\mathcal{H}}_{q_j} : j \in \cup_{l \leq i+1} J_l\}$ is a proper subgroup of every element of $\{U^{\mathcal{H}}_{q_j} : j \in J'_{i+1}\}$. So by induction, every element of $\{U^{\mathcal{H}}_{q_j} : j \in \cup_{l \leq i} J_l\}$ is a proper subgroup of every element of $\{U^{\mathcal{H}}_{q_j} : j \in J'_i\}$. Thus if $j \in J_i$ and $m \in J_{i+1} \subseteq J'_i$,

$$U^{\mathcal{H}}_{q_j} \subset U^{\mathcal{H}}_{q_m}$$

and

$$U^{\mathcal{H}}_{q_j} \subseteq L^{\mathcal{H}}_{q_m}$$

by the axioms for T^+. So (ii) holds, and since one may assume that all the congruence inequalities have the same parameters $\overline{P}, u$, the new system of congruence inequalities is g-normal of g-rank k. Clearly its g-image is satisfiable in $\mathcal{H}$ just in case the g-image of the original system is satisfiable in $\mathcal{H}$. □

The principal result of this section is

Proposition 4.3 *Assume that for any $\mathcal{L}_{\text{in}}$-homomorphism h of $\mathcal{N}$ into an $\aleph_1$-saturated model of T^+, the h-image of any h-normal system of h-rank less than k may be satisfied in the codomain of h. If $g : \mathcal{N} \restriction \mathcal{L}_{\text{in}} \to \mathcal{H} \restriction \mathcal{L}_{\text{in}}$ is as above and $\Psi = \{\psi_j(u) : j \in J\}$ is g-normal of g-rank k, then $g(\Psi)$ may be satisfied in $\mathcal{H}$.*

Proof If every $q_j = 0$, Ψ is a $(g,0)$-system and Proposition 3.2 makes $g(\Psi)$ satisfiable in $\mathcal{H}$. Assume, therefore, that some $q_j \neq 0$. If $J_k^{\neq 0} = \{j \in J_k : q_j \neq 0\}$, then $J_k^{\neq 0} \neq \emptyset$ and

$$j, l \in J_k^{\neq 0} \text{ and } m \in J - J_k^{\neq 0} \Rightarrow U_{q_m}^{\mathcal{H}} \subset U_{q_j}^{\mathcal{H}} = U_{q_l}^{\mathcal{H}};$$

so by the axioms for T^+,

$$j, l \in J_k^{\neq 0} \text{ and } m \in J - J_k^{\neq 0} \Rightarrow U_{q_m}^{\mathcal{H}} \subseteq L_{q_j}^{\mathcal{H}} = L_{q_l}^{\mathcal{H}}.$$

If $J_k^{\neq 0} = J$, then a similar argument shows that $\{0\} \subset U_{q_j}^{\mathcal{H}} = U_{q_l}^{\mathcal{H}}$ and $L_{q_j}^{\mathcal{H}} = L_{q_l}^{\mathcal{H}}$ for all $j, l \in J$. Let $\{G\} = \{L_{q_j}^{\mathcal{H}} : j \in J_k^{\neq 0}\}$, and apply Lemma 2.4 to $\mathcal{H}$ and G to obtain $\mathcal{S} \subseteq \mathcal{H}$ and the group-theoretic projections $\pi_1 : H \to S$ and $\pi_2 : H \to G$. $\pi_1 \circ g$ is an $\mathcal{L}_{\text{in}}$-homomorphism of $\mathcal{N}$ into $\mathcal{S}$, which is interpretable in $\mathcal{H}$ and so $\aleph_1$-saturated. $\pi_1(q_j) = 0$ when $j \in J - J_k^{\neq 0}$: so if $j_0 \in J_k^{\neq 0}$, Ψ is a $(\pi_1 \circ g, \pi_1(q_{j_0}))$-system with $L_{\pi_1(q_{j_0})}^{\mathcal{S}} = \{0\} \subset U_{\pi_1(q_{j_0})}^{\mathcal{S}}$. By Proposition 3.2, therefore, some $\mathbb{U} \in S$ satisfies $\pi_1 \circ g(\Psi)$ in $\mathcal{S}$. For each $j \in J$ pick a witness $\overline{\mathbb{X}}_j$ to the truth of $\pi_1 \circ g(\psi_j)(\mathbb{U})$ in $\mathcal{S}$; so

$$(\mathrm{I})_j^{\mathcal{S}} \qquad 0 \leq \overline{\mathbb{X}}_j \leq \pi_1 \circ g(B_j) \ \wedge\ D_j(\overline{\mathbb{X}}_j) \ \wedge\ H_j(\overline{\mathbb{X}}_j; \pi_1 \circ g(\overline{P}), \mathbb{U})$$

when $(\mathrm{I})_j$,

$$(\mathrm{II})_j^{\mathcal{S}} \qquad \pi_1 \circ g(A_j) \leq \overline{\mathbb{X}}_j \leq \mathbb{U} \ \wedge\ D_j(\overline{\mathbb{X}}_j) \ \wedge\ H_j(\overline{\mathbb{X}}_j; \pi_1 \circ g(\overline{P}), \mathbb{U})$$

when $(\mathrm{II})_j$, and

$$(\mathrm{III})_j^{\mathcal{S}} \qquad \mathbb{U} \leq \overline{\mathbb{X}}_j \leq \pi_1 \circ g(E_j) \ \wedge\ D_j(\overline{\mathbb{X}}_j) \ \wedge\ H_j(\overline{\mathbb{X}}_j; \pi_1 \circ g(\overline{P}), \mathbb{U})$$

when $(\mathrm{III})_j$. Apply the Chinese Remainder Theorem to obtain a conjunction $H(\overline{p}; u)$ of congruences equivalent to

$$\exists (\overline{x}_j)_{j \in J_k^{\neq 0}} \bigwedge_{l \in J_k^{\neq 0}} H_l(\overline{x}_l; \overline{p}, u)$$

in any torsion-free Abelian group. Letting $\widetilde{J} = J - J_k^{\neq 0}$, $\widetilde{\psi}(u)$ be

$$0 \leq_{\overline{P}, u}^{,H} 0$$

—so no conjunction D of extra inequalities appears—and

$$\widetilde{\Psi} = \{\psi_j(u) : j \in \widetilde{J}\} \cup \{\widetilde{\psi}(u)\},$$

one finds that $\widetilde{\Psi}$ is either a $(g,0)$-system (when $k = 0$) or a g-normal system of g-rank less than k (when $k > 0$). So by Proposition 3.2 or the induction hypothesis some $\widetilde{\mathbb{U}}$ satisfies $g(\widetilde{\Psi})$ in $\mathcal{H}$. For each $j \in \widetilde{J}$ pick a witness $\widetilde{\mathbb{X}}_j$ to the truth of $g(\psi_j)(\widetilde{\mathbb{U}})$ in $\mathcal{H}$: so

$$(\widetilde{\mathrm{I}})_j \qquad 0 \leq \widetilde{\mathbb{X}}_j \leq g(B_j) \wedge D_j(\widetilde{\mathbb{X}}_j) \wedge H_j(\widetilde{\mathbb{X}}_j; g(\overline{P}), \widetilde{\mathbb{U}})$$

when $(\mathrm{I})_j$,

$$(\widetilde{\mathrm{II}})_j \qquad g(A_j) \leq \widetilde{\mathbb{X}}_j \leq \widetilde{\mathbb{U}} \wedge D_j(\widetilde{\mathbb{X}}_j) \wedge H_j(\widetilde{\mathbb{X}}_j; g(\overline{P}), \widetilde{\mathbb{U}})$$

when $(\mathrm{II})_j$, and

$$(\widetilde{\mathrm{III}})_j \qquad \widetilde{\mathbb{U}} \leq \widetilde{\mathbb{X}}_j \leq g(E_j) \wedge D_j(\widetilde{\mathbb{X}}_j) \wedge H_j(\widetilde{\mathbb{X}}_j; g(\overline{P}), \widetilde{\mathbb{U}})$$

when $(\mathrm{III})_j$. Since $g(\widetilde{\psi})(\widetilde{\mathbb{U}})$, there are, for $j \in J_k^{\neq 0}$, $\widetilde{\mathbb{X}}_j$ from H with

$$(\widetilde{\mathrm{IV}})_j \qquad H_j(\widetilde{\mathbb{X}}_j; g(\overline{P}), \widetilde{\mathbb{U}}).$$

Applying the group homomorphism $\pi_2 : H \to G$ to the congruences from $(\widetilde{\mathrm{I}}) - (\widetilde{\mathrm{IV}})$, one concludes that

$$\bigwedge_{j \in J} H_j(\pi_2(\widetilde{\mathbb{X}}_j); \pi_2 \circ g(\overline{P}), \pi_2(\widetilde{\mathbb{U}})).$$

Let $\widehat{\mathbb{U}} = \mathbb{U} + \pi_2(\widetilde{\mathbb{U}})$ and

$$\widehat{\mathbb{X}}_j = \overline{\mathbb{X}}_j + \pi_2(\widetilde{\mathbb{X}}_j)$$

for $j \in J$. If one adds the congruences in the next-to-last display to corresponding ones in $(\mathrm{I})^{\mathcal{S}} - (\mathrm{III})^{\mathcal{S}}$, and remembers that $\pi_1 + \pi_2$ is the identity, one sees that

$$\bigwedge_{j \in J} H_j(\widehat{\mathbb{X}}_j; g(\overline{P}), \widehat{\mathbb{U}}).$$

Let B be the product of the moduli in all the H_j's. Arguments below will show that there are

$$\widehat{\mathbb{X}}'_j \equiv_B \widehat{\mathbb{X}}_j \text{ and } \widehat{\mathbb{U}}' \equiv_B \widehat{\mathbb{U}}$$

from $\mathcal{H}$ for which

$$0 \leq \widehat{\mathbb{X}}'_j \leq g(B_j) \wedge D_j(\widehat{\mathbb{X}}'_j)$$

when $(\mathrm{I})_j$,

$$g(A_j) \le \widehat{\mathbb{X}}'_j \le \widehat{\mathbb{U}}' \ \wedge\ D_j(\widehat{\mathbb{X}}'_j)$$

when $(\mathrm{II})_j$, and

$$\widehat{\mathbb{U}}' \le \widehat{\mathbb{X}}'_j \le g(E_j) \ \wedge\ D_j(\widehat{\mathbb{X}}'_j)$$

when $(\mathrm{III})_j$: so $\widehat{\mathbb{U}}'$ will satisfy $g(\Psi)$ in $\mathcal{H}$ as desired.

Suppose $(\mathrm{I})_j$ holds. If $j \in \widetilde{J}$, then $\pi_1 \circ g(B_j) = 0$, $\overline{\mathbb{X}}_j = \overline{0}$, and $\widehat{\mathbb{X}}_j = \pi_2(\widetilde{\mathbb{X}}_j)$. Since $0 \le \widetilde{\mathbb{X}}_j \le g(B_j) \in G$, $\pi_2(\widetilde{\mathbb{X}}_j) = \widetilde{\mathbb{X}}_j$; so by $(\widetilde{\mathrm{I}})_j$, $\widehat{\mathbb{X}}'_j = \widehat{\mathbb{X}}_j = \widetilde{\mathbb{X}}_j$ is a suitable witness. Suppose now that $j \notin \widetilde{J}$: i.e., $j \in J_k^{\neq 0}$. Then $0 < \pi_1 \circ g(B_j) \in S$, $\overline{\mathbb{X}}_j$ is from S, $0 \le \overline{\mathbb{X}}_j \le \pi_1 \circ g(B_j)$, $D_j(\overline{\mathbb{X}}_j)$, $\pi_1(\widehat{\mathbb{X}}_j - \overline{\mathbb{X}}_j) = \pi_1\pi_2(\widetilde{\mathbb{X}}_j) = 0$, and $\pi_1(\pi_1 \circ g(B_j) - g(B_j)) = 0$. If $\overline{\mathbb{X}}_{jk}$ is an entry of $\overline{\mathbb{X}}_j$ with

$$0 < \overline{\mathbb{X}}_{jk} < \pi_1 \circ g(B_j),$$

then any translation of $\overline{\mathbb{X}}_{jk}$ by an element of G will obey the same inequalities: so the corresponding entry $\widehat{\mathbb{X}}_{jk}$ of $\widehat{\mathbb{X}}_j$ has the same property, and one may replace all such $\widehat{\mathbb{X}}_{jk}$ by elements $\widehat{\mathbb{X}}'_{jk} \equiv_B \widehat{\mathbb{X}}_{jk}$, with $\widehat{\mathbb{X}}'_{jk} - \widehat{\mathbb{X}}_{jk} \in G$, ordered as D_j demands. For those k with

$$\overline{\mathbb{X}}_{jk} = 0,$$

$\widehat{\mathbb{X}}_{jk} \in G$: so by adding nonnegative B-divisible elements from G to these $\widehat{\mathbb{X}}_{jk}$, one may find nonnegative $\widehat{\mathbb{X}}'_{jk} \equiv_B \widehat{\mathbb{X}}_{jk}$ in G ordered as D_j demands and still less than any positive element of S. For those k with

$$\overline{\mathbb{X}}_{jk} = \pi_1 \circ g(B_j),$$

$\widehat{\mathbb{X}}_{jk} - g(B_j) \in G$, and by subtracting nonnegative B-divisible elements of G from these $\widehat{\mathbb{X}}_{jk}$ one may find $\widehat{\mathbb{X}}'_{jk} \equiv_B \widehat{\mathbb{X}}_{jk}$, with $\widehat{\mathbb{X}}'_{jk} \le g(B_j)$ and $\widehat{\mathbb{X}}'_{jk} - g(B_j) \in G$, ordered as D_j demands. $\widehat{\mathbb{X}}'_j$ is a witness to the truth of $g(\psi_j)(\widehat{\mathbb{U}})$ in $\mathcal{H}$, and any $\widehat{\mathbb{U}}' \equiv_B \widehat{\mathbb{U}}$ also will satisfy $g(\psi_j)(u)$.

When $(\mathrm{II})_j$ or $(\mathrm{III})_j$ holds, one may consider two cases.

Suppose first that $\widetilde{\Psi}$ is constrained. In this case $\mathbb{U} = 0$, $\widetilde{\mathbb{U}} \in G$, and $\widehat{\mathbb{U}} = \pi_2(\widetilde{\mathbb{U}}) = \widetilde{\mathbb{U}}$. If $j \in \widetilde{J}$, $\overline{\mathbb{X}}_j = \overline{0}$, $\widetilde{\mathbb{X}}_j$ comes from G, and $\widehat{\mathbb{X}}_j = \pi_2(\widetilde{\mathbb{X}}_j) = \widetilde{\mathbb{X}}_j$ serves as a witness for the truth of $g(\psi_j)(\widehat{\mathbb{U}})$ in $\mathcal{H}$. Assume now that $j \notin \widetilde{J}$. If $(\mathrm{II})_j$, then $\pi_1 \circ g(A_j) < 0$ in $\mathcal{S}$ and $\pi_1 \circ g(A_j) \le \overline{\mathbb{X}}_j \le \mathbb{U} = 0$; if $(\mathrm{III})_j$, then $0 = \mathbb{U} < \pi_1 \circ g(E_j)$ in $\mathcal{S}$ and $0 \le \overline{\mathbb{X}}_j \le \pi_1 \circ g(E_j)$. The argument above for $j' \in J$ with $(\mathrm{I})_{j'}$ thus still applies, though one endpoint of the interval is $\widetilde{\mathbb{U}} \in G$ rather than 0.

Assume now that $\widetilde{\Psi}$ is not constrained. If Ψ also is not constrained—say no $j \in J$ obeys $(\mathrm{III})_j$[5]—then by adding nonnegative B-divisible elements to the entries of $\widehat{\mathbb{X}}_j$ when $(\mathrm{II})_j$, one may obtain $\widehat{\mathbb{X}}'_j \equiv_B \widehat{\mathbb{X}}_j$, ordered as D_j demands, with $g(A_j) \leq \widehat{\mathbb{X}}'_j$. By then adding a nonnegative B-divisible element to $\widehat{\mathbb{U}}$, one obtains $\widehat{\mathbb{U}}' \equiv_B \widehat{\mathbb{U}}$ with $\widehat{\mathbb{X}}'_j \leq \widehat{\mathbb{U}}'$ for all j obeying $(\mathrm{II})_j$, and the argument is complete. Now suppose that Ψ is constrained. By Definition 4.1(3), some $\psi_j(u)$ is

$$0 \leq^{D_j,H_j}_{P,u} u;$$

because such a formula must belong to $\widetilde{\Psi}$, which is not constrained, no $j \in \widetilde{J}$ obeys $(\mathrm{III})_j$. So any $j \in J$ obeying $(\mathrm{III})_j$ has

$$0 < \pi_1 \circ g(E_j) \text{ in } \mathcal{S} \subseteq \mathcal{H}.$$

$\mathbb{U}$ and $\widetilde{\mathbb{U}}$ are nonnegative. If $\mathbb{U} = 0$, then $\widehat{\mathbb{U}} = \pi_2(\widetilde{\mathbb{U}}) \in G$. If $j \in J$ and $(\mathrm{II})_j$, then since

$$\pi_1 \circ g(A_j) \leq \overline{\mathbb{X}}_j \leq \mathbb{U} = 0,$$

$D_j(\overline{\mathbb{X}}_j)$, and $\widehat{\mathbb{X}}_j - \overline{\mathbb{X}}_j$ and $g(A_j) - \pi_1 \circ g(A_j)$ come from G, one may add nonnegative B-divisible elements from G to the entries of $\widehat{\mathbb{X}}_j$ to obtain $\widehat{\mathbb{X}}'_j \equiv_B \widehat{\mathbb{X}}_j$ greater than or equal to $g(A_j)$ and ordered as D_j demands; then by adding to $\widehat{\mathbb{U}} = \pi_2(\mathbb{U})$ a nonnegative B-divisible element from G, one may obtain $\widehat{\mathbb{U}}' \equiv_B \widehat{\mathbb{U}}$ in G greater than or equal to all the $\widehat{\mathbb{X}}'_j$ with $j \in J$ obeying $(\mathrm{II})_j$. Since $\widehat{\mathbb{U}}' \in G$, $\pi_1(\widehat{\mathbb{U}}') = 0 < \pi_1 \circ g(E_j)$ in $\mathcal{S} \subseteq \mathcal{H}$ whenever $j \in J$ obeys $(\mathrm{III})_j$. So since $0 = \mathbb{U} \leq \overline{\mathbb{X}}_j \leq \pi_1 \circ g(E_j)$, one may proceed as in the analysis of $j' \in J$ obeying $(\mathrm{I})_{j'}$—but replacing the entries $\widehat{\mathbb{X}}_{jk}$ of $\widehat{\mathbb{X}}_j$ corresponding to entries $\overline{\mathbb{X}}_{jk} = 0$ of $\overline{\mathbb{X}}_j$ by elements greater than or equal to $\widehat{\mathbb{U}}' \in G$—to obtain $\widehat{\mathbb{X}}'_j$ with the desired properties.

If $\mathbb{U} > 0$, then $\widehat{\mathbb{U}} = \mathbb{U} + \pi_2(\widetilde{\mathbb{U}}) > 0$. When $j \in J$ and $(\mathrm{III})_j$,

$$0 < \mathbb{U} \leq \overline{\mathbb{X}}_j \leq \pi_1 \circ g(E_j) \text{ in } \mathcal{S} \subseteq \mathcal{H},$$

$D_j(\overline{\mathbb{X}}_j)$, and $\widehat{\mathbb{X}}_j - \overline{\mathbb{X}}_j$ and $g(E_j) - \pi_1 \circ g(E_j)$ come from G. By subtracting nonnegative B-divisible elements of G from the entries of $\widehat{\mathbb{X}}_j$ when $(\mathrm{III})_j$, one may obtain $\widehat{\mathbb{X}}'_j \equiv_B \widehat{\mathbb{X}}_j$ less than or equal to $g(E_j)$ and ordered as D_j demands; then by subtracting from $\widehat{\mathbb{U}}$ a nonnegative B-divisible element from G, one obtains $\widehat{\mathbb{U}}' \equiv_B \widehat{\mathbb{U}}$ less than or equal to all the $\widehat{\mathbb{X}}'_j$ with $j \in J$

5 If instead no $j \in J$ obeys $(\mathrm{II})_j$, an analogous argument that subtracts rather than adds nonnegative B-divisible elements will reach the same conclusion.

obeying $(\mathrm{III})_j$. Since $\mathbb{U}$ is a positive element of S and $\widehat{\mathbb{U}}'-\mathbb{U} \in G$, $\widehat{\mathbb{U}}'$ and $\pi_1(\widehat{\mathbb{U}}')$ are positive. When $j \in J$ obeys $(\mathrm{II})_j$, Definition 4.1(3) makes every $\pi_1(\widehat{\mathbb{U}}' - g(A_j)) > 0$ in $\mathcal{S}$. Since

$$\pi_1 \circ g(A_j) \leq \overline{\mathbb{X}}_j \leq \mathbb{U},$$

$D_j(\overline{\mathbb{X}}_j)$, $\pi_1(\widehat{\mathbb{X}}_j - \overline{\mathbb{X}}_j) = \overline{0}$, $\overline{\mathbb{X}}_j$ is from S, and $\pi_1(\pi_1 \circ g(A_j) - g(A_j)) = 0$, one may finish the argument by proceeding as in the analysis of those $j' \in J$ obeying $(\mathrm{I})_{j'}$. Namely: if an entry $\overline{\mathbb{X}}_{jk}$ of $\overline{\mathbb{X}}_j$ obeys

$$\pi_1 \circ g(A_j) < \overline{\mathbb{X}}_{jk} < \mathbb{U},$$

then any translation of $\overline{\mathbb{X}}_{jk}$ by an element of G will be strictly between $g(A_j)$ and $\widehat{\mathbb{U}}'$: so the corresponding entry $\widehat{\mathbb{X}}_{jk}$ of $\widehat{\mathbb{X}}_j$ has the same property, and one may replace all such $\widehat{\mathbb{X}}_{jk}$ by elements $\widehat{\mathbb{X}}'_{jk} \equiv_B \widehat{\mathbb{X}}_{jk}$, with $\widehat{\mathbb{X}}'_{jk} - \widehat{\mathbb{X}}_{jk} \in G$, ordered as D_j demands. For those k with

$$\overline{\mathbb{X}}_{jk} = \pi_1 \circ g(A_j),$$

$\widehat{\mathbb{X}}_{jk} - g(A_j) \in G$: so by by adding to those $\widehat{\mathbb{X}}_{jk}$ nonnegative B-divisible elements from G, one finds elements $\widehat{\mathbb{X}}'_{jk} \equiv_B \widehat{\mathbb{X}}_{jk}$, with $\widehat{\mathbb{X}}'_{jk} \geq g(A_j)$ and $\widehat{\mathbb{X}}'_{jk} - g(A_j) \in G$, ordered as D_j demands and still less than $\widehat{\mathbb{U}}'$. For those k with

$$\overline{\mathbb{X}}_{jk} = \mathbb{U},$$

$\widehat{\mathbb{X}}_{jk} - \widehat{\mathbb{U}}' \in G$, and by subtracting nonnegative B-divisible elements from these $\widehat{\mathbb{X}}_{jk}$ one may find $\widehat{\mathbb{X}}'_{jk} \equiv_B \widehat{\mathbb{X}}_{jk}$, with $\widehat{\mathbb{X}}'_{jk} \leq \widehat{\mathbb{U}}'$ and $\widehat{\mathbb{X}}'_{jk} - \widehat{\mathbb{U}}' \in G$, ordered as D_j demands and still greater than $g(A_j)$. This $\widehat{\mathbb{X}}'_j$ is a witness to the truth of $g(\psi_j)(\widehat{\mathbb{U}}')$ when $j \in J$ obeys $(\mathrm{II})_j$, and the proof of Proposition 4.3 is complete. □

Course-of-values induction based on Proposition 4.3 implies

Corollary 4.4 *If $g : \mathcal{N} \restriction \mathcal{L}_{\mathrm{in}} \to \mathcal{H} \restriction \mathcal{L}_{\mathrm{in}}$, where $\mathcal{H} \models T^+$ is $\aleph_1$-saturated, then the g-image of any g-normal system of finite g-rank may be satisfied in $\mathcal{H}$.*

5 Projections

The machinery of Sections 2-4 allows one to prove

Proposition 5.1 *Modulo T_{in}, the class of conjunctions of atomic formulas is closed under existential quantification.*

Proof By Proposition 2.1 one need show merely that modulo T^+, the class of conjunctions of atomic $\mathcal{L}_{\text{in}}$-formulas is closed under existential quantification. So let $\psi(\overline{w}, \overline{z})$ be a conjunction of atomic $\mathcal{L}_{\text{in}}$-formulas in the disjoint lists $\overline{w}$, $\overline{z}$ of distinct variables. Because $T_{\text{in}} \models \psi(\overline{0}, \overline{0})$, one may apply Lemma 2.5 to show that $\varphi = \exists \overline{z} \psi(\overline{w}, \overline{z})$ is equivalent, modulo T^+, to a conjunction of atomic $\mathcal{L}_{\text{in}}$-formulas. Assume, therefore, that for each i in the nonempty set I, there is an n-tuple $\overline{a}_i$ from $\mathcal{M}_i \models T^+$ with $\mathcal{M}_i \models \varphi[\overline{a}_i]$, and that $\mathcal{N} \subseteq \prod_{i \in I} \mathcal{M}_i = \mathcal{M}$, $\overline{a} = (\overline{a}_i)_{i \in I} \in N^n$, $\mathcal{H} \models T^+$, and $f : \mathcal{N} \restriction \mathcal{L}_{\text{in}} \to \mathcal{H} \restriction \mathcal{L}_{\text{in}}$; the goal is to show that $\mathcal{H} \models \varphi[f(\overline{a})]$. Without loss of generality $f(N) \neq \{0\}$ and $\mathcal{H}$ is $|\mathcal{M}|^+$-saturated. Since $f(N) \neq \{0\}$, $\mathcal{N}$ is nontrivial, some $\mathcal{M}_i$ is nontrivial, and $\mathcal{M}$ is infinite: so $\mathcal{H}$ is also $\aleph_1$-saturated. By Zorn's Lemma there is a pair $(\mathcal{N}', g)$, with $\mathcal{N} \subseteq \mathcal{N}' \subseteq \mathcal{M}$, $g : \mathcal{N}' \restriction \mathcal{L}_{\text{in}} \to \mathcal{H} \restriction \mathcal{L}_{\text{in}}$, and $f \subseteq g$, maximal with respect to inclusion (in both coordinates). If $c \in M$, and $\Gamma(u)$ is the set of all congruence inequalities, over $\mathcal{N}'$, satisfied by c in $\mathcal{M}$, then if $g(\Gamma(u))$ is satisfiable in $\mathcal{H}$, g extends to an $\mathcal{L}_{\text{in}}$-homomorphism into $\mathcal{H}$ of the substructure of $\mathcal{M}$ with domain $N' + \mathbb{Z}c$, and so $c \in N'$ by the maximality of $(\mathcal{N}', g)$. Because $\mathcal{H}$ is $|\mathcal{M}|^+$-saturated, $g(\Gamma(u))$ is satisfiable in $\mathcal{H}$ just in case $g(\Delta(u))$ is satisfiable in $\mathcal{H}$ whenever $\Delta(u) \subseteq \Gamma(u)$ is finite. Lemma 4.2 provides a g-normal system $\Delta'(u)$, of finite g-rank, such that $g(\Delta(u))$ is satisfiable in $\mathcal{H}$ just in case $g(\Delta'(u))$ is. Since $\mathcal{H}$ is $\aleph_1$-saturated, Corollary 4.4 says that $g(\Delta'(u))$ is satisfiable in $\mathcal{H}$: so $c \in N'$ when $c \in M$, and $N' = M$. Because $g : \mathcal{M} \restriction \mathcal{L}_{\text{in}} \to \mathcal{H} \restriction \mathcal{L}_{\text{in}}$, $\varphi = \exists \overline{z} \psi$, and ψ is a conjunction of atomic $\mathcal{L}_{\text{in}}$-formulas, $\mathcal{H} \models \varphi[g(\overline{a})]$ and $\mathcal{H} \models \varphi[f(\overline{a})]$ as desired. □

Despite Proposition 5.1, T_{in} does not admit elimination of quantifiers[6]. In the lexicographically ordered Abelian group $\mathbb{Z}^{\mathbb{N}}$, let G be the ordered subgroup consisting of all $f \in \mathbb{Z}^{\mathbb{N}}$ with $f(i) = 0$ for all but finitely many i, and H be the ordered subgroup generated by G together with $\vec{5}$, the constant function with value 5. Expand both G and H to $\mathcal{L}_{\text{in}}$-structures $\mathcal{G}$ and $\mathcal{H}$ in the obvious way, and suppose A, B, and $\overline{P}$ come from G. If $\mathcal{G} \models A \leq_{\overline{P}}^{D,H} B$, then $\mathcal{H} \models A \leq_{\overline{P}}^{D,H} B$ because G is an ordered subgroup of H. Pick $n \in \mathbb{N}$ so that $A(i)$, $B(i)$, and every $P(i)$ is 0 when $i \geq n$. The function $\pi : H \to G$ given by

$$\pi(C)(j) = \begin{cases} C(j) & \text{if } j < n \\ 0 & \text{if } j \geq n \end{cases}$$

6 The following example is closely related to one in ([8], p. 393).

is a homomorphism of ordered Abelian groups which fixes A, B, and the P's: so if $\mathcal{H} \models A \leq_P^{D,H} B$, then $\mathcal{G} \models A \leq_P^{D,H} B$. Thus $\mathcal{G} \subseteq \mathcal{H}$ as $\mathcal{L}_{\text{in}}$-structures. Yet $\mathcal{G} \not\preccurlyeq \mathcal{H}$ because they disagree about the sentence

$$\forall x\,[\forall y > 0 \exists z\,(|x - z| < y \wedge z \equiv_5 0) \rightarrow x \equiv_5 0],$$

which is true in $\mathcal{G}$ but false in $\mathcal{H}$. Thus T_{in} does not admit elimination of quantifiers.

Though the manipulations behind Proposition 5.1 certainly exploit the complexity of congruence inequalities $\leq^{D,H}$, one might wonder whether one may get by with something simpler. In particular, Proposition 3.1 of [7] shows that in dense regular groups one may replace relations $\leq^{D,H}$ by simpler versions $\leq_k$ defined by

$$\forall x, y\,(x \leq_k y \leftrightarrow \exists z\,(x \leq kz \leq y))$$

for $k \geq 2$. However, an example mentioned in the Conclusion of [7] shows that

Proposition 5.2 *There is a congruence inequality $x \leq^{D,H} y$ inequivalent, over $\mathbb{Z}$, to every conjunction of congruence inequalities $r_i(x, y) \leq_{k_i} s_i(x, y)$.*

Proof Let $x \leq^{D,H} y$ be the congruence inequality defined by

$$\exists u, v(x \leq u, v \leq y \ \wedge \ u \leq v \ \wedge \ 3|u \ \wedge \ 2|v)$$

and let $S \subseteq \mathbb{Z}^2$ be the relation defined by $x \leq^{D,H} y$ in the ordered group $\mathbb{Z}$. The following argument will show that if every point of S obeys

$$ax + by \leq_m cx + dy,$$

where a, b, c, d, m are integers with $m \geq 2$, then $(1, 3)$ obeys this congruence inequality. Because $(1, 3) \notin S$, the desired result follows.

Since

$$x \leq_m y \text{ if and only if } nx \leq_{nm} ny$$

for all positive integers m, n with $m \geq 2$, one may assume that $m = 6^\alpha k$, where $\alpha \geq 2$ and $(6, k) = 1$.

If $x \leq y$ in $\mathbb{Z}$, then $6x \leq^{D,H} 6y$ and so

$$a(6x) + b(6y) \leq_m c(6x) + d(6y),$$

$6ax+6by \leq 6cx+6dy$, and $ax+by \leq cx+dy$. So by clearing denominators in $\mathbb{Q}$ one sees that

$$y - x \geq 0 \Rightarrow (d-b)y + (c-a)x \geq 0$$

in $\mathbb{Q}$. By Farkas' Lemma, therefore,

$$(d-b, c-a) = l(1,-1)$$

for some $l \geq 0$ in $\mathbb{Q}$, and of course $l \in \mathbb{Z}$. So every point of S obeys

$$(1) \qquad ax + by \leq_m ax + by + l(y-x),$$

and one wants to show that $(1,3)$ obeys this formula.

If $l = 0$, then (1) reduces to the congruence

$$ax + by \equiv_m 0.$$

Because $(-1,0) \in S$, $-a \equiv_m 0$; because $(0,1) \in S$, $b \equiv_m 0$. Thus $a + 3b \equiv_m 0$ and $(1,3)$ obeys (1) when $l = 0$.

Assume now that $l > 0$. Since one wants to show that

$$a + 3b \leq_m a + 3b + 2l,$$

one may without loss of generality assume that $2l < m$.

Since $(6,6) \in S$,

$$6^\alpha k = m | 6(a+b),$$

$6^{\alpha-1}k | a+b$, and

$$a + b \equiv_m 6^{\alpha-1}ki$$

for some integer i with $0 < i < 5$.

If $m | a+b$, then since $\leq_m$ is invariant under translation by m-divisible elements,

$$a + 3b \leq_m a + 3b + 2l \text{ if and only if } 2b \leq_m 2b + 2l.$$

Because $(0,2) \in S$, the right-hand side holds and one reaches the desired conclusion.

If $a + b \equiv_m 6^{\alpha-1}ki$ for $i \in \{2,4\}$, then $m | 3(a+b)$ and as above,

$$a + 3b \leq_m a + 3b + 2l \text{ if and only if } -2a \leq_m -2a + 2l;$$

because $(-2,0) \in S$, one again reaches the desired conclusion.

If $a + b \equiv_m 3 \cdot 6^{\alpha-1}k$, then $m | 2(a+b)$ and

$$a + 3b \leq_m a + 3b + 2l \text{ if and only if } -a + b \leq_m -a + b + 2l;$$

because $(-1, 1) \in S$, one again reaches the desired conclusion.

Suppose now that $a + b \equiv_m 6^{\alpha-1}k$ but $(1, 3)$ does not obey (1). As noted above,

$$2b \leq_m 2b + 2l :$$

there is a multiple of m between $2b$ and $2b + 2l$ (inclusive). Because

$$a + 3b \not\leq_m a + 3b + 2l$$

and $a + 3b = 2b + (a + b)$, $2b$ must be congruent mod m to an integer in $\{m - 6^{\alpha-1}k + j : 0 < j \leq 6^{\alpha-1}k\}$: otherwise, translating $2b$ and $2b + 2l$ by $a+b \equiv_m 6^{\alpha-1}k$ will yield integers still on opposite sides of a multiple of m. Because m, $2b$, and $6^{\alpha-1}k$ are even, $2b$ must be congruent mod m to an integer in $\{m - 6^{\alpha-1}k + 2j : 0 < 2j \leq 6^{\alpha-1}k\}$, and b must be congruent mod $\frac{1}{2}m$ $(= 3 \cdot 6^{\alpha-1}k)$ to an integer in $\{\frac{1}{2}m - \frac{1}{2}6^{\alpha-1}k + j : 0 < j \leq \frac{1}{2}6^{\alpha-1}k\}$ (note that $\frac{1}{2}6^{\alpha-1}k = 3 \cdot 6^{\alpha-2}k \in \mathbb{Z}$). So for some j with $0 < j \leq \frac{1}{2}6^{\alpha-1}k$, b is congruent mod m either to $\frac{1}{2}m - \frac{1}{2}6^{\alpha-1}k + j$ or to $\frac{1}{2}m + (\frac{1}{2}m - \frac{1}{2}6^{\alpha-1}k + j) = m - \frac{1}{2}6^{\alpha-1}k + j$.

If b is congruent mod m to $\frac{1}{2}m - \frac{1}{2}6^{\alpha-1}k + j$, then since $j \leq \frac{1}{2}6^{\alpha-1}k < \frac{1}{2}6^{\alpha}k = \frac{1}{2}m$,

$$0 < (1/2)m - (1/2)6^{\alpha-1}k + j < m$$

and b is not divisible by m. Since $(0, 1) \in S$,

$$b \leq_m b + l :$$

so

$$l \geq m - ((1/2)m - (1/2)6^{\alpha-1}k + j) = (1/2)m + (1/2)6^{\alpha-1}k - j \geq (1/2)m$$

and $2l \geq m$.

This contradiction implies that b is congruent mod m to $m - \frac{1}{2}6^{\alpha-1}k + j$. So

$$\begin{aligned} -3a - 2b &= -3(a + b) + b \\ &\equiv_m -3 \cdot 6^{\alpha-1}k + m - (1/2)6^{\alpha-1}k + j = (5/2)6^{\alpha-1}k + j. \end{aligned}$$

Since $0 < j \leq \frac{1}{2}6^{\alpha-1}k$, $-3a - 2b$ is not divisible by m. Yet since $(-3, -2) \in S$,

$$-3a - 2b \leq_m -3a - 2b + l$$

and

$$l \geq m - ((5/2)6^{\alpha-1}k + j) = (7/2)6^{\alpha-1}k - j \geq 3 \cdot 6^{\alpha-1}k = (1/2)m.$$

So again $2l \geq m$, and one reaches a contradiction showing that $(1,3)$ obeys (1) when $a+b \equiv_m 6^{\alpha-1}k$.

Assume finally that $a+b \equiv_m 5 \cdot 6^{\alpha-1}k$. Again one may argue by contradiction to show that $(1,3)$ obeys (1): so, assume

$$a + 3b \not\leq_m a + 3b + 2l.$$

Since $(2,4) \in S$,

$$2a + 4b \leq_m 2a + 4b + 2l,$$

and $a + 3b = 2a + 4b + (-a-b)$, where

$$-a - b \equiv_m -5 \cdot 6^{\alpha-1}k \equiv_m 6^{\alpha-1}k.$$

So much as before, $2a+4b$ must be congruent mod m to an integer in $\{m - 6^{\alpha-1}k + j : 0 < j \leq 6^{\alpha-1}k\}$. Again because $2a+4b$ is even, one argues as above that for some integer j with $0 < j \leq \frac{1}{2}6^{\alpha-1}k$, $a + 2b = \frac{1}{2}(2a+4b)$ is congruent mod m either to $\frac{1}{2}m - \frac{1}{2}6^{\alpha-1}k + j$ or to $m - \frac{1}{2}6^{\alpha-1}k + j$.

If $a+2b$ is congruent mod m to $\frac{1}{2}m - \frac{1}{2}6^{\alpha-1}k + j$, then

$$\begin{aligned} b &= (a+2b) + (-a-b) \\ &\equiv_m (1/2)m - (1/2)6^{\alpha-1}k + j + 6^{\alpha-1}k = (1/2)m + (1/2)6^{\alpha-1}k + j. \end{aligned}$$

Thus

$$\begin{aligned} -3a - 2b = 3(-a-b) + b &\equiv_m 3 \cdot 6^{\alpha-1}k + (1/2)m + (1/2)6^{\alpha-1}k + j \\ &\equiv_m (1/2)6^{\alpha-1}k + j. \end{aligned}$$

Since $0 < j \leq \frac{1}{2}6^{\alpha-1}k$ and $6^{\alpha-1}k = \frac{1}{6}m$, $-3a-2b$ is not divisible by m. As noted above,

$$-3a - 2b \leq_m -3a - 2b + l:$$

so

$$l \geq m - ((1/2)6^{\alpha-1}k + j) = (11/2)6^{\alpha-1}k - j \geq 5 \cdot 6^{\alpha-1}k = (5/6)m$$

and $2l \geq m$.

This contradiction implies that $a+2b$ is congruent mod m to $m - \frac{1}{2}6^{\alpha-1}k + j$. So as above

$$\begin{aligned} b = (-a-b) + (a+2b) &\equiv_m 6^{\alpha-1}k + m - (1/2)6^{\alpha-1}k + j \\ &\equiv_m (1/2)6^{\alpha-1}k + j \end{aligned}$$

and, since $0 < j \leq \frac{1}{2}6^{\alpha-1}k$, b is not divisible by m. As noted before,

$$b \leq_m b + l:$$

so

$$l \geq m - ((1/2)6^{\alpha-1}k + j),$$

and one reaches the same contradiction as before. Thus $(1, 3)$ obeys (1), no matter what the nature of $a + b \bmod m$, and the argument is complete. □

6 Uniformity

This section proves a version of Proposition 5.1 for theories of ordered modules viewed as two-sorted structures with a domain of scalars as well as a domain of vectors. In [2] van den Dries and Holly prove a related quantifier-elimination result for torsion-free modules over Bézout domains, and the present treatment borrows their axioms for Bézout domains. After pointing out that the kind of quantifier elimination established in [2] does not work in a more expressive language for ordered modules, Weispfenning shows how to eliminate quantifiers in favor of bounded quantifiers in this new context ([10], p. 51)[7]. If one views Proposition 5.1 as a quantifier-elimination result, this section aims to reveal how the coefficients of the quantifier-free formula depend on those of the given existential formula.

One may start with a two-sorted language $\mathcal{L}^{\mathrm{II}}$ for ordered modules. The two sorts—a module sort and a scalar sort—feature symbols $+$, $-$, $\leq$, 0 in the module sort (for addition, additive inverse, order, and additive identity); symbols $+$, $-$, $\cdot$, $\leq$, 0, 1, α, β, γ, g in the scalar sort (for addition, additive inverse, multiplication, order, additive identity, multiplicative identity, and four binary operations described below); a binary function symbol $\cdot$ (for scalar multiplication on the left); and a ternary relation symbol $\equiv$ (to represent congruences as described below). In the following description of of ordered modules over ordered Bézout domains, Greek letters ρ, σ, $\tau, \ldots$ will be used as scalar variables and letters x, y, $z, \ldots$ as module variables. T^{II} consists of $\mathcal{L}^{\mathrm{II}}$-sentences expressing axioms for ordered Abelian groups in the module sort; axioms for nontrivial ordered integral domains in the scalar sort; and the

7 If one admits a constant symbol for $1 \in \mathbb{Z}$, $\exists x_1 \ldots \exists x_m (u \leq \overline{x} \leq v \wedge D(\overline{x}) \wedge H(\overline{x}; \overline{y}))$ is equivalent over $\mathbb{Z}$ to a single-sorted version of one of Weispfenning's bounded-quantifier formulas: $\exists x_1 \ldots \exists x_m (\wedge_{i=1}^{m} |x_i| \leq (l+1)p \wedge \wedge_{j=1}^{m} (0 \leq x_j \wedge x_j + u \leq v) \wedge D(\overline{x} + u) \wedge H(\overline{x} + u; \overline{y}))$, where p is the least common multiple of the moduli in H and $l \geq 0$ is the greatest length of a chain in the graph $\{(i, j) : x_i \leq x_j$ is a conjunct of $D\}$.

universal closures of

$$\rho\cdot(x+y)=(\rho\cdot x)+(\rho\cdot y)\ \wedge\ (\rho+\sigma)\cdot x=(\rho\cdot x)+(\sigma\cdot x)\ \wedge\ 1\cdot x=x$$
$$\wedge\ (\rho\sigma)\cdot x=\rho\cdot(\sigma\cdot x)\ \wedge\ (\rho\cdot x=0\rightarrow\rho=0\ \vee\ x=0)$$
$$\wedge\ (\rho\leq\sigma\ \wedge\ 0\leq x\rightarrow\rho\cdot x\leq\sigma\cdot x)\wedge(0\leq\rho\wedge x\leq y\rightarrow\rho\cdot x\leq\rho\cdot y),$$

of

$$x\equiv_\rho y\leftrightarrow\exists z(x=y+\rho\cdot z),$$

and of

$$g(\rho,\sigma)=g(\sigma,\rho)\ \wedge\ \rho=\gamma(\rho,\sigma)g(\rho,\sigma)\ \wedge\sigma=\gamma(\sigma,\rho)g(\rho,\sigma)$$
$$\wedge\ 1=\alpha(\rho,\sigma)\gamma(\rho,\sigma)+\beta(\rho,\sigma)\gamma(\sigma,\rho).$$

This last sentence restricts attention to Bézout domains, any one of which may be expanded to a structure, with functions corresponding to g, α, β, and γ, in which g names a greatest-common-divisor function.

A formula $(*)=$

$$r\equiv_\theta s$$

is called a congruence of modulus θ. In torsion-free modules over Bézout domains, applying existential quantifiers to conjunctions of congruences with nonzero moduli yields conjunctions of congruences with nonzero moduli. To express this result exactly in the present context, let $k\in\mathbb{N}$, $\rho_1,\ldots,\rho_k$ be new constant symbols of scalar sort, $\mathcal{L}^{\mathrm{II}}_k=\mathcal{L}^{\mathrm{II}}\cup\{\rho_1,\ldots,\rho_k\}$ and $\mathcal{L}^{\mathrm{II,s}}$ ($\mathcal{L}^{\mathrm{II,s}}_k$) consist of the symbols of $\mathcal{L}^{\mathrm{II}}$ ($\mathcal{L}^{\mathrm{II}}_k$) purely of scalar sort. If R is an ordered Bézout domain viewed as an $\mathcal{L}^{\mathrm{II,s}}$-structure, and $y_1,\ldots,y_k\in R$, the substructure $\langle\overline{y}\rangle_R$ of R generated by $y_1,\ldots,y_k$ is a Bézout domain; letting $\rho_1,\ldots,\rho_k$ name $y_1,\ldots,y_k$ respectively, one may form the diagram $\Delta(\overline{y})$ of $\langle\overline{y}\rangle_R$. Say that a congruence $(*)$ is $\Delta(\overline{y})$-proper just in case θ is a closed $\mathcal{L}^{\mathrm{II,s}}_k$-term with $\Delta(\overline{y})\models\theta\neq0$. With the help of this jargon one may state

Lemma 6.1 *Modulo $T^{\mathrm{II}}\cup\Delta(\overline{y})$, the set of conjunctions of $\Delta(\overline{y})$-proper congruences is closed under existential quantification.*

Proof Let $r_1,\ldots,r_k,s_1,\ldots,s_k$ be closed $\mathcal{L}^{\mathrm{II,s}}_k$-terms with

$$\Delta(\overline{y})\models\bigwedge_{i=1}^{k}r_i,s_i\neq0,$$

and let $x,y_1,\ldots,y_k,z$ be distinct module variables. Because T's models are torsion-free, $T^{\mathrm{II}} \cup \Delta(\overline{y})$ implies that

$$\exists x(\bigwedge_{i=1}^{k} s_i x \equiv_{r_i} y_i)$$

is equivalent to

$$\exists z(\bigwedge_{i=1}^{k} z \equiv_{s_i' r_i} s_i' y_i \ \wedge \ z \equiv_s 0),$$

where s is $\prod_i s_i$ and s_i' is $\prod_{j \neq i} s_j$. So the given quantifier-elimination problem reduces to consideration of formulas $(\star) =$

$$\exists x(\bigwedge_{i=1}^{k} x \equiv_{r_i} y_i).$$

If a and b are nonzero elements of the Bézout domain $\langle \overline{y} \rangle_R$, then $l(a,b) = ab/g(a,b)$ is a least common multiple of a and b. Because Bézout domains are Prüfer domains ([4], Theorem 18.1), the principal ideals $(l(g(a,b), g(a,c)))$ and $(g(a, l(b,c)))$ are equal for all nonzero $a, b, c \in \langle \overline{y} \rangle_R$. So one may carry out the usual proof of the Chinese Remainder Theorem ([4], Theorems 21.1 and 21.2) to show that in any model of $T^{\mathrm{II}} \cup \Delta(\overline{y})$, $(\star)$ is equivalent to

$$\bigwedge_{i,j=1}^{k} y_i \equiv_{g(r_i, r_j)} y_j.$$

□

When $\overline{x} = (x_1, \ldots, x_m)$ and $\overline{y} = (y_1, \ldots, y_n)$ are disjoint lists of distinct module variables and $\overline{\sigma} = (\sigma_1, \ldots, \sigma_q)$ is a list of distinct scalar variables, an $\mathcal{L}^{\mathrm{II}}$-basic pair (D, H) consists of a conjunction $D(\overline{x})$ of inequalities

$$x_i \leq x_j$$

—where the graph $\{(i,j) : x_i \leq x_j$ is a conjunct of $D\}$ has no cycles—and a conjunction $H(\overline{x}; \overline{y}; \overline{\sigma})$ of congruences in $\overline{x}$ and $\overline{y}$ with scalar variables among the $\overline{\sigma}$. For each $\mathcal{L}^{\mathrm{II}}$-basic pair (D, H), adjoin an $(n+q+2)$-place relation symbol $\leq^{D,H}$ to $\mathcal{L}^{\mathrm{II}}$ to obtain $\mathcal{L}^{\mathrm{II}}_{\mathrm{in}}$, and let $T^{\mathrm{II}}_{\mathrm{in}}$ be the $\mathcal{L}^{\mathrm{II}}_{\mathrm{in}}$-theory consisting of T^{II} together with all axioms

$$\forall u, v, \overline{y}, \overline{\sigma} \, (u \leq^{D,H}_{\overline{y};\overline{\sigma}} v \leftrightarrow \exists \overline{x} \, (u \leq \overline{x} \leq v \wedge D(\overline{x}) \wedge H(\overline{x}; \overline{y}; \overline{\sigma}))).$$

Any formula $r \leq^{D,H}_{\overline{t};\overline{\theta}} s$ is called a congruence inequality, and its moduli are the moduli of the congruences in H. If $\Delta(\overline{y})$ is as above, a congruence inequality in $\mathcal{L}^{\mathrm{II}}_{\mathrm{in},k} = \mathcal{L}^{\mathrm{II}}_{\mathrm{in}} \cup \{\rho_1, \ldots, \rho_k\}$ is said to be $\Delta(\overline{y})$-proper just in case the moduli of the congruence inequality are closed $\mathcal{L}^{\mathrm{II}}_{\mathrm{in},k}$-terms δ with $\Delta(\overline{y}) \models \delta \neq 0$.

To establish an analogue of Proposition 5.1 for these new congruence inequalities, one needs new versions of $\mathcal{L}^+$ and T^+. Let $\mathcal{L}^{\mathrm{II},+} = \mathcal{L}^{\mathrm{II}}_{\mathrm{in}} \cup \{L, U\}$ and $T^{\mathrm{II},+}$ be the $\mathcal{L}^{\mathrm{II},+}$-theory extending $T^{\mathrm{II}}_{\mathrm{in}}$ by axioms written semi-formally as follows:

$$\forall x\, (L_x \text{ and } U_x \text{ are convex submodules})$$
$$\forall x\, (L_x \subseteq U_x)$$
$$\forall x\, (x \in U_x \ \wedge \ (x \in L_x \rightarrow x = 0))$$
$$U_0 = \{0\} = L_0$$
$$\forall x, y, \rho\, [(\rho \cdot x \in L_y \rightarrow \rho = 0 \vee x \in L_y) \wedge (\rho \cdot x \in U_y \rightarrow \rho = 0 \vee x \in U_y)]$$
$$\begin{aligned} \forall w\, [\forall x, y \in U_w \forall \rho\, (x + L_w < y + L_w \wedge 0 < \rho \\ \rightarrow \exists z \in U_w (x + L_w < \rho \cdot z + L_w < y + L_w)) \\ \vee \exists v \in U_w \{0 + L_w < v + L_w \\ \wedge \forall y \in U_w (0 + L_w < y + L_w \rightarrow v + L_w \leq y + L_w) \\ \wedge \forall x \in U_w \forall \rho (0 < \rho \rightarrow \\ \exists z \in U_w (x + L_w \leq \rho \cdot z + L_w < x + \rho \cdot v + L_w))]] \end{aligned}$$
$$\begin{aligned} \forall x, y\, [(L_x \subset U_y \rightarrow U_x \subseteq U_y) \wedge (L_x \subset L_y \rightarrow U_x \subseteq L_y) \\ \wedge (L_x \subset U_y \rightarrow L_x \subseteq L_y) \wedge (U_x \subset U_y \rightarrow U_x \subseteq L_y)]. \end{aligned}$$

One may view $T^{\mathrm{II},+}$ also as a theory in the language $\mathcal{L}^{\mathrm{II},+}_k = \mathcal{L}^{\mathrm{II},+} \cup \{\rho_1, \ldots, \rho_k\}$.

The proof of Proposition 2.1 shows that every ordered Abelian group, viewed as a $\mathbb{Z}$-module, may be expanded to a model of $T^{\mathrm{II},+}$; in this context, the fifth axiom comes for free, and the sixth states that each quotient U_x/L_x is dense regular or a $\mathbb{Z}$-group. So if one wants to generalize Proposition 5.1 to a result handling conjunctions of congruence inequalities, with scalar variables, in $\mathbb{Z}$-modules, there is no harm in assuming that $T^{\mathrm{II},+}$, rather than T^+, is the background theory. Yet one should note that when the ring of scalars is non-Archimedean, an ordered module need not be built out of convex submodules having quotients with the properties, akin to regularity, expressed by the axioms for $T^{\mathrm{II},+}$. It is not a conservative extension of $T^{\mathrm{II}}_{\mathrm{in}}$, as the following example shows[8]. Let $(\mathbb{Q}^*, +, -, \cdot, 0, 1, \leq, \mathbb{Z}^*)$ be a proper elementary extension of

8 See the Conclusion for more about this example.

the expanded ordered ring $(\mathbb{Q}, +, -, \cdot, 0, 1, \leq, \mathbb{Z})$. Because the extension is elementary, every element of $\mathbb{Q}^*$ is the ratio of elements of $\mathbb{Z}^*$, and one may restrict attention to the set $\mathbb{Q}^{\text{fin}}$ of elements of $\mathbb{Q}^*$ which may be written with denominator from $\mathbb{Z}$. $\mathbb{Q}^{\text{fin}}$ is an ordered $\mathbb{Z}^*$-submodule of $\mathbb{Q}^*$, and since every element of $\mathbb{Q}^{\text{fin}}$ is the ratio between an element of $\mathbb{Z}^*$ and an element of $\mathbb{Z}$, $\mathbb{Q}^{\text{fin}}$ has no nontrivial convex submodules. $\mathbb{Q}^{\text{fin}}$ is densely ordered, and in fact has $\mathbb{Q}$ as a convex subgroup: for an element of $\mathbb{Q}^*$, with denominator from $\mathbb{Z}$, which is bounded in absolute value by a rational number must be a rational number. So if $n > \mathbb{Z}$ belongs to $\mathbb{Z}^*$, there is no $a \in \mathbb{Q}^{\text{fin}}$ with $0 < na < 1$, and the ordered $\mathbb{Z}^*$-module $\mathbb{Q}^{\text{fin}}$, though a model of $T^{\text{II}}_{\text{in}}$, cannot be expanded to a model of $T^{\text{II},+}$.

Given these preliminaries, one may state

Lemma 6.2 *Let R be an ordered Bézout domain, $y_1, \ldots, y_k \in R$, and $\Delta(\overline{y})$ be the diagram of the $\mathcal{L}^{\text{II,s}}_k$-structure $\langle \overline{y} \rangle_R$, where ρ_i names y_i for $i = 1, \ldots, k$. Modulo $T^{\text{II},+} \cup \Delta(\overline{y})$, the existential quantification of a conjunction of congruence inequalities with no scalar variables is equivalent to a conjunction of $\Delta(\overline{y})$-proper congruence inequalities.*

Proof Each congruence inequality is equivalent, modulo $T^{\text{II}}_{\text{in}}$, to an existential quantification of a conjunction of congruence inequalities equivalent to weak inequalities or to congruences. Since $T^{\text{II}}_{\text{in}}$ implies that

$$\forall x, y, \sigma\, ((x \equiv_0 y \leftrightarrow x = y) \wedge (x \equiv_{-\sigma} y \leftrightarrow x \equiv_\sigma y)),$$

and since $\Delta(\overline{y})$ determines which closed $\mathcal{L}^{\text{II,s}}_k$-terms name zero, positive, or negative elements, the desired result holds if it holds for existential quantifications of conjunctions of $\Delta(\overline{y})$-proper congruence inequalities with positive moduli.

One establishes this result by repeating the arguments of Sections 2–5. Note first that since only $\Delta(\overline{y})$-proper congruence inequalities are under consideration, one may restrict the scalar sorts to the ordered Bézout domains of elements named by closed $\mathcal{L}^{\text{II,s}}_k$-terms. Because the $\mathcal{M}_i$'s and $\mathcal{H}$ are models of the diagram $\Delta(\overline{y})$ of $\langle \overline{y} \rangle_R$, all $\mathcal{L}_{\text{in}}$-homomorphisms in the proof are isomorphisms on these domains, and one may assume that all the $\mathcal{L}_{\text{in}}$-homomorphisms are homomorphisms of $D = \langle \overline{y} \rangle_R$-modules. To make the argument as close as possible to its predecessor one might eliminate the scalar sort: closed $\mathcal{L}^{\text{II,s}}_k$-terms would be eliminated in favor of function symbols for the corresponding scalar multiplications, and any congruences (or congruence inequalities) in the argument could be regarded as formulas without scalar-sort places because the moduli of these congruences (or congruence inequalities) would be closed $\mathcal{L}_k^{\mathrm{II,s}}$-terms τ for which $\Delta(\overline{y}) \models 0 < \tau$. Though the argument demands that one take quotients of $\mathcal{H}$ by definable submodules $U_a^{\mathcal{H}}$ or $L_a^{\mathcal{H}}$, the proofs of Lemmas 2.2–2.4 allow one to show that

Lemma 6.3 *Let $\mathcal{H} \models T^{\mathrm{II},+} \cup \Delta(\overline{y})$ be $\aleph_1$-saturated and G be $U_a^{\mathcal{H}}$ or $L_a^{\mathcal{H}}$. Restrict scalars to D, so that $\mathcal{H}$ and G become ordered D-modules. There is a D-submodule S of H such that as an ordered D-module, H is the lexicographic sum of S and G:*

$$H = S \overrightarrow{\times} G.$$

If $\mathcal{S}$ is the substructure of the D-module $\mathcal{H}$ determined by S, and $\pi : H \to H/G$ is the usual quotient map, then one may extend the D-module H/G to an $\mathcal{L}_k^{\mathrm{II},+}$-structure isomorphic via $(\pi \restriction S)^{-1}$ to $\mathcal{S}$, and both are models of $T^{\mathrm{II},+} \cup \Delta(\overline{y})$.

The earlier arguments often divide into cases, depending on whether an open interval $(A + L_x, B + L_x)$ in a quotient U_x/L_x is finite or infinite. When U_x/L_x is densely ordered, "finite" means that $A + L_x = B + L_x$. When U_x/L_x is discretely ordered with least element $1 + L_x$, "finite" means that $B - A + L_x$ is at most $\alpha \cdot 1 + L_x$, where α is a scalar named by some closed $\mathcal{L}_k^{\mathrm{II,s}}$-term. Subject to these qualifications, the earlier arguments generalize to the present context. □

A compactness argument now yields

Proposition 6.4 *Let $\psi(\overline{x}, \overline{y}, \overline{\sigma})$ be a conjunction of congruence inequalities with free variables among the module variables $\overline{x} = (x_1, \ldots, x_m)$ and $\overline{y} = (y_1, \ldots, y_n)$ and the scalar variables $\overline{\sigma} = (\sigma_1, \ldots, \sigma_k)$. There are $l \geq 1$, quantifier-free formulas $\delta_i(\overline{\sigma})$ of scalar sort (for $i = 1, \ldots, l$), scalar terms $\tau_{ij}(\overline{\sigma})$ (for $i = 1, \ldots, l$ and $j = 1, \ldots, m_i$), and conjunctions $\varphi_i(\overline{x}, \overline{\sigma})$ of congruence inequalities, with moduli among the $\tau_{ij}(\overline{\sigma})$'s, such that $T^{\mathrm{II},+}$ implies*

$$\forall \overline{\sigma}\, [(\bigvee_{i=1}^{l} \delta_i(\overline{\sigma})) \wedge \bigwedge_{i \neq j} \neg(\delta_i(\overline{\sigma}) \wedge \delta_j(\overline{\sigma}))$$

$$\wedge \bigwedge_{i=1}^{l} \{\delta_i(\overline{\sigma}) \to \wedge_{j=1}^{m_i} \tau_{ij}(\overline{\sigma}) \neq 0 \wedge \forall \overline{x}\, (\exists \overline{y} \psi(\overline{x}, \overline{y}, \overline{\sigma}) \leftrightarrow \varphi_i(\overline{x}, \overline{\sigma}))\}].$$

Proof Let S be the Stone space of $T^{\mathrm{II},+}$ in the language $\mathcal{L}_k^{\mathrm{II},+}$. When $t \in S$—i.e., when t is a complete $\mathcal{L}_k^{\mathrm{II},+}$-theory extending $T^{\mathrm{II},+}$—let Δ_t be the set of atomic and negated atomic $\mathcal{L}_k^{\mathrm{II,s}}$-sentences in t. Δ_t is the diagram of an ordered Bézout domain $\langle \overline{y}_t \rangle$ generated by k elements $y_{t1}, \ldots, y_{tk}$ named by $\rho_1, \ldots, \rho_k$, respectively. Lemma 6.2 provides scalar terms $\tau_{tj}(\overline{\rho})$, for $i = 1, \ldots, m_t$, and a conjunction $\theta_t(\overline{x}, \overline{\rho})$ of congruence inequalities, with moduli among the $\tau_{tj}(\overline{\rho})$, such that

$$T^{\mathrm{II},+} \cup \Delta_t \models \bigwedge_{j=1}^{m_t} \tau_{tj}(\overline{\rho}) \neq 0 \wedge \forall \overline{x}\, (\exists \overline{y} \psi(\overline{x}, \overline{y}, \overline{\rho}) \leftrightarrow \theta_t(\overline{x}, \overline{\rho})).$$

The compactness theorem yields a finite $\Delta_t' \subseteq \Delta_t$ with

$$T^{\mathrm{II},+} \models \bigwedge \Delta_t' \to \bigwedge_{j=1}^{m_t} \tau_{tj}(\overline{\rho}) \neq 0 \wedge \forall \overline{x}\, (\exists \overline{y} \psi(\overline{x}, \overline{y}, \overline{\rho}) \leftrightarrow \theta_t(\overline{x}, \overline{\rho})).$$

$\{\Delta_t'\}_{t \in S}$ corresponds to an open covering of the compact space S, and this open covering has a finite subcovering determined by some finite set $\{\Delta_{t_1}', \ldots, \Delta_{t_l}'\}$. Replacing the $\overline{\rho}$'s by the $\overline{\sigma}$'s, one may easily obtain suitable formulas $\delta_i(\overline{\sigma})$ as propositional combinations of the $\wedge \Delta_{t_i}'$'s. □

An immediate byproduct is

Corollary 6.5 [9] *Let $\psi(\overline{x}, \overline{y}, \overline{\sigma})$ be a conjunction of congruence inequalities with free variables among the module variables $\overline{x} = (x_1, \ldots, x_m)$ and $\overline{y} = (y_1, \ldots, y_n)$ and the scalar variables $\overline{\sigma} = (\sigma_1, \ldots, \sigma_k)$. There are $l \geq 1$, quantifier-free formulas $\delta_i(\overline{\sigma})$ of scalar sort (for $i = 1, \ldots, l$), scalar terms $\tau_{ij}(\overline{\sigma})$ (for $i = 1, \ldots, l$ and $j = 1, \ldots, m_i$), and conjunctions $\varphi_i(\overline{x}, \overline{\sigma})$ of congruence inequalities, with moduli among the $\tau_{ij}(\overline{\sigma})$'s, such that in any ordered $\mathbb{Z}$-module*

$$\forall \overline{\sigma}\, [(\bigvee_{i=1}^{l} \delta_i(\overline{\sigma})) \wedge \bigwedge_{i \neq j} \neg(\delta_i(\overline{\sigma}) \wedge \delta_j(\overline{\sigma}))$$

$$\wedge \bigwedge_{i=1}^{l} \{\delta_i(\overline{\sigma}) \to \wedge_{j=1}^{m_i} \tau_{ij}(\overline{\sigma}) \neq 0 \wedge \forall \overline{x}\, (\exists \overline{y}\, \psi(\overline{x}, \overline{y}, \overline{\sigma}) \leftrightarrow \varphi_i(\overline{x}, \overline{\sigma}))\}].$$

9 If when considering the $\mathbb{Z}$-module $\mathbb{Z}$ one introduces a module constant for $1 \in \mathbb{Z}$, one may follow the procedure mentioned in Note 7 to rewrite each $\varphi_i(\overline{x}, \overline{\sigma})$ as a formula using only bounded-existential quantifiers in the sense of [10].

7 Farkas' Lemma

On the basis of Propositions 5.1 and 6.4 one may generalize the analogues in [7] of Farkas' Lemma. Following Section 5 of [7], one may introduce a restricted kind of congruence inequality symmetric between scalars and variables. If in $\mathcal{L}_{\text{in}}$ $\overline{u} = (u_1, \ldots, u_k)$, $\overline{v} = (v_1, \ldots, v_k)$, and $\overline{x} = (x_1, \ldots, x_m)$ are disjoint lists of distinct variables and $(D(\overline{x}), H(\overline{x}; \overline{u}, \overline{v}))$ is a basic pair, $(\#) =$

$$\sum_{i=1}^{k} u_i \preccurlyeq^{D,H} \sum_{i=1}^{k} v_i$$

abbreviates the congruence inequality

$$\sum_{i=1}^{k} u_i \leq^{D,H}_{\overline{u},\overline{v}} \sum_{i=1}^{k} v_i,$$

and any formula

$$\sum_{i=1}^{k} t_i \preccurlyeq^{D,H} \sum_{i=1}^{k} w_i$$

obtained from $(\#)$ through replacement of the free variables $\overline{u}$, $\overline{v}$ by $\mathcal{L}$-terms $\overline{t}$, $\overline{w}$ is called a special congruence inequality. Following the proof of Lemma 5.1 in [7], one may prove

Lemma 7.1 *If $\overline{y} = (y_1, \ldots, y_k)^{\mathrm{T}}$ is a column of distinct variables, $a, b \in \mathbb{Z}^k$, and C is an integer matrix with k columns, then any congruence inequality*

$$a^{\mathrm{T}}\overline{y} \leq^{D,H}_{C\overline{y}} b^{\mathrm{T}}\overline{y}$$

is equivalent modulo T_{in} to a special congruence inequality

$$\sum_{i=1}^{k} f_i y_i \preccurlyeq^{D,K} \sum_{i=1}^{k} g_i y_i.$$

Then by the proof of Corollary 5.2 in [7]—but with Proposition 5.1 and Lemma 7.1 replacing Proposition 4.1 and Lemma 5.1 from [7]—one shows that

Corollary 7.2 *Let A be an m by n matrix over $\mathbb{Z}$. There are finitely many basic pairs (D_i, H_i) $(1 \leq i \leq l)$ such that for any ordered integral domain D and any $b \in D^m$ the following conditions are equivalent:*

(i) $b = Az$ *for some* $z \geq 0$ *in* D^n.

(ii) *For all* $y, w \in D^m$, *if* $y^{\mathrm{T}} A \preccurlyeq^{D_i,H_i} w^{\mathrm{T}} A$, *then* $y^{\mathrm{T}} b \preccurlyeq^{D_i,H_i} w^{\mathrm{T}} b$ *(for all* i, $1 \leq i \leq l$*).*

In (ii), $y^{\mathrm{T}} A \preccurlyeq^{D_i,H_i} w^{\mathrm{T}} A$ means that for each j, the jth column a_j of A obeys $y^{\mathrm{T}} a_j \preccurlyeq^{D_i,H_i} w^{\mathrm{T}} a_j$: i.e., $\sum_{i=1}^m y_i a_{ij} \preccurlyeq^{D_i,H_i} \sum_{i=1}^m w_i a_{ij}$.

Proposition 6.4 should permit the generalization of Corollary 7.2 to matrices with non-integer entries, and should allow one to say something about how the basic pairs (D_i, H_i) depend on A. To obtain the latter kind of result, one starts by generalizing the notion of special congruence inequality from $\mathcal{L}_{\text{in}}$ to $\mathcal{L}_{\text{in}}^{\text{II}}$. The definition may be stated exactly as above, though now the moduli in (#) may be scalar $\mathcal{L}_{\text{in}}^{\text{II}}$-terms, and the module terms $\bar{t}$, $\overline{w}$ may contain scalar terms. Proceeding as in the proof of Lemma 5.1 of [7] and the present Lemma 7.1, one shows that

Lemma 7.3 *Suppose* $\overline{y} = (y_1, \ldots, y_k)^{\mathrm{T}}$ *is a column of distinct module variables,* a *and* b *are each columns of* k *scalar terms with free variables among* $\overline{\sigma} = (\sigma_1, \ldots, \sigma_l)$*,* C *is an* m *by* k *matrix of scalar terms with free variables among* $\overline{\sigma}$*, and* $(D(\overline{x}), H(\overline{x}; z_1, \ldots, z_m; \overline{\sigma}))$ *is an* $\mathcal{L}^{\text{II}}$*-basic pair. There are* $q \in \mathbb{N}$*, quantifier-free* $\mathcal{L}^{\text{II,s}}$*-formulas* $\delta_j(\overline{\sigma})$ *(for* $j = 1, \ldots, q$*), and special congruence inequalities*

$$\sum_{i=1}^{k} f_{ij} y_i \preccurlyeq^{D,K_j} \sum_{i=1}^{k} g_{ij} y_i,$$

with scalar variables among $\overline{\sigma}$*, such that* $T^{\text{II},+}$ *implies*

$$\forall \overline{\sigma} \, [(\bigvee_{j=1}^{q} \delta_j(\overline{\sigma})) \wedge \bigwedge_{j \neq j'} \neg(\delta_j(\overline{\sigma}) \wedge \delta_{j'}(\overline{\sigma}))$$

$$\wedge \bigwedge_{j=1}^{q} \{\delta_j(\overline{\sigma}) \to \forall \overline{y} \, (a^{\mathrm{T}} \overline{y} \leq_{C\overline{y}}^{D,H} b^{\mathrm{T}} \overline{y} \leftrightarrow \sum_{i=1}^{k} f_{ij} y_i \preccurlyeq^{D,K_j} \sum_{i=1}^{k} g_{ij} y_i)\}].$$

The δ_j's are used to declare which entries of a, b are zero or nonzero, as one must do in the proof of Lemma 5.1 in [7].

Now using Proposition 6.4 instead of Proposition 5.1, one proves

Corollary 7.4 *Let* A *be an* m *by* n *matrix whose entries are terms of* $\mathcal{L}^{\text{II,s}}$ *in the scalar variables* $\overline{\sigma} = (\sigma_1, \ldots, \sigma_l)$*. There are finitely many quantifier-free* $\mathcal{L}^{\text{II,s}}$*-formulas* $\delta_i(\overline{\sigma})$ *(for* $i = 1, \ldots, q$*), and finitely many* $\mathcal{L}_{\text{in}}^{\text{II}}$*-basic pairs* (D_{ij}, H_{ij}) *(for* $i = 1, \ldots, q$ *and* $j = 1, \ldots, l_i$*)—with the* *moduli of H_{ij} among the $\mathcal{L}^{\mathrm{II}}$-terms $t_{ijk}(\overline{\sigma})$ (for $k = 1, \ldots, r_{ij}$)—having the following properties. Suppose D is an ordered integral domain which, as an ordered module over itself, expands to a model of $T^{\mathrm{II},+}$. Pick $d_1, \ldots, d_l \in D$, and let A^D be the m by n matrix denoted by A in D when $\sigma_1, \ldots, \sigma_l$ refer to $d_1, \ldots, d_l$. There is a unique $i = 1, \ldots, q$ such that $\overline{d} = (d_1, \ldots, d_l)$ obeys δ_i. All the corresponding $t_{ijk}(\overline{d})$'s are nonzero, and for every $b \in D^m$ the following conditions are equivalent:*

(i) *$b = A^D z$ for some $z \geq 0$ in D^n.*

(ii) *For all $y, w \in D^m$, if $y^{\mathrm{T}} A \preccurlyeq^{D_{ij},H_{ij}} w^{\mathrm{T}} A$, then $y^{\mathrm{T}} b \preccurlyeq^{D_{ij},H_{ij}} w^{\mathrm{T}} b$ (for all j, $1 \leq j \leq l_i$).*

Corollary 7.4 applies in particular to $\mathbb{Z}$ viewed as an ordered module over itself. If D is any ordered integral domain, but one picks $d_1, \ldots, d_l \in \mathbb{Z}$, the conclusion of Corollary 7.4 still holds.

8 Conclusion

One type of result, present in Section 5 of [7] but missing from the last section, asks that one start from a condition like 7.2(ii) and produce a corresponding condition 7.2(i). That is: given basic pairs (D_i, H_i) for $i = 1, \ldots, l$, one wants to expand the matrix A to a matrix $(A\ C)$ for which

$$b = (A\ C)z \text{ for some } z \geq 0 \text{ from } D$$

just in case

$$\text{For all } y, w \in D^m, \text{ if } y^{\mathrm{T}} A \preccurlyeq^{D_i,H_i} w^{\mathrm{T}} A, \text{ then } y^{\mathrm{T}} b \preccurlyeq^{D_i,H_i} w^{\mathrm{T}} b \quad (\text{for all } i, 1 \leq i \leq l).$$

$\mathbb{Z}$, as an ordered module over itself, obeys such a result ([7], Proposition 5.3), but the proof rests on the principle that the set of all nonnegative integer-vector solutions to a system of linear homogeneous equations with integer coefficients is the span, by nonnegative integer scalars, of finitely many integer vectors (see [5], pp. 102–103). I do not know which integral domains D other than $\mathbb{Z}$ obey such a principle, whose absence was noted in a related context in ([7], Section 5).

Because Corollary 7.2 generalizes Corollary 5.2 of [7] from $\mathbb{Z}$ to ordered integral domains—at least when A's entries are integers—one might hope that the integers obey a stronger analogue of Farkas' Lemma.

Though a modification of the present arguments allows one to prove a version of Lemma 6.2 for the $\mathbb{Z}^*$-module $\mathbb{Q}^{\mathrm{fin}}$, other examples resist treatment by the methods of this paper. $\mathbb{Q}^{\text{fin}}$ becomes a model of the theory, like $T^{\text{II},+}$, in which the sixth axiom about L and U is replaced by

$$\forall w \forall x, y, z \in U_w \forall \rho$$
$$[0 + L_w < z + L_w \wedge 0 < \rho \wedge (x + \rho \cdot z) + L_w < y + L_w \rightarrow$$
$$\exists v \in U_w (x + L_w < \rho \cdot v + L_w < y + L_w)].$$

If $\mathcal{H}$ is an ordered module over an ordered domain R, and ρ is a positive element of R, call an open interval (c, d) in H ρ-big just in case $d - c > \rho \cdot e$ for some positive e in H. The new axiom above guarantees that each ρ-big open interval in $U_a^{\mathcal{H}}/L_a^{\mathcal{H}}$ contains a ρ-divisible element. Call the interval (c, d) ρ-infinite just in case it is $n\rho$-big for every positive integer n. The argument for Lemma 6.2 may be based on the new axiom above if one replaces talk of infinite intervals in the module by talk of ρ-infinite intervals, where ρ depends on the system of congruence inequalities that one hopes to solve in $\mathcal{H}$. Section 6 uses its stronger axioms only because the arguments based on them mimic the arguments of [7] more closely.

Yet even this new approach will not handle certain easily described examples. Let R be the valuation ring of a non-Archimedean ordered field F: i.e., R is the set of elements of F bounded in absolute value by natural numbers. View R as an ordered module over itself, and let $e \in R$ be a positive infinitesimal. The interval $(1, 1 + e)$ in R is e^2-infinite, but contains no e^2-divisible element. And if $\{0\} \subset V \subset W \subseteq R$ are convex submodules of R, then W/V is a torsion module. If R obeys anything like Lemma 6.2, the proof may demand new techniques.

References

[1] C. C. Chang and H. J. Keisler, *Model Theory*, North-Holland Publishing Co., Amsterdam, 1973.

[2] L. van den Dries and J. Holly, Quantifier elimination for modules with scalar variables, *Ann. Pure Appl. Logic* **57** (1992), 161–179.

[3] J. Farkas, Theorie der einfachen Ungleichungen, *J. Reine Angew. Math.* **124** (1902), 1–27.

[4] R. W. Gilmer, *Multiplicative ideal theory*, Queen's Papers in Pure and Applied Mathematics, No. 12, Queen's University, Kingston, 1968.

[5] J. H. Grace and A. Young, *The algebra of invariants*, at the University Press, Cambridge, 1903.

[6] W. Hodges, *Model theory*, Encylopedia of Mathematics and its Applications, No. 42, Cambridge University Press, Cambridge, 1993.

[7] P. Scowcroft, Nonnegative solvability of linear equations in certain ordered rings, *Trans. Amer. Math. Soc.* **358** (2006), 3535–3570.

[8] P. Schmitt, Model- and substructure-complete theories of ordered Abelian groups, in G. H. Müller and M. M. Richter, eds., *Models and sets*, Lecture Notes in Math. 1103, Springer-Verlag, Berlin, 1984, pp. 389–418.

[9] V. Weispfenning, Model theory of abelian l-groups, in A. M. W. Glass and W. C. Holland, eds., *Lattice-ordered groups*, Mathematics and its Applications, 48, Kluwer Academic Publishers, Dordrecht, 1989, pp. 41–79.

[10] V. Weispfenning, Complexity and uniformity of elimination in Presburger arithmetic, *Proceedings of the 1997 International Symposium on Symbolic and Algebraic Computation (Kihei, HI)*, ACM, New York, 1997, pp. 48–53 (electronic).

Model theory for metric structures

Itaï Ben Yaacov
Université de Lyon, Université Lyon 1 and CNRS

Alexander Berenstein
Universidad de los Andes

C. Ward Henson
University of Illinois at Urbana-Champaign

Alexander Usvyatsov
University of California, Los Angeles

Contents

1 Introduction

A metric structure is a many-sorted structure in which each sort is a complete metric space of finite diameter. Additionally, the structure consists of some distinguished elements as well as some functions (of several variables) (a) between sorts and (b) from sorts to bounded subsets of $\mathbb{R}$, and these functions are all required to be uniformly continuous. Examples arise throughout mathematics, especially in analysis and geometry. They include metric spaces themselves, measure algebras, asymptotic cones of finitely generated groups, and structures based on Banach spaces (where one takes the sorts to be balls), including Banach lattices, C*-algebras, etc.

The usual first-order logic does not work very well for such structures, and several good alternatives have been developed. One alternative is the logic of *positive bounded formulas with an approximate semantics* (see [23, 25, 24]). This was developed for structures from functional analysis that are based on Banach spaces; it is easily adapted to the more general metric structure setting that is considered here. Another successful alternative is the setting of *compact abstract theories* (cats; see [1, 3, 4]). A recent development is the realization that for metric structures the frameworks of positive bounded formulas and of cats are equivalent. (The full cat framework is more general.) Further, out of this discovery has come a new *continuous* version of first-order logic that is suitable for metric structures; it is equivalent to both the positive bounded and cat approaches, but has many advantages over them.

The logic for metric structures that we describe here fits into the framework of continuous logics that was studied extensively in the 1960s and then dropped (see [12]). In that work, any compact Hausdorff space X was allowed as the set of truth values for a logic. This turned out to be too general for a completely successful theory.

We take the space X of truth values to be a closed, bounded interval of real numbers, with the order topology. It is sufficient to focus on the case where X is $[0, 1]$. In [12], a wide variety of quantifiers was allowed and studied. Since our truth value set carries a natural complete linear ordering, there are two canonical quantifiers that clearly deserve special attention; these are the operations sup and inf, and it happens that these are the only quantifiers we need to consider in the setting of continuous logic and metric structures.

The continuous logic developed here is strikingly parallel to the usual first-order logic, once one enlarges the set of possible truth values from $\{0,1\}$ to $[0,1]$. Predicates, including the equality relation, become functions from the underlying set A of a mathematical structure into the interval $[0,1]$. Indeed, the natural $[0,1]$-valued counterpart of the equality predicate is a metric d on A (of diameter at most 1, for convenience). Further, the natural counterpart of the assumption that equality is a congruence relation for the predicates and operations in a mathematical structure is the requirement that the predicates and operations in a metric structure be uniformly continuous with respect to the metric d. In the $[0,1]$-valued continuous setting, connectives are continuous functions on $[0,1]$ and quantifiers are sup and inf.

The analogy between this continuous version of first-order logic (CFO) for metric structures and the usual first-order logic (FOL) for ordinary structures is far reaching. In suitably phrased forms, CFO satisfies the compactness theorem, Löwenheim-Skolem theorems, diagram arguments, existence of saturated and homogeneous models, characterizations of quantifier elimination, Beth's definability theorem, the omitting types theorem, fundamental results of stability theory, and appropriate analogues of essentially all results in basic model theory of first-order logic. Moreover, CFO extends FOL: indeed, each mathematical structure treated in FOL can be viewed as a metric structure by taking the underlying metric d to be discrete ($d(a,b) = 1$ for distinct a, b). All these basic results true of CFO are thus framed as generalizations of the corresponding results for FOL.

A second type of justification for focusing on this continuous logic comes from its connection to applications of model theory in analysis and geometry. These often depend on an ultraproduct construction [11, 15] or, equivalently, the nonstandard hull construction (see [25, 24] and their references). This construction is widely used in functional analysis and also arises in metric space geometry (see [19], for example). The logic of positive bounded formulas was introduced in order to provide a model theoretic framework for the use of this ultraproduct (see [24]), which it does successfully. The continuous logic for metric structures that is presented here provides an equivalent background for this ultraproduct construction and it is easier to use. Writing positive bounded formulas to express statements from analysis and geometry is difficult and often feels unnatural; this goes much more smoothly in CFO. Indeed, continuous first-order logic provides model theorists and analysts with a common language; this is due to its being closely parallel to first-order logic while also using familiar constructs from analysis (*e.g.*, sup and inf in place of $\forall$ and $\exists$).

The purpose of this article is to present the syntax and semantics of this continuous logic for metric structures, to indicate some of its key theoretical features, and to show a few of its recent application areas.

In Sections 1 through 10 we develop the syntax and semantics of continuous logic for metric structures and present its basic properties. We have tried to make this material accessible without requiring any background beyond basic undergraduate mathematics. Sections 11 and 12 discuss imaginaries and omitting types; here our presentation is somewhat more brisk and full understanding may require some prior experience with model theory. Sections 13 and 14 sketch a treatment of quantifier elimination and stability, which are needed for the applications topics later in the paper; here we omit many proofs and depend on other articles for the details. Sections 15 through 18 indicate a few areas of mathematics to which continuous logic for metric structures has already been applied; these are taken from probability theory and functional analysis, and some background in these areas is expected of the reader.

The development of continuous logic for metric structures is very much a work in progress, and there are many open problems deserving of attention. What is presented in this article reflects work done over approximately the last three years in a series of collaborations among the authors. The material presented here was taught in two graduate topics courses offered during that time: a Fall 2004 course taught in Madison by Itaï Ben Yaacov and a Spring 2005 course taught in Urbana by Ward Henson. The authors are grateful to the students in those courses for their attention and help. The authors' research was partially supported by NSF Grants: Ben Yaacov, DMS-0500172; Berenstein and Henson, DMS-0100979 and DMS-0140677; Henson, DMS-0555904.

2 Metric structures and signatures

Let (M, d) be a complete, bounded metric space[1]. A *predicate* on M is a uniformly continuous function from M^n (for some $n \geq 1$) into some bounded interval in $\mathbb{R}$. A *function* or *operation* on M is a uniformly continuous function from M^n (for some $n \geq 1$) into M. In each case n is called the *arity* of the predicate or function.

A *metric structure* $\mathcal{M}$ based on (M, d) consists of a family $(R_i \mid i \in I)$ of predicates on M, a family $(F_j \mid j \in J)$ of functions on M, and a family $(a_k \mid k \in K)$ of distinguished elements of M. When we introduce such a metric structure, we will often denote it as

$$\mathcal{M} = (M, R_i, F_j, a_k \mid i \in I, j \in J, k \in K).$$

Any of the index sets I, J, K is allowed to be empty. Indeed, they might all be empty, in which case $\mathcal{M}$ is a pure bounded metric space.

The key restrictions on metric structures are: the metric space is *complete* and *bounded*, each predicate takes its values in a *bounded interval* of reals, and the functions and predicates are *uniformly continuous*. All of these restrictions play a role in making the theory work smoothly.

Our theory also applies to *many-sorted* metric structures, and they will appear as examples. However, in this article we will not explicitly bring them into our definitions and theorems, in order to avoid distracting notation.

2.1 Examples. We give a number of examples of metric structures to indicate the wide range of possibilities.

(1) A complete, bounded metric space (M, d) with no additional structure.

(2) A structure $\mathcal{M}$ in the usual sense from first-order logic. One puts the discrete metric on the underlying set ($d(a, b) = 1$ when a, b are distinct) and a relation is considered as a predicate taking values ("truth" values) in the set $\{0, 1\}$. So, in this sense the theory developed here is a generalization of first-order model theory.

(3) If (M, d) is an unbounded complete metric space with a distinguished element a, we may view (M, d) as a many-sorted metric structure $\mathcal{M}$; for example, we could take a typical sort to be a closed ball B_n of radius n around a, equipped with the metric obtained by restricting d. The inclusion mappings $I_{mn} \colon B_m \to B_n$

1 See the appendix to this section for some relevant basic facts about metric spaces.

($m < n$) should be functions in $\mathcal{M}$, in order to tie together the different sorts.

(4) The unit ball B of a Banach space X over $\mathbb{R}$ or $\mathbb{C}$: as functions we may take the maps $f_{\alpha\beta}$, defined by $f_{\alpha\beta}(x,y) = \alpha x + \beta y$, for each pair of scalars satisfying $|\alpha| + |\beta| \leq 1$; the norm may be included as a predicate, and we may include the additive identity 0_X as a distinguished element. Equivalently, X can be viewed as a many-sorted structure, with a sort for each ball of positive integer radius centered at 0, as indicated in the previous paragraph.

(5) Banach lattices: this is the result of expanding the metric structure corresponding to X as a Banach space (see the previous paragraph) by adding functions such as the absolute value operation on B as well as the positive and negative part operations. In section 17 of this article we discuss the model theory of some specific Banach lattices (namely, the L^p-spaces).

(6) Banach algebras: multiplication is included as an operation; if the algebra has a multiplicative identity, it may be included as a constant.

(7) C^*-algebras: multiplication and the *-map are included as operations.

(8) Hilbert spaces with inner product may be treated like the Banach space examples above, with the addition that the inner product is included as a binary predicate. (See section 15.)

(9) If $(\Omega, \mathcal{B}, \mu)$ is a probability space, we may construct a metric structure $\mathcal{M}$ from it, based on the metric space (M, d) in which M is the measure algebra of $(\Omega, \mathcal{B}, \mu)$ (elements of $\mathcal{B}$ modulo sets of measure 0) and d is defined to be the measure of the symmetric difference. As operations on M we take the Boolean operations $\cup, \cap, ^c$, as a predicate we take the measure μ, and as distinguished elements the 0 and 1 of M. In section 16 of this article we discuss the model theory of these metric structures.

Signatures

To each metric structure $\mathcal{M}$ we associate a *signature* L as follows. To each predicate R of $\mathcal{M}$ we associate a *predicate symbol* P and an integer $a(P)$ which is the arity of R; we denote R by $P^{\mathcal{M}}$. To each function F of $\mathcal{M}$ we associate a *function symbol* f and an integer $a(f)$ which is the arity of F; we denote F by $f^{\mathcal{M}}$. Finally, to each distinguished element a of $\mathcal{M}$ we associate a *constant symbol* c; we denote a by $c^{\mathcal{M}}$.

So, a signature L gives sets of predicate, function, and constant symbols, and associates to each predicate and function symbol its arity. In that respect, L is identical to a signature of first-order model theory. In addition, a signature for metric structures must specify more: for each predicate symbol P, it must provide a closed bounded interval I_P of real numbers and a modulus of uniform continuity[2] Δ_P. These should satisfy the requirements that $P^{\mathcal{M}}$ takes its values in I_P and that Δ_P is a modulus of uniform continuity for $P^{\mathcal{M}}$. In addition, for each function symbol f, L must provide a modulus of uniform continuity Δ_f, and this must satisfy the requirement that Δ_f is a modulus of uniform continuity for $f^{\mathcal{M}}$. Finally, L must provide a non-negative real number D_L which is a bound on the diameter of the complete metric space (M, d) on which $\mathcal{M}$ is based.[3] We sometimes denote the metric d given by $\mathcal{M}$ as $d^{\mathcal{M}}$; this would be consistent with our notation for the interpretation in $\mathcal{M}$ of the nonlogical symbols of L. However, we also find it convenient often to use the same notation "d" for the logical symbol representing the metric as well as for its interpretation in $\mathcal{M}$; this is consistent with usual mathematical practice and with the handling of the symbol $=$ in first-order logic.

When these requirements are all met and when the predicate, function, and constant symbols of L correspond exactly to the predicates, functions, and distinguished elements of which $\mathcal{M}$ consists, then we say $\mathcal{M}$ is an *L-structure*.

The key added features of a signature L in the metric structure setting are that L specifies (1) a bound on the diameter of the underlying metric space, (2) a modulus of uniform continuity for each predicate and function, and (3) a closed bounded interval of possible values for each predicate.

For simplicity, and without losing any generality, we will usually assume that our signatures L satisfy $D_L = 1$ and $I_P = [0, 1]$ for every predicate symbol P.

2.2 Remark. If $\mathcal{M}$ is an L-structure and A is a given closed subset of M^n, then $\mathcal{M}$ can be expanded by adding the predicate $x \mapsto \text{dist}(x, A)$, where x ranges over M^n and dist denotes the distance function with respect to the maximum metric on the product space M^n. Note that only in very special circumstances may A itself be added to $\mathcal{M}$ as a predicate (in the form of the characteristic function χ_A of A); this could

2 See the appendix to this section for a discussion of this notion.
3 If L is many-sorted, each sort will have its own diameter bound.

be done only if χ_A were uniformly continuous, which forces A to be a positive distance from its complement in M^n.

Basic concepts such as *embedding* and *isomorphism* have natural definitions for metric structures:

2.3 Definition. Let L be a signature for metric structures and suppose $\mathcal{M}$ and $\mathcal{N}$ are L-structures.

An *embedding* from $\mathcal{M}$ into $\mathcal{N}$ is a metric space isometry

$$T\colon (M, d^{\mathcal{M}}) \to (N, d^{\mathcal{N}})$$

that commutes with the interpretations of the function and predicate symbols of L in the following sense:

Whenever f is an n-ary function symbol of L and $a_1, \ldots, a_n \in M$, we have

$$f^{\mathcal{N}}(T(a_1), \ldots, T(a_n)) = T(f^{\mathcal{M}}(a_1, \ldots, a_n));$$

whenever c is a constant symbol c of L, we have

$$c^{\mathcal{N}} = T(c^{\mathcal{M}});$$

and whenever P is an n-ary predicate symbol of L and $a_1, \ldots, a_n \in M$, we have

$$P^{\mathcal{N}}(T(a_1), \ldots, T(a_n)) = P^{\mathcal{M}}(a_1, \ldots, a_n).$$

An *isomorphism* is a surjective embedding. We say that $\mathcal{M}$ and $\mathcal{N}$ are *isomorphic*, and write $\mathcal{M} \cong \mathcal{N}$, if there exists an isomorphism between $\mathcal{M}$ and $\mathcal{N}$. (Sometimes we say *isometric isomorphism* to emphasize that isomorphisms must be distance preserving.) An *automorphism* of $\mathcal{M}$ is an isomorphism between $\mathcal{M}$ and itself.

$\mathcal{M}$ is a *substructure* of $\mathcal{N}$ (and we write $\mathcal{M} \subseteq \mathcal{N}$) if $M \subseteq N$ and the inclusion map from M into N is an embedding of $\mathcal{M}$ into $\mathcal{N}$.

Appendix

In this appendix we record some basic definitions and facts about metric spaces and uniformly continuous functions; they will be needed when we develop the semantics of continuous first-order logic. Proofs of the results we state here are straightforward and will mostly be omitted.

Let (M, d) be a metric space. We say this space is *bounded* if there is a real number B such that $d(x, y) \leq B$ for all $x, y \in M$. The *diameter* of (M, d) is the smallest such number B.

Suppose (M_i, d_i) are metric spaces for $i = 1, \ldots, n$ and we take M to be the product $M = M_1 \times \cdots \times M_n$. In this article we will always regard M as being equipped with the maximum metric, defined for $x = (x_1, \ldots, x_n), y = (y_1, \ldots, y_n)$ by $d(x,y) = \max\{d_i(x_i, y_i) | i = 1, \ldots, n\}$.

A *modulus of uniform continuity* is any function $\Delta\colon (0,1] \to (0,1]$.

If (M,d) and (M',d') are metric spaces and $f\colon M \to M'$ is any function, we say that $\Delta\colon (0,1] \to (0,1]$ is a *modulus of uniform continuity for* f if for every $\epsilon \in (0,1]$ and every $x, y \in M$ we have

$$\text{(UC)} \qquad d(x,y) < \Delta(\epsilon) \implies d'(f(x), f(y)) \leq \epsilon.$$

We say f is *uniformly continuous* if it has a modulus of uniform continuity.

The precise way (UC) is stated makes the property *Δ is a modulus of uniform continuity for f* a topologically robust notion. For example, if $f\colon M \to M'$ is continuous and (UC) holds for a dense set of pairs (x,y), then it holds for all (x,y). In particular, if Δ is a modulus of uniform continuity for $f\colon M \to M'$ and we extend f in the usual way to a continuous function $\bar{f}\colon \overline{M} \to \overline{M'}$ (where $\overline{M}, \overline{M'}$ are completions of M, M', resp.), then, with this definition, Δ is a modulus of uniform continuity for the extended function $\bar{f}$.

If Δ is a function from $(0,\infty)$ to $(0,\infty)$ and it satisfies (UC) for all $\epsilon \in (0,\infty)$ and all $x, y \in M$, then we will often refer to Δ as a "modulus of uniform continuity" for f. In that case, f is uniformly continuous and the restriction of the function $\min(\Delta(\epsilon), 1)$ to $\epsilon \in (0,1]$ is a modulus of uniform continuity according to the strict meaning we have chosen to assign to this phrase, so no confusion should result.

2.4 Proposition. *Suppose $f\colon M \to M'$ and $f'\colon M' \to M''$ are functions between metrics spaces M, M', M''. Suppose Δ is a modulus of uniform continuity for f and Δ' is a modulus of uniform continuity for f'. Then the composition $f' \circ f$ is uniformly continuous; indeed, for each $r \in (0,1)$ the function $\Delta(r\Delta'(\epsilon))$ is a modulus of uniform continuity for $f' \circ f$.*

Let M, M' be metric spaces (with metrics d, d' resp.) and let f and $(f_n \mid n \geq 1)$ be functions from M into M'. Recall that $(f_n \mid n \geq 1)$ *converges uniformly to f on M* if

$$\forall \epsilon > 0\ \exists N\ \forall n > N\ \forall x \in M\ \big(d'(f_n(x), f(x)) \leq \epsilon\big).$$

2.5 Proposition. *Let M, M', f and $(f_n \mid n \geq 1)$ be as above, and suppose $(f_n \mid n \geq 1)$ converges uniformly to f on M. If each of the functions $f_n \colon M \to M'$ is uniformly continuous, then f must also be uniformly continuous. Indeed, a modulus of uniform continuity for f can be obtained from moduli Δ_n for f_n, for each $n \geq 1$, and from a function $N \colon (0,1] \to \mathbb{N}$ that satisfies*

$$\forall \epsilon > 0 \, \forall n > N(\epsilon) \, \forall x \in M \, (d'(f_n(x), f(x)) \leq \epsilon).$$

Proof A modulus Δ for f may be defined as follows: given $\epsilon > 0$, take $\Delta(\epsilon) = \Delta_n(\epsilon/3)$ where $n = N(\epsilon/3) + 1$. □

2.6 Proposition. *Suppose $f, f_n \colon M \to M'$ and $f', f'_n \colon M' \to M''$ are functions $(n \geq 1)$ between metric spaces M, M', M''. If $(f_n \mid n \geq 1)$ converges uniformly to f on M and $(f'_n \mid n \geq 1)$ converges uniformly to f' on M', then $(f'_n \circ f_n \mid n \geq 1)$ converges uniformly to $f' \circ f$ on M.*

Fundamental to the continuous logic described in this article are the operations sup and inf on bounded sets of real numbers. We use these to define new functions from old, as follows. Suppose M, M' are metric spaces and $f \colon M \times M' \to \mathbb{R}$ is a bounded function. We define new functions $\sup_y f$ and $\inf_y f$ from M to $\mathbb{R}$ by

$$(\sup_y f)(x) = \sup\{f(x,y) | y \in M'\}$$

$$(\inf_y f)(x) = \inf\{f(x,y) | y \in M'\}$$

for all $x \in M$. Note that these new functions map M into the same closed bounded interval in $\mathbb{R}$ that contained the range of f. Our perspective is that $\sup_y$ and $\inf_y$ are quantifiers that bind or eliminate the variable y, analogous to the way $\forall$ and $\exists$ are used in ordinary first-order logic.

2.7 Proposition. *Suppose M, M' are metric spaces and f is a bounded uniformly continuous function from $M \times M'$ to $\mathbb{R}$. Let Δ be a modulus of uniform continuity for f. Then $\sup_y f$ and $\inf_y f$ are bounded uniformly continuous functions from M to $\mathbb{R}$, and Δ is a modulus of uniform continuity for both of them.*

Proof Fix $\epsilon > 0$ and consider $u, v \in M$ such that $d(u,v) < \Delta(\epsilon)$. Then for every $z \in M'$ we have

$$f(v,z) \leq f(u,z) + \epsilon \leq (\sup_y f)(u) + \epsilon.$$

Taking the sup over $z \in M'$ and interchanging the role of u and v yields

$$|(\sup_y f)(u) - (\sup_y f)(v)| \leq \epsilon.$$

The function $\inf_y f$ is handled similarly. □

2.8 Proposition. *Suppose M is a metric space and $f_s \colon M \to [0,1]$ is a uniformly continuous function for each s in the index set S. Let Δ be a common modulus of uniform continuity for $(f_s \mid s \in S)$. Then $\sup_s f_s$ and $\inf_s f_s$ are uniformly continuous functions from M to $[0,1]$, and Δ is a modulus of uniform continuity for both of them.*

Proof In the previous proof, take M' to be S with the discrete metric, and define $f(x,s) = f_s(x)$. □

2.9 Proposition. *Suppose M, M' are metric spaces and let $(f_n \mid n \geq 1)$ and f all be bounded functions from $M \times M'$ into $\mathbb{R}$. If $(f_n \mid n \geq 1)$ converges uniformly to f on $M \times M'$, then $(\sup_y f_n \mid n \geq 1)$ converges uniformly to $\sup_y f$ on M and $(\inf_y f_n \mid n \geq 1)$ converges uniformly to $\inf_y f$ on M.*

Proof Similar to the proof of Proposition 2.7. □

In many situations it is natural to construct a metric space as the quotient of a pseudometric space (M_0, d_0); here we mean that M_0 is a set and $d_0 \colon M_0 \times M_0 \to \mathbb{R}$ is a pseudometric. That is,

$$\begin{aligned}
&d_0(x,x) = 0 \\
&d_0(x,y) = d_0(y,x) \geq 0 \\
&d_0(x,z) \leq d_0(x,y) + d_0(y,z)
\end{aligned}$$

for all $x, y, z \in M_0$; these are the same conditions as in the definition of a metric, except that $d_0(x,y) = 0$ is allowed even when x, y are distinct.

If (M_0, d_0) is a pseudometric space, we may define an equivalence relation E on M_0 by $E(x,y) \Leftrightarrow d_0(x,y) = 0$. It follows from the triangle inequality that d_0 is E-invariant; that is, $d_0(x,y) = d_0(x',y')$ whenever xEx' and yEy'. Let M be the quotient set M_0/E and $\pi \colon M_0 \to M$ the quotient map, so $\pi(x)$ is the E-equivalence class of x, for each $x \in M_0$. Further, define d on M by setting $d(\pi(x), \pi(y)) = d_0(x,y)$ for any $x, y \in M_0$. Then (M,d) is a metric space and π is a distance preserving function from (M_0, d_0) onto (M,d). We will refer to (M,d) as the *quotient* metric space induced by (M_0, d_0).

Suppose (M_0, d_0) and (M_0', d_0') are pseudometric spaces with quotient metric spaces (M, d) and (M', d') and quotient maps π, π', respectively. Let $f_0 \colon M_0 \to M_0'$ be any function. We say that f_0 is *uniformly continuous*, with modulus of uniform continuity Δ, if

$$d_0(x, y) < \Delta(\epsilon) \quad \Longrightarrow \quad d_0'(f_0(x), f_0(y)) \leq \epsilon$$

for all $x, y \in M_0$ and all $\epsilon \in (0, 1]$. In that case it is clear that $d_0(x, y) = 0$ implies $d_0'(f_0(x), f_0(y)) = 0$ for all $x, y \in M_0$. Therefore we get a well defined quotient function $f \colon M \to M'$ by setting $f(\pi(x)) = \pi'(f_0(x))$ for all $x \in M_0$. Moreover, f is uniformly continuous with modulus of uniform continuity Δ.

The following results are useful in many places for expressing certain kinds of implications in continuous logic, and for reformulating the concept of uniform continuity.

2.10 Proposition. *Let $F, G \colon X \to [0, 1]$ be arbitrary functions such that*

$$(\star) \qquad \forall \epsilon > 0\ \exists \delta > 0\ \forall x \in X\ (F(x) \leq \delta \Rightarrow G(x) \leq \epsilon)\,.$$

Then there exists an increasing, continuous function $\alpha \colon [0, 1] \to [0, 1]$ such that $\alpha(0) = 0$ and

$$(\star\star) \qquad \forall x \in X\ \ (G(x) \leq \alpha(F(x)))\,.$$

Proof Define a (possibly discontinuous) function $g \colon [0, 1] \to [0, 1]$ by

$$g(t) = \sup\{G(x) \mid F(x) \leq t\}$$

for $t \in [0, 1]$. It is clear that g is increasing and that $G(x) \leq g(F(x))$ holds for all $x \in X$. Moreover, statement $(\star)$ implies that $g(0) = 0$ and that $g(t)$ converges to 0 as $t \to 0$.

To complete the proof we construct an increasing, continuous function $\alpha \colon [0, 1] \to [0, 1]$ such that $\alpha(0) = 0$ and $g(t) \leq \alpha(t)$ for all $t \in [0, 1]$. Let $(t_n \mid n \in \mathbb{N})$ be any decreasing sequence in $[0, 1]$ with $t_0 = 1$ and $\lim_{n\to\infty} t_n = 0$. Define $\alpha \colon [0, 1] \to [0, 1]$ by setting $\alpha(0) = 0$, $\alpha(1) = 1$, and $\alpha(t_n) = g(t_{n-1})$ for all $n \geq 1$, and by taking α to be linear on each interval of the form $[t_{n+1}, t_n]$, $n \in \mathbb{N}$. It is easy to check that α has the desired properties. For example, if $t_1 \leq t \leq t_0 = 1$ we have that $\alpha(t)$ is a convex combination of $g(1)$ and 1 so that $g(t) \leq g(1) \leq \alpha(t)$. Similarly, if $t_{n+1} \leq t \leq t_n$ and $n \geq 1$, we have that $\alpha(t)$ is a convex combination of $g(t_n)$ and $g(t_{n-1})$ so that $g(t) \le g(t_n) \le \alpha(t)$. Together with $g(0) = \alpha(0) = 0$, this shows that $g(t) \le \alpha(t)$ for all $t \in [0,1]$. □

2.11 Remark. Note that the converse to Proposition 2.10 is also true. Indeed, if statement $(\star\star)$ holds and we fix $\epsilon > 0$, then taking

$$\delta = \sup\{s \mid \alpha(s) \le \epsilon\} > 0$$

witnesses the truth of statement $(\star)$.

2.12 Remark. The proof of Proposition 2.10 can be revised to show that the continuous function α can be chosen so that it only depends on the choice of an increasing function $\Delta\colon (0,1] \to (0,1]$ that witnesses the truth of statement $(\star)$, in the sense that

$$\forall x \in X\ (F(x) \le \Delta(\epsilon) \Rightarrow G(x) \le \epsilon)$$

holds for each $\epsilon \in (0,1]$. Given such a Δ, define $g\colon [0,1] \to [0,1]$ by $g(t) = \inf\{s \in (0,1] \mid \Delta(s) > t\}$. It is easy to check that $g(0) = 0$ and that g is an increasing function. Moreover, for any $\epsilon > 0$ we have from the definition that $g(t) < \epsilon$ for any t in $[0, \Delta(\epsilon))$; therefore $g(t)$ converges to 0 as t tends to 0. Finally, we claim that $G(x) \le g(F(x))$ holds for any $x \in X$. Otherwise we have $x \in X$ such that $g(F(x)) < G(x)$. The definition of g yields $s \in (0,1]$ with $s < G(x)$ and $\Delta(s) > F(x)$; this contradicts our assumptions.

Now α is constructed from g as in the proof of Proposition 2.10. This yields an increasing, continuous function $\alpha\colon [0,1] \to [0,1]$ with $\alpha(0) = 0$ such that whenever $F, G\colon X \to [0,1]$ are functions satisfying

$$\forall x \in X\ (F(x) \le \Delta(\epsilon) \Rightarrow G(x) \le \epsilon)$$

for each $\epsilon \in (0,1]$, then we have

$$\forall x \in X\ (G(x) \le \alpha(F(x))).$$

3 Formulas and their interpretations

Fix a signature L for metric structures, as described in the previous section. As indicated there (see page 321), we assume for simplicity of notation that $D_L = 1$ and that $I_P = [0,1]$ for every predicate symbol P.

Symbols of L

Among the *symbols* of L are the predicate, function, and constant symbols; these will be referred to as the *nonlogical* symbols of L and the remaining ones will be called the *logical* symbols of L. Among the logical symbols is a symbol d for the metric on the underlying metric space of an L-structure; this is treated formally as equivalent to a predicate symbol of arity 2. The logical symbols also include an infinite set V_L of *variables*; usually we take V_L to be countable, but there are situations in which it is useful to permit a larger number of variables. The remaining logical symbols consist of a symbol for each continuous function $u\colon [0,1]^n \to [0,1]$ of finitely many variables $n \geq 1$ (these play the role of connectives) and the symbols sup and inf, which play the role of quantifiers in this logic.

The *cardinality* of L, denoted card(L), is the smallest infinite cardinal number $\geq$ the number of nonlogical symbols of L.

Terms of L

Terms are formed inductively, exactly as in first-order logic. Each variable and constant symbol is an L-term. If f is an n-ary function symbol and $t_1, \ldots, t_n$ are L-terms, then $f(t_1, \ldots, t_n)$ is an L-term. All L-terms are constructed in this way.

Atomic formulas of L.

The *atomic formulas* of L are the expressions of the form $P(t_1, \ldots, t_n)$, in which P is an n-ary predicate symbol of L and $t_1, \ldots, t_n$ are L-terms; as well as $d(t_1, t_2)$, in which t_1 and t_2 are L-terms.

Note that the logical symbol d for the metric is treated formally as a binary predicate symbol, exactly analogous to how the equality symbol $=$ is treated in first-order logic.

Formulas of L

Formulas are also constructed inductively, and the basic structure of the induction is similar to the corresponding definition in first-order logic. Continuous functions play the role of connectives and sup and inf are used formally in the way that quantifiers are used in first-order logic. The precise definition is as follows:

3.1 Definition. The class of L-formulas is the smallest class of expressions satisfying the following requirements:

(1) Atomic formulas of L are L-formulas.
(2) If $u\colon [0,1]^n \to [0,1]$ is continuous and $\varphi_1, \dots, \varphi_n$ are L-formulas, then $u(\varphi_1, \dots, \varphi_n)$ is an L-formula.
(3) If φ is an L-formula and x is a variable, then $\sup_x \varphi$ and $\inf_x \varphi$ are L-formulas.

3.2 Remark. We have chosen to take all continuous functions on $[0,1]$ as our connectives. This is both too restrictive (see section 9 in which we want to close our set of formulas under certain kinds of limits, in order to develop a good notion of *definability*) and too general (see section 6). We made this choice in order to introduce formulas as early and as directly as possible.

An L-formula is *quantifier free* if it is generated inductively from atomic formulas without using the last clause; *i.e.*, neither $\sup_x$ nor $\inf_x$ are used.

Many syntactic notions from first-order logic can be carried over word for word into this setting. We will assume that this has been done by the reader for many such concepts, including *subformula* and *syntactic substitution* of a term for a variable, or a formula for a subformula, and so forth.

Free and *bound* occurrences of variables in L-formulas are defined in a way similar to how this is done in first-order logic. Namely, an occurrence of the variable x is bound if lies within a subformula of the form $\sup_x \varphi$ or $\inf_x \varphi$, and otherwise it is free.

An *L-sentence* is an L-formula that has no free variables.

When t is a term and the variables occurring in it are among the variables $x_1, \dots, x_n$ (which we always take to be distinct in this context), we indicate this by writing t as $t(x_1, \dots, x_n)$.

Similarly, we write an L-formula as $\varphi(x_1, \dots, x_n)$ to indicate that its free variables are among $x_1, \dots, x_n$.

Prestructures

It is common in mathematics to construct a metric space as the quotient of a pseudometric space or as the completion of such a quotient, and the same is true of metric structures. For that reason we need to consider what we will call *prestructures* and to develop the semantics of continuous logic for them.

As above, we take L to be a fixed signature for metric structures. Let (M_0, d_0) be a pseudometric space, satisfying the requirement that its diameter is $\leq D_L$. (That is, $d_0(x, y) \leq D_L$ for all $x, y \in M_0$.) An *L-prestructure* $\mathcal{M}_0$ based on (M_0, d_0) is a structure consisting of the following data:

(1) for each predicate symbol P of L (of arity n) a function $P^{\mathcal{M}_0}$ from M_0^n into I_P that has Δ_P as a modulus of uniform continuity;
(2) for each function symbol f of L (of arity n) a function $f^{\mathcal{M}_0}$ from M_0^n into M_0 that has Δ_f as a modulus of uniform continuity; and
(3) for each constant symbol c of L an element $c^{\mathcal{M}_0}$ of M_0.

Given an L-prestructure $\mathcal{M}_0$, we may form its *quotient* prestructure as follows. Let (M, d) be the quotient metric space induced by (M_0, d_0) with quotient map $\pi \colon M_0 \to M$. Then

(1) for each predicate symbol P of L (of arity n) define $P^{\mathcal{M}}$ from M^n into I_P by setting $P^{\mathcal{M}}(\pi(x_1), \ldots, \pi(x_n)) = P^{\mathcal{M}_0}(x_1, \ldots, x_n)$ for each $x_1, \ldots, x_n \in M_0$;
(2) for each function symbol f of L (of arity n) define $f^{\mathcal{M}}$ from M^n into M by setting $f^{\mathcal{M}}(\pi(x_1), \ldots, \pi(x_n)) = \pi(f^{\mathcal{M}_0}(x_1, \ldots, x_n))$ for each $x_1, \ldots, x_n \in M_0$;
(3) for each constant symbol c of L define $c^{\mathcal{M}} = \pi(c^{\mathcal{M}_0})$.

It is obvious that (M, d) has the same diameter as (M_0, d_0). Also, as noted in the appendix to section 2, for each predicate symbol P and each function symbol f of L, the predicate $P^{\mathcal{M}}$ is well defined and has Δ_P as a modulus of uniform continuity and the function $f^{\mathcal{M}}$ is well defined and has Δ_f as a modulus of uniform continuity. In other words, this defines an L-prestructure (which we will denote as $\mathcal{M}$) based on the (possibly not complete) metric space (M, d).

Finally, we may define an L-structure $\mathcal{N}$ by taking a *completion* of $\mathcal{M}$. This is based on a complete metric space (N, d) that is a completion of (M, d), and its additional structure is defined in the following natural way (made possible by the fact that the predicates and functions given by $\mathcal{M}$ are uniformly continuous):

(1) for each predicate symbol P of L (of arity n) define $P^{\mathcal{N}}$ from N^n into I_P to be the unique such function that extends $P^{\mathcal{M}}$ and is continuous;

(2) for each function symbol f of L (of arity n) define $f^{\mathcal{N}}$ from N^n into N to be the unique such function that extends $f^{\mathcal{M}}$ and is continuous;

(3) for each constant symbol c of L define $c^{\mathcal{N}} = c^{\mathcal{M}}$.

It is obvious that (N, d) has the same diameter as (M, d). Also, as noted in the appendix to section 2, for each predicate symbol P and each function symbol f of L, the predicate $P^{\mathcal{N}}$ has Δ_P as a modulus of uniform continuity and the function $f^{\mathcal{N}}$ has Δ_f as a modulus of uniform continuity. In other words, $\mathcal{N}$ is an L-structure.

Semantics

Let $\mathcal{M}$ be any L-prestructure, with $(M, d^{\mathcal{M}})$ as its underlying pseudometric space, and let A be a subset of M. We extend L to a signature $L(A)$ by adding a new constant symbol $c(a)$ to L for each element a of A. We extend the interpretation given by $\mathcal{M}$ in a canonical way, by taking the interpretation of $c(a)$ to be equal to a itself for each $a \in A$. We call $c(a)$ the *name* of a in $L(A)$. Indeed, we will often write a instead of $c(a)$ when no confusion can result from doing so.

Given an $L(M)$ term $t(x_1, \ldots, x_n)$, we define, exactly as in first-order logic, the interpretation of t in $\mathcal{M}$, which is a function $t^{\mathcal{M}} \colon M^n \to M$.

We now come to the key definition in continuous logic for metric structures, in which the semantics of this logic is defined. For each $L(M)$-sentence σ, we define *the value of σ in $\mathcal{M}$*. This value is a real number in the interval $[0, 1]$ and it is denoted $\sigma^{\mathcal{M}}$. The definition is by induction on formulas. Note that in the definition all terms mentioned are $L(M)$-terms in which no variables occur.

3.3 Definition. (1) $\left(d(t_1, t_2)\right)^{\mathcal{M}} = d^{\mathcal{M}}(t_1^{\mathcal{M}}, t_2^{\mathcal{M}})$ for any t_1, t_2;

(2) for any n-ary predicate symbol P of L and any $t_1, \ldots, t_n$,

$$\left(P(t_1, \ldots, t_n)\right)^{\mathcal{M}} = P^{\mathcal{M}}(t_1^{\mathcal{M}}, \ldots, t_n^{\mathcal{M}});$$

(3) for any $L(M)$-sentences $\sigma_1, \ldots, \sigma_n$ and any continuous function $u \colon [0, 1]^n \to [0, 1]$,

$$\left(u(\sigma_1, \ldots, \sigma_n)\right)^{\mathcal{M}} = u(\sigma_1^{\mathcal{M}}, \ldots, \sigma_n^{\mathcal{M}});$$

(4) for any $L(M)$-formula $\varphi(x)$,

$$\left(\sup_x \varphi(x)\right)^{\mathcal{M}}$$

is the supremum in $[0,1]$ of the set $\{\varphi(a)^{\mathcal{M}} \mid a \in M\}$;

(5) for any $L(M)$-formula $\varphi(x)$,

$$\left(\inf_x \varphi(x)\right)^{\mathcal{M}}$$

is the infimum in $[0,1]$ of the set $\{\varphi(a)^{\mathcal{M}} \mid a \in M\}$.

3.4 Definition. Given an $L(M)$-formula $\varphi(x_1,\dots,x_n)$ we let $\varphi^{\mathcal{M}}$ denote the function from M^n to $[0,1]$ defined by

$$\varphi^{\mathcal{M}}(a_1,\dots,a_n) = \left(\varphi(a_1,\dots,a_n)\right)^{\mathcal{M}}.$$

A key fact about formulas in continuous logic is that they define uniformly continuous functions. Indeed, the modulus of uniform continuity for the predicate does not depend on $\mathcal{M}$ but only on the data given by the signature L.

3.5 Theorem. *Let $t(x_1,\dots,x_n)$ be an L-term and $\varphi(x_1,\dots,x_n)$ an L-formula. Then there exist functions Δ_t and Δ_φ from $(0,1]$ to $(0,1]$ such that for any L-prestructure $\mathcal{M}$, Δ_t is a modulus of uniform continuity for the function $t^{\mathcal{M}}\colon M^n \to M$ and Δ_φ is a modulus of uniform continuity for the predicate $\varphi^{\mathcal{M}}\colon M^n \to [0,1]$.*

Proof The proof is by induction on terms and then induction on formulas. The basic tools concerning uniform continuity needed for the induction steps in the proof are given in the appendix to section 2. □

3.6 Remark. The previous result is the counterpart in this logic of the Perturbation Lemma in the logic of positive bounded formulas with the approximate semantics. See [24, Proposition 5.15].

3.7 Theorem. *Let $\mathcal{M}_0$ be an L-prestructure with underlying pseudometric space (M_0, d_0); let $\mathcal{M}$ be its quotient L-structure with quotient map $\pi\colon M_0 \to M$ and let $\mathcal{N}$ be the L-structure that results from completing $\mathcal{M}$ (as explained on page 330). Let $t(x_1,\dots,x_n)$ be any L-term and $\varphi(x_1,\dots,x_n)$ be any L-formula. Then:*

(1) $t^{\mathcal{M}}(\pi(a_1),\dots,\pi(a_n)) = t^{\mathcal{M}_0}(a_1,\dots,a_n)$ for all $a_1,\dots,a_n \in M_0$;
(2) $t^{\mathcal{N}}(b_1,\dots,b_n) = t^{\mathcal{M}}(b_1,\dots,b_n)$ for all $b_1,\dots,b_n \in M$.
(3) $\varphi^{\mathcal{M}}(\pi(a_1),\dots,\pi(a_n)) = \varphi^{\mathcal{M}_0}(a_1,\dots,a_n)$ for all $a_1,\dots,a_n \in M_0$;
(4) $\varphi^{\mathcal{N}}(b_1,\dots,b_n) = \varphi^{\mathcal{M}}(b_1,\dots,b_n)$ for all $b_1,\dots,b_n \in M$.

Proof The proofs are by induction on terms and then induction on formulas. In handling the quantifier cases in (3) the key is that the quotient map π is surjective. For the quantifier cases in (4), the key is that the functions $\varphi^{\mathcal{N}}$ are continuous and that $\mathcal{M}$ is dense in $\mathcal{N}$. □

3.8 Caution. Note that we only use words such as *structure* when the underlying metric space is *complete.* In some constructions this means that we must take a metric space completion at the end. Theorem 3.7 shows that this preserves all properties expressible in continuous logic.

Logical equivalence

3.9 Definition. Two L-formulas $\varphi(x_1,\ldots,x_n)$ and $\psi(x_1,\ldots,x_n)$ are said to be *logically equivalent* if

$$\varphi^{\mathcal{M}}(a_1,\ldots,a_n) = \psi^{\mathcal{M}}(a_1,\ldots,a_n)$$

for every L-structure $\mathcal{M}$ and every $a_1,\ldots,a_n \in M$.

If $\varphi(x_1,\ldots,x_n)$ and $\psi(x_1,\ldots,x_n)$ are L-formulas, we can extend the preceding definition by taking the *logical distance* between φ and ψ to be the supremum of all numbers

$$|\varphi^{\mathcal{M}}(a_1,\ldots,a_n) - \psi^{\mathcal{M}}(a_1,\ldots,a_n)|$$

where $\mathcal{M}$ is any L-structure and $a_1,\ldots,a_n \in M$. This defines a pseudometric on the set of all formulas with free variables among $x_1,\ldots,x_n$, and two formulas are logically equivalent if and only if the logical distance between them is 0.

3.10 Remark. Note that by Theorem 3.7, we could use L-prestructures in place of L-structures in the preceding Definition without changing the meaning of the concepts defined.

3.11 Remark. (Size of the space of L-formulas)
Some readers may be concerned that the set of L-formulas is be too large, because we allow all continuous functions as connectives. What matters, however, is the size of a set of L-formulas that is dense in the set of all L-formulas with respect to the logical distance defined in the previous paragraph. By Weierstrass's Theorem, there is a countable set of functions from $[0,1]^n$ to $[0,1]$ that is dense in the set of all continuous functions, with respect to the sup-distance between such functions. (The

sup-distance between f and g is the supremum of $|f(x) - g(x)|$ as x ranges over the common domain of f, g.) If we only use such connectives in building L-formulas, then (a) the total number of formulas that are constructed is card(L), and (b) any L-formula can be approximated arbitrarily closely in logical distance by a formula constructed using the restricted connectives. Thus the density character of the set of L-formulas with respect to logical distance is always $\leq$ card(L). (We explore this topic in more detail in section 6.)

Conditions of L

An *L-condition* E is a formal expression of the form $\varphi = 0$, where φ is an L-formula. We call E *closed* if φ is a sentence. If $x_1, \dots, x_n$ are distinct variables, we write an L-condition as $E(x_1, \dots, x_n)$ to indicate that it has the form $\varphi(x_1, \dots, x_n) = 0$ (in other words, that the free variables of E are among $x_1, \dots, x_n$).

If E is the $L(M)$-condition $\varphi(x_1, \dots, x_n) = 0$ and $a_1, \dots, a_n$ are in M, we say E is *true of $a_1, \dots, a_n$ in $\mathcal{M}$* and write $\mathcal{M} \models E[a_1, \dots, a_n]$ if $\varphi^{\mathcal{M}}(a_1, \dots, a_n) = 0$.

3.12 Definition. Let E_i be the L-condition $\varphi_i(x_1, \dots, x_n) = 0$, for $i = 1, 2$. We say that E_1 and E_2 are *logically equivalent* if for every L-structure $\mathcal{M}$ and every $a_1, \dots, a_n$ we have

$$\mathcal{M} \models E_1[a_1, \dots, a_n] \quad \text{iff} \quad \mathcal{M} \models E_2[a_1, \dots, a_n].$$

3.13 Remark. When φ and ψ are formulas, it is convenient to introduce the expression $\varphi = \psi$ as an abbreviation for the condition $|\varphi - \psi| = 0$. (Note that $u \colon [0,1]^2 \to [0,1]$ defined by $u(t_1, t_2) = |t_1 - t_2|$ is a connective.) Since each real number $r \in [0,1]$ is a connective (thought of as a constant function), expressions of the form $\varphi = r$ will thereby be regarded as conditions for any L-formula φ and $r \in [0,1]$. Note that the interpretation of $\varphi = \psi$ is semantically correct; namely for any L-structure $\mathcal{M}$ and elements a of M, $|\varphi - \psi|^{\mathcal{M}}(a) = 0$ if and only if $\varphi^{\mathcal{M}}(a) = \psi^{\mathcal{M}}(a)$.

Similarly, we introduce the expressions $\varphi \leq \psi$ and $\psi \geq \varphi$ as abbreviations for certain conditions. Let $\dot{-} \colon [0,1]^2 \to [0,1]$ be the connective defined by $\dot{-}(t_1, t_2) = \max(t_1 - t_2, 0) = t_1 - t_2$ if $t_1 \geq t_2$ and 0 otherwise. Usually we write $t_1 \dot{-} t_2$ in place of $\dot{-}(t_1, t_2)$. We take $\varphi \leq \psi$ and $\psi \geq \varphi$ to be abbreviations for the condition $\varphi \dot{-} \psi = 0$. (See section 6, where this connective plays a central role.) In $[0,1]$-valued logic, the condition $\varphi \leq \psi$ can be seen as family of implications, from the condition $\psi \leq r$ to the condition $\varphi \leq r$ for each $r \in [0,1]$.

4 Model theoretic concepts

Fix a signature L for metric structures. In this section we introduce several of the most fundamental model theoretic concepts and discuss some of their basic properties.

4.1 Definition. A *theory* in L is a set of closed L-conditions. If T is a theory in L and $\mathcal{M}$ is an L-structure, we say that $\mathcal{M}$ *is a model* of T and write $\mathcal{M} \models T$ if $\mathcal{M} \models E$ for every condition E in T. We write $\mathrm{Mod}_L(T)$ for the collection of all L-structures that are models of T. (If L is clear from the context, we write simply $\mathrm{Mod}(T)$.)

If $\mathcal{M}$ is an L-structure, the *theory of* $\mathcal{M}$, denoted $\mathrm{Th}(\mathcal{M})$, is the set of closed L-conditions that are true in $\mathcal{M}$. If T is a theory of this form, it will be called *complete*.

If T is an L-theory and E is a closed L-condition, we say E *is a logical consequence of* T and write $T \models E$ if $\mathcal{M} \models E$ holds for every model $\mathcal{M}$ of T.

4.2 Caution. Note that we only use words such as *model* when the underlying metric space is *complete*. Theorem 3.7 shows that whenever T is an L-theory and $\mathcal{M}_0$ is an L-prestructure such that $\varphi^{\mathcal{M}_0} = 0$ for every condition $\varphi = 0$ in T, then the completion of the canonical quotient of $\mathcal{M}_0$ is indeed a *model* of T.

4.3 Definition. Suppose that $\mathcal{M}$ and $\mathcal{N}$ are L-structures.

(1) We say that $\mathcal{M}$ and $\mathcal{N}$ are *elementarily equivalent*, and write $\mathcal{M} \equiv \mathcal{N}$, if $\sigma^{\mathcal{M}} = \sigma^{\mathcal{N}}$ for all L-sentences σ. Equivalently, this holds if $\mathrm{Th}(\mathcal{M}) = \mathrm{Th}(\mathcal{N})$.

(2) If $\mathcal{M} \subseteq \mathcal{N}$ we say that $\mathcal{M}$ is an *elementary substructure of* $\mathcal{N}$, and write $\mathcal{M} \preceq \mathcal{N}$, if whenever $\varphi(x_1, \ldots, x_n)$ is an L-formula and $a_1, \ldots, a_n$ are elements of M, we have

$$\varphi^{\mathcal{M}}(a_1, \ldots, a_n) = \varphi^{\mathcal{N}}(a_1, \ldots, a_n).$$

In this case, we also say that $\mathcal{N}$ is an *elementary extension* of $\mathcal{M}$.

(3) A function F from a subset of M into N is an *elementary map* from $\mathcal{M}$ into $\mathcal{N}$ if whenever $\varphi(x_1,\dots,x_n)$ is an L-formula and $a_1,\dots,a_n$ are elements of the domain of F, we have

$$\varphi^{\mathcal{M}}(a_1,\dots,a_n) = \varphi^{\mathcal{N}}(F(a_1),\dots,F(a_n)).$$

(4) An *elementary embedding* of $\mathcal{M}$ into $\mathcal{N}$ is a function from all of M into N that is an elementary map from $\mathcal{M}$ into $\mathcal{N}$.

4.4 Remark. (1) Every elementary map from one metric structure into another is distance preserving.
(2) The collection of elementary maps is closed under composition and formation of the inverse.
(3) Every isomorphism between metric structures is an elementary embedding.

In the following result we refer to sets $\mathcal{S}$ of L-formulas that are *dense with respect to logical distance.* (See page 333.) That is, such a set has the following property: for any L-formula $\varphi(x_1,\dots,x_n)$ and any $\epsilon > 0$ there is $\psi(x_1,\dots,x_n)$ in $\mathcal{S}$ such that for any L-structure $\mathcal{M}$ and any $a_1,\dots,a_n \in M$

$$|\varphi^{\mathcal{M}}(a_1,\dots,a_n) - \psi^{\mathcal{M}}(a_1,\dots,a_n)| \leq \epsilon.$$

4.5 Proposition. *(Tarski-Vaught Test for $\preceq$) Let $\mathcal{S}$ be any set of L-formulas that is dense with respect to logical distance. Suppose $\mathcal{M}$, $\mathcal{N}$ are L-structures with $\mathcal{M} \subseteq \mathcal{N}$. The following statements are equivalent:*

(1) $\mathcal{M} \preceq \mathcal{N}$;
(2) *For every L-formula $\varphi(x_1,\dots,x_n,y)$ in $\mathcal{S}$ and $a_1,\dots,a_n \in M$,*

$$\inf\{\varphi^{\mathcal{N}}(a_1,\dots,a_n,b) \mid b \in N\} = \inf\{\varphi^{\mathcal{N}}(a_1,\dots,a_n,c) \mid c \in M\}$$

Proof If (1) holds, then we may conclude (2) for the set of all L-formulas directly from the meaning of $\preceq$. Indeed, if $\varphi(x_1,\dots,x_n,y)$ is any L-formula and $a_1,\dots,a_n \in A$, then from (1) we have

$$\inf\{\varphi^{\mathcal{N}}(a_1,\dots,a_n,b) \mid b \in N\} = \big(\inf_y \varphi(a_1,\dots,a_n,y)\big)^{\mathcal{N}} =$$

$$\big(\inf_y \varphi(a_1,\dots,a_n,y)\big)^{\mathcal{M}} = \inf\{\varphi^{\mathcal{M}}(a_1,\dots,a_n,c) \mid c \in M\} =$$

$$\inf\{\varphi^{\mathcal{N}}(a_1,\dots,a_n,c) \mid c \in M\}.$$

For the converse, suppose (2) holds for a set S that is dense in the set of all L-formulas with respect to logical distance. First we will prove that (2) holds for the set of all L-formulas. Let $\varphi(x_1,\dots,x_n,y)$ be any L-formula. Given $\epsilon > 0$, let $\psi(x_1,\dots,x_n,y)$ be an element of S that approximates $\varphi(x_1,\dots,x_n,y)$ to within ϵ in logical distance. Let $a_1,\dots,a_n$ be elements of M. Then we have

$$\begin{aligned}
&\inf\{\varphi^{\mathcal{N}}(a_1,\dots,a_n,b) \mid b \in M\}\\
&\leq \inf\{\psi^{\mathcal{N}}(a_1,\dots,a_n,b) \mid b \in M\} + \epsilon\\
&= \inf\{\psi^{\mathcal{N}}(a_1,\dots,a_n,c) \mid c \in N\} + \epsilon\\
&\leq \inf\{\varphi^{\mathcal{N}}(a_1,\dots,a_n,c) \mid c \in N\} + 2\epsilon.
\end{aligned}$$

Letting ϵ tend to 0 and recalling $M \subseteq N$ we obtain the desired equality for $\varphi(x_1,\dots,x_n,y)$.

Now assume that (2) holds for the set of all L-formulas. One proves the equivalence

$$\psi^{\mathcal{M}}(a_1,\dots,a_n) = \psi^{\mathcal{N}}(a_1,\dots,a_n)$$

(for all $a_1,\dots,a_n$ in M) by induction on the complexity of ψ, using (2) to cover the case when ψ begins with sup or inf. □

5 Ultraproducts and compactness

First we discuss ultrafilter limits in topology. Let X be a topological space and let $(x_i)_{i\in I}$ be a family of elements of X. If D is an ultrafilter on I and $x \in X$, we write

$$\lim_{i,D} x_i = x$$

and say x *is the D-limit of* $(x_i)_{i\in I}$ if for every neighborhood U of x, the set $\{\, i \in I \mid x_i \in U \,\}$ is in the ultrafilter D. A basic fact from general topology is that X is a compact Hausdorff space if and only if for every family $(x_i)_{i\in I}$ in X and every ultrafilter D on I the D-limit of $(x_i)_{i\in I}$ exists and is unique.

The following lemmas are needed below when we connect ultrafilter limits and the semantics of continuous logic.

5.1 Lemma. *Suppose X, X' are topological spaces and $F\colon X \to X'$ is continuous. For any family $(x_i)_{i\in I}$ from X and any ultrafilter D on I, we have that*

$$\lim_{i,D} x_i = x \implies \lim_{i,D} F(x_i) = F(x)$$

where the ultrafilter limits are taken in X and X' respectively.

Proof Let U be an open neighborhood of $F(x)$ in X'. Since F is continuous, $F^{-1}(U)$ is open in X, and it contains x. If x is the D-limit of $(x_i)_{i\in I}$, there exists $A \in D$ such that for all $i \in A$ we have $x_i \in F^{-1}(U)$ and hence $F(x_i) \in U$. □

5.2 Lemma. *Let X be a closed, bounded interval in $\mathbb{R}$. Let S be any set and let $(F_i \mid i \in I)$ be a family of functions from S into X. Then, for any ultrafilter D on I*

$$\sup_x \big(\lim_{i,D} F_i(x)\big) \;\leq\; \lim_{i,D} \big(\sup_x F_i(x)\big), \text{ and}$$

$$\inf_x \big(\lim_{i,D} F_i(x)\big) \;\geq\; \lim_{i,D} \big(\inf_x F_i(x)\big).$$

where in both cases, $\sup_x$ and $\inf_x$ are taken over $x \in S$. Moreover, for each $\epsilon > 0$ there exist $(x_i)_{i\in I}$ and $(y_i)_{i\in I}$ in S such that

$$\lim_{i,D} F_i(x_i) + \epsilon \;\geq\; \lim_{i,D} \big(\sup_x F_i(x)\big), \text{ and}$$

$$\lim_{i,D} F_i(y_i) - \epsilon \;\leq\; \lim_{i,D} \big(\inf_x F_i(x)\big).$$

Proof We prove the statements involving sup; the inf statements are proved similarly (or by replacing each F_i by its negative).

Let $r_i = \sup_x F_i(x)$ for each $i \in I$ and let $r = \lim_{i,D} r_i$. For each $\epsilon > 0$, let $A(\epsilon) \in D$ be such that $r - \epsilon < r_i < r + \epsilon$ for every $i \in A(\epsilon)$.

First we show $\sup_x \lim_{i,D} F_i(x) \leq r$. For each $i \in A(\epsilon)$ and $x \in S$ we have $F_i(x) \leq r_i < r + \epsilon$. Hence the D-limit of $(F_i(x))_{i\in I}$ is $\leq r + \epsilon$. Letting ϵ tend to 0 gives the desired inequality.

For the other sup statement, fix $\epsilon > 0$ and for each $i \in I$ choose $x_i \in S$ so that $r_i \leq F_i(x_i) + \epsilon/2$. Then for $i \in A(\epsilon/2)$ we have $r \leq F_i(x_i) + \epsilon$. Taking the D-limit gives the desired inequality. □

Ultraproducts of metric spaces

Let $((M_i, d_i) \mid i \in I)$ be a family of bounded metric spaces, all having diameter $\leq K$. Let D be an ultrafilter on I. Define a function d on the Cartesian product $\prod_{i\in I} M_i$ by

$$d(x, y) = \lim_{i,D} d_i(x_i, y_i)$$

where $x = (x_i)_{i \in I}$ and $y = (y_i)_{i \in I}$. This D-limit is taken in the interval $[0, K]$. It is easy to check that d is a pseudometric on $\prod_{i \in I} M_i$.

For $x, y \in \prod_{i \in I} M_i$, define $x \sim_D y$ to mean that $d(x,y) = 0$. Then $\sim_D$ is an equivalence relation, so we may define

$$\Big(\prod_{i \in I} M_i\Big)_D = \Big(\prod_{i \in I} M_i\Big) / {\sim_D}.$$

The pseudometric d on $\prod_{i \in I} M_i$ induces a metric on this quotient space, and we also denote this metric by d.

The space $(\prod_{i \in I} M_i)_D$ with the induced metric d is called the *D-ultraproduct of* $((M_i, d_i) \mid i \in I)$. We denote the equivalence class of $(x_i)_{i \in I} \in \prod_{i \in I} M_i$ under $\sim_D$ by $((x_i)_{i \in I})_D$.

If $(M_i, d_i) = (M, d)$ for every $i \in I$, the space $(\prod_{i \in I} M_i)_D$ is called the *D-ultrapower of M* and it is denoted $(M)_D$. In this situation, the map $T \colon M \to (M)_D$ defined by $T(x) = ((x_i)_{i \in I})_D$, where $x_i = x$ for every $i \in I$, is an isometric embedding. It is called the *diagonal embedding* of M into $(M)_D$.

A particular case of importance is the D-ultrapower of a compact metric space (M, d). In that case the diagonal embedding of M into $(M)_D$ is surjective. Indeed, if $(x_i)_{i \in I} \in M^I$ and x is the D-limit of the family $(x_i)_{i \in I}$, which exists since (M, d) is compact, then it is easy to show that $((x_i)_{i \in I})_D = T(x)$. In particular, any ultrapower of a closed bounded interval may be canonically identified with the interval itself.

Since we require that structures are based on *complete* metric spaces, it is useful to note that every ultraproduct of such spaces is itself complete.

5.3 Proposition. *Let $((M_i, d_i) \mid i \in I)$ be a family of complete, bounded metric spaces, all having diameter $\leq K$. Let D be an ultrafilter on I and let (M, d) be the D-ultraproduct of $((M_i, d_i) \mid i \in I)$. The metric space (M, d) is complete.*

Proof Let $(x^k)_{k \geq 1}$ be a Cauchy sequence in (M, d). Without loss of generality we may assume that $d(x^k, x^{k+1}) < 2^{-k}$ holds for all $k \geq 1$; that is, to prove (M, d) complete it suffices to show that all such Cauchy sequences have a limit. For each $k \geq 1$ let x^k be represented by the family $(x_i^k)_{i \in I}$. For each $m \geq 1$ let A_m be the set of all $i \in I$ such that $d_i(x_i^k, x_i^{k+1}) < 2^{-k}$ holds for all $k = 1, \ldots, m$. Then the sets $(A_m)_{m \geq 1}$ form a decreasing chain and all of them are in D.

We define a family $(y_i)_{i \in I}$ that will represent the limit of the sequence

$(x^k)_{k\geq 1}$ in (M,d). If $i \notin A_1$, then we take y_i to be an arbitrary element of M_i. If for some $m \geq 1$ we have $i \in A_m \setminus A_{m+1}$, then we set $y_i = x_i^{m+1}$. If $i \in A_m$ holds for all $m \geq 1$, then $(x_i^m)_{m\geq 1}$ is a Cauchy sequence in the complete metric space (M_i, d_i) and we take y_i to be its limit.

An easy calculation shows that for each $m \geq 1$ and each $i \in A_m$ we have $d_i(x_i^m, y_i) \leq 2^{-m+1}$. It follows that $((y_i)_{i\in I})_D$ is the limit in the ultraproduct (M,d) of the sequence $(x^k)_{k\geq 1}$. □

Ultraproducts of functions

Suppose $((M_i, d_i) \mid i \in I)$ and $((M_i', d_i') \mid i \in I)$ are families of metric spaces, all of diameter $\leq K$. Fix $n \geq 1$ and suppose $f_i \colon M_i^n \to M_i'$ is a uniformly continuous function for each $i \in I$. Moreover, suppose the single function $\Delta\colon (0,1] \to (0,1]$ is a modulus of uniform continuity for all of the functions f_i. Given an ultrafilter D on I, we define a function

$$\Big(\prod_{i\in I} f_i\Big)_D : \Big(\prod_{i\in I} M_i\Big)_D^n \to \Big(\prod_{i\in I} M_i'\Big)_D$$

as follows. If for each $k = 1, \ldots, n$ we have $(x_i^k)_{i\in I} \in \prod_{i\in I} M_i$, we define

$$\Big(\prod_{i\in I} f_i\Big)_D(((x_i^1)_{i\in I})_D, \ldots, ((x_i^n)_{i\in I})_D) = \Big(\big(f_i(x_i^1, \ldots, x_i^n)\big)_{i\in I}\Big)_D.$$

We claim that this defines a uniformly continuous function that also has Δ as its modulus of uniform continuity. For simplicity of notation, suppose $n = 1$. Fix $\epsilon > 0$. Suppose the distance between $\big((x_i)_{i\in I}\big)_D$ and $\big((y_i)_{i\in I}\big)_D$ in the ultraproduct $\Big(\prod_{i\in I} M_i\Big)_D$ is $< \Delta(\epsilon)$. There must exist $A \in D$ such that for all $i \in A$ we have $d_i(x_i, y_i) < \Delta(\epsilon)$. Since Δ is a modulus of uniform continuity for all of the functions f_i, it follows that $d_i'(f_i(x_i), f_i(y_i)) \leq \epsilon$ for all $i \in A$. Hence the distance in the ultraproduct $\Big(\prod_{i\in I} M_i'\Big)_D$ between $\big((f(x_i))_{i\in I}\big)_D$ and $\big((f(y_i))_{i\in I}\big)_D$ must be $\leq \epsilon$. This shows that $\Big(\prod_{i\in I} f_i\Big)_D$ is well defined and that it has Δ as a modulus of uniform continuity. (Note that the precise form of our definition of "modulus of uniform continuity" played a role in this argument.)

Ultraproducts of L-structures

Let $(\mathcal{M}_i \mid i \in I)$ be a family of L-structures and let D be an ultrafilter on I. Suppose the underlying metric space of $\mathcal{M}_i$ is (M_i, d_i). Since there is a uniform bound on the diameters of these metric spaces, we may form their D-ultraproduct. For each function symbol f of L, the functions $f^{\mathcal{M}_i}$ all have the same modulus of uniform continuity Δ_f. Therefore the D-ultraproduct of this family of functions is well defined. The same is true if we consider a predicate symbol P of L. Moreover, the functions $P^{\mathcal{M}_i}$ all have their values in $[0,1]$, whose D-ultrapower can be identified with $[0,1]$ itself; thus the D-ultraproduct of $(P^{\mathcal{M}_i} \mid i \in I)$ can be regarded as a $[0,1]$-valued function on M.

Therefore we may define the *D-ultraproduct of the family* $(\mathcal{M}_i \mid i \in I)$ *of L-structures* to be the L-structure $\mathcal{M}$ that is specified as follows:

The underlying metric space of $\mathcal{M}$ is given by the ultraproduct of metric spaces

$$M = \Big(\prod_{i \in I} M_i\Big)_D.$$

For each predicate symbol P of L, the interpretation of P in $\mathcal{M}$ is given by the ultraproduct of functions

$$P^{\mathcal{M}} = \Big(\prod_{i \in I} P^{\mathcal{M}_i}\Big)_D$$

which maps M^n to $[0,1]$. For each function symbol f of L, the interpretation of f in $\mathcal{M}$ is given by the ultraproduct of functions

$$f^{\mathcal{M}} = \Big(\prod_{i \in I} f^{\mathcal{M}_i}\Big)_D$$

which maps M^n to M. For each constant symbol c of L, the interpretation of c in $\mathcal{M}$ is given by

$$c^{\mathcal{M}} = \big((c^{\mathcal{M}_i})_{i \in I}\big)_D.$$

The discussion above shows that this defines $\mathcal{M}$ to be a well-defined L-structure. We call $\mathcal{M}$ the *D-ultraproduct of the family* $(\mathcal{M}_i \mid i \in I)$ and denote it by

$$\mathcal{M} = \Big(\prod_{i \in I} \mathcal{M}_i\Big)_D.$$

If all of the L-structures $\mathcal{M}_i$ are equal to the same structure $\mathcal{M}_0$, then $\mathcal{M}$ is called the *D-ultrapower of* $\mathcal{M}_0$ and is denoted by

$$(\mathcal{M}_0)_D.$$

This ultraproduct construction finds many applications in functional analysis (see [24] and its references) and in metric space geometry (see

[19]). Its usefulness is partly explained by the following theorem, which is the analogue in this setting of the well known result in first-order logic proved by J. Łos. This is sometimes known as the *Fundamental Theorem of Ultraproducts.*

5.4 Theorem. *Let $(\mathcal{M}_i \mid i \in I)$ be a family of L-structures. Let D be any ultrafilter on I and let $\mathcal{M}$ be the D-ultraproduct of $(\mathcal{M}_i \mid i \in I)$. Let $\varphi(x_1, \ldots, x_n)$ be an L-formula. If $a_k = \big((a_i^k)_{i \in I}\big)_D$ are elements of $\mathcal{M}$ for $k = 1, \ldots, n$, then*

$$\varphi^{\mathcal{M}}(a_1, \ldots, a_n) = \lim_{i,D} \varphi^{\mathcal{M}_i}(a_i^1, \ldots, a_i^n).$$

Proof The proof is by induction on the complexity of φ. Basic facts about ultrafilter limits (discussed at beginning of this section) are used in the proof. □

5.5 Corollary. *If $\mathcal{M}$ is an L-structure and $T \colon M \to (M)_D$ is the diagonal embedding, then T is an elementary embedding of $\mathcal{M}$ into $(\mathcal{M})_D$.*

Proof From Theorem 5.4. □

5.6 Corollary. *If $\mathcal{M}$ and $\mathcal{N}$ are L-structures and they have isomorphic ultrapowers, then $\mathcal{M} \equiv \mathcal{N}$.*

Proof Immediate from the preceding result. □

The converse of the preceding corollary is also true, in a strong form:

5.7 Theorem. *If $\mathcal{M}$ and $\mathcal{N}$ are L-structures and $\mathcal{M} \equiv \mathcal{N}$, then there exists an ultrafilter D such that $(\mathcal{M})_D$ is isomorphic to $(\mathcal{N})_D$.*

The preceding result is an extension of the Keisler-Shelah Theorem from ordinary model theory. (See [36] and Chapter 6 in [13].) A detailed proof of the analogous result for normed space structures and the approximate logic of positive bounded formulas is given in [24, Chapter 10], and that argument can be readily adapted to continuous logic for metric structures.

There are characterizations of elementary equivalence that are slightly more complex to state than Theorem 5.7 but are much easier to prove, such as the following: $\mathcal{M} \equiv \mathcal{N}$ if and only if $\mathcal{M}$ and $\mathcal{N}$ have isomorphic elementary extensions that are each constructed as the union of an infinite sequence of successive ultrapowers. Theorem 5.7 has the positive feature that it connects continuous logic in a direct way to application areas in which the ultrapower construction is important, such as the theory of Banach spaces and other areas of functional analysis. Indeed, the result shows that the mathematical properties a metric structure and its ultrapowers share in common are exactly those that can be expressed by sentences of the continuous analogue of first-order logic. Theorem 5.7 also yields a characterization of axiomatizable classes of metric structures (5.14 below) whose statement is simpler than would otherwise be the case.

Compactness theorem

5.8 Theorem. *Let T be an L-theory and $\mathcal{C}$ a class of L-structures. Assume that T is finitely satisfiable in $\mathcal{C}$. Then there exists an ultraproduct of structures from $\mathcal{C}$ that is a model of T.*

Proof Let Λ be the set of finite subsets of T. Let $\lambda \in \Lambda$, and write $\lambda = \{E_1, \ldots, E_n\}$. By assumption there is an L-structure $\mathcal{M}_\lambda$ in $\mathcal{C}$ such that $\mathcal{M}_\lambda \models E_j$ for all $j = 1, \ldots, n$.

For each $E \in T$, let $S(E)$ be the set of all $\lambda \in \Lambda$ such that $E \in \lambda$. Note that the collection of sets $\{S(E) \mid E \in T\}$ has the finite intersection property. Hence there is an ultrafilter D on Λ that contains this collection.

Let

$$\mathcal{M} = \Big(\prod_{\lambda\in\Lambda} \mathcal{M}_\lambda\Big)_D.$$

Note that if $\lambda \in S(E)$, then $\mathcal{M}_\lambda \models E$. It follows from Theorem 5.4 that $\mathcal{M} \models E$ for every $E \in T$. In other words, the ultraproduct $\mathcal{M}$ of structures from $\mathcal{C}$ is a model of T. □

In many applications it is useful to note that the Compactness Theorem remains true even if the finite satisfiability hypothesis is weakened to an approximate version. This is an immediate consequence of basic properties of the semantics for continuous logic.

5.9 Definition. For any set Σ of L-conditions, Σ^+ is the set of all conditions $\varphi \leq 1/n$ such that $\varphi = 0$ is an element of Σ and $n \geq 1$.

5.10 Corollary. *Let T be an L-theory and $\mathcal{C}$ a class of L-structures. Assume that T^+ is finitely satisfiable in $\mathcal{C}$. Then there exists an ultraproduct of structures from $\mathcal{C}$ that is a model of T.*

Proof This follows immediately from Theorem 5.8, because T and T^+ obviously have the same models. □

The next result is a version of the Compactness Theorem for formulas. In it we allow an arbitrary family $(x_j \mid j \in J)$ of possible free variables.

5.11 Definition. Let T be an L-theory and $\Sigma(x_j \mid j \in J)$ a set of L-conditions. We say that Σ is *consistent with* T if for every finite subset F of Σ there exists a model $\mathcal{M}$ of T and elements a of M such that for every condition E in F we have $\mathcal{M} \models E[a]$. (Here a is a finite tuple suitable for the free variables in members of F.)

5.12 Corollary. *Let T be an L-theory and $\Sigma(x_j \mid j \in J)$ a set of L-conditions, and assume that Σ^+ is consistent with T. Then there is a model $\mathcal{M}$ of T and elements $(a_j \mid j \in J)$ of M such that*

$$\mathcal{M} \models E[a_j \mid j \in J]$$

for every L-condition E in Σ.

Proof Let $(c_j \mid j \in J)$ be new constants and consider the signature $L(\{c_j \mid j \in J\})$. This corollary is proved by applying the Compactness Theorem to the set $T \cup \Sigma^+(c_j \mid j \in J)$ of closed $L(\{c_j \mid j \in J\})$-conditions. As noted in the proof of the previous result, anything satisfying Σ^+ will also satisfy Σ. □

Axiomatizability of classes of structures

5.13 Definition. Suppose that $\mathcal{C}$ is a class of L-structures. We say that $\mathcal{C}$ is *axiomatizable* if there exists a set T of closed L-conditions such that $\mathcal{C} = \mathrm{Mod}_L(T)$. When this holds for T, we say that T is a set of *axioms* for $\mathcal{C}$ in L.

In this section we characterize axiomatizability in continuous logic using ultraproducts. The ideas are patterned after a well known characterization of axiomatizability in first-order logic due to Keisler [31]. (See Corollary 6.1.16 in [13].)

5.14 Proposition. *Suppose that $\mathcal{C}$ is a class of L-structures. The following statements are equivalent:*

(1) *$\mathcal{C}$ is axiomatizable in L;*
(2) *$\mathcal{C}$ is closed under isomorphisms and ultraproducts, and its complement, $\{\mathcal{M} \mid \mathcal{M}$ is an L-structure not in $\mathcal{C}\}$, is closed under ultrapowers.*

Proof (1)⇒(2) follows from the Fundamental Theorem of Ultraproducts.

To prove (2)⇒(1), we let T be the set of closed L-conditions that are satisfied by every structure in $\mathcal{C}$. We claim that T is a set of axioms for $\mathcal{C}$. To prove this, suppose $\mathcal{M}$ is an L-structure such that $\mathcal{M} \models T$.

We claim that $\mathrm{Th}(\mathcal{M})^+$ is finitely satisfiable in $\mathcal{C}$. If not, there exist L-sentences $\sigma_1, \ldots, \sigma_n$ and $\epsilon > 0$ such that $\sigma_j^{\mathcal{M}} = 0$ for all $j = 1, \ldots, n$, but such that for any $\mathcal{N} \in \mathcal{C}$, we have $\sigma_j^{\mathcal{N}} \geq \epsilon$ for some $j = 1, \ldots, n$. This means that the condition $\max(\sigma_1, \ldots, \sigma_n) \geq \epsilon$ is in T but is not satisfied in $\mathcal{M}$, which is a contradiction.

So $\mathrm{Th}(\mathcal{M})^+$ is finitely satisfiable in $\mathcal{C}$. By the Compactness Theorem this yields an ultraproduct $\mathcal{M}'$ of structures from $\mathcal{C}$ such that M' is a model of $\mathrm{Th}(\mathcal{M})^+$. One sees easily that this implies $\mathcal{M}' \equiv \mathcal{M}$. Theorem 5.7, the extension of the Keisler-Shelah theorem to this continuous logic, yields an ultrafilter D such that $(\mathcal{M}')_D$ and $(\mathcal{M})_D$ are isomorphic. Statement (2) implies that $\mathcal{M}$ is in $\mathcal{C}$. □

5.15 Remark. The proof of Proposition 5.14 contains the following useful elementary result: let $\mathcal{C}$ be a class of L-structures and let T be the set of all closed L-conditions E such that $\mathcal{M} \models E$ holds for all $\mathcal{M} \in \mathcal{C}$. Then, every model of T is elementarily equivalent to some ultraproduct of structures from $\mathcal{C}$.

6 Connectives

Recall that in our definition of *formulas* for continuous logic, we took a *connective* to be a continuous function from $[0,1]^n$ to $[0,1]$, for some $n \geq 1$. This choice is somewhat arbitrary; from one point of view it is too general, and from another it is too restrictive. We begin to discuss these issues in this section.

Here we continue to limit ourselves to *finitary* connectives, and our intention is to limit the connectives we use when building formulas. We consider a restricted set of formulas to be adequate if every formula can be "uniformly approximated" by formulas from the restricted set.

6.1 Definition. A system of connectives is a family $\mathcal{F} = (F_n \mid n \geq 1)$ where each F_n is a set of connectives $f\colon [0,1]^n \to [0,1]$. We say that $\mathcal{F}$ is *closed* if it is closed under arbitrary substitutions; more precisely:

(1) For each n, F_n contains the projection $\pi_j^n\colon [0,1]^n \to [0,1]$ onto the jth coordinate for each $j = 1,\ldots,n$.
(2) For each n and m, if $u \in F_n$, and $v_1,\ldots,v_n \in F_m$, then the function $w\colon [0,1]^m \to [0,1]$ defined by $w(t) = u(v_1(t),\ldots,v_n(t))$, where t denotes an element of $[0,1]^m$, is in F_m.

Note that each system $\mathcal{F}$ of connectives generates a smallest closed system of connectives $\overline{\mathcal{F}}$. We say that $\mathcal{F}$ is *full* if $\overline{\mathcal{F}}$ is uniformly dense in the system of all connectives; that is, for any $\epsilon > 0$ and any connective $f(t_1,\ldots,t_n)$, there is a connective $g \in \overline{F}_n$ such that

$$|f(t_1,\ldots,t_n) - g(t_1,\ldots,t_n)| \leq \epsilon$$

for all $(t_1,\ldots,t_n) \in [0,1]^n$.

6.2 Definition. Given a system $\mathcal{F}$ of connectives, we define the collection of $\mathcal{F}$*-restricted* formulas by induction:

(1) Atomic formulas are $\mathcal{F}$-restricted formulas.
(2) If $u \in F_n$ and $\varphi_1,\ldots,\varphi_n$ are $\mathcal{F}$-restricted formulas, then $u(\varphi_1,\ldots,\varphi_n)$ is an $\mathcal{F}$-restricted formula.
(3) If φ is an $\mathcal{F}$-restricted formula, so are $\sup_x \varphi$ and $\inf_x \varphi$.

The importance of full sets of connectives is that the restricted formulas made using them are dense in the set of all formulas, with respect to the logical distance between L-formulas, as the next result states. (See page 333.)

6.3 Theorem. *Assume that $\mathcal{F}$ is a full system of connectives. Then for any $\epsilon > 0$ and any L-formula $\varphi(x_1,\ldots,x_n)$, there is an $\mathcal{F}$-restricted L-formula $\psi(x_1,\ldots,x_n)$ such that for all L-structures $\mathcal{M}$ one has*

$$|\varphi^{\mathcal{M}}(a_1,\ldots,a_n) - \psi^{\mathcal{M}}(a_1,\ldots,a_n)| \leq \epsilon$$

for all $a_1,\ldots,a_n \in M$.

Proof By induction on formulas. □

Next we show that there is a very simple system of connectives that is full. In particular this system is countable, so there are at most card(L) many restricted L-formulas for this system of connectives. It follows from the previous result that the collection of all L formulas has density at most card(L) with respect to the logical distance between L-formulas.

6.4 Definition. We define a binary function $\dot{-}: \mathbb{R}^{\geq 0} \times \mathbb{R}^{\geq 0} \to \mathbb{R}^{\geq 0}$ by:

$$x \dot{-} y = \begin{cases} (x-y) & \text{if } x \geq y \\ 0 & \text{otherwise.} \end{cases}$$

Note that if $x, y \in [0,1]$, then $x \dot{-} y \in [0,1]$, so the restriction of $\dot{-}$ is a connective; we use $\dot{-}$ to denote this connective as well as its extension to all of $\mathbb{R}^{\geq 0}$.

There are many well known identities involving $\dot{-}$. For example, note that $x \dot{-} y = (x+z) \dot{-} (y+z)$ and $((x \dot{-} y) \dot{-} z) = x \dot{-} (y+z)$ for all $x, y, z \in \mathbb{R}^{\geq 0}$.

6.5 Definition. Let $\mathcal{F}_0 = (F_n \mid n \geq 1)$ where $F_1 = \{0, 1, x/2\}$ (with $0, 1$ treated as constant functions of one variable), $F_2 = \{\dot{-}\}$, and all other F_n are empty. We call the connectives in $\mathcal{F}_0$ *restricted*.

The following connectives belong to $\overline{F}_2$:

$$\min(t_1, t_2) = t_1 \dot{-} (t_1 \dot{-} t_2)$$

$$\max(t_1, t_2) = 1 \dot{-} (\min(1 \dot{-} t_1, 1 \dot{-} t_2))$$

$$|t_1 - t_2| = \max(t_1 \dot{-} t_2, t_2 \dot{-} t_1)$$

$$\min(t_1 + t_2, 1) = 1 \dot{-} ((1 \dot{-} t_1) \dot{-} t_2)$$

$$t_1 \dot{-} (mt_2) = \underbrace{((\ldots(t_1 \dot{-} t_2) \dot{-} \ldots) \dot{-} t_2)}_{m \text{ times}}.$$

Every dyadic fraction $m2^{-n}$ in $[0,1]$ is an element of $\overline{F}_1$.

6.6 Proposition. *The system of connectives $\mathcal{F}_0$ is full.*

Proof Let D be the set of dyadic fractions $m2^{-n}$ in $[0,1]$. First we show that for each distinct x, y in D, the set $\{(g(x), g(y)) \mid g \in \overline{F}_1\}$ includes all pairs (a, b) in D^2. Fix numbers $x, y, a, b \in D$ with $x < y$ and $a \geq b$.

Choose $m \in \mathbb{N}$ such that $a < m(y - x)$. Now let $g\colon [0,1] \to [0,1]$ be defined by

$$g(t) = \max(a \dot{-} m(t \dot{-} x), b).$$

Then $g \in \overline{F}_1$, and we also see that $g(x) = a$ and $g(y) = b$. If $a < b$ we can achieve the same result by using $1 \dot{-} a$ and $1 \dot{-} b$ in place of a, b in the construction of g, and then using the function $1 \dot{-} g(t)$.

Now we prove density by arguing exactly as in the proof of the lattice version of the Stone-Weierstrass Theorem on $[0,1]^n$. (See [17, pages 241–242].) □

6.7 Notation. By a *restricted formula* we mean an $\mathcal{F}_0$-restricted formula, where $\mathcal{F}_0$ is the system of connectives in Definition 6.5.

6.8 Definition. A formula is in *prenex form* if it is of the form

$$Q^1_{x_1} Q^2_{x_2} \cdots Q^n_{x_n} \psi$$

where ψ is a quantifier free formula and each Q^i is either sup or inf.

6.9 Proposition. *Every restricted formula is equivalent to a restricted formula in prenex form.*

Proof By induction on formulas. The proof proceeds like the usual proof in first-order logic. The main point of the proof is that the connective $\dot{-}$ is monotone in its arguments, increasing in the first and decreasing in the second. □

Existential conditions

Since the family of restricted formulas is uniformly dense in the family of all formulas, it follows from the previous result that every L-formula can be uniformly approximated by formulas in prenex form. This gives us a way of introducing analogues of the usual syntactic classes into continuous logic. For example, an L-formula is defined to be an *inf-formula* if it is approximated arbitrarily closely by prenex formulas of the form $\inf_{x_1} \ldots \inf_{x_n} \psi$, where ψ is quantifier free. A condition $\varphi = 0$ is defined to be *existential* if φ is an inf-formula. Other syntactic forms for L-conditions are defined similarly.

7 Constructions of models

Unions of chains

If Λ is a linearly ordered set, a Λ-*chain* of L-structures is a family of L-structures $(\mathcal{M}_\lambda \mid \lambda \in \Lambda)$ such that $\mathcal{M}_\lambda \subseteq \mathcal{M}_\eta$ for $\lambda < \eta$. If this holds, we can define the union of $(\mathcal{M}_\lambda \mid \lambda \in \Lambda)$ as an L-prestructure in an obvious way. (Note that for each function symbol or predicate symbol S, all of the interpretations $S^{\mathcal{M}_\lambda}$ have the same modulus of uniform continuity Δ_S, guaranteeing that the union of $(S^{\mathcal{M}_\lambda} \mid \lambda \in \Lambda)$ will also have Δ_S as a modulus of uniform continuity.) This union is based on a metric space, but it may not be complete. After taking the completion we get an L-structure that we will refer to as the *union* of the chain and that we will denote by $\bigcup_{\lambda\in\Lambda} \mathcal{M}_\lambda$.

Caution: In general, if the ordered set Λ has countable cofinality, the set-theoretic union $\bigcup_{\lambda\in\Lambda} M_\lambda$ will be a dense *proper* subset of the underlying set of the L-structure $\bigcup_{\lambda\in\Lambda} \mathcal{M}_\lambda$.

7.1 Definition. A chain of structures $(\mathcal{M}_\lambda \mid \lambda \in \Lambda)$ is called an *elementary chain* if $\mathcal{M}_\lambda \preceq \mathcal{M}_\eta$ for all $\lambda < \eta$.

7.2 Proposition. *If $(\mathcal{M}_\lambda \mid \lambda \in \Lambda)$ is an elementary chain and $\lambda \in \Lambda$, then $\mathcal{M}_\lambda \preceq \bigcup_{\lambda\in\Lambda} \mathcal{M}_\lambda$.*

Proof Use the Tarski-Vaught test (Proposition 4.5). □

Löwenheim-Skolem theorem

Recall that the *density character* of a topological space is the smallest cardinality of a dense subset of the space. For example, a space is separable if and only if its density character is $\leq \aleph_0$. If A is a topological space, we denote its density character by density(A).

7.3 Proposition. *(Downward Löwenheim-Skolem Theorem) Let κ be an infinite cardinal number and assume* card$(L) \leq \kappa$. *Let $\mathcal{M}$ be an L-structure and suppose $A \subseteq M$ has* density$(A) \leq \kappa$. *Then there exists a substructure $\mathcal{N}$ of $\mathcal{M}$ such that*

(1) $\mathcal{N} \preceq \mathcal{M}$;
(2) $A \subseteq N \subseteq M$;
(3) density$(N) \leq \kappa$.

Proof Let A_0 be a dense subset of A of cardinality at most κ. By suitably enlarging A_0, we may obtain a prestructure N_0 such that $A_0 \subseteq N_0 \subseteq M$ and $\mathrm{card}(N_0) \leq \kappa$ and such that the following closure property also holds: for every restricted L-formula $\varphi(x_1, \ldots, x_n, x_{n+1})$ and every rational $\epsilon > 0$, if $\varphi^{\mathcal{M}}(a_1, \ldots, a_n, c) \leq \epsilon$ with $a_k \in N_0$ for $k = 1, \ldots, n$ and $c \in M$, then there exists $b \in N_0$ such that $\varphi^{\mathcal{M}}(a_1, \ldots, a_n, b) \leq \epsilon$. It is possible to do this while maintaining the claimed cardinality bounds because L has at most κ many restricted formulas.

Let N be the closure of N_0 in M. By considering atomic formulas in the closure property above, one shows that there is a substructure $\mathcal{N}$ of $\mathcal{M}$ that is based on N. Continuity of formulas and uniform density of restricted formulas show that $\mathcal{N} \subseteq \mathcal{M}$ satisfy the Tarski-Vaught test (Proposition 4.5) and hence that $\mathcal{N} \preceq \mathcal{M}$. □

Saturated structures

7.4 Definition. Let $\Gamma(x_1, \ldots, x_n)$ be a set of L-conditions and let $\mathcal{M}$ be an L-structure. We say that $\Gamma(x_1, \ldots, x_n)$ is *satisfiable in* $\mathcal{M}$ if there exist elements $a_1, \ldots, a_n$ of $\mathcal{M}$ such that $\mathcal{M} \models \Gamma[a_1, \ldots, a_n]$.

7.5 Definition. Let $\mathcal{M}$ be an L-structure and let κ be an infinite cardinal. We say that $\mathcal{M}$ is *κ-saturated* if the following statement holds: whenever $A \subseteq M$ has cardinality $< \kappa$ and $\Gamma(x_1, \ldots, x_n)$ is a set of $L(A)$-conditions, if every finite subset of Γ is satisfiable in $(\mathcal{M}, a)_{a \in A}$, then the entire set Γ is satisfiable in $(\mathcal{M}, a)_{a \in A}$.

It is straightforward using ultraproducts to prove the existence of ω_1-saturated L-structures when L is countable, and we will do that next. This is the only degree of saturation that one needs for many applications.

Recall that an ultrafilter D is said to be *countably incomplete* if D is not closed under countable intersections. It is equivalent to require the existence of elements J_n of D for each $n \in \mathbb{N}$ such that the intersection $\bigcap_{n \in \mathbb{N}} J_n$ is the empty set. Evidently any non-principal ultrafilter on a countable set is countably incomplete.

7.6 Proposition. *Let L be a signature with* $\mathrm{card}(L) = \omega$ *and let D be a countably incomplete ultrafilter on a set Λ. Then for every family $(\mathcal{M}_\lambda \mid \lambda \in \Lambda)$ of L-structures, $(\prod_{\lambda \in \Lambda} \mathcal{M}_\lambda)_D$ is ω_1-saturated.*

Proof In order to simplify the notation, we verify that $(\prod_{\lambda\in\Lambda}\mathcal{M}_\lambda)_D$ satisfies the statement in Definition 7.5 for $n = 1$.

We have to prove the following statement: if $A \subseteq (\prod_{\lambda\in\Lambda}\mathcal{M}_\lambda)_D$ is countable and $\Gamma(x)$ is a set of $L(A)$-formulas such that every finite subset of $\Gamma(x)$ is satisfiable in $\big((\prod_{\lambda\in\Lambda}\mathcal{M}_\lambda)_D,\, a\big)_{a\in A}$, then the entire set $\Gamma(x)$ is satisfied in $\big((\prod_{\lambda\in\Lambda}\mathcal{M}_\lambda)_D,\, a\big)_{a\in A}$. For each $a \in A$, let

$$u(a) = (u_\lambda(a) \mid \lambda \in \Lambda) \in \prod_{\lambda\in\Lambda}\mathcal{M}_\lambda$$

be such that $a = \big((u_\lambda(a))_{\lambda\in\Lambda}\big)_D$. Note that

$$\Big((\prod_{\lambda\in\Lambda}\mathcal{M}_\lambda)_D,\, a\Big)_{a\in A} = \Big(\prod_{\lambda\in\Lambda}(\mathcal{M}_\lambda, u_\lambda(a))_{a\in A}\Big)_D.$$

Thus, since L is an arbitrary countable signature and A is also countable, it suffices to prove the following simpler statement:

If $\Gamma(x)$ is a set of L-conditions and every finite subset of $\Gamma(x)$ is satisfiable in $(\prod_{\lambda\in\Lambda}\mathcal{M}_\lambda)_D$, then $\Gamma(x)$ is satisfiable in $(\prod_{\lambda\in\Lambda}\mathcal{M}_\lambda)_D$.

So, suppose every finite subset of $\Gamma(x)$ is satisfiable in $(\prod_{\lambda\in\Lambda}\mathcal{M}_\lambda)_D$. Since L is countable, we may write

$$\Gamma(x) = \{\,\varphi_n(x) = 0 \mid n \in \mathbb{N}\}.$$

Since D is countably incomplete, we can fix a descending chain of elements of D

$$\Lambda = \Lambda_0 \supseteq \Lambda_1 \supseteq \dots$$

such that $\bigcap_{k\in\mathbb{N}}\Lambda_k = \emptyset$.

Let $X_0 = \Lambda$ and for each positive integer k define

$$X_k = \Lambda_k \cap \{\,\lambda \in \Lambda \mid \mathcal{M}_\lambda \models \inf_x \max(\varphi_1, \dots, \varphi_k) \le \frac{1}{k+1}\,\}.$$

Then $X_k \in D$ by Theorem 5.4, since

$$(\prod_{\lambda\in\Lambda}\mathcal{M}_\lambda)_D \models \inf_x \max(\varphi_1, \dots, \varphi_k) = 0.$$

We then have $X_k \supseteq X_{k+1}$ for every $k \in \mathbb{N}$ and $\bigcap_{k\in\mathbb{N}} X_k = \emptyset$, so for each $\lambda \in \Lambda$ there exists a largest positive integer $k(\lambda)$ such that $\lambda \in X_{k(\lambda)}$.

We now define an element $a = (a(\lambda))_{\lambda\in\Lambda} \in \prod_{\lambda\in\Lambda}\mathcal{M}_\lambda$ such that

$$\Big(\prod_{\lambda\in\Lambda}\mathcal{M}_\lambda\Big)_D \models \Gamma[(a)_D].$$

For $\lambda \in \Lambda$, if $k(\lambda) = 0$, let $a(\lambda)$ be any element in M_λ; otherwise, let $a(\lambda)$ be such that

$$\mathcal{M}_\lambda \models \{\max(\varphi_1, \ldots, \varphi_{k(\lambda)}) \leq \frac{1}{k(\lambda)}\}[a(\lambda)].$$

If $k \in \mathbb{N}$ and $\lambda \in X_k$, we have $k \leq k(\lambda)$, so $\mathcal{M}_\lambda \models (\varphi_k \leq \frac{1}{k(\lambda)})[a(\lambda)]$. It follows from Theorem 5.4 that $(\prod_{\lambda \in \Lambda} \mathcal{M}_\lambda)_D \models \Gamma[(a)_D]$. □

In saturated structures the meaning of L-conditions can be analyzed using the usual quantifiers $\forall$ and $\exists$, as the next result shows:

7.7 Proposition. *Let $\mathcal{M}$ be an L-structure and suppose $E(x_1, \ldots, x_m)$ is the L-condition*

$$(Q^1_{y_1} \cdots Q^n_{y_n} \varphi(x_1, \ldots, x_m, y_1, \ldots, y_n)) = 0$$

where each Q^i is either inf *or* sup *and φ is quantifier free.*

Let $\mathcal{E}(x_1, \ldots, x_m)$ be the mathematical statement

$$\widetilde{Q}^1 y_1 \ldots \widetilde{Q}^n y_n (\varphi(x_1, \ldots, x_m, y_1, \ldots, y_n) = 0)$$

where each $\widetilde{Q}^i y_i$ is $\exists y_i$ if $Q^i_{y_i}$ is $\inf_{y_i}$ *and is $\forall y_i$ if $Q^i_{y_i}$ is* $\sup_{y_i}$.

If $\mathcal{M}$ is ω-saturated, then for any elements $a_1, \ldots, a_m$ of M, we have

$$\mathcal{M} \models E[a_1, \ldots, a_m] \text{ if and only if } \mathcal{E}(a_1, \ldots, a_n) \text{ is true in } \mathcal{M}.$$

Proof By induction on n. For the induction step, suppose we are treating the condition $(\inf_y \psi(x_1, \ldots, x_n, y)) = 0$. If $\mathcal{M}$ is ω-saturated, then $(\inf_y \psi(a_1, \ldots, a_n, y)) = 0$ holds in $\mathcal{M}$ if and only if there exists some $b \in M$ such that $\psi(a_1, \ldots, a_n, b) = 0$ holds in $\mathcal{M}$. For the left to right direction, take $b \in M$ to satisfy the conditions $\psi(a_1, \ldots, a_n, b) \leq 1/n$ for $n \geq 1$. □

7.8 Definition. Let $\mathcal{M}$ be an L-structure and let $\mathcal{N}$ be an elementary extension of $\mathcal{M}$. We call $\mathcal{N}$ an *enlargement* of $\mathcal{M}$ if it has the following property: whenever $A \subseteq M$ and $\Gamma(x_1, \ldots, x_n)$ is a set of $L(A)$-conditions, if every finite subset of Γ is satisfiable in $(\mathcal{M}, a)_{a \in A}$, then the entire set Γ is satisfiable in $(\mathcal{N}, a)_{a \in A}$.

7.9 Lemma. *Every L-structure has an enlargement.*

Proof Let $\mathcal{M}$ be an L-structure and let J be a set of cardinality $\geq$ card$(L(M))$. Let I be the collection of finite subsets of J and let D be an ultrafilter on I that contains each of the sets $S_j = \{i \in I \mid j \in i\}$, for $j \in J$. Such an ultrafilter exists since this collection of sets has the finite intersection property. Let $\mathcal{N}$ be the D-ultrapower of $\mathcal{M}$, considered as an elementary extension of $\mathcal{M}$ via the diagonal map. We will show that $\mathcal{N}$ is an enlargement of $\mathcal{M}$.

Let $A \subseteq M$ and suppose $\Gamma(x_1,\dots,x_n)$ is a set of $L(A)$-conditions such that every finite subset of Γ is satisfiable in $(\mathcal{M},a)_{a\in A}$. Let α be a function from J onto Γ. Given $i = \{j_1,\dots,j_m\} \in I$, let $(a_i^1,\dots,a_i^n)$ be any n-tuple from M that satisfies the finite subset $\{\alpha(j_1),\dots,\alpha(j_m)\}$ of Γ in $(\mathcal{M},a)_{a\in A}$. For each $k = 1,\dots,n$ set $a_k = \left((a_i^k)_{i\in I}\right)_D$. Theorem 5.4 easily yields that $(a_1,\dots,a_n)$ satisfies Γ in $(\mathcal{N},a)_{a\in A}$. □

7.10 Proposition. *Let $\mathcal{M}$ be an L-structure. For every infinite cardinal κ, $\mathcal{M}$ has a κ-saturated elementary extension.*

Proof By increasing κ if necessary (for example, replacing κ by κ^+) we may assume κ is regular. By induction we construct an elementary chain $(\mathcal{M}_\alpha \mid \alpha < \kappa)$ such that $\mathcal{M}_0 = \mathcal{M}$ and for each $\alpha < \kappa$, $\mathcal{M}_{\alpha+1}$ is an enlargement of $\mathcal{M}_\alpha$. (At limit ordinals we take unions.) Let $\mathcal{N}$ be the union of the chain $(\mathcal{M}_\alpha \mid \alpha < \kappa)$. By Proposition 7.2, $\mathcal{M}_\alpha \preceq \mathcal{N}$ for all $\alpha < \kappa$; in particular, $\mathcal{M} \preceq \mathcal{N}$. We claim that $\mathcal{N}$ is κ-saturated. Let A be a subset of $\mathcal{N}$ of cardinality $< \kappa$. Since κ is regular, there exists $\alpha < \kappa$ such that A is a subset of $\mathcal{M}_\alpha$. The elements of $\mathcal{N}$ needed to verify Definition 7.5 for $(\mathcal{N},a)_{a\in A}$ can be found in $\mathcal{M}_{\alpha+1}$. □

Strongly homogeneous structures

7.11 Definition. Let $\mathcal{M}$ be an L-structure and let κ be an infinite cardinal. We say that $\mathcal{M}$ is *strongly κ-homogeneous* if the following statement holds: whenever $L(C)$ is an extension of L by constants with $\mathrm{card}(C) < \kappa$ and f,g are functions from C into M such that

$$(\mathcal{M}, f(c))_{c\in C} \equiv (\mathcal{M}, g(c))_{c\in C}$$

one has

$$(\mathcal{M}, f(c))_{c\in C} \cong (\mathcal{M}, g(c))_{c\in C}.$$

Note that an isomorphism from $(\mathcal{M}, f(c))_{c\in C}$ onto $(\mathcal{M}, g(c))_{c\in C}$ is an *automorphism* of $\mathcal{M}$ that takes $f(c)$ to $g(c)$ for each $c \in C$.

7.12 Proposition. *Let $\mathcal{M}$ be an L-structure. For every infinite cardinal κ, $\mathcal{M}$ has a κ-saturated elementary extension $\mathcal{N}$ such that each reduct of $\mathcal{N}$ to a sublanguage of L is strongly κ-homogeneous.*

Proof We may assume κ is regular, by increasing κ if necessary. Given any L-structure $\mathcal{M}$, we construct an elementary chain $(\mathcal{M}_\alpha \mid \alpha < \kappa)$ whose union has the desired properties. Let $\mathcal{M}_0 = \mathcal{M}$; for each $\alpha < \kappa$, let $\mathcal{M}_{\alpha+1}$ be an elementary extension of $\mathcal{M}_\alpha$ that is τ_α-saturated, where τ_α is a cardinal bigger than $\operatorname{card}(L)$ and bigger than the cardinality of $\mathcal{M}_\alpha$; take unions at limit ordinals. Let $\mathcal{N}$ be the union of $(\mathcal{M}_\alpha \mid \alpha < \kappa)$. By Proposition 7.2, $\mathcal{M} \preceq \mathcal{N}$. An argument such as in the proof of Proposition 7.10 shows that $\mathcal{N}$ is κ-saturated.

The fact that $\mathcal{N}$ is strongly κ-homogeneous follows from an inductive argument whose successor steps are based on the following easily proved fact:

Suppose $\mathcal{M} \preceq \mathcal{N}$ and $\mathcal{N}$ is τ-saturated, where τ is a cardinal satisfying $\tau > \operatorname{card}(L)$ and $\tau > \operatorname{card}(M)$. Let C be a set of new constants with $\operatorname{card}(C) < \tau$. Suppose f, g are functions from C into $\mathcal{M}$ such that $(\mathcal{M}, f(c))_{c\in C} \equiv (\mathcal{M}, g(c))_{c\in C}$. Then there exists an elementary embedding $T \colon \mathcal{M} \to \mathcal{N}$ such that for every $c \in C$, $T(f(c)) = g(c)$.

Finally, suppose L' is a sublanguage of L. For each $\alpha < \kappa$, the reduct of $\mathcal{M}_{\alpha+1}$ to L' is also τ_α-saturated. Hence an argument similar to the one given above for $\mathcal{N}$ shows that the reduct of $\mathcal{N}$ to L' is also strongly κ-homogeneous. □

Universal domains

7.13 Definition. Let T be a complete theory in L and let κ be an infinite cardinal number. A *κ-universal domain for T* is a κ-saturated, strongly κ-homogeneous model of T. If $\mathcal{U}$ is a κ-universal domain for T and $A \subseteq \mathcal{U}$, we will say A is *small* if $\operatorname{card}(A) < \kappa$.

By Proposition 7.12, every complete theory has a κ-universal domain for every infinite cardinal κ. Indeed, T has a model $\mathcal{U}$ such that not only $\mathcal{U}$ itself but also every reduct of $\mathcal{U}$ to a sublanguage of L is κ-saturated and strongly κ-homogeneous. Such models are needed for some arguments.

Implications

One of the subtleties of continuous first-order logic is that it is essentially a *positive* logic. In particular, there is no direct way to express an implication between conditions. This is inconvenient in applications, since many natural mathematical properties are stated using implications. However, when working in a saturated model or in all models of a theory, this obstacle can be clarified and often overcome in a natural way, which we explain here.

Let L be any signature for metric structures. For the rest of this section we fix two L-formulas $\varphi(x_1, \dots, x_n)$ and $\psi(x_1, \dots, x_n)$ and an L-theory T. For convenience we write x for $x_1, \dots, x_n$.

7.14 Proposition. *Let $\mathcal{M}$ be an ω-saturated model of T. The following statements are equivalent:*
(1) For all $a \in M^n$, if $\varphi^{\mathcal{M}}(a) = 0$ then $\psi^{\mathcal{M}}(a) = 0$.
(2) $\forall \epsilon > 0 \; \exists \delta > 0 \; \forall a \in M^n \; (\varphi^{\mathcal{M}}(a) < \delta \Rightarrow \psi^{\mathcal{M}}(a) \leq \epsilon)$.
(3) There is an increasing, continuous function $\alpha \colon [0,1] \to [0,1]$ with $\alpha(0) = 0$ such that $\psi^{\mathcal{M}}(a) \leq \alpha(\varphi^{\mathcal{M}}(a))$ for all $a \in M^n$.

Proof (1) $\Rightarrow$ (2): Suppose (1) holds. If (2) fails, then for some $\epsilon > 0$ the set of conditions $\psi(x) \geq \epsilon$ and $\varphi(x) \leq 1/n$ for $n \geq 1$ is finitely satisfiable in $\mathcal{M}$. Since $\mathcal{M}$ realizes every finitely satisfiable set of L-conditions, there exists $a \in M^n$ such that $\psi^{\mathcal{M}}(a) \geq \epsilon$ while $\varphi^{\mathcal{M}}(a) \leq 1/n$ for all $n \geq 1$. This contradicts (1).
(2) $\Rightarrow$ (3): This follows from Proposition 2.10.
(3) $\Rightarrow$ (1): This is trivial. □

The next results are variants of the previous one, with essentially the same proof. In stating them we translate (2) and (3) using conditions of continuous logic, as follows.

The statement in (2) holds for a given L-structure $\mathcal{M}$ and a given ϵ, δ if and only if the L-condition

$$\sup_x \min\left(\delta \dot{-} \varphi(x), \psi(x) \dot{-} \epsilon\right) = 0$$

is true in $\mathcal{M}$.

The statement in (3) holds for a given $\mathcal{M}$ and a given α if and only if the L-condition

$$\sup_x \left(\psi(x) \dot{-} \alpha(\varphi(x))\right) = 0$$

is true in $\mathcal{M}$.

7.15 Proposition. *The following statements are equivalent.*
(1) For all $\mathcal{M} \models T$ *and all* $a \in M^n$, *if* $\varphi^{\mathcal{M}}(a) = 0$ *then* $\psi^{\mathcal{M}}(a) = 0$.
(2) For all $\epsilon > 0$ *there exists* $\delta > 0$ *such that*

$$T \models \sup_x \min\left(\delta \dot- \varphi(x), \psi(x) \dot- \epsilon\right) = 0.$$

(3) There is an increasing, continuous function $\alpha\colon [0,1] \to [0,1]$ *with* $\alpha(0) = 0$ *such that*

$$T \models \sup_x \left(\psi(x) \dot- \alpha(\varphi(x))\right) = 0.$$

Proof It is obvious that (2) and (3) imply (1). If (2) fails, then for some $\epsilon > 0$ the set of conditions $\psi(x) \geq \epsilon$ and $\varphi(x) \leq 1/n$ for $n \geq 1$ is finitely satisfiable in models of T. The Compactness Theorem yields $\mathcal{M} \models T$ and $a \in M^n$ such that $\psi^{\mathcal{M}}(a) \geq \epsilon$ while $\varphi^{\mathcal{M}}(a) \leq 1/n$ for all $n \geq 1$. This contradicts (1).

Now assume that (1) and (equivalently) (2) hold. Define a function Δ on $(0,1]$ by setting each $\Delta(\epsilon)$ to be half the supremum of all $\delta \in (0,1]$ for which $T \models \sup_x \min\left(\delta \dot- \varphi(x), \psi(x) \dot- \epsilon\right) = 0$. Evidently Δ is an increasing function and (since (2) holds) it takes its values in $(0,1]$. For each $\mathcal{M} \models T$ and each $\epsilon \in (0,1]$ we have

$$\forall a \in M^n \left(\varphi^{\mathcal{M}}(a) \leq \Delta(\epsilon) \Rightarrow \psi^{\mathcal{M}}(a) \leq \epsilon\right).$$

Now we use the argument in Remark 2.12 to obtain an increasing, continuous function $\alpha\colon [0,1] \to [0,1]$ with $\alpha(0) = 0$ such that for each $\mathcal{M} \models T$ we have

$$\forall a \in M^n \left(\psi^{\mathcal{M}}(a) \leq \alpha(\varphi^{\mathcal{M}}(a))\right).$$

This proves that statement (3) holds for this α. □

7.16 Corollary. *Let* $\mathcal{C}$ *be the set of all models* $\mathcal{M}$ *of* T *that satisfy the requirement*

$$\forall a \in M^n \ \left[\varphi^{\mathcal{M}}(a) = 0 \ \Rightarrow \ \psi^{\mathcal{M}}(a) = 0\right].$$

Then the following statements are equivalent:
(1) $\mathcal{C}$ *is axiomatizable.*
(2) For each $\epsilon > 0$ *there exists* $\delta > 0$ *such that the L-condition*

$$\sup_x \min\left(\delta \dot- \varphi(x), \psi(x) \dot- \epsilon\right) = 0$$

is true in all members of $\mathcal{C}$.

(3) There exists an increasing, continuous function $\alpha\colon [0,1] \to [0,1]$ *with* $\alpha(0) = 0$ *such that the L-condition*

$$\sup_x \left(\psi(x) \dot{-} \alpha(\varphi(x))\right) = 0$$

is true in all members of $\mathcal{C}$.

Proof If (1) holds, apply the previous result to a theory T' that axiomatizes $\mathcal{C}$. If (2) or (3) holds, $\mathcal{C}$ can be axiomatized by adding conditions of the form displayed to T. □

8 Spaces of types

In this section we consider a fixed signature L for metric structures and a fixed L-theory T. Until further notice in this section we assume that T is a complete theory.

Suppose that $\mathcal{M}$ is a model of T and $A \subseteq M$. Denote the $L(A)$-structure $(\mathcal{M}, a)_{a\in A}$ by $\mathcal{M}_A$, and set T_A to be the $L(A)$-theory of $\mathcal{M}_A$. Note that any model of T_A is isomorphic to a structure of the form $(\mathcal{N}, a)_{a\in A}$, where $\mathcal{N}$ is a model of T.

8.1 Definition. Let T_A be as above and let $x_1, \ldots, x_n$ be distinct variables.

A set p of $L(A)$-conditions with all free variables among $x_1, \ldots, x_n$ is called an *n-type over A* if there exists a model $(\mathcal{M}, a)_{a\in A}$ of T_A and elements $e_1, \ldots, e_n$ of M such that p is the set of all $L(A)$-conditions $E(x_1, \ldots, x_n)$ for which $\mathcal{M}_A \models E[e_1, \ldots, e_n]$.

When this relationship holds, we denote p by $\mathrm{tp}_{\mathcal{M}}(e_1, \ldots, e_n/A)$ and we say that $(e_1, \ldots, e_n)$ *realizes p in* $\mathcal{M}$. (The subscript $\mathcal{M}$ will be omitted if doing so causes no confusion; A will be omitted if it is empty.)

The collection of all such n-types over A is denoted by $S_n(T_A)$, or simply by $S_n(A)$ if the context makes the theory T_A clear.

8.2 Remark. Let $\mathcal{M}, A$ be as above, and let e, e' be n-tuples from M.
(1) $\mathrm{tp}_{\mathcal{M}}(e/A) = \mathrm{tp}_{\mathcal{M}}(e'/A)$ if and only if $(\mathcal{M}_A, e) \equiv (\mathcal{M}_A, e')$.
(2) If $\mathcal{M} \preceq \mathcal{N}$, then $\mathrm{tp}_{\mathcal{M}}(e/A) = \mathrm{tp}_{\mathcal{N}}(e/A)$.

8.3 Remark. Suppose $\mathcal{M}$ is a κ-saturated L-structure. It is easy to show that for any subset $A \subseteq M$ with $\mathrm{card}(A) < \kappa$, every type in $S_n(T_A)$ is realized in $\mathcal{M}$. Indeed, this property (even just with $n = 1$) is equivalent to κ-saturation of $\mathcal{M}$.

The logic topology on types

Fix T_A as above. If $\varphi(x_1,\dots,x_n)$ is an $L(A)$-formula and $\epsilon > 0$, we let $[\varphi < \epsilon]$ denote the set

$$\{q \in S_n(T_A) \mid \text{for some } 0 \leq \delta < \epsilon \text{ the condition } (\varphi \leq \delta) \text{ is in } q\}.$$

8.4 Definition. The *logic topology* on $S_n(T_A)$ is defined as follows. If p is in $S_n(T_A)$, the basic open neighborhoods of p are the sets of the form $[\varphi < \epsilon]$ for which the condition $\varphi = 0$ is in p and $\epsilon > 0$.

Note that the logic topology is Hausdorff. Indeed, if p, q are distinct elements of $S_n(T_A)$, then there exists an $L(A)$-formula $\varphi(x_1,\dots,x_n)$ such that the condition $\varphi = 0$ is in one of the types but not the other. Therefore, for some positive r the condition $\varphi = r$ is in that other type. Taking $\epsilon = r/2 > 0$ we see that $[\varphi < \epsilon]$ and $[(r \dot{-} \varphi) < \epsilon]$ are disjoint basic open sets for the logic topology, one containing p and the other containing q.

It is also useful to introduce notation such as the following (where $\varphi(x_1,\dots,x_n)$ is an $L(A)$-formula and $\epsilon \geq 0$):

$$[\varphi \leq \epsilon] = \{q \in S_n(T_A) \mid \text{the condition } (\varphi \leq \epsilon) \text{ is in } q\}.$$

Each set $[\varphi \leq \epsilon]$ is closed in the logic topology; indeed, its complement is $\emptyset$ if $\epsilon \geq 1$ and it is $[1 \dot{-} \varphi < \delta]$ if $\epsilon < 1$ and $\delta = 1 - \epsilon$.

8.5 Lemma. *The closed subsets of $S_n(T_A)$ for the logic topology are exactly the sets of the form $C_\Gamma = \{p \in S_n(T_A) \mid \Gamma(x_1,\dots,x_n) \subseteq p\}$ where $\Gamma(x_1,\dots,x_n)$ is a set of $L(A)$-conditions.*

Proof Given such a set $\Gamma(x_1,\dots,x_n)$, note that C_Γ is the intersection of all sets $[\varphi \leq 0]$ where $\varphi = 0$ is any condition in Γ. Hence C_Γ is closed. Conversely, suppose C is a subset of $S_n(T_A)$ that is closed in the logic topology and let p be any element of $S_n(T_A) \setminus C$. By the definition of the logic topology there exists an $L(A)$-condition $\varphi = 0$ in p and $\epsilon > 0$ such that $[\varphi < \epsilon]$ is disjoint from C. Without loss of generality we may assume $\epsilon \leq 1$. Then the closed set $[(\epsilon \dot{-} \varphi) \leq 0]$ contains C and does not have p as an element. To represent C in the desired form it suffices to take Γ to be the set of all conditions of the form $(\epsilon \dot{-} \varphi) = 0$ that arise in this way. □

8.6 Proposition. *For any $n \geq 1$, $S_n(T_A)$ is compact with respect to the logic topology.*

Proof In light of the preceding discussion, one sees that this is just a restatement of the Compactness Theorem (Corollary 5.12). □

The d-metric on types

Let T_A be as above. For each $n \geq 1$ we define a natural metric on $S_n(T_A)$; it is induced as a quotient of the given metric d on M^n, where $(\mathcal{M}, a)_{a\in A}$ is a suitable model of T_A, so we also denote this metric on types by d.

To define this metric, let $\mathcal{M}_A = (\mathcal{M}, a)_{a\in A}$ be any model of T_A in which each type in $S_n(T_A)$ is realized, for each $n \geq 1$. (Such a model exists by Proposition 7.10.) Let (M, d) be the underlying metric space of $\mathcal{M}$. For $p, q \in S_n(T_A)$ we define $d(p, q)$ to be

$$\inf \{ \max_{1\leq j\leq n} d(b_j, c_j) \mid \mathcal{M}_A \models p[b_1, \ldots, b_n], \ \mathcal{M}_A \models q[c_1, \ldots, c_n] \}.$$

Note that this expression for $d(p, q)$ does not depend on $\mathcal{M}_A$, since $\mathcal{M}_A$ realizes every type of a $2n$-tuple $(b_1, \ldots, b_n, c_1, \ldots, c_n)$ over A. It follows easily that d is a pseudometric on $S_n(T_A)$. Note that if $p, q \in S_n(A)$, then by the Compactness Theorem and our assumptions about $\mathcal{M}_A$, there exist realizations $(b_1, \ldots, b_n)$ of p and $(c_1, \ldots, c_n)$ of q in $\mathcal{M}_A$, such that $\max_j d(b_j, c_j) = d(p, q)$. In particular, if $d(p, q) = 0$, then $p = q$; so d is indeed a metric on $S_n(T_A)$.

8.7 Proposition. *The d-topology is finer than the logic topology on $S_n(T_A)$.*

Proof This follows from the uniform continuity of formulas. □

8.8 Proposition. *The metric space $(S_n(T_A), d)$ is complete.*

Proof Let $(p_k)_{k\geq 1}$ be a Cauchy sequence in $(S_n(T_A), d)$. Without loss of generality we may assume $d(p_k, p_{k+1}) \leq 2^{-k}$ for all k; that is, for completeness it suffices to show that every such Cauchy sequence has a limit. Let $\mathcal{N}$ be an ω-saturated and strongly ω-homogeneous model of T_A. Without loss of generality we may assume that $\mathcal{N} = \mathcal{M}_A$ for some model $\mathcal{M}$ of T. Using our saturation and homogeneity assumptions, we see that for any $a \in M^n$ realizing p_k there exists $b \in M^n$ realizing p_{k+1} such that $d(a, b) = d(p_k, p_{k+1})$. Therefore, proceeding inductively we may generate a sequence (b_k) in M^n such that b_k realizes p_k and $d(b_k, b_{k+1}) = d(p_k, p_{k+1}) \leq 2^{-k}$ for all k. This implies that (b_k) is a Cauchy sequence in M^n so it converges in M^n to some $b \in M^n$. It follows that the type realized by b in $(\mathcal{M}, a)_{a \in A}$ is the limit of the sequence (p_k) in the metric space $(S_n(T_A), d)$. □

Functions on type spaces defined by formulas

Let T_A be as above. Let $\mathcal{M}_A = (\mathcal{M}, a)_{a \in A}$ be any model of T_A in which each type in $S_n(T_A)$ is realized, for each $n \geq 1$.

Let $\varphi(x_1, \ldots, x_n)$ be any $L(A)$-formula. For each type $p \in S_n(T_A)$ we let $\widetilde{\varphi}(p)$ denote the unique real number $r \in [0,1]$ for which the condition $\varphi = r$ is in p. Equivalently, $\widetilde{\varphi}(p) = \varphi^{\mathcal{M}}(b)$ when b is any realization of p in $\mathcal{M}_A$.

8.9 Lemma. *Let $\varphi(x_1, \ldots, x_n)$ be any $L(A)$-formula. The function $\widetilde{\varphi}\colon S_n(T_A) \to [0,1]$ is continuous for the logic topology and uniformly continuous for the d-metric distance on $S_n(T_A)$.*

Proof For any $r \in [0,1]$ and $\epsilon > 0$, note that

$$\widetilde{\varphi}^{-1}(r - \epsilon, r + \epsilon) = [|\varphi \dot{-} r| < \epsilon].$$

This shows that $\widetilde{\varphi}$ is continuous for the logic topology. For the uniform continuity, use Theorem 3.5 to obtain a modulus of uniform continuity Δ_φ for $\varphi^{\mathcal{M}}$ on M^n. It is easy to show that Δ_φ is a modulus of uniform continuity for $\widetilde{\varphi}$. Indeed, suppose $\epsilon \in (0,1]$ and let $\delta = \Delta_\varphi(\epsilon)$. Suppose $p, q \in S_n(T_A)$ have $d(p,q) < \delta$. Suppose $r = \widetilde{\varphi}(p)$ and $s = \widetilde{\varphi}(q)$. We need to show $|r - s| \leq \epsilon$. Choose $a, b \in M^n$ to realize p, q respectively in $\mathcal{M}_A$ with $d(a,b) = d(p,q)$. Our choice of Δ_φ ensures that $|r - s| = |\varphi^{\mathcal{M}}(a) - \varphi^{\mathcal{M}}(b)| \leq \epsilon$, as desired. □

8.10 Proposition. *For any function $\Phi\colon S_n(T_A) \to [0,1]$ the following statements are equivalent:*
(1) Φ is continuous for the logic topology on $S_n(T_A)$;
(2) There is a sequence $(\varphi_k(x_1, \ldots, x_n))_{k \geq 1}$ of $L(A)$-formulas such that $(\widetilde{\varphi_k})_{k \geq 1}$ converges to Φ uniformly on $S_n(T_A)$;
(3) Φ is continuous for the logic topology and uniformly continuous for the d-metric on $S_n(T_A)$.

Proof (1) ⇒ (2): Note that the set of functions of the form $\widetilde{\varphi}$, where $\varphi(x_1, \ldots, x_n)$ is an $L(A)$-formula, separates points of $S_n(T_A)$ and is closed under the pointwise lattice operations max and min. Applying the lattice version of the Stone-Weierstrass Theorem to the compact, Hausdorff space $S_n(T_A)$ with the logic topology yields (2).

(2) ⇒ (3): Each $\widetilde{\varphi}$ is continuous for the logic topology and uniformly continuous for the d-metric on $S_n(T_A)$. These properties are preserved under uniform convergence.

(3) ⇒ (1): trivial. □

There are many connections between the type spaces $S_n(T_A)$ as n and A vary. We conclude this subsection with a result that summarizes some of them.

8.11 Proposition. *Let $\mathcal{M} \models T$ and $A \subseteq B \subseteq M$. Let π be the restriction map from $S_n(T_B)$ to $S_n(T_A)$ (defined by letting $\pi(p)$ be the set of $L(A)$-conditions in p). Then:*
(1) π is surjective;
(2) π is continuous (hence closed) for the logic topologies;
(3) π is contractive (hence uniformly continuous) for the d-metrics;
(4) if A is d-dense in B, then π is a homeomorphism for the logic topologies and a surjective isometry for the d-metrics.

Proof (1) Let $p \in S_n(T_A)$. The set p^+ of $L(A)$-formulas is finitely satisfied in $(\mathcal{M}, b)_{b\in B}$, so p itself is satisfied in some elementary extension of $(\mathcal{M}, b)_{b\in B}$, by the Compactness Theorem. If $(e_1, \ldots, e_n)$ realizes p in such an elementary extension, then $p = \pi(\text{tp}(e_1, \ldots, e_n/B))$.
(2) If $\varphi(x_1, \ldots, x_n)$ is any $L(A)$-formula and $\epsilon > 0$, then π obviously maps $[\varphi < \epsilon]$ as a basic neighborhood in $S_n(T_B)$ into $[\varphi < \epsilon]$ as a basic neighborhood in $S_n(T_A)$. Therefore π is continuous for the logic topologies; hence π is also a closed map, since those topologies are compact and Hausdorff.
(3) Any realization of $p \in S_n(T_B)$ is also a realization of $\pi(p)$.
(4) This follows from the fact that formulas define uniformly continuous functions. Hence any $L(B)$-formula can be uniformly approximated by a sequence of $L(A)$-formulas, when A is dense in B. In particular, if $p \in S_n(T_B)$, then any realization of $\pi(p)$ is a realization of p itself. □

Types over an arbitrary theory

For the rest of this section we consider a theory T that is satisfiable but not necessarily complete. For some purposes one needs to consider the spaces of types over $\emptyset$ that are consistent with T.

Fix $n \geq 0$ and let $x_1, \ldots, x_n$ be distinct variables. We let $S_n(T)$ denote the set of all n-types over $\emptyset$ of the form $\mathrm{tp}_{\mathcal{M}}(e_1, \ldots, e_n)$ where $\mathcal{M}$ is any model of T and $e_1, \ldots, e_n$ are in M. We equip $S_n(T)$ with the *logic topology*, whose basic open sets are of the form

$$[\varphi < \epsilon] = \{q \in S_n(T) \mid \text{condition } (\varphi \leq \delta) \text{ is in } q \text{ for some } 0 \leq \delta < \epsilon\}$$

where $\epsilon > 0$ and $\varphi(x_1, \ldots, x_n)$ is an L-formula. (Note that $S_0(T)$ is simply the space of all complete L-theories that extend T.)

For each L-formula $\varphi(x_1, \ldots, x_n)$ we define $\widetilde{\varphi}\colon S_n(T) \to [0,1]$ by setting $\widetilde{\varphi}(p)$ to be the unique real number $r \in [0,1]$ for which the condition $\varphi = r$ is in p, for each $p \in S_n(T)$. Equivalently, $\widetilde{\varphi}(p) = \varphi^{\mathcal{M}}(a)$ where a is any realization of p in any $\mathcal{M} \models T$. The proof of Lemma 8.9 shows that each $\widetilde{\varphi}$ is continuous as a function into $[0,1]$ from $S_n(T)$ with the logic topology.

8.12 Theorem. *Let T be any satisfiable L-theory; $S_n(T)$ equipped with the logic topology has the following properties:*

(1) *The closed subsets of $S_n(T)$ are exactly the sets of the form*

$$C_\Gamma = \{p \in S_n(T) \mid \Gamma(x_1, \ldots, x_n) \subseteq p\}$$

where $\Gamma(x_1, \ldots, x_n)$ is a set of L-conditions.

(2) *$S_n(T)$ is a compact, Hausdorff space.*

(3) *$\Phi\colon S_n(T) \to [0,1]$ is continuous if and only if there is a sequence $(\varphi_k(x_1, \ldots, x_n))_{k \geq 1}$ of L-formulas such that $(\widetilde{\varphi_k})_{k\geq 1}$ converges to Φ uniformly on $S_n(T)$.*

Proof The proofs of Lemma 8.5 and Propositions 8.6 and 8.10 apply without change to the current situation. □

9 Definability in metric structures

In this section we discuss some issues around *definability*, which is arguably the central topic in model theory and its applications.

Let $\mathcal{M}$ be a metric structure, and L a signature for $\mathcal{M}$. First we consider definability of predicates $P\colon M^n \to [0,1]$. Then we use definable predicates to discuss definability of subsets of M^n and functions from M^n into M.

Let A be any subset of M, which we think of as a set of possible parameters to use in definitions.

Definable predicates

9.1 Definition. A predicate $P\colon M^n \to [0,1]$ is definable in $\mathcal{M}$ over A if and only if there is a sequence $(\varphi_k(x) \mid k \geq 1)$ of $L(A)$-formulas such that the predicates $\varphi_k^{\mathcal{M}}(x)$ converge to $P(x)$ uniformly on M^n; *i.e.*,

$$\forall \epsilon > 0\, \exists N\, \forall k \geq N\, \forall x \in M^n \quad (|\varphi_k^{\mathcal{M}}(x) - P(x)| \leq \epsilon)\,.$$

In other words, a predicate is definable over A if it is in the uniform closure of the set of functions from M^n to $[0,1]$ that are obtained by interpreting $L(A)$-formulas in $\mathcal{M}$. We will give various results to show why we think this is the "right" notion of definability for predicates in metric structures.

9.2 Remark. Suppose $P\colon M^n \to [0,1]$ is definable in $\mathcal{M}$ over A. By taking $A_0 \subseteq A$ to be the set of elements of A whose names appear in the sequence of $L(A)$-formulas $(\varphi_k(x) \mid k \geq 1)$ witnessing that P is definable, we get a countable set A_0 such that P is definable in $\mathcal{M}$ over A_0. In contrast to what happens in ordinary first-order logic, it need not be possible to do this with a *finite* set of parameters.

It can be useful to have a specific representation of definable predicates. To do this we broaden our perspective of "connectives". Consider the product space $[0,1]^{\mathbb{N}}$ of infinite sequences; this is a compact metrizable space; for example, its topology is given by the metric ρ defined by

$$\rho((a_k),(b_k)) = \sum_{k=0}^{\infty} 2^{-k}|a_k - b_k|$$

for any pair of sequences $(a_k \mid k \in \mathbb{N})$ and $(b_k \mid k \in \mathbb{N})$ in $[0,1]^{\mathbb{N}}$. In representing definable predicates we will regard any continuous function $u\colon [0,1]^{\mathbb{N}} \to [0,1]$ as a kind of connective.

9.3 Proposition. *Let $\mathcal{M}$ be an L-structure with $A \subseteq M$, and suppose $P\colon M^n \to [0,1]$ is a predicate. Then P is definable in $\mathcal{M}$ over A if and only if there is a continuous function $u\colon [0,1]^{\mathbb{N}} \to [0,1]$ and $L(A)$-formulas $(\varphi_k \mid k \in \mathbb{N})$ such that for all $x \in M^n$*

$$P(x) = u\left(\varphi_k^{\mathcal{M}}(x) \mid k \in \mathbb{N}\right).$$

Proof Suppose P has the specified form. Fix $\epsilon > 0$; we need to show that P can be approximated uniformly to within ϵ by the interpretation of some $L(A)$-formula.

Since $([0,1]^{\mathbb{N}}, \rho)$ is compact, the function u is uniformly continuous with respect to ρ. Hence there exists $m \in \mathbb{N}$ such that

$$|u((a_k)) - u((b_k))| \leq \epsilon$$

holds whenever $(a_k \mid k \in \mathbb{N})$ and $(b_k \mid k \in \mathbb{N})$ are sequences such that $a_k = b_k$ for all $k = 0, \ldots, m$. Let $u_m\colon [0,1]^{m+1} \to [0,1]$ be the continuous function obtained from u by setting

$$u_m(a_0, \ldots, a_m) = u(a_0, \ldots, a_m, 0, 0, 0, \ldots)$$

for all $a_0, \ldots, a_m \in \mathbb{N}$. Let $\varphi(x)$ be the $L(A)$-formula given by

$$\varphi(x) = u_m(\varphi_0(x), \ldots, \varphi_m(x)).$$

Then we have immediately that

$$|P(x) - \varphi^{\mathcal{M}}(x)| = |u\left(\varphi_k^{\mathcal{M}}(x) \mid k \in \mathbb{N}\right) - u_m\left(\varphi_0^{\mathcal{M}}(x), \ldots, \varphi_m^{\mathcal{M}}(x)\right)| \leq \epsilon$$

for all $x \in M^n$. Hence P is definable over A in $\mathcal{M}$.

For the converse direction, assume that $P\colon M^n \to [0,1]$ is definable over A in $\mathcal{M}$. For each $k \in \mathbb{N}$ let $\varphi_k(x)$ be an $L(A)$-formula such that

$$|\varphi_k^{\mathcal{M}}(x) - P(x)| \leq 2^{-k}$$

holds for all $x \in M^n$.

Now consider the set C of all sequences $(a_k \mid k \in \mathbb{N})$ in $[0,1]^{\mathbb{N}}$ such that $|a_k - a_l| \leq 2^{-N}$ holds whenever $N \in \mathbb{N}$ and $k, l \geq N+1$. Each sequence (a_k) in C is a Cauchy sequence in $[0,1]$, so it converges to a limit that we denote by $\lim(a_k)$. It is easy to check that C is a closed subset of $[0,1]^{\mathbb{N}}$ and that the sequence $(\varphi_k^{\mathcal{M}}(x) \mid k \in \mathbb{N})$ is in C for every $x \in M^n$. Moreover, it is easy to check that the function $\lim\colon C \to [0,1]$ is continuous with respect to the restriction of the product topology on $[0,1]^{\mathbb{N}}$ to C. By the Tietze Extension Theorem, there is a continuous function $u\colon [0,1]^{\mathbb{N}} \to [0,1]$ that agrees with lim on C. From this we conclude immediately that

$$P(x) = u\left(\varphi_k^{\mathcal{M}}(x) \mid k \in \mathbb{N}\right)$$

for all $x \in M^n$, as desired. □

9.4 Remark. Note that in our proof of Proposition 9.3, the continuous function $u\colon [0,1]^{\mathbb{N}} \to [0,1]$, in terms of which the definable predicate P was represented, is completely independent of P. It is not too difficult to give a constructive proof of this result, in which u is described concretely, and doing so can be useful. See, for example, the *forced limit function* discussed in [6, section 3].

9.5 Remark. Proposition 9.3 shows one way to represent definable predicates so that they become meaningful in every L-structure. This suggests how the notion of L-formula could be expanded by allowing continuous *infinite* connectives, without expanding the notion of *definability* for predicates, in order to have an exact correspondence between *formulas* and *definable predicates.* There is the complication that, as noted above, a definable predicate may depend on *infinitely many* of the parameters that are used in its definition. We will not explore this direction here.

9.6 Lemma. *Suppose* $P\colon M^n \to [0,1]$ *is definable in* $\mathcal{M}$ *over* A *and consider* $\mathcal{N} \preceq \mathcal{M}$ *with* $A \subseteq N$. *Then* $\inf_x P(x)$ *and* $\sup_x P(x)$ *have the same value in* $\mathcal{N}$ *as in* $\mathcal{M}$.

Proof Let $(\varphi_k \mid k \geq 1)$ be $L(A)$-formulas such that for all $x \in M^n$ and all $k \geq 1$ we have

$$|P(x) - \varphi_k^{\mathcal{M}}(x)| \leq \frac{1}{k}.$$

Since $\mathcal{N} \preceq \mathcal{M}$, we also have for all $x \in N^n$ and all $k \geq 1$

$$|P(x) - \varphi_k^{\mathcal{N}}(x)| \leq \frac{1}{k}.$$

We conclude that

$$\begin{aligned} \inf_x P(x) &= \lim_{k\to\infty} \inf_x \varphi_k^{\mathcal{M}}(x) && (\text{inf over } x \in M^n) \\ &= \lim_{k\to\infty} \inf_x \varphi_k^{\mathcal{N}}(x) && (\text{inf over } x \in N^n) \\ &= \inf_x P(x) && (\text{inf over } x \in N^n) \end{aligned}$$

as claimed. □

This Lemma is a special case of a more general result:

9.7 Proposition. *Let* $P_i\colon M^n \to [0,1]$ *be definable in* $\mathcal{M}$ *over* A *for* $i = 1, \ldots, m$ *and consider* $\mathcal{N} \preceq \mathcal{M}$ *with* $A \subseteq N$. *Let* Q_i *be the restriction of* P_i *to* N^n *for each* i. *Then* $(\mathcal{N}, Q_1, \ldots, Q_m) \preceq (\mathcal{M}, P_1, \ldots, P_m)$.

Proof This is proved using an elaboration of the ideas above. The proof is by induction on formulas, using the tools concerning uniform convergence that were developed in the appendix to section 2. □

9.8 Proposition. *Let $P\colon M^n \to [0,1]$ be definable in $\mathcal{M}$ over A and consider an elementary extension $\mathcal{N}$ of $\mathcal{M}$. There is a unique predicate $Q\colon N^n \to [0,1]$ such that Q is definable in $\mathcal{N}$ over A and P is the restriction of Q to M^n. This predicate satisfies $(\mathcal{M}, P) \preceq (\mathcal{N}, Q)$.*

Proof Let $(\varphi_k(x) \mid k \geq 1)$ be a sequence of $L(A)$-formulas such that the functions $(\varphi_k^{\mathcal{M}} \mid k \geq 1)$ converge uniformly to P on M^n. Note that for any $k, l \geq 1$ we have

$$\sup\{|\varphi_k^{\mathcal{N}}(b) - \varphi_l^{\mathcal{N}}(b)| \mid b \in N^n\} = \sup\{|\varphi_k^{\mathcal{M}}(a) - \varphi_l^{\mathcal{M}}(a)| \mid a \in M^n\}.$$

Therefore the functions $(\varphi_k^{\mathcal{N}} \mid k \geq 1)$ converge uniformly on N^n to some function $Q\colon N^n \to [0,1]$. Evidently Q extends P and the construction of Q ensures that it is definable in $\mathcal{N}$ over A. The last statement follows from Proposition 9.7.

It remains to prove Q is unique. Suppose Q_1, Q_2 are predicates definable in $\mathcal{M}$ over A whose restriction to M^n equals P. Applying Proposition 9.7 we conclude that $(\mathcal{N}, Q_1, Q_2)$ is an elementary extension of $(\mathcal{M}, P, P)$ and hence $\sup\{|Q_1(x) - Q_2(x)| \mid x \in N^n\} = \sup\{|P(x) - P(x)| \mid x \in M^n\} = 0$. Therefore $Q_1 = Q_2$. □

The following result gives a useful and conceptually appealing characterization of definable predicates. As above we have an L-structure $\mathcal{M}$ with $A \subseteq M$ and we set $\mathcal{M}_A = (\mathcal{M}, a)_{a \in M}$; let $T = \mathrm{Th}(\mathcal{M})$ be the complete L-theory of which $\mathcal{M}$ is a model and let $T_A = \mathrm{Th}(\mathcal{M}_A)$.

9.9 Theorem. *Let $P\colon M^n \to [0,1]$ be a function. Then P is a predicate definable in $\mathcal{M}$ over A if and only if there exists $\Phi\colon S_n(T_A) \to [0,1]$ that is continuous with respect to the logic topology on $S_n(T_A)$ such that $P(a) = \Phi(\mathrm{tp}_{\mathcal{M}}(a/A))$ for all $a \in M^n$.*

Proof First suppose that there is a continuous $\Phi\colon S_n(T_A) \to [0,1]$ such that $P(a) = \Phi(\mathrm{tp}_{\mathcal{M}}(a/A))$ for all $a \in M^n$. By Proposition 8.10 there is a sequence $(\varphi_k(x) \mid k \geq 1)$ of $L(A)$-formulas such that the functions $(\widetilde{\varphi}_k \mid k \geq 1)$ converge uniformly to Φ on $S_n(T_A)$. For any $a \in M^n$ let $p = \mathrm{tp}_{\mathcal{M}}(a/A)$ and note that $|\varphi_k^{\mathcal{M}}(a) - P(a)| = |\widetilde{\varphi}_k(p) - \Phi(p)|$. Therefore the functions $(\varphi_k^{\mathcal{M}} \mid k \geq 1)$ converge uniformly to P on M^n, from which it follows that P is a predicate (*i.e.*, that it is uniformly continuous) and that it is definable in $\mathcal{M}$ over A.

For the converse, suppose $(\varphi_k(x) \mid k \geq 1)$ is a sequence of $L(A)$-formulas such that the functions $(\varphi_k^{\mathcal{M}} \mid k \geq 1)$ converge uniformly to

P on M^n. Let $\mathcal{N}$ be a κ-saturated elementary extension of $\mathcal{M}$, where $\kappa > \mathrm{card}(A)$. Arguing as in the proof of Proposition 9.8 we see that the functions $(\varphi_k^{\mathcal{N}} \mid k \geq 1)$ converge uniformly on N^n to some function $Q\colon N^n \to [0,1]$. Evidently Q extends P and the construction of Q ensures that it is definable in $\mathcal{N}$ over A. Let p be any type in $S_n(T_A)$. Define $\Phi(p) = Q(b)$ where $b \in N^n$ realizes p; since $Q(b)$ is the limit of $(\varphi_k^{\mathcal{N}}(b) \mid k \geq 1)$ and $(\varphi_k \mid k \geq 1)$ are $L(A)$-formulas, the value of $Q(b)$ depends only on the type of b over A. Moreover, our construction of Q ensures that Φ is the uniform limit of the functions $(\widetilde{\varphi}_k \mid k \geq 1)$ on $S_n(T_A)$. Therefore Φ is continuous with respect to the logic topology by Proposition 8.10. For any $a \in M^n$ we have $P(a) = Q(a) = \Phi(\mathrm{tp}_{\mathcal{M}}(a/A))$ as desired. □

The next result provides a characterization of definability for predicates in saturated models that proves to be technically helpful in many situations. If $\mathcal{M}$ is an L-structure and $A \subseteq M$, a subset $S \subseteq M^n$ is called *type-definable* in $\mathcal{M}$ over A if there is a set $\Sigma(x_1,\dots,x_n)$ of $L(A)$-formulas such that for any $a \in M^n$ we have $a \in S$ if and only if $\varphi^{\mathcal{M}}(a) = 0$ for every $\varphi \in \Sigma$. In this case we will say S is *type-defined by* Σ.

9.10 Corollary. *Let $\mathcal{M}$ be a κ-saturated L-structure and $A \subseteq M$ with $\mathrm{card}(A) < \kappa$; let $P\colon M^n \to [0,1]$ be a function. Then P is a predicate definable in $\mathcal{M}$ over A if and only if the sets $\{a \in M^n \mid P(a) \leq r\}$ and $\{a \in M^n \mid P(a) \geq r\}$ are type-definable in $\mathcal{M}$ over A for every $r \in [0,1]$.*

Proof Suppose P is a predicate definable in $\mathcal{M}$ over A. Using Theorem 9.9 we get a continuous function $\Phi\colon S_n(T_A) \to [0,1]$ such that $P(a) = \Phi(\mathrm{tp}_{\mathcal{M}}(a/A))$ for all $a \in M^n$. For $r \in [0,1]$, $\Phi^{-1}([0,r])$ is a closed subset of $S_n(A)$ for the logic topology. By Lemma 8.5, it is of the form $\{p \in S_n(T_A) \mid \Gamma(x_1,\dots,x_n) \subseteq p\}$ where $\Gamma(x_1,\dots,x_n)$ is some set of $L(A)$-conditions. It follows that $\{a \in M^n \mid P(a) \leq r\}$ is type-defined by Γ. A similar argument applies to $\{a \in M^n \mid P(a) \geq r\}$.

Conversely, suppose P is a function such that $\{a \in M^n \mid P(a) \leq r\}$ and $\{a \in M^n \mid P(a) \geq r\}$ are type-definable in $\mathcal{M}$ over A for every $r \in [0,1]$. This allows us to define $\Phi\colon S_n(T_A) \to [0,1]$ by setting $\Phi(p) = P(a)$ whenever $p \in S_n(T_A)$ and $a \in M^n$ realizes p in $\mathcal{M}_A$. Note that such an a exists for every p because of our saturation assumption. Further, our type-definability assumption ensures that Φ is well defined and, moreover, that it is continuous for the logic topology. It follows by Theorem 9.9 that P is in $\mathcal{M}$ over A. □

As another corollary to Theorem 9.9 we get a characterization of definability for predicates that is a generalization of the Theorem of Svenonius from ordinary model theory. Note that we here assume that the given function P is a predicate, *i.e.*, that it is uniformly continuous; this allows us to consider the expansion $(\mathcal{M}, P)$ as a metric structure.

9.11 Corollary. *Let $\mathcal{M}$ be an L-structure with $A \subseteq M$, and suppose $P\colon M^n \to [0,1]$ is a predicate. Then P is definable in $\mathcal{M}$ over A if and only if whenever $(\mathcal{N}, Q) \succeq (\mathcal{M}, P)$, the predicate Q is invariant under all automorphisms of $\mathcal{N}$ that leave A fixed pointwise.*

Proof For the left to right direction, assume that P is definable in $\mathcal{M}$ over A. So there is a sequence of $L(A)$-formulas $(\varphi_k \mid \kappa \geq 1)$ such that P is the uniform limit of $(\varphi_k^{\mathcal{M}} \mid k \geq 1)$ on M^n. Suppose $(\mathcal{N}, Q)$ is any elementary extension of $(\mathcal{M}, P)$. As discussed in the previous proof, we have that Q is the uniform limit of $(\varphi_k^{\mathcal{N}} \mid k \geq 1)$ N^n. Since each function $\varphi_k^{\mathcal{N}}$ is the interpretation of an $L(A)$-formula, it must be invariant under all automorphisms of $\mathcal{N}$ that leave A pointwise fixed. Hence the same is true of its uniform limit Q.

For the right to left direction, let $(\mathcal{N}, Q) \succeq (\mathcal{M}, P)$ be such that $\mathcal{N}$ is strongly κ-homogeneous and $(\mathcal{N}, Q)$ is κ-saturated, where $\kappa > \operatorname{card}(A)$, obtained using Proposition 7.12. Then define $\Phi\colon S_n(T_A) \to [0,1]$ by $\Phi(p) = Q(b)$ for $b \in N^n$ realizing p. We first need to show Φ is well-defined. On the one hand, $\mathcal{N}_A$ realizes every type $p \in S_n(T_A)$ (so b exists). On the other hand, Q is assumed to be $\operatorname{Aut}_A(\mathcal{N})$-invariant and this group acts transitively on the set of realizations of any given $p \in S_n(T_A)$.

Next we show that Φ is continuous with respect to the logic topology on $S_n(T_A)$. Fix $p \in S_n(T_A)$ and let $r = \Phi(p)$. For any $b \in N^n$ we have the implication

$$b \text{ realizes } p \text{ in } \mathcal{N}_A \quad \Rightarrow \quad Q(b) = r.$$

Since $(\mathcal{N}, Q)$ is κ-saturated, it follows that for each $\epsilon > 0$ there exists a condition $\varphi = 0$ in p and $\delta > 0$ so that for any $b \in N^n$ we have the implication

$$\varphi^{\mathcal{N}}(b) < \delta \quad \Rightarrow \quad |Q(b) - r| \leq \epsilon/2.$$

Therefore Φ maps $[\varphi < \delta]$, which is a logic neighborhood of p, into the open interval $(r-\epsilon, r+\epsilon)$. Hence Φ is continuous.

By Theorem 9.9 we conclude that P is definable in $\mathcal{M}$ over A. $\square$

Distance predicates

Let $\mathcal{M}$ be an L-structure and $D \subseteq M^n$. The predicate giving the distance in M^n to D is given by

$$\operatorname{dist}(x, D) = \inf\{d(x,y) \mid y \in D\}.$$

Here $x = (x_1, \ldots, x_n)$ and $y = (y_1, \ldots, y_n)$ range over M^n and we consider the metric d on M^n defined by

$$d(x,y) = \max\big(d(x_1,y_1), \ldots, d(x_n,y_n)\big).$$

Predicates of the form $\operatorname{dist}(x, D)$ are important in the model theory of metric structures. We show next that they can be characterized by axioms in continuous logic.

Consider a predicate $P\colon M^n \to [0,1]$ and the following conditions.

(E_1) $$\sup_x \inf_y \max\big(P(y), |P(x) - d(x,y)|\big) = 0;$$

(E_2) $$\sup_x \big| P(x) - \inf_y \min\big(P(y) + d(x,y), 1\big)\big| = 0.$$

Observe that for any $D \subseteq M^n$, $P(x) = \operatorname{dist}(x, D)$ satisfies E_1 and E_2.

9.12 Theorem. *Let $(\mathcal{M}, F)$ be an L structure satisfying conditions E_1 and E_2. Let $D = \{x \in M^n \mid F(x) = 0\}$ be the zeroset of F. Then $F(x) = \operatorname{dist}(x, D)$ for all $x \in M^n$.*

Proof By E_2, $F(x) \leq F(y) + d(x,y)$ for all y. So for $y \in D$, $F(x) \leq d(x,y)$. So $F(x) \leq \operatorname{dist}(x, D)$.

Fix $\epsilon > 0$. We will show that $\operatorname{dist}(x, D) \leq F(x) + \epsilon$ for all $x \in M^n$. Letting ϵ go to 0 will complete the proof. We generate a sequence (y_k) in $\mathcal{M}$ using E_1. Set $y_1 = x$, any fixed element of M^n. Choose y_2 such that $F(y_2) \leq \frac{\epsilon}{8}$ and $|F(x) - d(x, y_2)| \leq \frac{\epsilon}{8}$. Continue by induction, satisfying

$$F(y_{k+1}) \leq \frac{\epsilon}{2^{k+2}} \qquad |F(y_k) - d(y_k, y_{k+1})| \leq \frac{\epsilon}{2^{k+2}}.$$

Therefore, $d(y_k, y_{k+1}) \leq F(y_k) + |F(y_k) - d(y_k, y_{k+1})| \leq \epsilon 2^{-k}$. So (y_k)

is a Cauchy sequence, and hence it converges to some $y \in M^n$. By the continuity of F, $F(y) = 0$. Moreover,

$$d(x, y) = \lim_{k\to\infty} d(y_1, y_{k+1}) \leq d(y_1, y_2) + \sum_{k=2}^{\infty} d(y_k, y_{k+1}) \leq F(x) + \epsilon.$$

Since $y \in D$, this shows $\mathrm{dist}(x, D) \leq F(x) + \epsilon$, as desired. □

Zerosets

Next we turn to definability for *sets* (*i.e.*, subsets of M^n). An obvious way to carry over definability for sets from first-order logic would be to regard a condition $\varphi(x) = 0$ as defining the set of x that satisfy it. This is indeed an important kind of definability for sets, but it turns out to be somewhat weak, due to fact that the rules for constructing formulas in continuous logic are rather generous. This is especially true if one puts zerosets of definable predicates on the same footing as zerosets of formulas (as one should).

In this subsection we briefly explore this notion of definability, and in the next subsection we introduce a stronger, less obvious, but very important kind of definability for sets in metric structures.

9.13 Definition. Let $D \subseteq M^n$. We say that D is a *zeroset* in $\mathcal{M}$ over A if there is a predicate $P\colon M^n \to [0,1]$ definable in $\mathcal{M}$ over A such that $D = \{x \in M^n \mid P(x) = 0\}$.

The next result shows that zerosets are the same as type-definable sets, with the restriction that the partial type is countable.

9.14 Proposition. *For $D \subseteq M^n$, the following are equivalent.*

(1) *D is a zeroset in $\mathcal{M}$.*
(2) *there is a sequence $(\varphi_m \mid m \geq 1)$ of L-formulas such that*

$$D = \left\{x \in M^n \mid \varphi_m^{\mathcal{M}}(x) = 0 \textit{ for all } m \in \mathbb{N}\right\} = \bigcap_{m=1}^{\infty} \textit{Zeroset of } \varphi_m^{\mathcal{M}}.$$

Proof $(1 \Rightarrow 2)$: Suppose $D = \{x \in M^n \mid P(x) = 0\}$ and $(\varphi_m^{\mathcal{M}} \mid m \geq 1)$ are L-formulas such that for all $x \in M^n$ and all $m \geq 1$

$$|P(x) - \varphi_m^{\mathcal{M}}(x)| \leq \frac{1}{m}.$$

Then $D = \bigcap_m D_m$, where D_m is the zero set of the interpretation of the L-formula $(\varphi_m(x) \dot{-} \frac{1}{m})$ in $\mathcal{M}$.

$(2 \Rightarrow 1)$: If $D = \{x \in M^n \mid \varphi_m^{\mathcal{M}}(x) = 0 \text{ for all } m \in \mathbb{N}\}$, then the definable predicate

$$P(x) = \sum_{m=1}^{\infty} 2^{-m} \varphi_m^{\mathcal{M}}(x).$$

has D as its zeroset. □

9.15 Corollary. *The collection of zerosets in $\mathcal{M}$ over A is closed under countable intersections.*

Definability of sets

The next definition gives what we believe is the correct concept of first-order definability for sets in metric structures.

9.16 Definition. A closed set $D \subseteq M^n$ is *definable* in $\mathcal{M}$ over A if and only if the distance predicate $\operatorname{dist}(x, D)$ is definable in $\mathcal{M}$ over A.

The importance of this concept of definability for sets is shown by the following result. It says essentially that in continuous first-order logic we will retain definability of predicates if we quantify (using sup or inf) over sets that are definable in this sense, but not if we quantify over other sets.

9.17 Theorem. *For a closed set $D \subseteq M^n$ the following are equivalent:*
(1) D is definable in $\mathcal{M}$ over A.
(2) For any predicate $P\colon M^m \times M^n \to [0,1]$ that is definable in $\mathcal{M}$ over A, the predicate $Q\colon M^m \to [0,1]$ defined by

$$Q(x) = \inf\{P(x,y) \mid y \in D\}$$

is definable in $\mathcal{M}$ over A.

Proof To prove (1) we only need to assume (2) for the case in which $P(x,y) = d(x,y)$ (so P is the interpretation of a quantifier-free formula); here $m = n$.

So assume D is definable in $\mathcal{M}$ over A. Let $P\colon M^m \times M^n \to [0,1]$ be any predicate that is definable in $\mathcal{M}$ over A. This ensures that P is uniformly continuous, so (see Proposition 2.10) there is an increasing, continuous function $\alpha\colon [0,1] \to [0,1]$ with $\alpha(0) = 0$ such that

$$|P(x,y) - P(x,z)| \le \alpha(d(y,z))$$

for any $x \in M^m$ and $y, z \in M^n$. Let Q be defined on M^m by $Q(x) = \inf\{P(x,y) \mid y \in D\}$.

We will show that $Q(x) = \inf\{P(x,z) + \alpha(\mathrm{dist}(z,D)) \mid z \in M^n\}$ for all $x \in M^m$. This shows that Q is definable in $\mathcal{M}$ over A, given that P and $\mathrm{dist}(z,D)$ are definable in $\mathcal{M}$ over A. (Notice that we replaced the inf over D by the inf over M^n and it is expressible by one of the basic constructs of continuous logic.)

By our choice of α we have $P(x,y) \le P(x,z) + \alpha(d(y,z))$ for all $x \in M^m$ and $y, z \in M^n$. Taking the inf over $y \in D$ and using the fact that α is continuous and increasing, we conclude that

$$Q(x) \le P(x,z) + \alpha(\mathrm{dist}(z,D))$$

for all $x \in M^m$ and $z \in M^n$. Taking the inf of the right side over $z \in D$ yields $Q(x)$, and the inf over $z \in M^n$ is even smaller, but is bounded below by $Q(x)$. This shows

$$Q(x) = \inf\{P(x,z) + \alpha(\mathrm{dist}(z,D)) \mid z \in M^n\}$$

for all $x \in M^m$ and completes the proof of (1) $\Rightarrow$ (2). □

The following result shows another useful property of definable sets that need not be true of zerosets.

9.18 Proposition. *Let $\mathcal{N} \preceq \mathcal{M}$ be L-structures, and let $D \subseteq M^n$ be definable in $\mathcal{M}$ over A, where $A \subseteq N$. Then:*
(1) For any $x \in N^n$, $\mathrm{dist}(x,D) = \mathrm{dist}(x, D \cap N^n)$. Thus $D \cap N^n$ is definable in $\mathcal{N}$ over A.
(2) $(\mathcal{N}, \mathrm{dist}(\cdot, D \cap N^n)) \preceq (\mathcal{M}, \mathrm{dist}(\cdot, D))$.
(3) If $D \neq \emptyset$, then $D \cap N^n \neq \emptyset$.

Proof Statement (1) is an immediate consequence of Theorem 9.12 and (2) follows from (1) by Proposition 9.7. For (3), note that if $D \neq \emptyset$, then $\inf_x \mathrm{dist}(x,D) = 0$ in $\mathcal{M}$. Therefore $\inf_x \mathrm{dist}(x, D \cap N^n) = 0$ in $\mathcal{N}$ by (2). Hence there exists $a \in N$ such that $\mathrm{dist}(a, D \cap N^n) < 1$. This implies $D \cap N^n \neq \emptyset$, since otherwise $\mathrm{dist}(a, D \cap N^n) = 1$. □

If $D \subseteq M^n$ is definable in $\mathcal{M}$, then it is certainly a zeroset (it is the set of zeros of the definable predicate $\mathrm{dist}(x,D)$). However, the converse is not generally true. The next result explores the distinction between these two concepts.

9.19 Proposition. *For a closed set $D \subseteq M^n$, the following are equivalent:*

(1) *D is definable in $\mathcal{M}$ over A.*
(2) *There is a predicate $P\colon M^n \to [0,1]$, definable in $\mathcal{M}$ over A, such that $P(x) = 0$ for all $x \in D$ and*

$$\forall \epsilon > 0\ \exists \delta > 0\ \forall x \in M^n\ \ (P(x) \le \delta \Rightarrow \operatorname{dist}(x, D) \le \epsilon).$$

(3) *There is a sequence $(\varphi_m \mid m \ge 1)$ of $L(A)$-formulas and a sequence $(\delta_m \mid m \ge 1)$ of positive real numbers such that for all $m \ge 1$ and $x \in M^n$,*

$$(x \in D \Rightarrow \varphi_m^{\mathcal{M}}(x) = 0) \ \ \textit{and}$$

$$\left(\varphi_m^{\mathcal{M}}(x) \le \delta_m \Rightarrow \operatorname{dist}(x, D) \le \frac{1}{m}\right).$$

Proof (1 ⇒ 3): Let $F(x) = \operatorname{dist}(x, D)$ and assume it is definable in $\mathcal{M}$ over A. So there exists a sequence $(\psi_m(x) \mid m \ge 1)$ of $L(A)$-formulas such that for all $x \in M^n$ and $m \ge 1$ we have

$$|F(x) - \psi_m^{\mathcal{M}}(x)| \le \frac{1}{3m}.$$

If $x \in D$, then $F(x) = 0$ and so $\psi_m^{\mathcal{M}}(x) \le \frac{1}{3m}$. Also, if $\psi_m^{\mathcal{M}}(x) \le \frac{2}{3m}$ we have

$$F(x) \le \psi_m^{\mathcal{M}}(x) + |F(x) - \psi_m^{\mathcal{M}}(x)| \le \frac{2}{3m} + \frac{1}{3m} = \frac{1}{m}.$$

Hence the $L(A)$-formulas $\varphi_m(x) = (\psi_m^{\mathcal{M}}(x) \dot{-} \frac{1}{3m})$ have the desired property (with $\delta_m = \frac{2}{3m}$.)

(3 ⇒ 2): Set $P(x) = \sum_{m=1}^{\infty} 2^{-m} \varphi_m^{\mathcal{M}}(x)$.

(2 ⇒ 1): We use Proposition 2.10. This gives us a continuous, increasing function $\alpha\colon [0,1] \to [0,1]$ such that $\alpha(0) = 0$ and for all $x \in M^n$, $\operatorname{dist}(x, D) \le \alpha(P(x))$.

Consider the function

$$F(x) = \inf_y \min\left(\alpha\left(P(y)\right) + d(x,y), 1\right).$$

First of all, this predicate is definable in $\mathcal{M}$ over A, since P is definable in $\mathcal{M}$ over A. Indeed, if $(\varphi_n \mid n \geq 1)$ are $L(A)$-formulas such that $\varphi_n^{\mathcal{M}}$ converges to P on M^n, then

$$\psi_n = \inf_y \min\left(\alpha(\varphi_n(y)) + d(x,y), 1\right)$$

gives a sequence of $L(A)$-formulas such that $\psi_n^{\mathcal{M}}$ converges uniformly to F.

Second, we observe that $F(x) = \operatorname{dist}(x, D)$ for all $x \in M^n$. If y is any element of D (so $P(y) = 0$), we see that

$$F(x) \leq \min\left(\alpha(0) + d(x,y), 1\right) = d(x,y)$$

and hence $F(x) \leq \operatorname{dist}(x, D)$.

On the other hand, we have $\alpha\left(P(y)\right) \geq \operatorname{dist}(y, D)$ for all y, and hence

$$\begin{aligned} F(x) &\geq \inf_y \min\left(\operatorname{dist}(y,D) + d(x,y), 1\right) \\ &\geq \min\left(\operatorname{dist}(x,D), 1\right). \end{aligned}$$

Since the metric d is bounded by 1, this shows $F(x) \geq \operatorname{dist}(x, D)$, as desired. □

9.20 Remark. Suppose $\mathcal{M}$ is an ω_1-saturated L-structure and A is a countable subset of M.

If $P, Q \colon M^n \to [0,1]$ are definable in $\mathcal{M}$ over A and P, Q have the same zeroset, then an easy saturation argument shows that

$$\forall \epsilon > 0\ \exists \delta > 0\ \forall x \in M^n\ \ (P(x) \leq \delta \Rightarrow Q(x) \leq \epsilon)$$

and

$$\forall \epsilon > 0\ \exists \delta > 0\ \forall x \in M^n\ \ (Q(x) \leq \delta \Rightarrow P(x) \leq \epsilon)\,.$$

Now suppose $P \colon M^n \to [0,1]$ is definable in $\mathcal{M}$ over A. The proof of $(2 \Rightarrow 1)$ in Proposition 9.19 together with the observation in the previous paragraph yields the following:

The zeroset D of P is definable in $\mathcal{M}$ over A if and only if P satisfies the statement in Proposition 9.19(2):

$$\forall \epsilon > 0\ \exists \delta > 0\ \forall x \in M^n\ \ (P(x) \leq \delta \Rightarrow \operatorname{dist}(x, D) \leq \epsilon)\,.$$

As an example, we use this to give a geometric criterion for definability of closed balls $B(a,r) = \{x \in M^n \mid d(a,x) \leq r\}$; namely, $B(a,r)$ is definable in $\mathcal{M}$ over $\{a\}$ if and only if the family $(B(a, r+\delta) \mid \delta > 0)$ of closed balls converges to $B(a,r)$, in the sense of the Hausdorff metric, as $\delta \to 0$. (To prove this, apply the preceding statement to the predicate

$P(x) = (d(a,x) \dot{-} r)$, which is definable in $\mathcal{M}$ over $\{a\}$ and has $B(a,r)$ as its zeroset. Note that $P(x) \leq \delta$ iff $d(a,x) \leq r + \delta$.)

9.21 Remark. Let $\mathcal{M}$ be a discrete structure from ordinary first-order logic. We explore the meaning of these notions of definability for sets in $\mathcal{M}$, regarding it as a metric structure in which we take the distance between distinct elements always to be 1.

Note that in the metric setting there are more formulas than the usual first-order ones. For example, $\psi = \sum_{k=1}^{N} 2^{-k}\varphi_k$ is a formula whenever $(\varphi_k \mid 1 \leq k \leq N)$ are first-order formulas; for $x \in M^n$, the possible values of $\psi^{\mathcal{M}}$ are $\frac{k}{2^N}$, where $k = 0, \ldots, 2^N - 1$.

What one can easily prove here, by induction on formulas, is the following: for any L-formula φ, $\{\varphi^{\mathcal{M}}(x) \mid x \in M^n\}$ is a finite set. Moreover, for any $r \in [0,1]$, $\{x \in M^n \mid \varphi^{\mathcal{M}}(x) = r\}$ is definable in $\mathcal{M}$ by an ordinary first-order formula.

For any $D \subseteq M^n$ one has the following characterizations:

(1) D is definable in $\mathcal{M}$ over A if and only if D is definable in $\mathcal{M}$ over A by a first-order formula, in the usual sense of model theory.

(2) D is a zeroset in $\mathcal{M}$ over A if and only if D is the intersection of countably many sets definable in $\mathcal{M}$ over A by first-order formulas.

To prove (1) above, suppose $\operatorname{dist}(x, D)$ is definable in $\mathcal{M}$ over A. In this setting, $\operatorname{dist}(x, D)$ is simply $1 - \chi_D$, where χ_D denotes the characteristic function of D. Let φ be an $L(A)$-formula such that for all $x \in M^n$

$$|\operatorname{dist}(x, D) - \varphi^{\mathcal{M}}(x)| \leq \frac{1}{3}.$$

Then $D = \left\{x \in M^n \mid \varphi^{\mathcal{M}}(x) \leq \frac{1}{2}\right\}$ and this is first-order definable in $\mathcal{M}$ over A.

Definability of functions

We give a brief introduction to definability of functions (from M^n into M) in a metric structure $\mathcal{M}$.

9.22 Definition. Let $\mathcal{M}$ be an L-structure and $A \subseteq M$ and consider a function $f\colon M^n \to M$. We say f is definable in $\mathcal{M}$ over A if and only if the function $d(f(x), y)$ on M^{n+1} is a predicate definable in $\mathcal{M}$ over A.

9.23 Proposition. *If the function $f\colon M^n \to M$ is definable in $\mathcal{M}$ over A, then f is uniformly continuous; indeed, any modulus of uniform continuity for the predicate $d(f(x), y)$ is a modulus of uniform continuity for f.*

Proof Suppose $\Delta\colon (0,1] \to (0,1]$ is a modulus of uniform continuity for $d(f(x), y)$. That is, for any $x, x' \in M^n$ and $y, y' \in M$ and for any $\epsilon \in (0,1]$ we know that if $d(x, x') < \Delta(\epsilon)$ and $d(y, y') < \Delta(\epsilon)$, then $|d(f(x), y) - d(f(x'), y')| \leq \epsilon$. Taking $y' = y = f(x')$ we get that $d(x, x') < \Delta(\epsilon)$ implies $d(f(x), f(x')) \leq \epsilon$. □

For any function $f\colon M^n \to M$ we denote the graph of f by $\mathcal{G}_f$. We regard $\mathcal{G}_f$ as a subset of M^{n+1}.

Note that if f is definable in $\mathcal{M}$ over A then its graph $\mathcal{G}_f$ is definable in $\mathcal{M}$ over A as a subset of M^{n+1}. This follows from the identity

$$\operatorname{dist}\big((x,y), \mathcal{G}_f\big) = \inf_z \max\big(d(x,z), d(f(z), y)\big)$$

(in which x, z range over M^n and y ranges over M).

The converse of this observation is true in a strong form, if we work in a sufficiently saturated model.

9.24 Proposition. *Let $\mathcal{M}$ be κ-saturated, where κ is uncountable, and let $A \subseteq M$ have cardinality $< \kappa$. Let $f\colon M^n \to M$ be any function. The following are equivalent:*
(1) f is definable in $\mathcal{M}$ over A.
(2) $\mathcal{G}_f$ is type-definable in $\mathcal{M}$ over A.

Proof It remains to prove (2) $\Rightarrow$ (1). Write $P(x, y) = d(f(x), y)$ for $x \in M^n$ and $y \in M$. Let $\Gamma(x, y)$ be a set of $L(A)$-conditions that type-defines $\mathcal{G}_f$ in $\mathcal{M}$. Fix $r \in [0,1]$ and note that

$$P(x,y) \leq r \Leftrightarrow \exists z\big((x,z) \in \mathcal{G}_f \wedge d(z,y) \leq r\big).$$

This shows that the set $\{(x,y) \in M^{n+1} \mid P(x,y) \leq r\}$ is type-defined in $\mathcal{M}$ by the set of $L(A)$-conditions of the form

$$\inf_z \max\big(\varphi(x,z), d(z,y) \dot{-} r\big) = 0$$

where $\varphi = 0$ is any condition in Γ. A similar argument shows that the set $\{(x,y) \in M^{n+1} \mid P(x,y) \geq r\}$ is type-definable in $\mathcal{M}$ over A for each $r \in [0,1]$. By Corollary 9.10 this shows that P is definable and hence that f is definable in $\mathcal{M}$ over A. □

9.25 Proposition. *Let $\mathcal{M}$ be an L-structure and $A \subseteq M$. Suppose the function $f\colon M^n \to M$ is definable in $\mathcal{M}$ over A. Then:*
(1) If $\mathcal{N} \preceq \mathcal{M}$ and $A \subseteq N$, then f maps N^n into N and the restriction of f to N^n is definable in $\mathcal{N}$ over A.
(2) If $\mathcal{N} \succeq \mathcal{M}$ then there is a function $g\colon N^n \to N$ such that g extends f and g is definable in $\mathcal{N}$ over A.

Proof (1) Fix any $(a_1, \ldots, a_n) \in N^n$ and let $b = f(a_1, \ldots, a_n) \in M$. Let $P\colon M \to [0,1]$ be the predicate defined by $P(y) = d(b, y)$ for all $y \in M$; note that P is definable in $\mathcal{M}$ over $A \cup \{a_1, \ldots, a_n\} \subseteq N$. Let $Q\colon N \to [0,1]$ be the restriction of P to N; evidently Q is definable in $\mathcal{N}$ over $A \cup \{a_1, \ldots, a_n\}$ and, by Proposition 9.7, $(\mathcal{N}, Q) \preceq (\mathcal{M}, P)$. Note that P satisfies $\inf_y P(y) = 0$ and $d(x,y) \leq P(x) + P(y)$, where x, y range over M. Transferring these conditions from $(\mathcal{M}, P)$ to $(\mathcal{N}, Q)$ we may find a sequence $(c_k \mid k \geq 1)$ in N satisfying $Q(c_k) \leq 1/k$ for all $k \geq 1$ and hence also $d(c_k, c_l) \leq 1/k + 1/l$ for all $k, l \geq 1$. It follows that $(c_k \mid k \geq 1)$ converges to an element of N which is a zero of Q and hence a zero of P. This limit must be b, since b is the only zero of P.
(2) By making $\mathcal{N}$ larger we may assume that it is ω_1-saturated. Using (1) it suffices to prove (2) for the larger elementary extension. Let $P\colon M^{n+1} \to [0,1]$ be the predicate $P(x,y) = d(f(x), y)$. Using Proposition 9.8 let $Q\colon N^{n+1} \to [0,1]$ be the predicate that extends P and is definable in $\mathcal{N}$ over A, so we have $(\mathcal{N}, Q) \succeq (\mathcal{M}, P)$. Note that P satisfies $\sup_x \inf_y P(x,y) = 0$. Hence the same is true of Q. Using this and the fact that $\mathcal{N}$ is ω_1-saturated it follows that for all $x \in N^n$ there exists at least one $y \in N$ such that $Q(x,y) = 0$. Note also that from the definition of P and the triangle inequality for d it follows that P satisfies

$$\sup_x \sup_y \sup_{y'} \left(|d(y', y) - P(x, y')| \dot{-} P(x,y) \right) = 0.$$

Hence the same is true of Q. From this it follows that for each $x \in N^n$ there is at most one $y \in N$ such that $Q(x,y) = 0$. Therefore the zero set of Q is the graph of some function $g\colon N^n \to N$. Moreover, it also follows that if $Q(x,y) = 0$ then $Q(x, y') = d(y, y')$ for all $y' \in N$. That is, for all $x \in N^n$ and $y' \in N$ we have $Q(x, y') = d(g(x), y')$. This shows that the function g is definable in $\mathcal{N}$ over A as desired. □

The preceding results permit one to show easily that the collection of functions definable in a given structure is closed under composition and that the result of substituting definable functions into a definable predicate is a definable predicate.

Extension by definition

Here we broaden our perspective to consider our definability concepts in a *uniform* way, relative to the class of all models of a satisfiable L-theory T. The earlier parts of this section can be interpreted as dealing with the case where T is *complete*; now we consider more general theories.

Let L be any signature for metric structures and let T be any satisfiable L-theory; let L_0 be a signature contained in L.

If T_0 is an L_0-theory, we say that T_0 is the *restriction of* T *to* L_0 (or, equivalently, that T is a *conservative extension of* T_0) if for every closed L_0-condition E we have

$$T \models E \quad \text{iff} \quad T_0 \models E.$$

Note that for T to be a conservative extension of T_0 it suffices (but need not be necessary) to require that T is an extension of T_0 and that every model of T_0 has an expansion that is a model of T.

9.26 Definition. An L-formula $\varphi(x_1, \ldots, x_n)$ is *defined in* T *over* L_0 if for each $\epsilon > 0$ there exists an L_0-formula $\psi(x_1, \ldots, x_n)$ such that

$$T \models \big(\sup_{x_1} \ldots \sup_{x_n} |\varphi - \psi|\big) \leq \epsilon.$$

If P is an n-ary predicate symbol in L, we say P is *defined in* T *over* L_0 if the formula $P(x_1, \ldots, x_n)$ is defined in T over L_0. If f is an n-ary function symbol in L, we say f is *defined in* T *over* L_0 if the formula $d(f(x_1, \ldots, x_n), y)$ is defined in T over L_0. In particular, if c is a constant symbol in L, we say c is *defined in* T *over* L_0 if the formula $d(c, y)$ is defined in T over L_0.

9.27 Definition. Let T, L, L_0 be as above and let T_0 be an L_0-theory. We say T is an *extension by definitions of* T_0 if T is a conservative extension of T_0 and every nonlogical symbol in L is defined in T over L_0.

Let T, L, L_0 be as above and let T_0 be the restriction of T to L_0. For each $n \geq 0$ define $\pi_n \colon S_n(T) \to S_n(T_0)$ to be the restriction map: $\pi_n(p)$ is the set of all L_0-conditions in p, for any $p \in S_n(T)$. Evidently each π_n is continuous with respect to the logic topologies. Using Theorem 8.12(1,2) we see that π_n is surjective for each n. (Since $\pi_n(S_n(T))$ is a closed subset of $S_n(T_0)$, there is a set $\Gamma(x_1, \ldots, x_n)$ of L_0-conditions such that $\pi_n(S_n(T))$ is the set of all $q \in S_n(T_0)$ such that $\Gamma \subseteq q$. If

$\varphi(x_1,\ldots,x_n)=0$ is any condition in Γ we have

$$T \models (\sup_{x_1}\ldots\sup_{x_n}\varphi(x_1,\ldots,x_n))=0$$

and therefore

$$T_0 \models (\sup_{x_1}\ldots\sup_{x_n}\varphi(x_1,\ldots,x_n))=0.$$

It follows that every $q\in S_n(T_0)$ contains Γ.)

9.28 Proposition. *Let T,L,L_0 be as above and suppose T is an extension by definitions of the L_0-theory T_0. Then every L-formula is defined in T over L_0.*

Proof Note that every model $\mathcal{M}$ of T is completely determined by its reduct $\mathcal{M}|L_0$ to L_0. We will use this to show that the restriction map π_n is injective for each n. Therefore π_n^{-1} is a homeomorphism from $S_n(T_0)$ onto $S_n(T)$ for each n. Using Theorem 8.12(3) we conclude that every L-formula is defined in T over L_0.

To complete the proof, suppose $p_1,p_2\in S_n(T)$ have $\pi_n(p_1)=\pi_n(p_2)$. For each $j=1,2$ let $\mathcal{M}_j$ be a model of T and a_j an n-tuple in M_j that realizes p_j in $\mathcal{M}_j$. It follows that $(\mathcal{M}_1|L_0,a_1)$ and $(\mathcal{M}_2|L_0,a_2)$ are elementarily equivalent. By Theorem 5.7 there is an ultrafilter D such that the ultrapowers $(\mathcal{M}_1|L_0,a_1)_D$ and $(\mathcal{M}_2|L_0,a_2)_D$ are isomorphic, say by the function f from $(\mathcal{M}_1|L_0)_D$ onto $(\mathcal{M}_2|L_0)_D$. Let $\mathcal{N}$ be the unique L-structure for which f is an isomorphism from $(\mathcal{M}_1)_D$ onto $\mathcal{N}$. This ensures that $\mathcal{N}$ is a model of T and $\mathcal{N}|L_0=(\mathcal{M}_2|L_0)_D=(\mathcal{M}_2)_D|L_0$. Therefore $\mathcal{N}$ and $(\mathcal{M}_2)_D$ are identical, and hence f is an isomorphism from $(\mathcal{M}_1)_D$ onto $(\mathcal{M}_2)_D$. Since f maps a_1 onto a_2, this shows $p_1=p_2$, as desired. □

9.29 Corollary. *The property of being an extension by definitions is transitive. That is, if T_1 is an extension by definitions of T_0 and T_2 is an extension by definitions of T_1, then T_2 is an extension by definitions of T_0.*

Proof Immediate from the previous result. □

9.30 Remark. The following observation is useful when constructing extensions by definition in the metric setting. Suppose γ is an ordinal and $(T_\alpha \mid \alpha<\gamma)$ are theories in continuous logic such that (a) $T_{\alpha+1}$

is an extension by definitions of T_α whenever $\alpha + 1 < \gamma$ and (b) $T_\lambda = \cup(T_\alpha \mid \alpha < \lambda)$ whenever λ is a limit ordinal $< \gamma$. Let $T = \cup(T_\alpha \mid \alpha < \gamma)$. Then T is an extension by definitions of T_0. (The proof is by induction on γ; Corollary 9.29 takes care of the case when γ is a successor ordinal and the limit ordinal case is trivial.)

9.31 Corollary. *Let T, L, L_0 be as above and let T_0 be an L_0-theory. If T is an extension by definitions of T_0, then every model of T_0 has a (unique) expansion to a model of T.*

Proof We have already noted the uniqueness of the expansion, so only its existence needs to be proved. Let $\mathcal{M}_0$ be any model of T_0. As shown in the proof of Proposition 9.28, there is a unique complete L-theory T_1 that contains T and $\mathrm{Th}(\mathcal{M}_0)$. ($T_1$ is the unique element of $S_0(T)$ that satisfies $\pi_0(T_1) = \mathrm{Th}(\mathcal{M}_0)$.) Let $\mathcal{M}$ be a $\mathrm{card}(\mathcal{M}_0)^+$-saturated model of T_1. Since $\mathcal{M}|L_0 \equiv \mathcal{M}_0$, we may assume without loss of generality that $\mathcal{M}_0 \preceq \mathcal{M}|L_0$. A simple modification of the proof of Proposition 9.25(1) shows that the universe of $\mathcal{M}_0$ is closed under $f^{\mathcal{M}}$ for every function symbol f of L and contains $c^{\mathcal{M}}$ for every constant symbol c of L. Hence there exists a substructure $\mathcal{M}_0^*$ of $\mathcal{M}$ whose reduct to L_0 equals $\mathcal{M}_0$. It follows from Proposition 9.28 and the fact that $\mathcal{M}_0 \preceq \mathcal{M}|L_0$ that $\mathcal{M}_0^*$ is an elementary substructure of $\mathcal{M}$. In particular, $\mathcal{M}_0^*$ is a model of T and an expansion of $\mathcal{M}_0$, as desired. □

Now we describe certain standard ways of obtaining extensions by definition of a given L_0-theory T_0. Typically this is done via a sequence of steps, in each of which we add a definable predicate, or a definable constant, or a definable function. As the basis of each step we have in hand a previously constructed theory T that is an extension by definitions of T_0, with L being the signature of T.

First, consider the case where we want to add a definable n-ary predicate. Here we have a sequence $(\varphi_k(x_1, \ldots, x_n) \mid k \geq 1)$ of L-formulas that is uniformly Cauchy in all models of T, in the sense that the following statement holds:

$$\forall \epsilon > 0\ \exists N\ \forall k, l > N\ T \models \big(\sup_{x_1} \ldots \sup_{x_n} |\varphi_k - \varphi_l|\big) \leq \epsilon.$$

Choose an increasing sequence of positive integers $(N_m \mid m \geq 1)$ such that

$$\forall m \geq 1\ \forall k, l > N_m\ T \models \big(\sup_{x_1} \ldots \sup_{x_n} |\varphi_k - \varphi_l|\big) \leq 2^{-m}.$$

We then let P be a new n-ary predicate symbol and take T' to be the $L(P)$-theory obtained by adding to T the conditions

$$(\sup_{x_1} \dots \sup_{x_n} |\varphi_{k(m)}(x_1, \dots, x_n) - P(x_1, \dots, x_n)|) \leq 2^{-m}$$

for every $m \geq 1$ and $k(m) = N_m + 1$. Then every model of T has an expansion that is a model of T', and P is defined in T' over L, by construction. Therefore T' is an extension by definitions of T. Hence T' is an extension by definitions of T_0 by Corollary 9.29. Given any model $\mathcal{M}$ of T', this implies that the predicate $P^{\mathcal{M}}$ is definable (over $\emptyset$) in the reduct $\mathcal{M}|L_0$. More precisely, $P^{\mathcal{M}}$ is the uniform limit of the predicates $\varphi_m^{\mathcal{M}|L}$ as $m \to \infty$, and each of these predicates is definable in $\mathcal{M}|L_0$.

Note that in defining the signature $L(P)$ we need to specify a modulus of uniform continuity for the predicate symbol P. Such a modulus can be defined from the sequence $(N_m \mid m \geq 1)$ together with moduli for the formulas $(\varphi_k \mid k \geq 1)$ as indicated in the proof of Proposition 2.5.

Next consider the case where we want to add a definable constant. Without loss of generality we may assume that we have an L-formula $\varphi(y)$ such that

$$T \models (\inf_z \sup_y |d(z,y) - \varphi(y)|) = 0.$$

This implies that in every model $\mathcal{M}$ of T the zeroset of $\varphi^{\mathcal{M}}$ has a single element, by Proposition 9.25(1).

We then let c be a new constant symbol and take T' to be the $L(c)$-theory obtained by adding to T the condition $(\sup_y |d(c,y) - \varphi(y)|) = 0$. Again we have that every model of T has an expansion that is a model of T' and c is definable in T' over L. Hence T' is an extension by definitions of T_0.

Finally, generalizing the case of adding a definable constant, consider the case where we want to add a definable function. Without loss of generality we may assume that we have an L-formula $\varphi(x_1, \dots, x_n, y)$ such that

$$T \models (\sup_{x_1} \dots \sup_{x_n} \inf_z \sup_y |d(z,y) - \varphi(x_1, \dots, x_n, y)|) = 0.$$

This implies that in every model $\mathcal{M}$ of T the zeroset of $\varphi^{\mathcal{M}}$ is the graph of a total function from M^n to M, by Proposition 9.25(1).

We then let f be a new n-ary function symbol and take T' to be the $L(f)$-theory obtained by adding to T the condition

$$(\sup_{x_1} \dots \sup_{x_n} \sup_y |d(f(x_1, \dots, x_n), y) - \varphi(x_1, \dots, x_n, y)|) = 0.$$

Again we have that every model of T has an expansion that is a model of T' and f is definable in T' over L. Hence T' is an extension by definitions of T_0. Note that when introducing the signature $L(f)$ we must specify a modulus of uniform continuity for f. This can be taken to be the modulus of uniform continuity of the L-formula φ. (See the proof of Proposition 9.23.)

In order to simplify matters, we have described the addition of a constant c or function f only in the apparently restricted situation where the definitions of $d(c, y)$ or $d(f(x_1, \ldots, x_n), y)$ are given by *formulas* of L rather than by definable predicates. We want to emphasize that this is not a real limitation; it would be overcome by first adding the needed definable predicate and then using it to add the desired constant or function.

We conclude this section with a result that generalizes Beth's Definability Theorem to continuous logic.

9.32 Theorem. *Let T be an L-theory and let L_0 be a signature contained in L. Let S be any nonlogical symbol in L. Assume that S is implicitly defined in T over L_0; that is, assume that if $\mathcal{M}, \mathcal{N}$ are models of T for which $\mathcal{M}|L_0 = \mathcal{N}|L_0$, then one always has $S^{\mathcal{M}} = S^{\mathcal{N}}$. Then S is defined in T over L_0.*

Proof Let φ be the $L_0(S)$-formula $S(x_1, \ldots, x_n)$ if S is an n-ary predicate symbol, $d(S(x_1, \ldots, x_n), x_{n+1})$ if S is an n-ary function symbol, and $d(S, x_1)$ if S is a constant symbol. Denote the list of variables in φ simply as x. To show that S is defined in T over L_0 we need to show that the formula $\varphi(x)$ is defined in T over L_0.

Let T_1 be the restriction of T to the signature $L_0(S)$. Our first step is to prove that T_1 implicitly defines S over L_0. To prove this we will show that for each $\epsilon > 0$ there exist finitely many L_0-formulas, $\psi_1(x), \ldots, \psi_k(x)$ such that

$$T \models \big(\min_{1 \le j \le k} (\sup_x |\varphi(x) - \psi_j(x)|) \big) \le \epsilon.$$

Suppose this fails for some specific $\epsilon > 0$. Then there exists a model $\mathcal{M}$ of T such that for every L_0-formula $\psi(x)$ we have

$$\mathcal{M} \models (\sup_x |\varphi(x) - \psi_j(x)|) \ge \epsilon.$$

This implies that the predicate $Q = \varphi^{\mathcal{M}}$ is not definable (over $\emptyset$) in $\mathcal{M}|L_0$. By Proposition 7.12 we may assume that $\mathcal{M}$ is ω-saturated and that $\mathcal{M}|L_0$ is strongly ω-homogeneous. As shown in the proof of Corollary 9.11, there is an automorphism τ of $\mathcal{M}|L_0$ such that $Q \neq Q \circ \tau$. Let $\mathcal{N}$ be the unique L-structure for which τ is an isomorphism from $\mathcal{M}$ onto $\mathcal{N}$. Evidently $\mathcal{N} \models T$. Since τ is an automorphism of $\mathcal{M}|L_0$, we have $\mathcal{N}|L_0 = \mathcal{M}|L_0$. However, $\varphi^{\mathcal{N}} = Q \circ \tau \neq Q = \varphi^{\mathcal{M}}$. This contradicts the assumption that S is implicitly defined in T over L_0.

So, we now have that S is implicitly defined in T_1 over L_0. Let T_0 be the restriction of T_1 to L_0 (which is the same as the restriction of T to L_0). For each $n \geq 0$, let $\pi_n \colon S_n(T_1) \to S_n(T_0)$ be the restriction map: $\pi_n(p)$ is equal to the set of all L_0-conditions in p. Arguing as in the proof of Proposition 9.28 we see that each π_n is a surjective homeomorphism and we conclude that the formula $\varphi(x)$ is defined in T_1 over L_0. Hence $\varphi(x)$ is defined in T over L_0, as desired. □

9.33 Corollary. *Let T be an L-theory and T_0 an L_0-theory, both satisfiable, where L_0 is contained in L. Then T is an extension by definitions of T_0 if and only if every model of T_0 has a unique expansion that is a model of T.*

Proof This is immediate from Corollary 9.31 (for the left to right direction) and Theorem 9.32 (for the other direction). □

10 Algebraic and definable closures

In this section we introduce the concepts of *definable* and *algebraic* closure of a set in a metric structure. There are several reasonable choices for the definitions, but they turn out to be equivalent.

10.1 Definition. Let $\mathcal{M}$ be an L-structure and A a subset of M, and let $a \in M^n$. We say that a is *definable* in $\mathcal{M}$ over A if the set $\{a\}$ is definable in $\mathcal{M}$ over A (that is, if the predicate $d(\cdot, a)$ is definable in $\mathcal{M}$ over A). We say that a is *algebraic* in $\mathcal{M}$ over A if there exists a compact set $C \subseteq M^n$ such that $a \in C$ and C is definable in $\mathcal{M}$ over A.

As in the usual first-order setting, these properties of tuples reduce to the corresponding properties of their coordinates:

10.2 Proposition. *Let $\mathcal{M}$ be an L-structure and A a subset of M. Let $a = (a_1, \ldots, a_n) \in M^n$. Then a is definable (resp., algebraic) in $\mathcal{M}$ over A if and only if a_j is definable (resp., algebraic) in $\mathcal{M}$ over A for each $j = 1, \ldots, n$.*

Proof We treat the algebraic case. To prove the left to right direction it suffices to prove that if $C \subseteq M^n$ is compact and definable in $\mathcal{M}$ over A, then its projection C_i onto the ith coordinates is definable for each $i = 1, \dots, n$. (It is obviously compact.) For this we first note that for each i the predicate P_i on M^n defined by

$$P_i(x_1, \dots, x_n) = \inf\{d(x_i, y_i) \mid (y_1, \dots, y_n) \in C\}$$

is definable in $\mathcal{M}$ over A, by Theorem 9.17. Then we have $\mathrm{dist}(x_i, C_i) = P_i(x_i, \dots, x_i)$, showing that C_i is definable in $\mathcal{M}$ over A for each i.

For the right to left direction, suppose $C_i \subseteq M$ is a compact set witnessing that a_i is algebraic in $\mathcal{M}$ over A. Then the product $C = C_1 \times \dots \times C_n$ witnesses the same property for a. Note that the distance function for such a product is given by

$$\mathrm{dist}\big((x_1, \dots, x_n), C\big) = \inf_{y_1 \in C_1} \dots \inf_{y_n \in C_n} \max\big(d(x_1, y_1), \dots, d(y_n, x_n)\big)$$

and the right side is definable by n uses of Theorem 9.17. □

Notation 10.3 We let $\mathrm{dcl}_{\mathcal{M}}(A)$ denote the set of all elements of M that are definable in $\mathcal{M}$ over A and we call it the *definable closure* of A in $\mathcal{M}$. Similarly, we let $\mathrm{acl}_{\mathcal{M}}(A)$ denote the set of all elements of M that are algebraic in $\mathcal{M}$ over A and call it the *algebraic closure* of A in $\mathcal{M}$.

We first want to show that the definable and algebraic closures depend only on A and not on the structure in which they are defined. For that we apply some definability results from the previous section.

10.4 Proposition. *Let $\mathcal{M} \preceq \mathcal{N}$ and $A \subseteq M$. Suppose $C \subseteq N^n$ is definable in $\mathcal{N}$ over A and that $C \cap M^n$ is compact. Then C is contained in M^n.*

Proof Let $Q \colon N^n \to [0, 1]$ be the predicate given by $Q(x) = \mathrm{dist}(x, C)$, and assume that Q is definable in $\mathcal{N}$ over A. Let P be the restriction of Q to M^n. By Proposition 9.18 we have that $(\mathcal{M}, P) \preceq (\mathcal{N}, Q)$ and that $P(x) = \mathrm{dist}(x, C \cap M^n)$ for all $x \in M^n$. Fix $\epsilon > 0$ and suppose $c_1, \dots, c_m$ is an ϵ-net in $C \cap M^n$. It follows that whenever $P(x) < \epsilon$ we have $d(x, c_j) \leq 2\epsilon$ for some $j = 1, \dots, m$. In other words, the condition

$$\sup_x \min\big(\epsilon \dot{-} P(x), \min(d(x, c_1), \dots, d(x, c_m)) \dot{-} 2\epsilon\big) = 0$$

holds in $(\mathcal{M}, P)$. Hence the condition

$$\sup_x \min\left(\epsilon \dot{-} Q(x), \min(d(x,c_1),\ldots,d(x,c_m)) \dot{-} 2\epsilon\right) = 0$$

holds in $(\mathcal{N}, Q)$. It follows that $c_1,\ldots,c_m$ is a 2ϵ-net in C. Therefore every element of C is the limit of a sequence from M^n and hence C is contained in M^n. □

10.5 Corollary. *Let $\mathcal{M} \preceq \mathcal{N}$ be L-structures and let A be a subset of M. Then*

$$\mathrm{dcl}_{\mathcal{M}}(A) = \mathrm{dcl}_{\mathcal{N}}(A) \subseteq \mathrm{acl}_{\mathcal{N}}(A) = \mathrm{acl}_{\mathcal{M}}(A).$$

Proof Proposition 10.4 shows that if $C \subseteq N^n$ is compact and is definable in $\mathcal{N}$ over A, then $C \subseteq M^n$. Obviously C is then definable in $\mathcal{M}$ over A. It follows that $\mathrm{dcl}_{\mathcal{N}}(A) \subseteq \mathrm{dcl}_{\mathcal{M}}(A)$ and $\mathrm{acl}_{\mathcal{N}}(A) \subseteq \mathrm{acl}_{\mathcal{M}}(A)$.

For the opposite containment, suppose $C \subseteq M^n$ is compact and definable in $\mathcal{M}$ over A. We want to show that C is definable in $\mathcal{N}$ over A. Let $P(x) = \mathrm{dist}(x, C)$ for all $x \in M^n$. Since P is definable in $\mathcal{M}$ over A, by Proposition 9.8 there is a predicate $Q\colon N^n \to [0,1]$ that is definable in $\mathcal{N}$ over A and that extends P. Moreover we have $(\mathcal{M}, P) \preceq (\mathcal{N}, Q)$. Let $D \subseteq N^n$ be the zero set of Q. By Theorem 9.12 we have that $Q(x) = \mathrm{dist}(x, D)$ for all $x \in N^n$. It follows from Proposition 10.4 that $D = C$. Therefore C is definable in $\mathcal{N}$ over A. It follows that $\mathrm{dcl}_{\mathcal{M}}(A) \subseteq \mathrm{dcl}_{\mathcal{N}}(A)$ and $\mathrm{acl}_{\mathcal{M}}(A) \subseteq \mathrm{acl}_{\mathcal{N}}(A)$. □

An alternative way to approach *definable closure* and *algebraic closure* in metric structures would be to consider compact *zerosets* rather than definable sets. The next result shows that the concepts would be the same, as long as one takes care to work in saturated models.

10.6 Proposition. *Let $\mathcal{M}$ be an ω_1-saturated L-structure and A a subset of M. For compact subsets C of M^n the following are equivalent:*
(1) C is a zeroset in $\mathcal{M}$ over A.
(2) C is definable in $\mathcal{M}$ over A.

Proof Obviously (2) implies (1). For the converse, let $P\colon M^n \to [0,1]$ be a predicate definable in $\mathcal{M}$ over A whose zero set is C.

Given $\epsilon > 0$, let $F \subseteq C$ be a finite $\epsilon/2$-net in C. We claim there exists $\delta > 0$ such that any a satisfying $P(a) \leq \delta$ must lie within ϵ of some element of F. Otherwise we may use the ω_1-saturation of $\mathcal{M}$ to obtain an element a of M^n such that $P(a) \leq 1/k$ for every $k \geq 1$ while $d(a, c) \geq \epsilon$ for all $c \in F$, which is impossible.

The existence of such a $\delta > 0$ for each $\epsilon > 0$ shows that P verifies (2) in Proposition 9.19. Hence C is definable in $\mathcal{M}$ over A. $\square$

The following result gives alternative characterizations of definability.

10.7 Exercise. *Let $\mathcal{M}$ be an L-structure, $A \subseteq M$, and $a \in M^n$. Statements (1),(2) and (3) are equivalent:*
(1) a is definable in $\mathcal{M}$ over A.
(2) For any $\mathcal{N} \succeq \mathcal{M}$ the only realization of $\mathrm{tp}_{\mathcal{M}}(a/A)$ *in $\mathcal{N}$ is a.*
(3) For any $\epsilon > 0$ there is an $L(A)$-formula $\varphi(x)$ and $\delta > 0$ such that $\varphi^{\mathcal{M}}(a) = 0$ and the diameter of $\{b \in M^n \mid \varphi^{\mathcal{M}}(b) < \delta\}$ is $\leq \epsilon$.
If $\mathcal{N}$ is any fixed ω_1-saturated elementary extension of $\mathcal{M}$, then (1) is equivalent to:
(4) The only realization of $\mathrm{tp}_{\mathcal{M}}(a/A)$ *in $\mathcal{N}$ is a.*

The next result gives alternative characterizations of algebraicity:

10.8 Exercise. *Let $\mathcal{M}$ be an L-structure, $A \subseteq M$, and $a \in M^n$. Statements (1),(2) and (3) are equivalent:*
(1) a is algebraic in $\mathcal{M}$ over A.
(2) For any $\mathcal{N} \succeq \mathcal{M}$, every realization of $\mathrm{tp}_{\mathcal{M}}(a/A)$ *in $\mathcal{N}$ is in M^n.*
(3) For any $\epsilon > 0$ there is an $L(A)$-formula $\varphi(x)$ and $\delta > 0$ such that $\varphi^{\mathcal{M}}(a) = 0$ and the set $\{b \in M^n \mid \varphi^{\mathcal{M}}(b) < \delta\}$ has a finite ϵ-net.
(4) For any $\mathcal{N} \succeq \mathcal{M}$, the set of realizations of $\mathrm{tp}_{\mathcal{M}}(a/A)$ *in $\mathcal{N}$ is compact.*
If $\mathcal{N}$ is any fixed ω_1-saturated elementary extension of $\mathcal{M}$, then (1) is equivalent to:
(5) The set of realizations of $\mathrm{tp}_{\mathcal{M}}(a/A)$ *in $\mathcal{N}$ is compact.*
If $\mathcal{N}$ is any fixed κ-saturated elementary extension of $\mathcal{M}$, with κ uncountable, then (1) is equivalent to:
(6) The set of realizations of $\mathrm{tp}_{\mathcal{M}}(a/A)$ *in $\mathcal{N}$ has density character $< \kappa$.*

10.9 Remark. The previous result shows that $\mathrm{acl}_{\mathcal{M}}(A)$ is the same as the *bounded* closure of A in $\mathcal{M}$. (Equivalence of (1) and (6) in highly saturated models.)

The usual basic properties of algebraic and definable closure in first-order logic hold in this more general setting. There is one (unsurprising) modification needed: if a is algebraic (resp., definable) over A in

$\mathcal{M}$, there is a *countable* subset A_0 of A such that a is algebraic (resp., definable) over A_0 in $\mathcal{M}$. However, it may be impossible to satisfy this property with A_0 finite. (For example, let A consist of the elements of a Cauchy sequence whose limit is a. Then a is definable over A but it is not necessarily even algebraic over a finite subset of A.) There is also one new feature: the algebraic (resp., definable) closure of a set is the same as the algebraic closure of any dense subset.

In what follows we fix an L-structure $\mathcal{M}$ and A, B are subsets of M. We write acl instead of $\mathrm{acl}_{\mathcal{M}}$.

10.10 Exercise. *(Properties of* dcl*)*
(1) $A \subseteq \mathrm{dcl}(A)$*;*
(2) if $A \subseteq \mathrm{dcl}(B)$ *then* $\mathrm{dcl}(A) \subseteq \mathrm{dcl}(B)$*;*
(3) if $a \in \mathrm{dcl}(A)$ *then there exists a countable set* $A_0 \subseteq A$ *such that* $a \in \mathrm{dcl}(A_0)$*;*
(4) if A *is a dense subset of* B*, then* $\mathrm{dcl}(A) = \mathrm{dcl}(B)$.

10.11 Proposition. *(Properties of* acl*)*
(1) $A \subseteq \mathrm{acl}(A)$*;*
(2) if $A \subseteq \mathrm{acl}(B)$ *then* $\mathrm{acl}(A) \subseteq \mathrm{acl}(B)$*;*
(3) if $a \in \mathrm{acl}(A)$ *then there exists a countable set* $A_0 \subseteq A$ *such that* $a \in \mathrm{acl}(A_0)$*;*
(4) if A *is a dense subset of* B*, then* $\mathrm{acl}(A) = \mathrm{acl}(B)$.

10.12 Proposition. *If* $\alpha \colon A \to B$ *is an elementary map, then it extends to an elementary map* $\alpha' \colon \mathrm{acl}(A) \to \mathrm{acl}(B)$*. Moreover, if* α *is surjective, then so is* α'.

Proofs of the last two results are given in [26].

11 Imaginaries

In this section we explain how to add *finitary imaginaries* to a metric structure. This construction is the first stage of a natural generalization of the $\mathcal{M}^{eq}$ construction in ordinary first-order model theory.

There are several different ways to look at this construction. We first take a "geometric" point of view, based on forming the quotient of a pseudometric space. This is a generalization to the metric setting of taking the quotient of a set X by an equivalence relation E and connecting X to X/E by the quotient map π_E which takes each element of X to its E-class.

Let L be any signature for metric structures and let $\rho(x,y)$ be an L-formula in which x and y are n-tuples of variables. If $\mathcal{M}$ is any L-structure, we think of $\rho^{\mathcal{M}}$ as being a function of two variables on M^n, and we are interested in the case where this function is a pseudometric on M^n. (Note: to consider a pseudometric that is definable in $\mathcal{M}$ over $\emptyset$, but is not itself the interpretation of a formula, one should first pass to an extension by definitions in which it is the interpretation of a predicate.)

We define a signature L_ρ that extends L by adding a new sort, which we denote by M' with metric ρ', and a new n-ary function symbol π_ρ, which is to be interpreted by functions from M^n into M'. The modulus of uniform continuity specified by L_ρ for π_ρ is Δ_ρ. (Recall that Δ_ρ is a modulus of uniform continuity for $\rho^{\mathcal{M}}$ in every L-structure $\mathcal{M}$. See Theorem 3.5.)

Now suppose $\mathcal{M}$ is any L-structure in which $\rho^{\mathcal{M}}$ is a pseudometric on M^n. We expand $\mathcal{M}$ to an L_ρ-structure by interpreting (M',ρ') to be the completion of the quotient metric space of $(M^n,\rho^{\mathcal{M}})$ and by interpreting π_ρ to be the canonical quotient mapping from M^n into M'. This expansion of $\mathcal{M}$ will be denoted by $\mathcal{M}_\rho$. Saying that $\mathcal{M}_\rho$ is an L_ρ-structure requires checking that Δ_ρ is a modulus of uniform continuity for the interpretation of π_ρ. We indicate why this is true: suppose $\epsilon > 0$ and $x, x' \in M^n$ satisfy $d(x,x') < \Delta_\rho(\epsilon)$. Then

$$\rho'(\pi_\rho(x),\pi_\rho(x')) = \rho(x,x') = |\rho(x,x') - \rho(x',x')| \leq \epsilon$$

as desired.

Let T_ρ be the L_ρ-theory consisting of the following conditions:

(1) $\sup_x \rho(x,x) = 0$;
(2) $\sup_x \sup_y |\rho(x,y) - \rho(y,x)| = 0$;
(3) $\sup_x \sup_y \sup_{y'} (\rho(x,y) \dot{-} \min(\rho(x,y') + \rho(y',y), 1)) = 0$;
(4) $\sup_x \sup_y |\rho'(\pi_\rho(x),\pi_\rho(y)) - \rho(x,y)| = 0$;
(5) $\sup_z \inf_x \rho'(z,\pi_\rho(x)) = 0$.

11.1 Theorem. *(1) For every L-structure $\mathcal{M}$ in which $\rho^{\mathcal{M}}$ is a pseudometric, the expansion $\mathcal{M}_\rho$ is a model of T_ρ.*
(2) If $\mathcal{N}$ is any model of T_ρ, with $\mathcal{M}$ its reduct to L, then $\rho^{\mathcal{M}}$ is a pseudometric and $\mathcal{N}$ is isomorphic to $\mathcal{M}_\rho$ by an isomorphism that is the identity on $\mathcal{M}$.

Proof (1) is obvious. If $\mathcal{N}$ is any model of T_ρ, with $\mathcal{M}$ its reduct to L, the first three conditions in T_ρ ensure that $\rho^{\mathcal{M}}$ is a pseudometric.

Statement (4) ensures that the interpretation of π_ρ together with the metric ρ' on its range is isomorphic to the canonical quotient of the pseudometric space $(M^n, \rho^{\mathcal{M}})$ by an isomorphism that is the identity on M^n. Statement (5) ensures that the range of the interpretation of π_ρ is dense in (M', ρ'). □

To keep notation simple we have limited our discussion to the situation where L is a 1-sorted signature. If it is many-sorted, we may replace M^n by any finite cartesian product of sorts and carry out exactly the same construction. In particular, the process of adding imaginary sorts can be iterated.

Canonical parameters for formulas

Now we use the quotient construction above to add canonical parameters for any L-formula $\varphi(u, x)$, where u is an m-tuple of variables and x is an n-tuple of variables (thought of as the parameters). We take $\rho(x, y)$ to be the L-formula $\sup_u |\varphi(u, x) - \varphi(u, y)|$. Note that for any L-structure $\mathcal{M}$ we have that $\rho^{\mathcal{M}}$ is a pseudometric on M^n. Thus we may carry out the construction above, obtaining the signature L_ρ and the expansion $\mathcal{M}_\rho$ of any L-structure $\mathcal{M}$ to a uniquely determined model of T_ρ. Let $\widehat{\varphi}(u, z)$ be the L_ρ-formula

$$\inf_y \left(\varphi(u, y) + \rho'(z, \pi_\rho(y))\right).$$

11.2 Proposition. *If $\mathcal{N}$ is any model of T_ρ, then*

$$\mathcal{N} \models \sup_u \sup_x |\widehat{\varphi}(u, \pi_\rho(x)) - \varphi(u, x)| = 0 \text{ and}$$

$$\mathcal{N} \models \sup_w \sup_z \left|\rho'(w, z) - \sup_u |\widehat{\varphi}(u, w) - \widehat{\varphi}(u, z)|\right| = 0.$$

Proof Let $\mathcal{N}$ be any model of T_ρ and let $\mathcal{M}$ be its reduct to L. Let (M', ρ') be the sort of $\mathcal{N}$ that is added when expanding $\mathcal{M}$ to an L_ρ-structure.

To prove the first statement, take any $x, y \in M^n$. Then for any $u \in M^m$ we have

$$\begin{aligned} \varphi^{\mathcal{M}}(u, y) + \rho'(\pi_\rho^{\mathcal{N}}(x), \pi_\rho^{\mathcal{N}}(y)) &= \varphi^{\mathcal{M}}(u, y) + \rho^{\mathcal{M}}(x, y) \\ &\geq \varphi^{\mathcal{M}}(u, y) + |\varphi^{\mathcal{M}}(u, x) - \varphi^{\mathcal{M}}(u, y)| \\ &\geq \varphi^{\mathcal{M}}(u, x). \end{aligned}$$

Taking the inf over $y \in M^n$ shows that $\widehat{\varphi}^{\mathcal{N}}(u, \pi_\rho^{\mathcal{N}}(x)) \geq \varphi^{\mathcal{M}}(u,x)$. On the other hand, taking $y = x$ in the definition of $\widehat{\varphi}^{\mathcal{N}}$ shows that $\widehat{\varphi}^{\mathcal{N}}(u, \pi_\rho^{\mathcal{N}}(x)) \leq \varphi^{\mathcal{M}}(u,x)$.

For the second statement, note that it suffices to consider w, z in the range of $\pi_\rho^{\mathcal{N}}$, since it is dense in M'. When $w = \pi_\rho^{\mathcal{N}}(x)$ and $z = \pi_\rho^{\mathcal{N}}(y)$, the equality to be proved follows easily from the first statement in the Proposition and the definition of $\rho(x,y)$. □

11.3 Remark. Let $\mathcal{M}$ be any L-structure and $\mathcal{N} = \mathcal{M}_\rho$ its canonical expansion to a model of T_ρ. As before, denote the extra sort of this expansion by (M', ρ'). The results above show that M' is a space of canonical parameters for the predicate $\varphi^{\mathcal{M}}(u,x)$ and that $\widehat{\varphi}^{\mathcal{N}}$ is the predicate resulting from $\varphi^{\mathcal{M}}$ by the identification of each $x \in M^n$ with its associated canonical parameter $z = \pi_\rho^{\mathcal{N}}(x)$.

11.4 Remark. Note that in any model $\mathcal{N}$ of T_ρ the function $\pi_\rho^{\mathcal{N}}$ is definable from $\widehat{\varphi}^{\mathcal{N}}$ and $\varphi^{\mathcal{N}}$ in the sense of Definition 9.22. Indeed, when $\mathcal{N} \models T_\rho$ we have

$$\mathcal{N} \models \sup_x \sup_z \big|\rho'(\pi_\rho(x), z) - \sup_u |\varphi(u,x) - \widehat{\varphi}(u,z)|\big| = 0.$$

(Proof: Specialize w to $\pi_\rho(x)$ in the second statement in Proposition 11.2 and then use the first statement to replace $\widehat{\varphi}(u, \pi_\rho(x))$ by $\varphi(u,x)$.)

In the metric setting, the construction of $\mathcal{M}^{eq}$ should allow expansions corresponding to extensions by definitions as well as those corresponding to quotients by definable (over $\emptyset$) pseudometrics.

In fact, the full construction of $\mathcal{M}^{eq}$ requires the addition of more sorts than are described here. In particular, for some uses in stability theory one needs to add imaginaries that provide canonical parameters for definable predicates that depend on *countably many* parameters. (What we describe here only covers the case where the predicate depends on *finitely many* parameters.) See [6, end of Section 5] for a sketch of how this is done.

12 Omitting types and ω-categoricity

In this section we assume that the signature L has only a countable number of nonlogical symbols.

Let T be a complete L-theory. We emphasize here our point of view that models of T are always complete for their underlying metric(s). That is especially significant for the meaning of properties such as categoricity and omitting types.

12.1 Definition. Let κ be a cardinal $\geq \omega$. We say T is κ-categorical if whenever $\mathcal{M}$ and $\mathcal{N}$ are models of T having density character equal to κ, one has that $\mathcal{M}$ is isomorphic to $\mathcal{N}$.

One of the main goals of this section is to state a characterization of ω-categoricity for complete theories in continuous logic, extending the Ryll-Nardzewski Theorem from first-order model theory. We note that Ben Yaacov has proved the analogue for this setting of Morley's Theorem concerning uncountable categoricity. (See [4].)

Let p be an n-type over $\emptyset$ for T; that is, $p \in S_n(T)$. If $\mathcal{M}$ is a model of T, we let $p(\mathcal{M})$ denote the set of realizations of p in $\mathcal{M}$.

12.2 Definition. Let $p \in S_n(T)$. We say that p is *principal* if for every model $\mathcal{M}$ of T, the set $p(\mathcal{M})$ is definable in $\mathcal{M}$ over $\emptyset$.

The following Lemma shows that for $p \in S_n(T)$ to be principal, it suffices for there to exist *some* model of T in which the set of realizations of p is a nonempty definable set. It also shows that a principal type is realized in every model of T.

12.3 Lemma. *Let $p \in S_n(T)$. Suppose there exists a model $\mathcal{M}$ of T such that $p(\mathcal{M})$ is nonempty and definable in $\mathcal{M}$ over $\emptyset$. Then for any model $\mathcal{N}$ of T, the set $p(\mathcal{N})$ is nonempty and definable in $\mathcal{N}$ over $\emptyset$.*

Proof Since T is complete, any two models of T have isomorphic elementary extensions. Therefore it suffices to consider the case in which one of the structures is an elementary extension of the other.

First suppose $\mathcal{M} \preceq \mathcal{N}$ and that $p(\mathcal{N})$ is nonempty and definable in $\mathcal{N}$ over $\emptyset$. Since $p(\mathcal{M}) = p(\mathcal{N}) \cap M^n$, Theorem 9.12 and Proposition 9.18 yield that $p(\mathcal{M})$ is nonempty and definable in $\mathcal{M}$ over $\emptyset$.

Now suppose $\mathcal{M} \preceq \mathcal{N}$ and that $p(\mathcal{M})$ is nonempty and definable in $\mathcal{M}$ over $\emptyset$. Let $P(x) = \text{dist}(x, p(\mathcal{M}))$, so P is a definable predicate in $\mathcal{M}$ (over $\emptyset$). Use Proposition 9.8 to obtain a predicate Q on N^n so that $(\mathcal{M}, P) \preceq (\mathcal{N}, Q)$. By Theorem 9.12 the predicate Q satisfies $Q(x) = \text{dist}(x, D)$ for all $x \in N^n$, where D is the zeroset of Q. Therefore, it suffices to show that $p(\mathcal{N})$ is the zeroset of Q.

First we prove that the zeroset of Q is contained in $p(\mathcal{N})$. Suppose

$\varphi(x)$ is any L-formula for which the condition $\varphi = 0$ is in p. Fix $\epsilon > 0$; by Theorem 3.5 there exists $\delta > 0$ such that whenever $x, y \in M^n$ satisfy $d(x,y) < \delta$, then $|\varphi^{\mathcal{M}}(x) - \varphi^{\mathcal{M}}(y)| \leq \epsilon$. We may assume $\delta < 1$. If $x \in M^n$ satisfies $\text{dist}(x, p(\mathcal{M})) < \delta$, there must exist $y \in M^n$ realizing p with $d(x,y) < \delta$ and hence

$$\varphi^{\mathcal{M}}(x) = |\varphi^{\mathcal{M}}(x) - \varphi^{\mathcal{M}}(y)| \leq \epsilon.$$

That is,

$$(\mathcal{M}, P) \models \sup_x (\min(\delta \dot{-} P(x), \varphi(x) \dot{-} \epsilon)) = 0.$$

It follows that

$$(\mathcal{N}, Q) \models \sup_x (\min(\delta \dot{-} Q(x), \varphi(x) \dot{-} \epsilon)) = 0.$$

Since ϵ was arbitrary and $\varphi = 0$ was an arbitrary condition in p, this shows that the zeroset of Q is contained in $p(\mathcal{N})$.

Finally, we need to show that Q is identically zero on $p(\mathcal{N})$. By construction, there is a sequence $(\varphi_n(x))$ of L-formulas such that

$$|Q(x) - \varphi_n^{\mathcal{N}}(x)| \leq 1/n$$

for all $x \in N^n$ and all $n \geq 1$. Since $p(\mathcal{M})$ is nonempty, we may take $x \in M^n$ realizing p; since $Q(x) = P(x) = \text{dist}(x, p(\mathcal{M})) = 0$, we see that the condition $\varphi_n(x) \leq 1/n$ is in $p(x)$ for all $n \geq 1$. For any $x \in p(\mathcal{N})$ we therefore have

$$Q(x) \leq |Q(x) - \varphi_n^{\mathcal{N}}(x)| + \varphi_n^{\mathcal{N}}(x) \leq 2/n$$

for all $n \geq 1$. Therefore $Q(x) = 0$ for any x in $p(\mathcal{M})$. □

12.4 Proposition. *Let $p \in S_n(T)$. Then p is principal if and only if the logic topology and the d-metric topology agree at p.*

Proof Let $\mathcal{M}$ be an ω-saturated model of T. Since p is realized in $\mathcal{M}$, Lemma 12.3 implies that p is principal if and only if $p(\mathcal{M})$ is definable in $\mathcal{M}$ over $\emptyset$. We apply Proposition 9.19, taking $D = p(\mathcal{M})$ and $A = \emptyset$. This yields that p is principal if and only if for each $m \geq 1$ there is an L-formula $\varphi_m(x)$ and $\delta_m > 0$ such that $\varphi_m = 0$ is in p and any $q \in S_n(T)$ that contains the condition $\varphi_m \leq \delta_m$ must satisfy $d(q,p) \leq 1/m$.

So, when p is principal and $m \geq 1$, we have a logic open neighborhood of p, namely $[\varphi_m < \delta_m]$, that is contained in the $1/m$-ball around p.

On the other hand, suppose $[\psi < \delta]$ is a basic logic neighborhood of

p that is contained in the $1/m$-ball around p. There exists $0 < \eta < \delta$ such that the condition $\psi \leq \eta$ is in p. Taking φ_m to be the formula $\psi \dot{-} \eta$ and δ_m to satisfy $0 < \delta_m < \delta - \eta$, we have that $\varphi_m = 0$ is in p and that any $q \in S_n(T)$ that contains the condition $\varphi_m \leq \delta_m$ must satisfy $d(q,p) \leq 1/m$. If this is possible for every $m \geq 1$, then p must be principal. □

The fact that the metric is included as a predicate in L allows us to characterize principal types by a formally weaker topological property than the one in the previous Proposition:

12.5 Proposition. *Let $p \in S_n(T)$. Then p is principal if and only if the ball $\{q \in S_n(T) \mid d(q,p) \leq \epsilon\}$ has nonempty interior in the logic topology, for each $\epsilon > 0$.*

Proof ($\Rightarrow$) This follows from Proposition 12.4.

($\Leftarrow$) Suppose $[\psi < \delta]$ is a nonempty basic open set contained in the ϵ-ball around $p \in S_n(T)$. We may assume $\delta \leq \epsilon$, since $[\frac{1}{k}\psi < \frac{1}{k}\delta] = [\psi < \delta]$ for all k. Choose η such that $0 < \eta < \delta$ and $[\psi \leq \eta]$ is nonempty. Consider the formula

$$\varphi(x) = \inf_y \max(\psi(y) \dot{-} \eta, d(x,y) \dot{-} \epsilon).$$

The condition $\varphi = 0$ is in p. This is because $[\psi \leq \eta]$ is nonempty and is contained in the ϵ-ball around p. Furthermore, the basic open set $[\varphi < \delta - \eta]$ is contained in the $(2\epsilon + \delta)$-ball around p. Taking ϵ arbitrarily small gives the desired result, by Proposition 12.4. □

12.6 Theorem. *(Omitting Types Theorem, local version) Let T be a complete theory in a countable signature, and let $p \in S_n(T)$. The following statements are equivalent:*

(1) *p is principal.*
(2) *p is realized in every model of T.*

Proof (1) $\Rightarrow$ (2). Since p must be realized in *some* model of T, Lemma 12.3 shows that a principal type is realized in *all* models of T.

(2) $\Rightarrow$ (1). We sketch a proof of the contrapositive. Suppose p is not principal. By Proposition 12.5, there exists $\epsilon > 0$ such that the ϵ-ball $\{q \in S_n(T) \mid d(q,p) \leq \epsilon\}$ has empty interior in the logic topology. That is, for any L-formula $\varphi(x)$ and any $\delta > 0$, the logic neighborhood $[\varphi < \delta]$ is either empty or contains a type q such that $d(q,p) > \epsilon$. An argument

as in the usual proof of the omitting types theorem in classical first-order model theory yields a countable L-prestructure $\mathcal{M}_0$ satisfying the theory T^+, such that any n-type q realized in $\mathcal{M}_0$ satisfies $d(q,p) > \epsilon$. Let $\mathcal{M}$ be the completion of $\mathcal{M}_0$. It follows that any type q realized in $\mathcal{M}$ satisfies $d(q,p) \geq \epsilon$, and hence $\mathcal{M}$ is a model of T in which p is not realized. □

12.7 Definition. A model $\mathcal{M}$ of T is *atomic* if every n-type realized in $\mathcal{M}$ is principal.

12.8 Proposition. *Let $\mathcal{M}$ be a model of T and let $p(x_1,\ldots,x_m) = \mathrm{tp}_{\mathcal{M}}(a_1,\ldots,a_m)$ for a sequence $a_1,\ldots,a_m$ in M. Let $n > m$ and suppose $q(x_1,\ldots,x_n) \in S_n(T)$ extends p. If q is principal, then for each $\epsilon > 0$ there exist $(b_1,\ldots,b_n)$ realizing q in M and satisfying $d(a_j,b_j) \leq \epsilon$ for all $j = 1,\ldots,m$.*

Proof Let $D \subseteq M^m$ and $E \subseteq M^n$ be the sets of realizations of p and q respectively in $\mathcal{M}$. Because q is principal, both D and E are definable in $\mathcal{M}$ over $\emptyset$. Hence the function F defined for $x \in M^m$ by

$$F(x) = \inf_y |\mathrm{dist}(x,D) - \mathrm{dist}((x,y),E)|$$

is a predicate definable in $\mathcal{M}$ over $\emptyset$. Here y ranges over M^{n-m}. If $\mathcal{N}$ is an ω_1-saturated elementary extension of $\mathcal{M}$, the sequence $(a_1,\ldots,a_m)$ can be extended in $\mathcal{N}$ to a realization of q. If we extend F to N^m using the same definition, this shows that $F(a_1,\ldots,a_m) = 0$ in $\mathcal{N}$. By Lemma 9.6 we have that $F(a_1,\ldots,a_m) = 0$ in $\mathcal{M}$. So there exist $c \in M^{n-m}$ such that the n-tuple $(a_1,\ldots,a_m,c)$ has distance $\leq \epsilon$ to E in M^n. This completes the proof. □

12.9 Corollary. *Let $\mathcal{M}$ be a separable atomic model of T and let $\mathcal{N}$ be any other model of T. Let $(a_1,\ldots,a_m)$ realize the same type in $\mathcal{M}$ that $(b_1,\ldots,b_m)$ realizes in $\mathcal{N}$. Then, for each $\epsilon > 0$ there exists an elementary embedding F from $\mathcal{M}$ into $\mathcal{N}$ such that $d(b_j,F(a_j)) \leq \epsilon$ for all $j = 1,\ldots,m$. Furthermore, if $\mathcal{N}$ is also separable and atomic, then F can be taken to be an isomorphism from $\mathcal{M}$ onto $\mathcal{N}$.*

Proof Extend $a_1,\ldots,a_m$ to an infinite sequence (a_k) that is dense in M. Let (δ_k) be a sequence of positive real numbers whose sum is less than ϵ. By induction on $n \geq 0$ we use the previous Proposition to generate sequences c_n in N^{m+n} with the following properties: (1) $c_0 =$

$(b_1, \ldots, b_m)$; (2) c_n realizes the same type in $\mathcal{N}$ that $(a_1, \ldots, a_{m+n})$ realizes in $\mathcal{M}$; (3) the first $m+n$ coordinates of c_{n+1} are at a distance less than δ_n away from the corresponding coordinates of c_n.

It follows that for each j, the sequence of jth coordinates of c_n is a Cauchy sequence in N. Let its limit be d_j. Continuity of formulas ensures that the map taking a_j to d_j for each j is an elementary map. Therefore it extends to the desired elementary embedding of $\mathcal{M}$ into $\mathcal{N}$.

A "back-and-forth" version of the same argument proves the final statement in the Corollary. □

The following result is the analogue of the Ryll-Nardzewski Theorem in this setting:

12.10 Theorem. *Let T be a complete theory in a countable signature. The following statements are equivalent:*

(1) *T is ω-categorical;*
(2) *For each $n \geq 1$, every type in $S_n(T)$ is principal;*
(3) *For each $n \geq 1$, the metric space $(S_n(T), d)$ is compact.*

Proof (1) ⇒ (2) This is immediate from the Omitting Types Theorem.

(2) ⇒ (1) Condition (2) implies that every model of T is atomic. Therefore, Corollary 12.9 (especially, the last sentence) yields that any two separable models of T are isomorphic.

(2) ⇔ (3) By Proposition 12.4, statement (2) is equivalent to saying that the logic topology and the d-metric topology are identical on $S_n(T)$ for every n. Since the logic topology is compact, this shows that (2) implies (3). On the other hand, if the metric space $(S_n(T), d)$ is compact, then its topology must agree with the logic topology, since both topologies are compact and Hausdorff, and one is coarser than the other. □

12.11 Corollary. *Suppose T is ω-categorical and $\mathcal{M}$ is the separable model of T. Then $\mathcal{M}$ is* strongly ω-near-homogeneous *in the following sense: if $a, b \in M^n$, then for every $\epsilon > 0$ there is an automorphism F of $\mathcal{M}$ such that*

$$d(F(a), b) \leq d(\mathrm{tp}(a), \mathrm{tp}(b)) + \epsilon.$$

Proof By Theorem 12.10 we see that $\mathcal{M}$ is atomic. Let

$$r = d(\mathrm{tp}(a), \mathrm{tp}(b)).$$

First consider the case where $r = 0$. Corollary 12.9 yields the existence of an automorphism F of $\mathcal{M}$ with the desired properties. Now suppose $r > 0$. Since $\mathcal{M}$ is the unique separable model of T, there exist $a', b' \in M^n$ such that $\text{tp}(a') = \text{tp}(a)$, $\text{tp}(b') = \text{tp}(b)$, and $d(a', b') = r$. Applying the $r = 0$ case to a, a' and to b, b', for each $\epsilon > 0$ we get automorphisms F_a, F_b of $\mathcal{M}$ such that $d(F_a(a), a') < \epsilon/2$ and $d(F_b(b), b') < \epsilon/2$. Taking $F = F_b^{-1} \circ F_a$ gives an automorphism with

$$\begin{aligned} d(F(a), b) &= d(F_a(a), F_b(b)) \\ &\leq d(F_a(a), a') + d(a', b') + d(F_b(b), b') \\ &< r + \epsilon \end{aligned}$$

as desired. □

12.12 Remark. In the previous result, note that for any automorphism F of $\mathcal{M}$, $d(F(a), b) \geq d(\text{tp}(a), \text{tp}(b))$ since $F(a)$ and a realize the same type. Moreover, Example 17.7 shows that in the setting of the previous result, we need not be able to find an automorphism F of $\mathcal{M}$ that satisfies $d(F(a), b) = d(\text{tp}(a), \text{tp}(b))$; so this result gives the strongest possible kind of homogeneity for ω-categorical metric structures, in general. (Example 17.7 notes that the theory of the Banach lattice L^p is ω-categorical, but that there exist elements f, g realizing the same type but are such that there is no automorphism of L^p taking f to g.)

12.13 Corollary. *Suppose $L \subseteq L'$ are countable signatures, T' is a complete theory in L' and T is its restriction to L. If T' is ω-categorical, then so is T.*

Proof The restriction map (discarding formulas not in L) defines a map from $S_n(T')$ onto $S_n(T)$ that is contractive with respect to the d-metrics. Hence it preserves compactness. □

12.14 Remark. Example 17.7 shows that ω-categoricity is *not* necessarily preserved under the addition of designated elements to the language, in contrast to what happens in classical first-order model theory. There can exist pairs (a, b) realizing a principal type in some model, but such that $\text{tp}(b/a)$ is not principal. This indicates a complication in the model theory of metric structures that is not completely understood, and that affects a number of important aspects of the theory (including, for example, superstability).

13 Quantifier elimination

Fix a signature L and an L-theory T. We give basic definitions and state (but do not prove) some results around quantifier elimination in continuous logic. The proofs are similar to those in [24, pages 84–91].

13.1 Definition. An L-formula $\varphi(x_1, \ldots, x_n)$ is *approximable in T by quantifier-free formulas* if for every $\epsilon > 0$ there is a quantifier-free L-formula $\psi(x_1, \ldots, x_n)$ such that for all $\mathcal{M} \models T$ and all $a_1, \ldots, a_n \in M$, one has

$$|\varphi^{\mathcal{M}}(a_1, \ldots, a_n) - \psi^{\mathcal{M}}(a_1, \ldots, a_n)| \leq \epsilon.$$

13.2 Proposition. *Let $\varphi(x_1, \ldots, x_n)$ be an L-formula. The following statements are equivalent.*

(1) *φ is approximable in T by quantifier-free formulas;*

(2) *Whenever we are given*
- *models $\mathcal{M}$ and $\mathcal{N}$ of T;*
- *substructures $\mathcal{M}_0 \subseteq \mathcal{M}$ and $\mathcal{N}_0 \subseteq \mathcal{N}$;*
- *an isomorphism Φ from $\mathcal{M}_0$ onto $\mathcal{N}_0$; and*
- *elements $a_1, \ldots, a_n$ of $\mathcal{M}_0$;*

we have

$$\varphi^{\mathcal{M}}(a_1, \ldots, a_n) = \varphi^{\mathcal{N}}(\Phi(a_1), \ldots, \Phi(a_n)).$$

Moreover, for the implication (2)⇒(1) it suffices to assume (2) only for the cases in which $\mathcal{M}_0$ and $\mathcal{N}_0$ are finitely generated.

13.3 Definition. An L-theory T *admits quantifier elimination* if every L-formula is approximable in T by quantifier-free formulas.

13.4 Remark. (1) Let T be an L-theory, and let $L(C)$ be an extension of L by constants. If T admits quantifier elimination in L, then T admits quantifier elimination in $L(C)$.
(2) Let $T \subseteq T'$ be theories in a signature L. If T admits quantifier elimination in L, then T' admits quantifier elimination in L.

13.5 Lemma. *Suppose that T is an L-theory and that every restricted L-formula of the form $\inf_x \varphi$, with φ quantifier-free, is approximable in T by quantifier-free formulas. Then T admits quantifier elimination.*

13.6 Proposition. *Let T be an L-theory. Then the following statements are equivalent:*

(1) *T admits quantifier elimination;*
(2) *If $\mathcal{M}$ and $\mathcal{N}$ are models of T, then every embedding of a substructure of $\mathcal{M}$ into $\mathcal{N}$ can be extended to an embedding of $\mathcal{M}$ into an elementary extension of $\mathcal{N}$.*

Moreover, if $\operatorname{card}(L) \leq \kappa$*, then in statement (2) it suffices to consider models $\mathcal{M}$ of density character $\leq \kappa$.*

14 Stability and independence

In this section we sketch three general approaches to stability in metric structures. The first one is based on measuring the size of the type spaces $S_1(T_A)$ in various ways. The second one comes from properties of the notion of independence obtained from non-dividing. The third one comes from definability of types. In all three cases we give clear statements of the definitions and the basic results (which we need in later sections), but we only give some of the proofs. In spite of having these different approaches to stability, there is, in the end, only one notion of stability, as we discuss.

Throughout this section T is a complete L-theory, κ is a cardinal $> \operatorname{card}(L)$, and λ is an infinite cardinal. As is usual, we often denote $\operatorname{card}(L)$ also by $|T|$; recall that this is the least *infinite* cardinal $\geq$ the number of nonlogical symbols in L. When $\mathcal{M}$ is a model of T and $A \subseteq M$, recall that T_A is the theory of $(\mathcal{M}, a)_{a \in A}$. We take x and y to be finite sequences of distinct variables; usually $x = x_1, \ldots, x_n$.

We begin with an approach to stability based simply on the cardinality of the type spaces $S_1(T_A)$.

14.1 Definition. We say that T is *λ-stable with respect to the discrete metric* if for any $\mathcal{M} \models T$ and any $A \subseteq M$ of cardinality $\leq \lambda$, the set $S_1(T_A)$ has cardinality $\leq \lambda$. We say that T is *stable with respect to the discrete metric* if T is λ-stable with respect to the discrete metric for some λ.

14.2 Definition. Let $\mathcal{U}$ be a κ-universal domain for T. Let $\varphi(x, y)$, $\psi(x, y)$ be formulas such that the conditions $\varphi(x, y) = 0$ and $\psi(x, y) = 0$ are contradictory in $\mathcal{U}$. Since $\mathcal{U}$ is κ-saturated, there exists some $\epsilon > 0$ such that $\{\varphi(x, y) = 0, \psi(x', y) = 0, d(x, x') \leq \epsilon\}$ is not satisfiable in $\mathcal{U}$.

If $p(x)$ is any satisfiable partial type over a small subset of $\mathcal{U}$, we define the rank $R(p,\varphi,\psi,2)$ inductively, in the usual manner. First we define a relation $R(p,\varphi,\psi,2) \geq \alpha$ by induction on the ordinal α, as follows:

- $R(p,\varphi,\psi,2) \geq 0$ for any satisfiable p;
- for λ a limit ordinal, $R(p,\varphi,\psi,2) \geq \lambda$ if $R(p,\varphi,\psi,2) \geq \alpha$ for all $\alpha < \lambda$;
- $R(p,\varphi,\psi,2) \geq \alpha+1$ if there are satisfiable extensions p_1, p_2 of p and $b \in \mathcal{U}$ such that $\varphi(x,b) = 0$ is in p_1, $\psi(x,b) = 0$ is in p_2, $R(p_1,\varphi,\psi,2) \geq \alpha$, and $R(p_2,\varphi,\psi,2) \geq \alpha$.

We write $R(p,\varphi,\psi,2) = \infty$ if $R(p,\varphi,\psi,2) \geq \alpha$ for all ordinals α. Otherwise there is an ordinal γ such that $R(p,\varphi,\psi,2) \geq \alpha$ holds iff $\alpha \leq \gamma$; in that case we set $R(p,\varphi,\psi,2) = \gamma$. By compactness, if $R(p,\varphi,\psi,2) \geq \omega$, then $R(p,\varphi,\psi,2) = \infty$, so the values of R lie in $\mathbb{N} \cup \{\infty\}$.

14.3 Proposition. *Let $\mathcal{U}$ be a κ-universal domain for T. Then T is stable with respect to the discrete metric if and only if for all pairs of conditions $\varphi(x,y) = 0$, $\psi(x,y) = 0$ that are contradictory in $\mathcal{U}$, we have $R(\{d(x,x) = 0\},\varphi,\psi,2) < \omega$.*

Proof The argument is similar to the proof of the corresponding result in classical first-order model theory. See Proposition 2.2 in [3] for this proof in the cat framework. □

Next we introduce a notion of stability based on measuring the size of $S_1(T_A)$ by its density character with respect to the d-metric.

14.4 Definition. We say that T is *λ-stable* if for any $\mathcal{M} \models T$ and $A \subseteq M$ of cardinality $\leq \lambda$, there is a subset of $S_1(T_A)$ of cardinality $\leq \lambda$ that is dense in $S_1(T_A)$ with respect to the d-metric. We say that T is *stable* if T is λ-stable for some infinite λ.

14.5 Remark. There are two reasons why we have chosen to associate *stability* most closely with topological properties of the type spaces $S_1(T_A)$ expressed in terms of the d-metric. First, as will be seen in later sections, many theories of interest turn out to be ω-stable in this sense (but not necessarily ω-stable with respect to other natural topologies on $S_1(T_A)$ including the discrete topology). The second (and main) reason for this choice is that theories in continuous first-order logic that are λ-stable in this sense (*i.e.*, with respect to the d-metric) have properties analogous to those of λ-stable theories in classical first-order logic. (See

[3], [6].) For example ω-stable theories have prime models and ω-stable theories are λ-stable for all infinite λ. (See Remark 14.8 below.)

If we drop the quantitative aspect (*i.e.*, we drop λ) then the distinction between these two notions of stability disappears:

14.6 Theorem. *A theory T is stable if and only if T is stable with respect to the discrete metric.*

Proof Clearly if T is λ-stable with respect to the discrete metric then T is λ-stable.

Assume now that T is not stable with respect to the discrete metric. Let $\mathcal{U}$ be a κ-universal domain for T. By Proposition 14.3 there are formulas $\varphi_0(x, y)$, $\varphi_1(x, y)$ and $\epsilon > 0$ such that

$$\{\varphi_0(x, y) = 0, \varphi_1(x', y) = 0, d(x, x') \leq \epsilon\}$$

is not satisfiable in $\mathcal{U}$, and

$$R(\{d(x, x) = 0\}, \varphi_0(x, y), \varphi_1(x, y), 2) = \infty.$$

Given an infinite cardinal λ, let μ be a cardinal such that $2^{<\mu} \leq \lambda < 2^{\mu}$. We may assume that $\kappa > \lambda$. By the definition of the rank and the saturation of $\mathcal{U}$ we can find a sequence $\{b_\sigma \mid \sigma \in 2^{<\mu}\}$ of elements in $\mathcal{U}$ such that $\{\varphi_{\sigma(\alpha)}(x, b_{\sigma\restriction\alpha}) = 0 \mid \alpha < \mu\}$ is realized by some c_σ in $\mathcal{U}$, for every $\sigma \in 2^{\mu}$. Let $B = \{b_\sigma \mid \sigma \in 2^{<\mu}\}$. Then $\mathrm{card}(B)$ is $\leq \lambda$, yet $d(\mathrm{tp}(c_\sigma/B), \mathrm{tp}(c_\tau/B)) \geq \epsilon$ for all distinct $\sigma, \tau \in 2^{\mu}$. Therefore T is not λ-stable. □

For many theories there are several natural topologies on type spaces that can be used as the basis of alternative notions of "λ-stability". This approach was considered by Iovino in [28, 29] for metric structures based on Banach spaces; he proved a generalization of Theorem 14.6 for such notions, in the setting of positive bounded formulas.

We continue the discussion of these type-counting notions of stability by quoting the Stability Spectrum Theorem. The proof is very much like the one in classical first-order logic. See Theorem 4.12 in [4] for this proof in the cat framework.

14.7 Theorem. *Let T be a stable theory and let $\mu(T)$ be the first cardinal in which T is stable. Then there exists a cardinal $\kappa = \kappa(T)$ such that T is λ-stable if and only if $\lambda = \mu(T) + \lambda^{<\kappa(T)}$.*

14.8 Remark. The previous result yields that any ω-stable theory is λ-stable for all infinite λ, because $\kappa(T)$ must be $\aleph_0$ for such theories. Most of the examples we treat in later sections are ω-stable.

14.9 Definition. A theory T is *superstable* if $\kappa(T) = \aleph_0$.

By the remark above, any ω-stable theory is superstable, but the converse implication is not true. For example, the theory of infinite dimensional Hilbert spaces with a generic automorphism is superstable but not ω-stable. (See [7].) There is a characterization of superstability in terms of the stability spectrum:

14.10 Proposition. *T is superstable if and only if it is λ-stable for all $\lambda \geq 2^{|T|}$.*

Proof See [6]. □

Now we begin discussing our second approach to stability, which is based on non-dividing:

14.11 Definition. Let $\mathcal{U}$ be a κ-universal domain for T and let B, C be small subsets of $\mathcal{U}$. Let $p(X, B)$ be a partial type over B in a possibly infinite tuple of variables X (so $p(X, Y)$ is a partial type without parameters). We say that $p(X, B)$ *divides* over C if there exists a C-indiscernible sequence $(B_i \mid i < \omega)$ in $\mathrm{tp}(B/C)$ such that $\bigcup_{i<\omega} p(X, B_i)$ is inconsistent with T.
Furthermore, if A, B, C are small sets in $\mathcal{U}$ such that $\mathrm{tp}(A/BC)$ does not divide over C, then we say that A *is independent from* B *over* C and we write $A \mathop{\smile\!\!\!\!\!\!|}_C B$.

This notion of independence has good properties in every stable theory, just as it does in classical first-order model theory:

14.12 Theorem. *Let $\mathcal{U}$ be a κ-universal domain for T. If T is stable, then the independence relation $\mathop{\smile\!\!\!\!\!\!|}$ defined using non-dividing satisfies the following properties (here A, B, etc., are any small subsets of $\mathcal{U}$ and M is a small elementary submodel of $\mathcal{U}$):*

(1) *Invariance under automorphisms of $\mathcal{U}$.*
(2) *Symmetry: $A \mathop{\smile\!\!\!\!\!\!|}_C B \Longleftrightarrow B \mathop{\smile\!\!\!\!\!\!|}_C A$.*
(3) *Transitivity: $A \mathop{\smile\!\!\!\!\!\!|}_C BD$ if and only if $A \mathop{\smile\!\!\!\!\!\!|}_C B$ and $A \mathop{\smile\!\!\!\!\!\!|}_{B \cup C} D$.*

(4) *Finite Character:* $A \mathop{\smile\!\!\!\!\!\!|}_C B$ *if and only if* $a \mathop{\smile\!\!\!\!\!\!|}_C B$ *for all finite tuples* a *from* A.
(5) *Extension: For all* A, B, C *there exists* A' *such that* $A' \mathop{\smile\!\!\!\!\!\!|}_C B$ *and* $\mathrm{tp}(A/C) = \mathrm{tp}(A'/C)$.
(6) *Local Character: If* a *is any finite tuple, then there is* $B_0 \subseteq B$ *of cardinality* $\leq |T|$ *such that* $a \mathop{\smile\!\!\!\!\!\!|}_{B_0} B$.
(7) *Stationarity of types: If* $\mathrm{tp}(A/M) = \mathrm{tp}(A'/M)$, $A \mathop{\smile\!\!\!\!\!\!|}_M B$, *and* $A' \mathop{\smile\!\!\!\!\!\!|}_M B$, *then* $\mathrm{tp}(A/B \cup M) = \mathrm{tp}(A'/B \cup M)$.

Proof The proof follows the ideas of the corresponding result in classical first-order model theory. See, for example, [39].

A similar result has been proved in the more general framework of cats. Most of the proof can be found in [3, Theorems 1.51,2.8]), except that the extension property may fail in the setting considered there. In [4, Theorem 1.15] it is shown that in a cat that is *thick* and stable (more generally, if it is thick and simple), every type has a non-dividing extension. This applies to the setting considered in this paper, since any complete theory of metric structures gives rise to a Hausdorff cat, and every Hausdorff cat is thick. □

Stability can be characterized by the existence of an independence relation with suitable properties, just as in the classical first-order setting.

14.13 Definition. Let $\mathcal{U}$ be a κ-universal domain for T. A relation satisfying properties (1)–(7) in Theorem 14.12 is called a *stable independence relation* on $\mathcal{U}$.

14.14 Theorem. *Let* $\mathcal{U}$ *be a* κ*-universal domain for* T. *If* T *is stable, there is precisely one stable independence relation on* $\mathcal{U}$. *Moreover, if there exists a stable independence relation* $A \mathop{\smile\!\!\!\!\!\!|}^*_C B$ *on triples of small subsets of* $\mathcal{U}$, *then* T *is stable.*

Proof If T is a stable theory, the existence of a stable independence relation on $\mathcal{U}$ is given by Theorem 14.12. The rest of this Theorem is proved as in the classical first-order case. See, for example, [39]. □

Theorem 14.14 will play an important role in our treatment of application areas in sections 15–18. As is often true, the theories treated in those sections are already equipped with natural independence relations; for example, one has *orthogonality* in Hilbert spaces and *probabilistic independence* in probability spaces. (See sections 15 and 16 below.) Showing that such a relation is a stable independence relation is often not hard. Once this has been done, Theorem 14.14 implies that the theory in question is stable and that the natural candidate is indeed *the* relation of model-theoretic independence on $\mathcal{U}$ (yielding a complete understanding of non-dividing).

A third (equivalent) approach to stability is given by the notion of definability of types, which plays a central role in stability theory:

14.15 Definition. We say that a complete type $p \in S_n(B)$ is *definable* over a set A if for each formula $\varphi(x, y)$ with $x = x_1, \ldots, x_n$, there exists a predicate $\Psi(y)$ definable over A such that for all suitable tuples b in B we have that $\Psi(b)$ is the unique $r \in [0, 1]$ such that the condition $\varphi(x, b) = r$ is in p.

The following theorem characterizes stability in terms of definability of types:

14.16 Theorem. *The theory T is stable if and only if every type over a model $\mathcal{M}$ of T is definable over M.*

Proof Similar to the classical proof for first-order theories, but requiring a slightly more delicate analysis. See Theorem 8.5 in [6]. □

We conclude this section with the following result, which gives an alternative proof for Theorem 14.6 and which provides additional information about the stability spectrum.

14.17 Corollary. *Suppose T is stable. Then T is λ-stable with respect to the discrete metric for all λ that satisfy $\lambda = \lambda^{|T|}$.*

Proof Assume T is stable and that $\lambda = \lambda^{|T|}$. Let $\mathcal{M}$ be a model of T of density character λ; then M has cardinality at most $\lambda^{\aleph_0}$, which equals λ because of our special assumptions. Note that the number of definitions of types over M is at most $\lambda^{|T|} = \lambda$. Therefore, Theorem 14.16 yields that $S_1(T_M)$ has cardinality at most λ, as desired. □

The study of stability theory and its applications in analysis started with the work of Krivine and Maurey [32] around quantifier free formulas in Banach spaces. A study of stability and ω-stability for metric structures based on normed spaces is carried out in [28, 29, 30]. Theorem 14.6 was first proved (in a more general form) in Corollary 7.2 and

Corollary 7.3 in [28, page 88]. The proof provided here comes from [4, Remark 4.11]. A more general approach to stability in the setting of cats can be found in [3].

Stability of a general function (in particular, of a formula in continuous logic, *i.e.*, *local* stability) was introduced in [34] in the setting of functional analysis. Local stability of a formula in continuous logic is developed in [6].

There are several known examples of stable theories in the setting of metric structures, such as the theories of Hilbert spaces, probability spaces and L^p spaces. (See the next three sections.) Expansions of Hilbert spaces and probability spaces by generic automorphisms also turn out to be stable. (See [7, 8, 10] and the last section of this article.)

15 Hilbert spaces

A pre-Hilbert space H over $\mathbb{R}$ is a vector space over $\mathbb{R}$ with an inner product $\langle\ \rangle$ that satisfies the following properties:

(1) $\langle rx + sy, z\rangle = r\langle x, z\rangle + s\langle y, z\rangle$ for all $x, y, z \in H$ and all $r, s \in \mathbb{R}$;
(2) $\langle x, y\rangle = \langle y, x\rangle$ for all $x, y \in H$;
(3) $\langle x, x\rangle \in (0, \infty)$ for all nonzero $x \in H$; $\langle 0, 0\rangle = 0$.

On each pre-Hilbert space we define a norm by $\|x\| = \sqrt{\langle x, x\rangle}$. A pre-Hilbert space that is complete with respect to this norm is called a *Hilbert space*.

Let H be a pre-Hilbert space over $\mathbb{R}$. We treat H in continuous logic by identifying it with the many-sorted metric prestructure

$$\mathcal{M}(H) = ((B_n(H) \mid n \geq 1), 0, \{I_{mn}\}_{m<n}, \{\lambda_r\}_{r\in\mathbb{R}}, +, -, \langle\ \rangle)$$

where $B_n(H) = \{x \in H \mid \|x\| \leq n\}$ for $n \geq 1$; 0 is the zero vector in $B_1(H)$; for $m < n$, $I_{mn}\colon B_m \to B_n$ is the inclusion map; for $r \in \mathbb{R}$ and $n \geq 1$, $\lambda_r\colon B_n(H) \to B_{nk}(H)$ is scalar multiplication by r, with k the unique integer satisfying $k \geq 1$ and $k - 1 \leq |r| < k$; furthermore, $+, -\colon B_n(H) \times B_n(H) \to B_{2n}(H)$ are vector addition and subtraction and $\langle\ \rangle\colon B_n(H) \to [-n^2, n^2]$ is the inner product, for each $n \geq 1$. The metric on each sort is given by $d(x, y) = \|x - y\|$.

There is an obvious continuous signature L such that for each pre-Hilbert space H, the many-sorted prestructure described above is an L-prestructure; the necessary bounds and moduli of uniform convergence are easy to specify. We see easily that if H is a pre-Hilbert space, then the completion of the L-prestructure $\mathcal{M}(H)$ is equal to $\mathcal{M}(\overline{H})$, where $\overline{H}$ is the Hilbert space obtained by completing H.

It is not difficult to show that there is an L-theory, which we will denote by HS, such that $\mathcal{M} \models HS$ if and only if there is a Hilbert space H such that $\mathcal{M} \cong \mathcal{M}(H)$. (One way to verify this is using Proposition 5.14; it is well known and easy to check that the class of L-structures isomorphic to some $\mathcal{M}(H)$, with H a Hilbert space, is closed under ultraproducts and under elementary substructures. It is also not difficult to extract a set of axioms for this class by directly translating the requirements that $\mathcal{M}$ comes from a vector space over $\mathbb{R}$ and that the inner product satisfies (1),(2),(3) above.)

For any Hilbert space H, we see that H is infinite dimensional if and only if $\mathcal{M}(H)$ satisfies the conditions

$$\big(\inf_{x_1}\dots\inf_{x_n}\max_{1\le i,j\le n}(|\langle x_i,x_j\rangle - \delta_{ij}|)\big) = 0$$

for $n \geq 1$; here $\delta_{ij} = 1$ if $i = j$ and $\delta_{ij} = 0$ if $i \neq j$, and the variables $x_1, \dots, x_n$ range over the sort $B_1(H)$. Let IHS be the L-theory obtained by adding this infinite set of conditions to HS. An L-structure $\mathcal{M}$ is a model of IHS if and only if $\mathcal{M}$ is isomorphic to $\mathcal{M}(H)$ for some infinite dimensional Hilbert space H. In what follows we will identify H with $\mathcal{M}(H)$ when applying model theoretic techniques and concepts, such as types.

Note that IHS is κ-categorical for every infinite cardinal κ. Therefore IHS is a complete theory. In the rest of this section we will show that IHS admits quantifier elimination and is ω-stable.

Let $x, y \in H$ and let $A \subset H$. By $\overline{A}$ we mean the norm closure of the linear span of A. Let $P_{\bar{A}}(x)$ be the projection of x on the subspace $\overline{A}$. We denote by $A^{\perp}$ the set $\{z \in H : \langle a, z\rangle = 0 \text{ for all } a \in A\}$; it is a closed subspace of H known as the *orthogonal complement* of $\overline{A}$, since H is the Hilbert space direct sum of $\overline{A}$ and $A^{\perp}$.

15.1 Lemma. *Let H be an infinite dimensional Hilbert space, with $c_1, \dots, c_n, d_1, \dots, d_n \in H$. Then $(c_1, \dots, c_n)$ and $(d_1, \dots, d_n)$ realize the same type over $A \subset H$ if and only if $P_{\bar{A}}(c_i) = P_{\bar{A}}(d_i)$ and $\langle c_i, c_j\rangle = \langle d_i, d_j\rangle$ for all $1 \leq i, j \leq n$.*

Proof If $\mathrm{tp}(c_1, \dots, c_n/A) = \mathrm{tp}(d_1, \dots, d_n/A)$ then $\langle c_i, c_j\rangle = \langle d_i, d_j\rangle$ for $i, j \leq n$ and for every $a, b \in A$, $\langle c_i - b, a\rangle = \langle d_i - b, a\rangle$; thus $P_{\bar{A}}(c_i) = P_{\bar{A}}(d_i)$.

Conversely, assume that $P_{\bar{A}}(c_i) = P_{\bar{A}}(d_i)$ and $\langle c_i, c_j \rangle = \langle d_i, d_j \rangle$ for $i, j \leq n$. Then $c_i - P_{\bar{A}}(c_i), d_i - P_{\bar{A}}(d_i) \in A^{\perp}$ and

$$\langle c_i - P_{\bar{A}}(c_i), c_j - P_{\bar{A}}(c_j) \rangle = \langle d_i - P_{\bar{A}}(d_i), d_j - P_{\bar{A}}(d_j) \rangle.$$

Using the Gram-Schmidt process we can build an automorphism of H that fixes $\bar{A}$ pointwise and takes $c_i - P_{\bar{A}}(c_i)$ to $d_i - P_{\bar{A}}(d_i)$ for all $1 \leq i \leq n$. □

15.2 Corollary. *The theory IHS admits quantifier elimination.*

Proof We apply Proposition 13.6. Suppose $\mathcal{M}_1, \mathcal{M}_2 \models IHS$; let $\mathcal{N}$ be a substructure of M_1 and $f : \mathcal{N} \to \mathcal{M}_2$ an embedding. Let $\mathcal{M}_2' \succeq \mathcal{M}_2$ be such that the orthogonal complement of M_2 in M_2' has dimension $\geq \dim(M_1)$. We can extend f so it maps an orthonormal basis of $M_1 \cap N^{\perp}$ into an orthonormal subset of $M_2' \cap M_2^{\perp}$ and then extend f linearly to all of M_1. By Lemma 15.1, such a map is an embedding. □

15.3 Lemma. *Let H be an infinite dimensional Hilbert space and let $A \subset H$. Then the definable closure of A equals $\bar{A}$.*

Proof By passing to an elementary extension of H, which does not change $\mathrm{dcl}(A)$, we may assume that $\bar{A}$ is a proper subspace of H.

We first show that if $c \in \bar{A}$, then $c \in \mathrm{dcl}(A)$. Given $c \in \bar{A}$, there is a Cauchy sequence $\{c_n : n \geq 1\}$ of elements in the space spanned by A such that $\lim_{n\to\infty} c_n = c$. We may assume that $\|c_n - c\| \leq 1/(2n)$ for $n \geq 1$. Let $\varphi_n(x) = \|x - c_n\| \dot{-} 1/(2n)$. Then the family of formulas $\{\varphi_n(x) \mid n \geq 1\}$ and numbers $\{\delta_n = 1/(2n) \mid n \geq 1\}$ shows that $\{c\}$ is A-definable.

Assume now that $c \notin \bar{A}$, so $c - P_{\bar{A}}(c) \neq 0$. Take any $y \in A^{\perp}$ such that $\|y\| = \|c - P_{\bar{A}}(c)\|$. Then $\mathrm{tp}(c/A) = \mathrm{tp}(P_{\bar{A}}(c) + y/A)$. Since $A^{\perp}$ is not the 0 subspace, this shows that $\mathrm{tp}(c/A)$ has realizations in H that are different from c, and thus $c \notin \mathrm{dcl}(A)$. □

15.4 Proposition. *Let H be an infinite dimensional Hilbert space. For each $x, y \in H$ and $A \subset H$ we have*

$$d(\mathrm{tp}(x/A), \mathrm{tp}(y/A))^2 = \|P_{\bar{A}}(x) - P_{\bar{A}}(y)\|^2 + \big|\|x - P_{\bar{A}}(x)\| - \|y - P_{\bar{A}}(y)\|\big|^2$$

Proof Let $x, y \in H$ and let $A \subset H$. If $\mathrm{tp}(x'/A) = \mathrm{tp}(x/A)$ and

$\mathrm{tp}(y'/A) = \mathrm{tp}(y/A)$, then

$$\begin{aligned}\|x' - y'\|^2 &= \|P_{\bar{A}}(x') - P_{\bar{A}}(y')\|^2 + \|(x' - P_{\bar{A}}(x')) - (y' - P_{\bar{A}}(y'))\|^2 \\ &\geq \|P_{\bar{A}}(x) - P_{\bar{A}}(y)\|^2 + \big|\|x - P_{\bar{A}}(x)\| - \|y - P_{\bar{A}}(y)\|\big|^2.\end{aligned}$$

For the reverse inequality, let $x_\perp = x - P_{\bar{A}}(x)$ and $y_\perp = y - P_{\bar{A}}(y)$. If $x_\perp = 0$ the result is clear, so we may assume that $x_\perp \neq 0$. Let $\alpha = \|y_\perp\|/\|x_\perp\|$ and let $z = \alpha x_\perp$. Then we have $\mathrm{tp}(y/A) = \mathrm{tp}(P_{\bar{A}}(y) + z/A)$ by Lemma 15.1 and

$$\begin{aligned}\|x - (P_A(y) + z)\|^2 &= \|P_{\bar{A}}(x) - P_{\bar{A}}(y)\|^2 + \big|\|x_\perp - \alpha x_\perp\|\big|^2 \\ &= \|P_{\bar{A}}(x) - P_{\bar{A}}(y)\|^2 + \big|\|x_\perp\| - \|y_\perp\|\big|^2\end{aligned}$$

by the Pythagorean theorem. □

15.5 Proposition. *The theory IHS is ω-stable.*

Proof Let H be an infinite dimensional Hilbert space and let $A \subset H$ be countable. Let $T_A = \mathrm{Th}(\mathcal{M}(H), a)_{a \in A}$. We may assume that the dimension of $A^\perp$ (in H) is ω, so H is separable. Using Lemma 15.1 it is easy to show that every 1-type over A is realized in H. Therefore $S_1(A)$ is separable with respect to the d-metric. □

We close this section by giving a concrete description in terms of familiar Hilbert space concepts of the independence relation that is associated to the theory IHS. As usual, this gives an alternate proof that IHS is stable, although it does not identify the values of λ for which IHS is λ-stable.

In what follows, we fix a cardinal number $\kappa > 2^{\aleph_0}$ and we fix a κ-universal domain H for the theory IHS.

15.6 Definition. Whenever A, B, C are small subsets of H, we write $A \mathop{\smile\!\!\!\!\!\!\mid}^*_C B$ to mean that $P_{\bar{C}}(a) = P_{\overline{C \cup B}}(a)$ for all $a \in A$.

15.7 Lemma. *Let $A, B, C \subset H$ be small. Then $A \mathop{\smile\!\!\!\!\!\!\mid}^*_C B$ if and only if $a - P_{\bar{C}}(a) \perp b - P_{\bar{C}}(b)$ for all $a \in A$ and $b \in B$.*

Proof Assume first that $A \mathop{\smile\!\!\!\!\!\!\mid}^*_C B$, so for all $a \in A$, $P_{\bar{C}}(a) = P_{\overline{C \cup B}}(a)$. Then $a - P_{\bar{C}}(a) \in (C \cup B)^\perp$, so $a - P_{\bar{C}}(a) \perp b - P_{\bar{C}}(b)$ for any $b \in B$. Now assume that $a - P_{\bar{C}}(a) \perp b - P_{\bar{C}}(b)$ for all $a \in A$ and $b \in B$. Then $a - P_{\bar{C}}(a) \perp b$ for all $a \in A$ and $b \in B$ and thus $P_{\bar{C}}(a) = P_{\overline{C \cup B}}(a)$ for all $a \in A$. □

15.8 Theorem. *The relation $\mathop{\smile\!\!\!\!\!\!|}^*$ is a stable independence relation on H. Therefore, $\mathop{\smile\!\!\!\!\!\!|}^*$ is identical to the independence relation $\mathop{\smile\!\!\!\!\!\!|}$ based on non-dividing for the stable theory IHS.*

Proof We prove directly that $\mathop{\smile\!\!\!\!\!\!|}^*$ has all seven properties of a stable independence relation:

(1) Invariance: Let $f \in \mathrm{Aut}(H)$. Then for every $u, v \in H$, $\langle u|v\rangle = \langle f(u)|f(v)\rangle$. Therefore, if E is any subspace of H, it is easy to see that f carries P_E to $P_{f(E)}$ and carries $E^\perp$ to $f(E)^\perp$. This makes it clear (from the definition) that $f(A) \mathop{\smile\!\!\!\!\!\!|}^*_{f(C)} f(B)$ is equivalent to $A \mathop{\smile\!\!\!\!\!\!|}^*_C B$ for any small subsets A, B, C of H.

(2) Symmetry: This follows from Lemma 15.7.

(3) Transitivity: This follows from the definition.

(4) Finite character: This is immediate from the definition.

(5) Extension: By finite character and compactness, it suffices to prove the property for finite tuples. Let $a_1, \ldots, a_n \in H$ and let $B, C \subseteq H$ be small. Since $B \cup C$ is small, $(B \cup C)^\perp$ is an infinite dimensional subspace of H. Hence there are $c_1, \ldots, c_n \in (C \cup B)^\perp$ such that $\mathrm{tp}(c_1, \ldots, c_n) = \mathrm{tp}(a_1 - P_{\bar{C}}(a_1), \ldots, a_n - P_{\bar{C}}(a_n))$. Let $a'_i = P_{\bar{C}}(a_i) + c_i$, for $i = 1, \ldots, n$. Then

$$\mathrm{tp}(a'_1, \ldots, a'_n/C) = \mathrm{tp}(a_1, \ldots, a_n/C)$$

and $\{a'_1, \ldots, a'_n\} \mathop{\smile\!\!\!\!\!\!|}^*_C B$.

(6) Local Character: Given $a_1, \ldots, a_n$ and B there exists a countable subset B_0 of B such that each of $P_{\bar{B}}(a_i)$ is an element of $\overline{B_0}$. Then we have $a_1, \ldots, a_n \mathop{\smile\!\!\!\!\!\!|}^*_{B_0} B$.

(7) Stationarity of types: We will show that the property holds for general sets, that is, we do not need to assume that the underlying set C is an elementary substructure of H. By finite character, it suffices to prove the property when A is a finite tuple. So let $a = (a_1, \ldots, a_n), a' = (a'_1, \ldots, a'_n) \in H^n$ and let $C, B \subseteq H$ be small. Assume that $\mathrm{tp}(a/C) = \mathrm{tp}(a'/C)$ and that $a \mathop{\smile\!\!\!\!\!\!|}^*_C B$, $a' \mathop{\smile\!\!\!\!\!\!|}^*_C B$. Then for every $i = 1, \ldots, n$,

$$P_{\overline{B \cup C}}(a_i) = P_{\bar{C}}(a_i) = P_{\bar{C}}(a'_i) = P_{\overline{B \cup C}}(a'_i).$$

Thus $\mathrm{tp}(a/B \cup C) = \mathrm{tp}(a'/B \cup C)$.

The second statement follows from the first by Theorem 14.14. □

Note that the proof of Theorem 15.8 shows that all types of tuples in IHS are stationary.

Model theoretic studies of infinite dimensional Hilbert spaces are carried out in [9] and in [1]. A direct proof to characterize non-dividing in terms of orthogonality can be found in Corollary 2 and Lemma 8 in [9]. Proposition 15.5 appears in [28].

16 Probability spaces

In this section we give an introduction to the model theory of probability spaces using their measure algebras as metric structures.

A *probability space* is a triple $(X, \mathcal{B}, \mu)$, where X is a set, $\mathcal{B}$ is a σ-algebra of subsets of X and μ is a σ-additive measure on $\mathcal{B}$ such that $\mu(X) = 1$.

We say that $B \in \mathcal{B}$ is an *atom* if $\mu(B) > 0$ and there does not exist any $B' \in \mathcal{B}$ satisfying $B' \subseteq B$ and $0 < \mu(B') < \mu(B)$. Further, $B \in \mathcal{B}$ is *atomless* if there is no atom that is a subset of B. The probability space $(X, \mathcal{B}, \mu)$ is *atomless* if X itself is atomless. It is well known that if B is an atomless element of $\mathcal{B}$ and $0 \leq r \leq 1$ then there exists $B' \in \mathcal{B}$ satisfying $B' \subseteq B$ and $\mu(B') = r \cdot \mu(B)$. (See [20, Section 41] for a discussion.) This uniformity in the property of being atomless plays a role in axiomatizing the property in continuous logic. (See below.)

We write $A_1 \sim_\mu A_2$, and say that $A_1, A_2 \in \mathcal{B}$ determine the same *event*, if the symmetric difference of the sets, denoted by $A_1 \triangle A_2$, has measure zero. Clearly $\sim_\mu$ is an equivalence relation. We denote the class of $A \in \mathcal{B}$ under the equivalence relation $\sim_\mu$ by $[A]_\mu$. The collection of equivalence classes of $\mathcal{B}$ modulo $\sim_\mu$ is called the set of *events* and it is denoted by $\widehat{\mathcal{B}}$. The operations of complement, union and intersection are well defined for events and make $\widehat{\mathcal{B}}$ a σ-algebra; in addition, μ induces a well-defined, strictly positive, countably additive probability measure on $\widehat{\mathcal{B}}$. We refer to $\widehat{\mathcal{B}}$ as the *measure algebra* and to $(\widehat{\mathcal{B}}, \mu)$ as the *measured algebra* associated to $(X, \mathcal{B}, \mu)$.

Given $(X, \mathcal{B}, \mu)$, we build a 1-sorted metric structure (called a *probability structure*)

$$\mathcal{M} = (\widehat{\mathcal{B}}, 0, 1, \cdot^c, \cap, \cup, \mu)$$

whose metric is given by $d([A]_\mu, [B]_\mu) = \mu(A \triangle B)$. Here 0 is the event of measure zero, 1 the event of measure one, and $\cdot^c, \cap, \cup$ are the Boolean operations induced on $\widehat{\mathcal{B}}$. The modulus of uniform continuity for $\cdot^c$ is the identity $\Delta(\epsilon) = \epsilon$ and the moduli of uniform continuity for $\cup$ and $\cap$ are given by $\Delta'(\epsilon) = \epsilon/2$. We sometimes write a^{-1} for a^c and a^{+1} for a, when a is an element of $\widehat{\mathcal{B}}$.

Let L be the signature associated to these probability structures. The following L-conditions are true in all probability structures. Indeed, Theorem 16.1 shows they axiomatize that class of structures.

(1) Boolean algebra axioms:
Each of the usual axioms for Boolean algebras is the closure under universal quantifiers of an equation between terms (see [27, page 38]) and thus it can be expressed in continuous logic as a condition. For example, the axiom $\forall x \forall y (x \cup y = y \cup x)$ is equivalent to $\sup_x \sup_y \big(d(x \cup y, y \cup x)\big) = 0$.

(2) Measure axioms:
$\mu(0) = 0$ and $\mu(1) = 1$;
$\sup_x \sup_y \big(\mu(x \cap y) \dot{-} \mu(x)\big) = 0$;
$\sup_x \sup_y \big(\mu(x) \dot{-} \mu(x \cup y)\big) = 0$;
$\sup_x \sup_y |(\mu(x) \dot{-} \mu(x \cap y)) - (\mu(x \cup y) \dot{-} \mu(y))| = 0$.
The last three axioms express that $\mu(x \cup y) + \mu(x \cap y) = \mu(y) + \mu(x)$ for all x, y.

(3) Connection between d and μ:
$\sup_x \sup_y |d(x, y) - \mu(x \Delta y)| = 0$ where $x \Delta y$ denotes the Boolean term giving the symmetric difference: $(x \cap y^c) \cup (x^c \cap y)$.

We denote the set of L-conditions above by PrA.

16.1 Theorem. *Let $\mathcal{M}$ be an L-structure with underlying metric space (M, d). Then $\mathcal{M}$ is a model of PrA if and only if $\mathcal{M}$ is the probability structure associated to a probability space $(X, \mathcal{B}, \mu)$ as above.*

Proof It is clear that probability structures satisfy the conditions in PrA. A proof that such structures are metrically complete is given in [38, Chapter 7]. For the converse, let $\mathcal{M} \models PrA$; recall this implies that underlying metric space of $\mathcal{M}$ is complete. In [38, Chapter 7] it is discussed how to realize $\mathcal{M}$ as a probability structure. □

To say that a model is *atomless* we need the following axiom; it states that every set of positive measure can be cut nearly "in half" measurably. As noted above, this is well known to hold in a probability space if and only if the space is atomless.

(4) $\sup_x \inf_y |\mu(x \cap y) - \mu(x \cap y^c)| = 0$.

We denote by APA the set of axioms PrA together with (4). Its models are exactly the probability structures obtained from atomless probability spaces.

16.2 Proposition. *The theory APA is separably categorical (and therefore APA is complete).*

Proof Let $\mathcal{M} \models APA$ be separable. As is shown in [38, Chapter 7], $\mathcal{M}$ is the probability structure associated to a countably generated probability space that is necessarily atomless. A familiar back and forth argument shows that any two such probability spaces are isomorphic in a measure preserving manner. □

Next we characterize the d-metric on spaces of types for APA. To do this, we need the following special case of the Radon-Nikodym theorem:

16.3 Theorem. *(Radon-Nikodym; see [16, Theorem 3.8]) Let $(X, \mathcal{B}, \mu)$ be a probability space, let $\mathcal{C} \subseteq \mathcal{B}$ be a σ-subalgebra and let $A \in \mathcal{B}$. Let a be the event corresponding to A. Then there is a unique $g_a \in L^1(X, \mathcal{C}, \mu)$ such that for any $B \in \mathcal{C}$, $\int_B g_a d\mu = \int_B \chi_A d\mu$. Such an element g_a is called the* conditional probability *of a with respect to $\mathcal{C}$ and it is denoted by $\mathbb{P}(a|\mathcal{C})$.*

The next lemma provides an explicit form for the d-metric on probability structures (this formula was known to analysts [37, Lemma 6.3] in the case $C = \emptyset$).

16.4 Lemma. *Let $\mathcal{M} \models APA$ be a κ-universal domain, with $\kappa \geq \omega_1$. Assume that $\mathcal{M}$ is the probability structure associated to the probability space $(Y, \mathcal{D}, m)$. Let $a = (a_1, \ldots, a_n) \in M^n$, $b = (b_1, \ldots, b_n) \in M^n$ be partitions of the probability structure. Let $C \subseteq M$ be small, and let $\mathcal{C}$ be the σ-subalgebra of $\mathcal{D}$ generated by the measurable sets A such that the event of A is in C. Then*

$$d(\mathrm{tp}(a/C), \mathrm{tp}(b/C)) = \max_{1 \leq i \leq n} \|\mathbb{P}(a_i|\mathcal{C}) - \mathbb{P}(b_i|\mathcal{C})\|_1$$

where $\| \ \|_1$ is the L_1-norm.

Proof This is Lemma 3.14 in [10]. □

16.5 Corollary. *(Ben Yaacov [2]) Let $\mathcal{M} \models APA$ be a κ-universal domain, with $\kappa \geq \omega_1$, and let $C \subseteq M$ be small. Let $\mathcal{C}$ be obtained from C as in the previous result. Let $a = (a_1, \ldots, a_n) \in M^n$, $b = (b_1, \ldots, b_n) \in M^n$ be arbitrary. Then $\mathrm{tp}(a/C) = \mathrm{tp}(b/C)$ iff*

$$\mathbb{P}(a_1^{i_1} \cap \cdots \cap a_n^{i_n}|\mathcal{C}) = \mathbb{P}(b_1^{i_1} \cap \cdots \cap b_n^{i_n}|\mathcal{C})$$

for all n-tuples $(i_1, \dots, i_n)$ *from* $\{+1, -1\}$.

Proof Apply the previous result to the tuples of atoms in the finite subalgebras generated by $a_1, \dots, a_n$ and $b_1, \dots, b_n$ respectively. □

16.6 Proposition. *The theory APA admits quantifier elimination.*

Proof If $\mathcal{M} \models APA$ and $a_1, \dots, a_n \in M$, then the previous result shows that $\mathrm{tp}(a_1, \dots, a_n)$ is determined by the measures of the atoms of the finite Boolean subalgebra generated by $a_1, \dots, a_n$. This together with Proposition 13.2 implies that APA admits quantifier elimination. □

16.7 Proposition. *Let* $\mathcal{M}$ *be a model of APA and* $C \subseteq M$. *The* definable closure of C *is the smallest* σ*-algebra of events containing* C.

Proof This follows from quantifier elimination plus an analysis of restricted quantifier free formulas. □

Non-dividing in probability structures has a natural characterization:

16.8 Proposition. *(Ben Yaacov [2, Theorem 2.10]) Let* $\mathcal{M} \models APA$ *be a* κ*-universal domain, with* $\kappa \geq \omega_1$, *and let* $C \subseteq M$ *be small. Assume that* $\mathcal{M}$ *is the probability structure associated to the probability space* $(X, \mathcal{B}, \mu)$. *Let* $a_1, \dots, a_n; b_1, \dots, b_m \in M$. *Let* $\mathcal{C}$ *be the* σ*-subalgebra of* $\mathcal{B}$ *generated by a collection of measurable sets whose events make up* C *and let* $\mathcal{C}_b$ *be the* σ*-subalgebra of* $\mathcal{B}$ *generated by a larger collection of measurable sets whose events make up* $C \cup \{b_1, \dots, b_m\}$. *Then* $\mathrm{tp}(a_1, \dots, a_n / C \cup \{b_1, \dots, b_m\})$ *does not divide over* C *if and only if*

$$\mathbb{P}(a_1^{i_1} \wedge \cdots \wedge a_n^{i_n} | \mathcal{C}_b) = \mathbb{P}(a_1^{i_1} \wedge \cdots \wedge a_n^{i_n} | \mathcal{C})$$

for all n-tuples $(i_1, \dots, i_n)$ *from* $\{-1, 1\}$.

16.9 Proposition. *(1) The theory APA is* ω*-stable.*
(2) Let $\mathcal{N} \models APA$ *and consider* $a \in N^n$ *and* $C \subseteq N$. *Then* $\mathrm{tp}(a/C)$ *is stationary.*

Proof Let $\mathcal{M} \models APA$ be a κ-universal domain, with $\kappa \geq \omega_1$. Let $(X, \mathcal{B}, \mu)$ be a probability space whose events correspond to the elements of M. For part (1), let $C \subseteq M$ be countable. We may assume that C is closed under finite intersections, unions and complements. Let $\mathcal{C}$ be a set of measurable sets whose set of events is C. We may choose $\mathcal{C}$ so that it is a countable Boolean algebra. Let $\langle \mathcal{C} \rangle$ be the σ-algebra generated by $\mathcal{C}$, let Step(C) be the set of step functions in $L^1(X, \mathcal{C}, m)$ with coefficients in $\mathbb{Q}$ and let $\mathcal{F} = \{\text{tp}(a/C) \mid \mathbb{P}(a|\langle \mathcal{C} \rangle) \in \text{Step}(C)\}$. Then $\mathcal{F}$ is a countable set of types. It follows from Lemma 16.4 that $\mathcal{F}$ is a dense subset of the space of 1-types over C with respect to the d-metric. Therefore APA is ω-stable. The proof of (2) follows from 16.8. □

A model theoretic study of (atomless) probability spaces is carried out in [2]. The author shows that they give rise to a compact abstract theory (see [1] for the definition) and that, in this cat, the notion of independence obtained from non-dividing (Proposition 16.8) agrees with probabilistic independence. A characterization of the d-metric (Lemma 16.4) is derived in [10], which also contains the proof of ω-stability given above. In general, the presentation of the material in this section follows [10] closely.

17 L^p Banach lattices

Let X be a set, U a σ-algebra on X and μ a σ-additive measure on U, and let $p \in [1, \infty)$. We denote by $L^p(X, U, \mu)$ the space of (equivalence classes of) U-measurable functions $f : X \to \mathbb{R}$ such that $\|f\| = (\int |f|^p d\mu)^{1/p} < \infty$. We consider this space as a Banach lattice (complete normed vector lattice) over $\mathbb{R}$ in the usual way; in particular, the lattice operations $\wedge, \vee$ are given by pointwise maximum and minimum.

We will work on models of the form

$$((B_n \mid n \geq 1), 0, \{I_{mn}\}_{m<n}, \{\lambda_r\}_{r \in \mathbb{R}}, +, -, \wedge, \vee, \| \ \|)$$

where $B_n = B_n(L^p(X, U, \mu)) = \{f \in L^p(X, U, \mu) \mid \|f\| \leq n\}$ and $I_{mn} : B_m \to B_n$ is the inclusion map for $m < n$. The metric on each B_n is given by $d(f, g) = \|f - g\|$. The diameter of B_n is $2n$ and the values of the predicate $\| \ \|$ on B_n are in $[0, n]$. The operations $+, -, \wedge, \vee$ map B_n into B_{2n}. For $r \in \mathbb{R}$ with $k - 1 < |r| \leq k$, where $k \geq 1$ is an integer, the operation λ_r (of scalar multiplication by r) maps B_n into B_{kn}.

The moduli of uniform continuity for the norm and for the inclusion maps I_{mn} are all given by $\Delta(\epsilon) = \epsilon$. The moduli of uniform continuity for $+, -, \wedge, \vee$ are all given by $\Delta'(\epsilon) = \epsilon/2$. For $r \in \mathbb{R}$ with $k-1 < |r| \leq k$, where k is an integer ≥ 1, the modulus of uniform continuity of λ_r is given by $\Delta_{\lambda_r}(\epsilon) = \epsilon/k$.

Let L denote the signature just described.

Basic analysis and probability

We start with a review of some results from analysis that we use to approach L^p spaces model theoretically.

A measure space (X, U, μ) is called *decomposable* (also called *strictly localizable*) if there exists a partition $\{X_i \mid i \in I\} \subseteq U$ of X into measurable sets such that $\mu(X_i) < \infty$ for all $i \in I$ and such that for any subset A of X, $A \in U$ iff $A \cap X_i \in U$ for all $i \in I$ and, in that case, $\mu(A) = \sum_{i \in I} \mu(A \cap X_i)$.

17.1 Convention. Throughout this section we require that all measure spaces are decomposable.

There is no loss of generality in adopting this convention: the representation theorem for abstract L^p spaces shows that every such space (and, in particular therefore, every concrete $L^p(X, U, \mu)$ space) can be represented in this way with (X, U, μ) being decomposable. (See the proof of Theorem 3 in [11], for example.)

Let E be any Banach lattice and $f \in E$. The *positive part* of f is $f \vee 0$, and it is denoted f^+. The *negative part* of f is $f^- = (-f)^+$, and one has $f = f^+ - f^-$ and the *absolute value* of f is given by $|f| = f^+ + f^-$. Further, f is *positive* if $f = f^+$ and f is *negative* if $-f$ is positive. For $f, g \in E$, one has $f \geq g$ iff $f - g$ is positive.

Let (X, U, μ) be a measure space. A measurable set $S \in U$ is an *atom* if $\mu(S) > 0$ and there does not exist any $S' \in U$ satisfying $S' \subseteq S$ and $0 < \mu(S') < \mu(S)$. One calls (X, U, μ) *atomless* if it has no atoms.

If E is a Banach lattice and $x \in E$, a *component* of x is $y \in E$ such that $|y| \wedge |x - y| = 0$. If (X, U, μ) is a measure space and $E = L^p(X, U, \mu)$, then the components of $x \in E$ are the results of restricting x to some measurable subset of the support of x.

17.2 Notation. Let (X, U, μ) and (Y, V, μ) be a measure spaces. We write $(Y, V, \mu) \subseteq (X, U, \mu)$ to mean that $Y \in U$ and $V \subseteq U$.

Model theory of L^p Banach lattices

In this section, unless stated otherwise, we work on the unit ball. So all elements under consideration and all quantifiers range over B_1.

It is routine to write down L-conditions expressing the following axioms, which are true in $L^p(X, U, \mu)$, where (X, U, μ) is a measure space.

(1) The Banach lattice axioms, described in [35, pages 47–49, 81].

(2) Axioms for abstract L^p spaces, which state that

$$\|x \wedge y\|^p \leq \|x\|^p + \|y\|^p \leq \|x + y\|^p$$

whenever x and y are positive.

We write LpL for the theory axiomatized above.

17.3 Proposition. *(Axiomatizability) Let $\mathcal{M}$ be an L-structure with underlying metric space (M, d). Then $\mathcal{M}$ is a model of LpL if and only if there is a measure space (X, U, μ) such that $\mathcal{M}$ is isomorphic to $L^p(X, U, \mu)$.*

Proof See the proof of Theorem 3 in [11], for example. □

To ensure that the measure space representing the structure under consideration is atomless we need an additional axiom:

(3) $\sup_x \inf_y \big(\max(\big|\|y\| - \|x^+ - y\|\big|, \|y \wedge (x^+ - y)\|\big) = 0.$

It is obvious that this condition is satisfied in any $L^p(X, U, \mu)$ for which (X, U, μ) is atomless. For the converse, note that (3) states that for every positive function $u = x^+$ and every $\epsilon > 0$ there is some y such that $\|y\|$ and $\|u - y\|$ differ by at most ϵ and $\|y \wedge (u - y)\| \leq \epsilon$. Assume $u \neq 0$. By taking ϵ small enough and subtracting $y \wedge (u - y)$ from y, we get a nontrivial component of u. Hence, in models of LpL that also satisfy condition (3), there are no atoms.

We denote by $ALpL$ the set of conditions $LpL + (3)$. We just proved:

17.4 Proposition. *If $\mathcal{M}$ is an L-structure, then $\mathcal{M}$ is a model of the theory $ALpL$ if and only if there is an atomless measure space (X, U, μ) such that $\mathcal{M}$ is isomorphic to $L^p(X, U, \mu)$.*

17.5 Fact. (Quantifier elimination) Let $\mathcal{M}$ be the L^p space of an atomless measure space and let $a, b \in M^n$. If a and b have the same quantifier free type in $\mathcal{M}$, then a and b have the same type in $\mathcal{M}$. That is, if $\|t(a)\| = \|t(b)\|$ in $\mathcal{M}$ for every term $t(x_1, \ldots, x_n)$, then a and b have the same type in $\mathcal{M}$. From this observation and Proposition 13.6 it follows that $ALpL$ admits quantifier elimination.

For a proof of Fact 17.5, see Example 13.18 in [24].

Note that Fact 17.5 is not true without the assumption that $\mathcal{U}$ is atomless; atoms and non-atoms can have the same quantifier free type but never have the same type.

17.6 Fact. (Separable categoricity) If $\mathcal{M} \models ALpL$ be separable, then $\mathcal{M}$ is isomorphic to $L^p([0,1], \mathcal{B}, m)$, where $\mathcal{B}$ is the σ-algebra of Lebesgue measurable sets and m is Lebesgue measure.

Note that ω-categoricity need not be preserved when we add constants to the language:

17.7 Example. Let f and g be any two norm 1, positive elements of $L^p([0,1], \mathcal{B}, m)$. By Fact 17.5, we get

$$(L^p([0,1], \mathcal{B}, m), f) \equiv (L^p([0,1], \mathcal{B}, m), g).$$

However, there are two possible isomorphism types of such structures, depending on whether or not the support of the adjoined function has measure 1 or not.

17.8 Fact. Let $\mathcal{M}$ be a model of $ALpL$ and $A \subseteq M$. The definable closure of A in $\mathcal{M}$ is the closed linear sublattice of $\mathcal{M}$ generated by A.

Proof This follows from quantifier elimination plus an analysis of restricted quantifier free formulas. □

17.9 Remark. Let $\mathcal{M}$ be the L^p space of a measure space and let C be a closed linear sublattice of $\mathcal{M}$. One can use the representation theorem for abstract L^p spaces (see [33, pages 15–16], part(2) of the axiomatization) to show that there exist measure spaces (X, U, μ) and (Y, V, μ) satisfying $(Y, V, \mu) \subseteq (X, U, \mu)$, as well as an isomorphism Φ from $L^p(X, U, \mu)$ onto $\mathcal{M}$ such that Φ maps $L^p(Y, V, \mu)$ exactly onto C.

The previous remark has interesting consequences. Let $\mathcal{M}$ and C be as in Remark 17.9 and let $f \in M^n$. In [5] it is proved that the type over C realized by f in $\mathcal{M}$ is characterized by the joint conditional distribution of f over the σ-algebra associated to C (V in the notation above). The proof of this result is beyond the scope of this paper. However the ideas behind it are illustrated by the special case of a single characteristic function:

17.10 Proposition. *Let $\mathcal{M}$ and C be as in Remark 17.9. Suppose $(Y, V, \mu) \subset (X, U, \mu)$ are measure spaces such that $C = L^p(Y, V, \mu)$ and $M = L^p(X, U, \mu)$. Let $A, B \subset Y$ be such that $\chi_A, \chi_B \in M$. Then $\mathbb{P}(A|V) = \mathbb{P}(B|V)$ implies $\mathrm{tp}(\chi_A/C) = \mathrm{tp}(\chi_B/C)$.*

Proof Assume that $\mathbb{P}(A|V) = \mathbb{P}(B|V)$. By quantifier elimination, to show that $\mathrm{tp}(\chi_A/C) = \mathrm{tp}(\chi_B/C)$ it suffices to prove that for any $g \in C^l$ and any lattice term $t(x, y)$, we have $\|t(\chi_A, g)\|^p = \|t(\chi_B, g)\|^p$.

Let ν be the measure on Borel subsets D of $\mathbb{R}^{1+l}$ defined by $\nu(D) = \mu\{x \in X : (\chi_A, g)(x) \in D\}$. Since $\mathbb{P}(A|V) = \mathbb{P}(B|V)$, we have that $\nu(D) = \mu\{x \in X : (\chi_B, g)(x) \in D\}$ for any Borel $D \subset \mathbb{R}^{1+l}$. Then, by the change of variable formula,

$$\int_X |t(\chi_A(x), g(x))|^p d\mu(x) = \int_{\mathbb{R}^{1+l}} |t(r, s)|^p d\nu(r, s)$$
$$= \int_X |t(\chi_B(x), g(x))|^p d\mu(x)$$

and hence $\|t(\chi_A, g)\|^p = \|t(\chi_B, g)\|^p$ as desired. □

Finally we show stability.

17.11 Theorem. *(Henson [22]) The theory ALpL is ω-stable.*

Proof Let $\mathcal{U}$ be a κ-universal domain for *ALpL*, with $\kappa \geq \omega_1$, and let $A \subsetneq \mathcal{U}$ be countably infinite. Then $\mathrm{dcl}(A)$ is a closed linear sublattice of $\mathcal{U}$ and thus we can find measure spaces $(Y, V, \mu) \subseteq (X, U, \mu)$ such that $\mathrm{dcl}(A) = L^p(Y, V, \mu)$ and $\mathcal{U} = L^p(X, U, \mu)$. Let $T_A = \mathrm{Th}((\mathcal{U}, a)_{a \in A})$.

Any function $h \in \mathcal{U}$ can be written as $f + g$, where the support of f is contained in Y and the support of g is disjoint from Y. Moreover, $\mathrm{tp}((f + g)/A)$ is determined by $\mathrm{tp}(f/A)$ and $\mathrm{tp}(g/A)$.

Therefore, to find the density character of $S_1(T_A)$ it suffices to consider the following two cases:

Let $f \in \mathcal{U}$ be an element supported on Y. It suffices to consider the case where f is a simple function, since every function is a limit of simple functions. We identify $L^p(Y, V, \mu)$ with its canonical image in the space $L^p((Y, V, \mu) \otimes ([0, 1], \mathcal{B}, m))$, where $([0, 1], \mathcal{B}, m)$ is the standard Lebesgue space. Since $\mathcal{U}$ is sufficiently saturated, we may assume that $L^p((Y, V, \mu) \otimes ([0, 1], \mathcal{B}, m))$ is a closed linear sublattice of $\mathcal{U}$. Using quantifier elimination (see Fact 17.5), the fact that f is a simple function, and Proposition 17.10, we can find $f' \in L^p((Y, V, \mu) \otimes ([0, 1], \mathcal{B}, m))$ such that $\mathrm{tp}(f/A) = \mathrm{tp}(f'/A)$. Since $L^p((Y, V, \mu) \otimes ([0, 1], \mathcal{B}, m))$ is separable, the density character of the space of types of functions supported on Y is ω.

Let $g \in \mathcal{U}$ be an element whose support is disjoint from Y. The type $\mathrm{tp}(g/A)$ is determined by $\|g^+\|$ and $\|g^-\|$. Let $B, C \in U$ be disjoint from Y, each of measure one. The set $\{\mathrm{tp}(c_1\chi_B - c_2\chi_C) \mid c_1, c_2 \in \mathbb{Q}^+\}$

is a countable dense subset of the space of types of functions disjoint to $\mathrm{dcl}(A)$. □

The basic model-theoretic properties of L^p spaces, such as axiomatizability and quantifier elimination (see Fact 17.5) are proved in [24]. Further results about the model theory of these structures are given in [5], including a characterization of the d-metric and an analysis of non-dividing in terms of conditional expectation, as well as stability theoretic properties such as stationarity of types over sets.

18 Probability spaces with generic automorphism

In this section we will study the existentially closed structures of the form $(\mathcal{M}, \tau)$, where $\mathcal{M}$ is a probability structure and τ is an automorphism of $\mathcal{M}$. We show this class is axiomatizable and its theory is stable. We also discuss the model-theoretic meaning of some results in ergodic theory.

Lebesgue spaces and their automorphisms

There are two approaches to isomorphisms on probability spaces. On the one hand, we have measure preserving point maps between the spaces; on the other, we have measure preserving maps between measured algebras.

18.1 Definition. Let $(X_1, \mathcal{B}_1, \mu_1)$, $(X_2, \mathcal{B}_2, \mu_2)$ be probability spaces and let $\widehat{\mathcal{B}}_1$, $\widehat{\mathcal{B}}_2$ be their measure algebras. By an *isomorphism* of the measured algebras we mean a bijection $\Phi\colon \widehat{\mathcal{B}}_1 \to \widehat{\mathcal{B}}_2$ that preserves complements, countable unions and intersections and satisfies $\mu_2(\Phi(b)) = \mu_1(b)$ for all $b \in \widehat{\mathcal{B}}_1$. The probability spaces are said to be *conjugate* if their measured algebras are isomorphic.

18.2 Definition. Let $(X_1, \mathcal{B}_1, \mu_1)$, $(X_2, \mathcal{B}_2, \mu_2)$ be probability spaces and let $\widehat{\mathcal{B}}_1$, $\widehat{\mathcal{B}}_2$ be their measure algebras. Let $C_1 \in \mathcal{B}_1$, $C_2 \in \mathcal{B}_2$ with $\mu_1(C_1) = 1 = \mu_2(C_2)$. An invertible measure preserving transformation $\Phi\colon C_1 \to C_2$ is called an *isomorphism* between $(X_1, \mathcal{B}_1, \mu_1)$ and $(X_2, \mathcal{B}_2, \mu_2)$. If $(X_1, \mathcal{B}_1, \mu_1) = (X_2, \mathcal{B}_2, \mu_2)$, we call Φ an *automorphism*. For $b \in \widehat{\mathcal{B}}_1$, let $B \in \mathcal{B}_1$ be such that $[B]_{\mu_1} = b$. Let $\widehat{\Phi}(b) = [\Phi(B \cap C_1)]_{\mu_2}$. The induced map $\widehat{\Phi}\colon \widehat{\mathcal{B}}_1 \to \widehat{\mathcal{B}}_2$ is an isomorphism and it is called an *induced isomorphism* of the measured algebras.

Clearly any two isomorphic probability spaces are conjugate; however, the converse does not hold in general. The next definition concerns a well-known special class of probability spaces where the converse does hold.

18.3 Definition. A probability space $(I, \mathcal{L}, m)$ is a *Lebesgue space* if it is isomorphic to a probability space that is the disjoint union of two spaces:

(1) One that is a countable (or finite) set of points $\{y_1, y_2, \ldots\}$, each of positive measure.
(2) The space $([0,s], \mathcal{L}([0,s]), l)$, where $\mathcal{L}([0,s])$ is the Lebesgue σ-algebra on $[0,s]$ and l is Lebesgue measure.

Here $s = 1 - \sum_{i=1}^{\infty} p_i$, where $p_i > 0$ is the measure of $\{y_i\}$.

On Lebesgue spaces the notion of isomorphism and conjugacy coincide. (See Theorem 2.2 in [40].)

18.4 Definition. Let $(Y, \mathcal{C}, \mu)$ be an atomless probability space and let τ_Y be an automorphism of $(Y, \mathcal{C}, \mu)$. We say that τ_Y is *aperiodic* if for every $n \in \mathbb{N}^+$, the set $\{y \in Y \mid \tau_Y^n(y) = y\}$ has measure zero.

For the rest of this section we will study aperiodic maps and their properties. A good source for this material is the book of Halmos [21, pages 69–76] on ergodic theory. One of the key tools for studying aperiodic automorphisms is:

18.5 Theorem. *(Rokhlin's Lemma [21, page 71]) Let $(Y, \mathcal{C}, \mu)$ be an atomless probability space and τ_Y an aperiodic automorphism of this space. Then for every positive integer n and $\epsilon > 0$, there exists a measurable set $E \in \mathcal{C}$ such that the sets $E, \tau_Y(E), \ldots, \tau_Y^{n-1}(E)$ are disjoint and $\mu(\cup_{i<n} \tau_Y^i(E)) > 1 - \epsilon$.*

18.6 Remark. Let $(Y, \mathcal{C}, \mu)$ be an atomless probability space and let τ_Y be an automorphism of this space. Let $\mathcal{N}$ be the probability structure induced by $(Y, \mathcal{C}, \mu)$ and let τ be the automorphism of $\mathcal{N}$ induced by τ_Y. Then τ_Y is aperiodic iff

$$\inf_e \max\left(|1/n - \mu(e)|, \mu(e \cap \tau(e)), \mu(e \cap \tau^2(e)), \ldots, \mu(e \cap \tau^{n-1}(e))\right) = 0$$

for all $n \geq 1$.

For the rest of this section we fix an atomless Lebesgue space $(I, \mathcal{L}, m)$.

18.7 Fact. [21, page 74] Let $A, B \in \mathcal{L}$ be such that $m(A) = m(B)$. Then there is an automorphism η of $(I, \mathcal{L}, m)$ such that $m(\eta(A) \triangle B) = 0$.

From now on, G denotes the group of measure preserving automorphisms on $(I, \mathcal{L}, m)$, where we identify two maps if they agree on a set of measure one. There is a natural representation of G in $\mathbb{B}(L^2(I, \mathcal{L}, m))$ (the space of bounded linear operators on $L^2(I, \mathcal{L}, m)$); it sends $\tau \in G$ to the unitary operator U_τ defined for all $f \in L^2(I, \mathcal{L}, m)$ by $U_\tau(f) = f \circ \tau$. The norm topology on $\mathbb{B}(L^2(I, \mathcal{L}, m))$ pulls back to a group topology on G, which is called the *uniform topology* on G in [21, page 69]. For $\tau, \eta \in G$, let $\rho(\tau, \eta) = m(\{x \in X \mid \tau(x) \neq \eta(x)\})$. It is shown in [21, pages 72–73] that ρ is a metric for the uniform topology.

18.8 Definition. We call a map $\eta \in G$ a *cycle* of period n if there is a set $E \in \mathcal{L}$ of measure $1/n$ such that $E, \eta(E), \ldots, \eta^{n-1}(E)$ are pairwise disjoint, and $\eta^n = id \restriction_X$ (up to measure zero).

18.9 Remark. (1) Let $\tau \in G$ be aperiodic. For every $n > 0$ there is a cycle $\eta \in G$ of period n such that $\rho(\tau, \eta) \leq 2/n$. (By Rokhlin's Lemma.)
(2) Given any two cycles $\eta_1, \eta_2 \in G$ of period n, there is $\gamma \in G$ such that $\gamma^{-1}\eta_1\gamma = \eta_2$. (This follows from Fact 18.7.)
(3) Let $\tau_1, \tau_2 \in G$ be aperiodic. Then for every $\epsilon > 0$, there is $\gamma \in G$ such that $\rho(\tau_1, \gamma^{-1}\tau_2\gamma) \leq \epsilon$. (This follows from (1) and (2).)

Existentially closed structures

We denote by L be the language of probability structures and by APA the theory of atomless probability structures. Write L_τ for the language L expanded by a unary function with symbol τ and let APA_τ be the theory $APA \cup$ "τ is an automorphism". We can axiomatize APA_τ by adding to APA the following conditions:

(1) $\sup_x \left|\mu(x) - \mu(\tau(x))\right| = 0$
(2) $\sup_x \inf_y \left|\mu(x \triangle \tau(y))\right| = 0$
(3) $\sup_x \sup_y \left|\mu(\tau(x \cup y) \triangle (\tau(x) \cup \tau(y)))\right| = 0$
(4) $\sup_x \sup_y \left|\mu(\tau(x \cap y) \triangle (\tau(x) \cap \tau(y)))\right| = 0$

(Note that (1) expresses that τ is measure preserving, (2) that τ is surjective, and (3,4) that τ is a Boolean homomorphism.)

We write tp for types in the language L and tp_τ for types in the language L_τ. Let $\mathcal{M}$ be the probability structure associated to $(I, \mathcal{L}, m)$.

Recall that G is the group of automorphisms of $(I, \mathcal{L}, m)$, where we identify two maps if they agree on a set of measure one.

18.10 Proposition. *Let τ_I, τ'_I be aperiodic automorphisms of $(\mathcal{I}, L, m)$ and let τ, τ' be the corresponding induced automorphisms of $\mathcal{M}$. Then $(\mathcal{M}, \tau) \equiv (\mathcal{M}, \tau')$.*

Proof An application of Remark 18.9(3) with the values for ϵ ranging over the sequence $\{1/n \mid n \in \mathbb{N}^+\}$ together with the uniform continuity of formulas shows that $(\mathcal{M}, \tau) \equiv (\mathcal{M}, \tau')$. □

Thus any two models induced by aperiodic transformations on an atomless *Lebesgue space* have the same elementary theory. The aim of this section is to study this theory.

18.11 Definition. Let $(X, \mathcal{B}, \mu)$ be a probability space, let $\widehat{\mathcal{B}}$ be the corresponding measure algebra of events and let τ be an automorphism of the measure algebra $\widehat{\mathcal{B}}$. The map τ is called *aperiodic* if it satisfies the following condition corresponding to the conclusion of Rokhlin's lemma:

$$\inf_e \max\left(|1/n - \mu(e)|, \mu(e \cap \tau(e)), \mu(e \cap \tau^2(e)), \ldots, \mu(e \cap \tau^{n-1}(e))\right) = 0$$

for all $n \geq 1$.

Let $APAA$ be the theory APA_τ together with the conditions in L_τ describing that τ is an aperiodic automorphism.

18.12 Lemma. *The theory APAA is complete.*

Proof Let (M_1, τ_1) and (M_2, τ_2) be two models of $APAA$. Then there are separable models $(M'_i, \tau'_i) \models APAA$ that are elementarily equivalent to (M_i, τ_i) for $i = 1, 2$ respectively. By separable categoricity of APA, for $i = 1, 2$ we may assume that M'_i is the probability structure associated to a Lebesgue space and that τ'_i is induced by an automorphism η_i of the corresponding Lebesgue space. Then η_1 and η_2 are aperiodic automorphisms of the Lebesgue spaces and thus by Proposition 18.10 we have $(M'_1, \tau'_1) \equiv (M'_2, \tau'_2)$. □

18.13 Remark. The theory of aperiodic automorphisms on probability structures is the limit, as n goes to infinity, of the theory of a probability structure formed by n atoms $\{a_1, \ldots, a_n\}$ each of measure $1/n$, equipped with a cycle τ of period n; that is, τ is a permutation of $\{a_1, \ldots, a_n\}$ such that

$$\{a_1, \tau(a_1), \ldots, \tau^{n-1}(a_1)\} = \{a_1, \ldots, a_n\}.$$

Indeed, for each sentence φ in the language of $APAA$, let $v(n, \varphi)$ denote the value φ gets in the n-point probability space with a cycle of period n. On the other hand, let $v(\varphi)$ be the value φ gets in any model of $APAA$. (This value is well defined since $APAA$ is complete.) Then, $\lim_{n\to\infty} v(n, \varphi) = v(\varphi)$. Otherwise there would be a non-principal ultrafilter $\mathcal{U}$ on the positive integers such that if $\mathcal{M}$ is the $\mathcal{U}$-ultraproduct of the family consisting of the n-point probability structures with cycles of period n, then $\varphi^{\mathcal{M}} \neq v(\varphi)$. But this is a contradiction, because any such $\mathcal{M}$ is a model of $APAA$.

Our next aim is to show that $APAA$ is model complete. The techniques used for the proof are similar to the ones used in proving the completeness of the theory $APAA$, but now we need to include parameters.

18.14 Proposition. *The theory APAA is model complete. That is, if* $(N_0, \tau_0) \subseteq (N_1, \tau_1)$ *are models of APAA, then* $(N_0, \tau_0) \preceq (N_1, \tau_1)$.

Proof Let $(\mathcal{M}, \tau) \subseteq (\mathcal{N}, \tau)$ be separable models of $APAA$. We may assume that there is an atomless Lebesgue space $(X, \mathcal{B}, m)$ such that $\mathcal{M}$ is the model induced by $(X, \mathcal{B}, m)$ and that there is an atomless Lebesgue space $(Y, \mathcal{C}, m)$ such that $\mathcal{N}$ is the structure induced by $(Y, \mathcal{C}, m)$. Furthermore, we may assume that there is an aperiodic automorphism τ_X on $(X, \mathcal{B}, m)$ that induces the action of τ on $\mathcal{M}$ and that there is an aperiodic automorphism τ_Y on $(Y, \mathcal{C}, m)$ that induces the action of τ on $\mathcal{N}$. Note that both maps τ_X and τ_Y are aperiodic.

Let $a_1, \ldots, a_p \in M$. Take $A_1^X, \ldots, A_p^X \in \mathcal{B}$ such that $[A_j^X]_m = a_j$ for $j = 1, \ldots, p$, and $A_1^Y, \ldots, A_p^Y \in \mathcal{C}$ such that $[A_j^Y]_m = a_j$ for $j = 1, \ldots, p$.

Let $\epsilon > 0$ and let n be a positive integer such that $1/n < \epsilon$. By Rokhlin's Lemma, there is $B \in \mathcal{B}$ such that $B, \tau_X(B), \ldots, \tau_X^{n-1}(B)$ are disjoint and $m(\cup_{i<n} \tau_X^i(B)) \geq 1 - \epsilon$. Let P^X be the partition of B generated by $\tau_X^{-i}(\tau_X^i(B) \cap A_j^X)$ for $1 \leq j \leq p$ and $0 \leq i < n$. Since $(\mathcal{M}, \tau) \subseteq (\mathcal{N}, \tau)$, there is $C \in \mathcal{C}$ such that $[C]_m = [B]_m$. Let P^Y be the partition of C generated by $\tau_Y^{-i}(\tau_Y^i(C) \cap A_j^Y)$ for $1 \leq j \leq p$ and $0 \leq i < n$.

Since $(X, \mathcal{B}, m)$ and $(Y, \mathcal{C}, m)$ are Lebesgue spaces, by Fact 18.7 there is $h_0 \colon B \to C$ a measure preserving bijection such that

$$h_0(\tau_X^{-i}(\tau_X^i(B) \cap A_j^X)) = \tau_Y^{-i}(\tau_Y^i(C) \cap A_j^Y)$$

(up to measure zero) for $1 \leq j \leq p$ and $0 \leq i < n$. Extend h_0 to $\cup_{1<i<n}\tau_X^i(B)$ by setting $h_0(\tau_X^i(x)) = \tau_Y^i(h_0(x))$ for $x \in B$, $0 < i < n$.

Let $Z^X = X \setminus \cup_{1<i<n}\tau_X^i(B)$ and let $Z^Y = Y \setminus \cup_{1<i<n}\tau_Y^i(C)$. Extend h_0 to a measure preserving bijection h from $(X, \mathcal{B}, m)$ to $(Y, \mathcal{C}, m)$ by defining $h \colon Z^X \to Z^Y$ so that $h(Z^X \cap A_j^X) = Z^Y \cap A_j^Y$ for all $1 \leq j \leq p$.

Note that h induces a map $\widehat{h}$ from $\mathcal{M}$ to $\mathcal{N}$ satisfying $\widehat{h}(a_j) = a_j$ for all j and that $\rho(\tau^X, h^{-1}\tau^Y h) \leq 2\epsilon$. The proposition follows from the uniform continuity of formulas. □

18.15 Definition. We say that $(\mathcal{M}, \tau) \models APA_\tau$ is *existentially closed* if whenever $(\mathcal{N}, \tau) \models APA_\tau$, $(\mathcal{N}, \tau) \supseteq (\mathcal{M}, \tau)$, $a \in M^n$ and $\varphi(x, y)$ is a quantifier free formula such that $(\mathcal{N}, \tau) \models \inf_x \varphi(x, a) = 0$, then $(\mathcal{M}, \tau) \models \inf_x \varphi(x, a) = 0$. (Here x, y are disjoint finite sequences of distinct variables, with y of length n.)

18.16 Lemma. *Let $\mathcal{M}$ be an L_τ-structure. Then $\mathcal{M} \models APAA$ if and only if $\mathcal{M}$ is an existentially closed model of APA_τ.*

Proof Any model of APA_τ can be embedded in a model of $APAA$. Indeed, if $\mathcal{M} \models APA_\tau$, then $\mathcal{M}$ can be embedded in the product of $\mathcal{M}$ and the L_τ-structure that is based on the unit circle with normalized Lebesgue measure plus the rotation through an irrational multiple of π. It is easy to see that this product is a model of $APAA$. Since $APAA$ is axiomatized by adding a set of inf-conditions to APA_τ, this yields that any existentially closed model of APA_τ is a model of $APAA$. The other direction follows from the previous proposition. □

In ergodic theory, *joinings* (see [18, page 125]) give different ways of amalgamating two probability structures with automorphisms into a common extension. In particular, the *relative independent joining over a common factor* (described in [18, page 127]) corresponds to a free amalgamation of two models $\mathcal{M}_1$, $\mathcal{M}_2$ of APA_τ over a common substructure $\mathcal{N}$. Call this new structure $\mathcal{M}_1 \oplus_\mathcal{N} \mathcal{M}_2$.

18.17 Theorem. *The theory APAA has elimination of quantifiers.*

Proof We apply Proposition 13.6. Let $\mathcal{M}_1$, $\mathcal{M}_2 \models APAA$, let $\mathcal{N}$ be a substructure of $\mathcal{M}_1$ and let $f : \mathcal{N} \to \mathcal{M}_2$ be an embedding. We may assume that f is the identity. Since $APAA$ is model complete, $\mathcal{M}_1 \oplus_{\mathcal{N}} \mathcal{M}_2$ is an elementary extension of $\mathcal{M}_2$ and the canonical embedding of $\mathcal{M}_1$ into this space gives the desired extension of f. □

18.18 Proposition. *Let (M, τ) be a κ-universal domain of $APAA$, with $\kappa \geq \omega_1$, and let $A \subseteq M$. Then the definable closure of A in L_τ is the smallest σ-algebra of events containing $\tau^i(A)$ for all $i \in \mathbb{Z}$.*

Proof This follows from quantifier elimination plus an analysis of restricted quantifier free formulas. □

Independence and stability

In this section we introduce an abstract notion of independence, which we call τ-independence, and show that it agrees with non-dividing. This idea follows the approach used in [14, Section 3] to characterize non-dividing inside a first-order stable structure expanded by a generic automorphism. We reserve the word "independence" here for independence of events in the sense of probability structures. Fix a κ-universal domain $(\mathcal{U}, \tau) \models APAA$, where $\kappa \geq \omega_1$.

18.19 Definition. Let $A, B, C \subset \mathcal{U}$ be small. We write $A \mathop{\smile\!\!\!\!\!\mid}^{\tau}_{C} B$, and say that A *is τ-independent from B over C*, if $\mathrm{dcl}_\tau(A)$ is independent from $\mathrm{dcl}_\tau(B)$ over $\mathrm{dcl}_\tau(C)$.

We will show that $\mathop{\smile\!\!\!\!\!\mid}^{\tau}$ is a stable independence relation. We start by proving a strong form of stationarity:

18.20 Proposition. *Let $a, b \in \mathcal{U}^n$ and let $C \subseteq D \subseteq \mathcal{U}$. Suppose that $\mathrm{tp}_\tau(a/C) = \mathrm{tp}_\tau(b/C)$ and that $a \mathop{\smile\!\!\!\!\!\mid}^{\tau}_{C} D$ and $b \mathop{\smile\!\!\!\!\!\mid}^{\tau}_{C} D$. Then $\mathrm{tp}_\tau(a/D) = \mathrm{tp}_\tau(b/D)$.*

Proof Let a, b, C, D be as above. Then for every $k < \omega$,

$$\mathrm{tp}(\tau^{-k}(a), \ldots, \tau^k(a))/\mathrm{dcl}_\tau(C)) = \mathrm{tp}(\tau^{-k}(b), \ldots, \tau^k(b))/\mathrm{dcl}_\tau(C)).$$

By stationarity of types in models of APA, we get

$$\mathrm{tp}(\tau^{-k}(a), \ldots, \tau^k(a))/\mathrm{dcl}_\tau(D)) = \mathrm{tp}(\tau^{-k}(b), \ldots, \tau^k(b))/\mathrm{dcl}_\tau(D)).$$

Since this equality holds for all $k < \omega$, by quantifier elimination of $APAA$, $\mathrm{tp}_\tau(a/D) = \mathrm{tp}_\tau(b/D)$. □

18.21 Corollary. *The theory APAA is stable and the relation of τ-independence agrees with non-dividing.*

Proof We note that τ-independence is a stable independence relation on κ-universal domains for $APAA$, where $\kappa \geq \omega_1$. Using the properties of independence in a κ-universal domain $\mathcal{U}$ for APA, it is clear that τ-independence satisfies: invariance, symmetry, transitivity, extension, local character and finite character. By the previous proposition it also satisfies stationarity. Applying Theorem 14.14, this shows that τ-independence agrees with non-dividing and $APAA$ is stable. □

18.22 Remark. The theory $APAA$ is not ω-stable. For every irrational $\alpha \in [0, 1]$, consider the model of APAA that is based on the unit circle with normalized Lebesgue measure and is equipped with the aperiodic automorphism corresponding to rotation by the angle $2\pi\alpha$. Let p_α be the 1-type realized in this model by the event corresponding to the semicircle in the upper half plane.

Assume now that α, β are irrational and that $\alpha - \beta$ is also irrational. Let $a \models p_\alpha$ and let $b \models p_\beta$. For every $\epsilon > 0$ there is $n < \omega$ such that $d(a, \sigma^n(a)) = \mu(a \triangle \sigma^n(a)) < \epsilon$ while $d(b, \sigma^n(b)) > 1 - \epsilon$. It follows that $d(a, b) + d(\sigma^n(a), \sigma^n(b)) \geq 1 - 2\epsilon$. Since $d(a, b) = d(\sigma^n(a), \sigma^n(b))$, we conclude that $d(a, b) \geq \frac{1}{2} - \epsilon$. Therefore $d(p_\alpha, p_\beta) \geq \frac{1}{2}$.

Since we can choose $2^{\aleph_0}$ irrationals in $[0, 1]$, each two of which are linearly independent over $\mathbb{Q}$, this yields $2^{\aleph_0}$ types over $\emptyset$ each two of which have distance $\geq \frac{1}{2}$ from each other.

Further results about probability algebras with a generic automorphism can be found in [10]. In particular, it is shown there how *entropy* can be seen as a model-theoretic rank.

References

[1] I. Ben Yaacov, Positive model theory and compact abstract theories, *Journal of Mathematical Logic* **3**, 2003, 85–118.

[2] I. Ben Yaacov, Schroedinger's cat, *Israel Journal of Mathematics* **153**, 2006, 157–191.

[3] I. Ben Yaacov, Simplicity in compact abstract theories, *Journal of Mathematical Logic* **3**, 2003, 163–191.

[4] I. Ben Yaacov, Uncountable dense categoricity in cats, *The Journal of Symbolic Logic* **70**, 2005, 829-860.

[5] I. Ben Yaacov, Alexander Berenstein and C. Ward Henson, Model-theoretic independence in the Banach lattices $L^p(\mu)$, submitted.

[6] I. Ben Yaacov and Alex Usvyatsov, Continuous first order logic and local stability, submitted.

[7] I. Ben Yaacov, Alex Usvyatsov and Moshe Zadka, The theory of a Hilbert space with a generic automorphism, in preparation.

[8] A. Berenstein, Hilbert spaces with generic groups of automorphisms, *Archive for Mathematical Logic* **46**, 2007, 289–299.

[9] A. Berenstein and S. Buechler, Homogeneous expansions of Hilbert spaces, *Annals of Pure and Applied Logic* **128**, 2004, 75–101.

[10] A. Berenstein and C. W. Henson, Model theory of probability spaces with an automorphism, submitted.

[11] J. Bretagnolle, D. Dacunha-Castelle, J.-L. Krivine, Lois stables et espaces L^p, *Annals Inst. Henry Poincaré Sect. B (N.S.)* **2**, 1965/1966, 231–259.

[12] C. C. Chang and H. J. Keisler, *Continuous Model Theory*, Princeton University Press, 1966.

[13] C. C. Chang and H. J. Keisler. *Model theory*, volume 73 of Studies in Logic and the Foundations of Mathematics, North-Holland, third edition, 1990.

[14] Z. Chatzidakis and A. Pillay, Generic structures and simple theories, *Annals of Pure and Applied Logic* **95**, 1998, 71–92.

[15] D. Dacunha-Castelle and J.-L. Krivine, *Applications des ultraproduits à l'étude des espaces et des algèbres de Banach*, Studia Math. **41** (1972), 315–334.

[16] G. B. Folland, *Real Analysis*, John Wiley and Sons, 1984.

[17] L. Gillman and M. Jerison, *Rings of Continuous Functions*, Van Nostrand, 1960.

[18] E. Glasner, *Ergodic Theory via Joinings*, Mathematical Surveys and Monographs, 101, American Mathematical Society, 2003.

[19] M. Gromov, *Metric Structures for Riemannian and Non-Riemannian Spaces*, Birkhäuser, 1999.

[20] P. A. Halmos, *Measure Theory*, Van Nostrand, 1950.

[21] P. A. Halmos, *Lectures on Ergodic Theory*, Chelsea, 1956.

[22] C. W. Henson, Model theory of separable Banach spaces (abstract), *The Journal of Symbolic Logic* **52**, 1987, 1059–1060.

[23] C. W. Henson, Nonstandard hulls of Banach spaces, *Israel Journal of Mathematics* **25**, 1976, 108–144.

[24] C. W. Henson and J. Iovino, Ultraproducts in analysis, in *Analysis and Logic*, London Mathematical Society Lecture Notes Series, vol. 262, 2002, 1–113.

[25] C. W. Henson and L. C. Moore, Nonstandard analysis and the theory of Banach spaces, in *Nonstandard Analysis – Recent Developments (Victoria, B.C., 1980)*, Springer Lecture Notes in Mathematics No. 983, 1983, 27–112.

[26] C. W. Henson and H. Tellez, Algebraic closure in continuous logic, *Revista Colombiana de Matemáticas*, to appear.

[27] W. Hodges, *Model Theory*, Cambridge University Press, 1993.

[28] J. Iovino, Stable Banach space structures, I: Fundamentals, in *Models, algebras and proofs* (Bogotá, 1995), Lecture Notes in Pure and Applied Mathematics, 203, Marcel Dekker, 1999, 77–95.

[29] J. Iovino, Stable Banach space structures, II: Forking and compact topologies, in *Models, algebras and proofs* (Bogotá, 1995), Lecture Notes in Pure and Applied Mathematics, 203, Marcel Dekker, 1999, 97–117.

[30] J. Iovino, The Morley rank of a Banach space, *The Journal of Symbolic

Logic **61**, 1996, 85–118.
[31] H. J. Keisler, Ultraproducts and elementary classes, *Proceedings of the Koninklijke Nederlandse Akademie van Wetenschappen (Indagationes mathematicae)*, Series A, **23**, 1961, 477–495.
[32] J.-L. Krivine and B. Maurey, Espaces de Banach stables, *Israel Journal of Mathematics* **39**, 1981, 273–295.
[33] J. Lindenstrauss and L. Tzafriri, *Classical Banach Spaces, II; Function Spaces*, Springer Verlag, 1979.
[34] Y. Raynaud, Espaces de Banach superstables, distances stables et homéomorphismes uniformes, *Israel Journal of Mathematics* **44**, 1983, 33–52.
[35] H. H. Schaefer, *Banach Lattices and Positive Operators*, Springer Verlag, 1974.
[36] S. Shelah, Every two elementarily equivalent models have isomorphic ultrapowers, *Israel Journal of Mathematics* **10**, 1971, 224–233.
[37] P. Shields, *The Theory of Bernoulli Shifts*, Chicago Lectures in Mathematics, The University of Chicago Press, 1973.
[38] D. A. Vladimirov, *Boolean Algebras in Analysis*, Kluwer Academic Publishers, 2002.
[39] F. O. Wagner, *Simple Theories*, Kluwer Academic Publishers, 2000.
[40] P. Walters, *An Introduction to Ergodic Theory*, Springer Verlag, 1982.

Printed in the United States
By Bookmasters

Printed in the United States
By Bookmasters